T0317607

**Thermal Spreading and Contact Resistance**

# Wiley-ASME Press Series

Thermal Spreading and Contact Resistance: Fundamentals and Applications
*Yuri S. Muzychka and M. Michael Yovanovich*

Analysis of ASME Boiler, Pressure Vessel, and Nuclear Components in the Creep Range
*Maan H. Jawad, Robert I. Jetter*

Voltage-Enhanced Processing of Biomass and Biochar
*Gerardo Diaz*

Pressure Oscillation in Biomedical Diagnostics and Therapy
*Ahmed Al-Jumaily, Lulu Wang*

Robust Control: Youla Parameterization Method
*Farhad Assadian, Kevin R. Mellon*

Metrology and Instrumentation: Practical Applications for Engineering and Manufacturing
*Samir Mekid*

Fabrication of Process Equipment
*Owen Greulich, Maan H. Jawad*

Engineering Practice with Oilfield and Drilling Applications
*Donald W. Dareing*

Flow-Induced Vibration Handbook for Nuclear and Process Equipment
*Michel J. Pettigrew, Colette E. Taylor, Nigel J. Fisher*

Vibrations of Linear Piezostructures
*Andrew J. Kurdila, Pablo A. Tarazaga*

Bearing Dynamic Coefficients in Rotordynamics: Computation Methods and Practical Applications
*Lukasz Brenkacz*

Advanced Multifunctional Lightweight Aerostructures: Design, Development, and Implementation
*Kamran Behdinan, Rasool Moradi-Dastjerdi*

Vibration Assisted Machining: Theory, Modelling and Applications
*Li-Rong Zheng, Dr. Wanqun Chen, Dehong Huo*

Two-Phase Heat Transfer
*Mirza Mohammed Shah*

Computer Vision for Structural Dynamics and Health Monitoring
*Dongming Feng, Maria Q Feng*

Theory of Solid-Propellant Nonsteady Combustion
*Vasily B. Novozhilov, Boris V. Novozhilov*

Introduction to Plastics Engineering
*Vijay K. Stokes*

Fundamentals of Heat Engines: Reciprocating and Gas Turbine Internal Combustion Engines
*Jamil Ghojel*

Offshore Compliant Platforms: Analysis, Design, and Experimental Studies
*Srinivasan Chandrasekaran, R. Nagavinothini*

Computer Aided Design and Manufacturing
*Zhuming Bi, Xiaoqin Wang*

Pumps and Compressors
*Marc Borremans*

Corrosion and Materials in Hydrocarbon Production: A Compendium of Operational and Engineering Aspects
*Bijan Kermani and Don Harrop*

Design and Analysis of Centrifugal Compressors
*Rene Van den Braembussche*

Case Studies in Fluid Mechanics with Sensitivities to Governing Variables
*M. Kemal Atesmen*

The Monte Carlo Ray-Trace Method in Radiation Heat Transfer and Applied Optics
*J. Robert Mahan*

Dynamics of Particles and Rigid Bodies: A Self-Learning Approach
*Mohammed F. Daqaq*

Primer on Engineering Standards, Expanded Textbook Edition
*Maan H. Jawad and Owen R. Greulich*

Engineering Optimization: Applications, Methods and Analysis
*R. Russell Rhinehart*

Compact Heat Exchangers: Analysis, Design and Optimization using FEM and CFD Approach
*C. Ranganayakulu and Kankanhalli N. Seetharamu*

Robust Adaptive Control for Fractional-Order Systems with Disturbance and Saturation
*Mou Chen, Shuyi Shao, and Peng Shi*

Robot Manipulator Redundancy Resolution
*Yunong Zhang and Long Jin*

Stress in ASME Pressure Vessels, Boilers, and Nuclear Components
*Maan H. Jawad*

Combined Cooling, Heating, and Power Systems: Modeling, Optimization, and Operation
*Yang Shi, Mingxi Liu, and Fang Fang*

Applications of Mathematical Heat Transfer and Fluid Flow Models in Engineering and Medicine
*Abram S. Dorfman*

Bioprocessing Piping and Equipment Design: A Companion Guide for the ASME BPE Standard
*William M. (Bill) Huitt*

Nonlinear Regression Modeling for Engineering Applications: Modeling, Model Validation, and Enabling Design of Experiments
*R. Russell Rhinehart*

Geothermal Heat Pump and Heat Engine Systems: Theory and Practice
*Andrew D. Chiasson*

Fundamentals of Mechanical Vibrations
*Liang-Wu Cai*
Introduction to Dynamics and Control in Mechanical Engineering Systems
*Cho W.S. To*

This Work is a co-publication between John Wiley & Sons, Inc. and ASME Press.

# Thermal Spreading and Contact Resistance

## Fundamentals and Applications

*Yuri S. Muzychka*
Department of Mechanical and Mechatronics Engineering
Faculty of Engineering and Applied Science
Memorial University of Newfoundland
St. John's, NL, Canada

*M. Michael Yovanovich*
University of Waterloo
Department of Mechanical and Mechatronics Engineering
Faculty of Engineering
University of Waterloo
Waterloo, Ontario, Canada

WILEY

*Library of Congress Cataloging-in-Publication Data applied for.*

Hardback ISBN: 9781394187522

Cover Design: Wiley
Cover Image: © DrPixel/Getty Images; Courtesy of Y.S. Muzychka and M.M. Yovanovich

Set in 9.5/12.5pt STIXTwoText by Straive, Chennai, India

# Contents

# About the Authors

**Yuri S. Muzychka** is a professor of mechanical engineering at Memorial University of Newfoundland (Canada). He joined the Faculty of Engineering and Applied Science at Memorial University of Newfoundland in 2000. He completed his PhD in 1999 at the University of Waterloo. Since joining Memorial University, he has focused his research efforts in several areas of heat transfer and fluid dynamics, namely Fundamentals of Convection and Conduction Heat Transfer, Thermal Management in Electronics, Transport Phenomena in Internal Flows, Multiphase Phase Flow, and most recently Marine Icing Phenomena. He has published over 250 papers in high-quality journals and international conference proceedings, and five book chapters. He is a Fellow of the ASME, CSME, and the Engineering Institute of Canada (EIC).

**M. Michael Yovanovich** is a distinguished professor emeritus at the University of Waterloo (Canada). He joined the University of Waterloo in 1969 after a short period as an associate professor at the University of Poitiers (France). He completed his mechanical engineering studies at Queens University (Canada) and obtained his ScD from the Massachusetts Institute of Technology in 1967. He devoted his early research in the area of thermal contact resistance and constriction resistance analysis. Later in his career, he was active in thermal management in electronics, conduction heat ransfer, convective heat transfer, microfluidics, and thermal contact resistance. He has published seven book chapters, given over 150 Keynote Lectures, and published over 350 papers in international journals and conference proceedings. He is a fellow of ASME, CSME, AIAA, AAAS, and RSC.

# Preface

Thermal spreading and contact resistance are fundamentally important topics in heat transfer. The topic appears in virtually all thermal management applications in engineering. Applications are found in mechanical, aerospace, chemical, nuclear, and countless other engineering fields. The fundamentals of this topic have grown out of research needs in electronics cooling, aerospace systems, and tribology applications over the past 60+ years. Much of the information in this field is found in research papers, compendiums, handbooks, unpublished reports, and other reference books in varying degrees of coverage. We have set out to provide the necessary fundamentals and applications in one comprehensive reference book which will enable current practitioners to apply the state-of-the-art knowledge to their problems and also enable them to go further with new problems as technology demands lead to new areas of research.

Thermal engineering covers a broad area of applications. In nearly all applications, heat enters or leaves a system through a smaller area of contact. This area of contact may be perfect or imperfect. If it is perfect, thermal spreading (or constriction) resistance occurs. If it is imperfect, then thermal transport occurs over a much smaller region of real contact within the apparent area of contact. This is contact resistance. The topics covered in this book have been drawn from over 60 years of research and development applications in many engineering fields. Each decade has brought new applications and new solutions. Early applications focused on thermal interfaces of mechanical joints, followed by electronics cooling using heat sinks for discrete heat sources. Today, we see micro-devices such as GaN high electron mobility transistors (HEMTs), LED lighting, and three-dimensional chip stacks. Other areas such as conduction in roller bearings and mass transfer from pharmaceutical patches also contain spreading resistance. Presently, the emerging area of super-hydrophobic (nonwetting) surfaces requires knowledge of thermal spreading at small scales. Even common problems such as heat loss from buildings and homes into the earth (a half-space) are modeled using thermal spreading resistance. We expect there will always be a need for the useful knowledge collected in this unique manuscript.

The book is presented primarily as a research reference. It compiles historical solutions and summarizes key results from over 60 years of published works in the field of thermal constriction (spreading) and contact resistance studies. The book is divided into 12 chapters covering the fundamentals and applications in problems where heat transfer from discrete heat sources is prevalent. The book also provides numerous simple predictive models for determining thermal spreading resistance and contact conductance of mechanical joints and interfaces. The book also discusses advanced applications in contact resistance, mass transfer, transport from super-hydrophobic surfaces, droplet/surface phase change problems, tribology applications such as sliding surfaces and roller bearings, heat transfer in micro-devices, micro-electronics, and thermal spreaders used in thermal management of electronics equipment. Solutions to fundamental problems are presented with extensive details which are often left out in handbook references. Fundamental concepts are explained clearly and concisely.

Much of the material that is presented in Chapters 10–12, along with key results in Chapters 2–4, was presented in a large number of short courses and book chapters dealing with Thermal Contact Resistance. The second author (M. Michael Yovanovich) presented several 3–5 day courses on Thermal Contact Resistance at various research labs in the 1970s, 1980s, 1990s, and early 2000s. He presented his first five-day course on *Theory and Applications of Thermal Contact Resistance* to the engineers and scientists at IBM Research Laboratory in Paris-Saclay, France, in 1968. Over the next 30 years, he presented numerous two to three-day and five-day courses at many companies and research establishments. Over 500 engineers and physicists attended these courses with most short courses attended by 35–40 engineers from industry and government research labs. Five-day long courses were presented to IBM Research and Manufacturing at Burlington, Vermont, Kingston and Poughkeepsie, New York; Los Alamos National Laboratory, Los Alamos, New Mexico; Nokia Research, Helsinki, Finland; and Sandia National Laboratories, Albuquerque, New Mexico. Ten 3-day courses were sponsored by Cooling Zone of Marlborough, Massachusetts, in the late 1990s and early 2000s. Finally, numerous two to three-day short courses presented during conferences sponsored by the American Society of Mechanical Engineers (ASME) and the American Institute of Aeronautics and Aeronautics (AIAA). The content of the courses was based on research results of about 60 graduate students (masters and PhD), two-to-three postdoctoral fellows and faculty members.

The funded research topics arose from the Nuclear Industry (Atomic Energy of Canada Limited in Pinawa, Manitoba and Chalk River, Ontario); Aerospace Industry (Lockheed Missiles and Space Corporation, in Palo, Alto, California; Los Alamos Laboratory in Los Alamos, New Mexico; Sandia Laboratories in Albuquerque, New Mexico); Microelectronics industries (such as Nokia Research in Helsinki, Finland, and IBM Research in Binghamton, Kingston, Poughkeepsie in New York, and Burlington in Vermont). Much of the theoretical and practical materials found in this book are based on the aforementioned sponsored research.

Though it is written primarily as a research reference, a limited number of examples are included throughout each chapter. As a research reference its purpose is to collect, summarize, and disseminate the most useful results that engineers use in the many fields of application. The analysis is detailed, allowing readers to appreciate the limitations of the theory as well as adapt it to new problems as they arise in practice. We have specifically targeted researchers and engineers working in the fields of heat transfer, thermal management of electronics, and tribology. Practitioners in other fields such as mechanical, aerospace, or nuclear engineering will also benefit, as many applications occur in these broad areas of engineering. Readers can also use it to learn solution techniques to extend and adapt theory to new problems.

The topics are distributed over 12 chapters and 3 appendices. Chapter 1 discusses the fundamental concepts and definitions used throughout the text. Chapters 2 through 8 present a thorough discussion of thermal spreading resistance fundamentals and solutions for a wide array of applications including the semi-infinite (or half-space) domain, flux tubes and disks, rectangular flux channels, orthotropic systems, multisource systems, transient spreading solutions, and finally problems with variable sink conductance. Chapter 9 provides extensive discussion on a number of special applications including moving heat sources, mass diffusion, superhydrophobic surfaces, temperature dependent thermal conductivity, and problems in spherical domains. The remaining three chapters, Chapters 10–12, provide comprehensive coverage of thermal contact resistance theory and its applications. Finally, we include three appendices. The first contains a concise overview of the special mathematical functions that appear in many of the solutions. The second presents useful materials pertaining to hardness which is needed in the application of contact resistance models. Finally, the third provides useful data for thermal properties of common metals, nonmetals, gases, and thermal interface materials.

Throughout each chapter, there are a handful of illustrative examples that are collected from our research experience as well as the published literature. These are provided to assist the reader in the appropriate application of the concepts and to provide some calculation benchmarks. Over 400 works spanning six decades of research have been cited. While we have made every effort to review all of the relevant historical literature, we no doubt have missed some excellent works. We strived to present the most useful and recurring material, i.e. the results used most often in practice or applications that can be easily extended using the methods outlined in the book.

Yuri S. Muzychka, St. John's, NL, Canada, 2023
M. Michael Yovanovich, Waterloo, ON, Canada, 2023

# Acknowledgments

The authors would like to acknowledge the contributions of many individuals whose work contributed immensely in the preparation of this text.

The first author is deeply indebted to the second author M. Michael Yovanovich for 30 years of mentorship and collaboration that ensued from the beginning of my research career as a graduate student at the University of Waterloo. I would also like to acknowledge the support and collaborations of R. Culham (University of Waterloo), M. Hodes (Tufts University), K. Bagnall (Device Lab at MIT), M. Razavi (Memorial University), B. Al-Khamaiseh (Memorial University), L. Lam (Tufts University, Memorial University), and S. Goudarzi (Memorial University) on the many problems in thermal spreading resistance we tackled and solved together. Additional thanks to A. Etminan for the preparation of many excellent figures, plots, and drawings appearing throughout the text.

The second author would like to thank all the students who collaborated in thermal spreading and contact resistance research over the past five decades. These include G. Schneider, S. Burde, V. Antonetti, A. Hegazy, J. DeVaal, S. Song, K. Negus, R. Culham, K. Kno, M. Sridhar, M. Mantelli, Y. Muzychka, P. Teerstra, H. Attia, V. Cecco, C. Tien, K. Martin, W. Kitscha, G. McGee, J. Zwart, J. Saabas, N. Fisher, P. Turyk, T. Lemczyk, M. Stevanovic, I. Savija, and M. Bahrami. Their contributions to this text are numerous.

Additionally, the many faculty and industry collaborators through the last five decades: H. Cordier and J. Coutenceau (University of Poitiers, France); H. Fenech, W. Rohsenow, B. Mikic, M. Cooper, and J. Henry (Massachusetts Institute of Technology, USA); L. Fletcher, G. Peterson (Texas A&M University, USA); V. Antonetti (Manhattan College, USA); J. Beck (Michigan State University, USA); G. Schneider, L. Chow, G. Gladwell, A. Strong, J. Thompson, J. Tevaarwerk, R. Culham, and D. Roulston (University of Waterloo, Canada); J. Dryden (Western University, Canada); Y. Muzychka (Memorial University of Newfoundland, Canada); M. Bahrami (Simon Fraser University, Canada); M. Mantelli and H. Milanez (Federal University of Santa Catarina, Brazil); M. Shankula (AECL Canada); and D. Wesley (Babcock and Wilcox, USA).

# Nomenclature

| | |
|---|---|
| $a, b, c, d$ | linear dimensions (m) |
| $a, b, c$ | radial dimensions (m) |
| $a, b$ | semi-axes of an ellipse or rectangle (m) |
| $a$ | contact spot radius (m) |
| $a_e$ | elastic contact spot radius (m) |
| $a_{ep}$ | elastic–plastic contact spot radius (m) |
| $a_H$ | Hertz contact spot radius (m) |
| $a_p$ | plastic contact spot radius (m) |
| $a_L$ | contact spot radius for layer (m) |
| $a_S$ | contact spot radius for substrate (m) |
| $A$ | area (m$^2$) |
| $A_a$ | apparent contact area (m$^2$) |
| $A_c$ | contact area (m$^2$) |
| $A_g$ | gap area (m$^2$) |
| $A_r$ | real contact area (m$^2$) |
| $A_t$ | flux tube cross-sectional area (m$^2$) |
| $A_0, A_m, A_n, A_{mn}$ | Fourier coefficients |
| $b$ | hydrodynamic slip length |
| $b_t$ | thermal slip length |
| $b_t$ | effective CLA roughness ($\equiv 2(CLA_1 + CLA_2)$) |
| $B_0, B_m, B_n, B_{mn}$ | modified Fourier coefficients |
| $Bi$ | Biot number ($\equiv h\mathcal{L}/k$) |
| $c_p$ | specific heat (J/kg K) |
| $C_0, C_m, C_n, C_{mn}$ | Fourier coefficients |
| $C_0, C_2, C_2, C_3 \ldots$ | equation coefficients |
| $C$ | concentration (mass/m$^3$) |
| $C_c$ | dimensionless contact conductance ($\equiv h_c\sigma/k_s m$) |
| $CLA$ | Center Line Average roughness (m) |
| $d$ | plate separation (m) |
| $D$ | diameter of contacting sphere (m) |
| $D_{AB}$ | mass diffusivity of $A$ in $B$ (m$^2$/s) |
| $\hat{f}, \bar{f}$ | influence coefficient (K/W) |
| $e^x$ | exponential function ($\equiv \exp(\cdot)$) |
| $e$ | eccentricity (m) |
| $\mathrm{erf}(\cdot)$ | Gaussian error function |

| | |
|---|---|
| $\mathrm{erfc}(\cdot)$ | complementary error function |
| $\mathrm{erfc}^{-1}(\cdot)$ | inverse complementary error function |
| $E(\cdot)$ | complete elliptic integral of the second kind |
| $E, E_1, E_2$ | modulus of elasticity (GPa) |
| $E'$ | effective modulus of elasticity (GPa) |
| $Ei(\cdot)$ | exponential integral |
| $f_{ep}$ | elastic–plastic contact parameter |
| $F$ | applied load (force) (N) |
| $F(\cdot)$ | incomplete elliptic integral of the first kind |
| $\mathcal{F}_{12}$ | effective radiative surface factor |
| $Fo$ | Fourier number $(\equiv \alpha t / \mathcal{L}^2)$ |
| $G$ | superposition functions |
| $h$ | convection film coefficient or conductance (W/(m$^2$ K)) |
| $h_s$ | sink plane conductance (W/(m$^2$ K)) |
| $h_c$ | contact conductance (W/(m$^2$ K)) |
| $h_e$ | edge conductance (W/(m$^2$ K)) |
| $h_g$ | gap conductance (W/(m$^2$ K)) |
| $h_j$ | joint conductance (W/(m$^2$ K)) |
| $h_0$ | maximum conductance (W/(m$^2$ K)) |
| $H_e$ | equivalent elastic micro-hardness (GPa) |
| $H_{ep}$ | elasto-plastic micro-hardness (GPa) |
| $H_B$ | Brinell hardness (GPa) |
| $H_L$ | layer hardness (GPa) |
| $H_S$ | micro-hardness of softer substrate (GPa) |
| $H_V$ | Vickers micro-hardness (GPa) |
| $H_p$ | equivalent plastic micro-hardness (GPa) |
| $H'$ | effective micro-hardness of coated surface (GPa) |
| $\mathrm{ierfc}(\cdot)$ | integrated complementary error function |
| $I$ | polar second moment of area |
| $I_0(\cdot), I_1(\cdot)$ | modified Bessel functions of the first kind of order 0 and 1 |
| $I_g$ | gap conductance integral |
| $I_{g,l}$ | gap conductance integral for line contact |
| $I_{g,p}$ | gap conductance integral for point contact |
| $J_z$ | mass flux in $z$-direction (mass/(m$^2$ s)) |
| $J_0(\cdot), J_1(\cdot)$ | Bessel functions of the first kind of order 0 and 1 |
| $K$ | thermal conductivity ratio |
| $K_0(\cdot), K_1(\cdot)$ | modified Bessel functions of the second kind of order 0 and 1 |
| $k, k_1, k_2$ | thermal conductivities (W/m K) |
| $k_g$ | gas thermal conductivity (W/(m K)) |
| $k_p$ | polymer thermal conductivity (W/(m K)) |
| $k_s$ | effective contact spot thermal conductivity $(\equiv 2k_1 k_2/(k_1 + k_2))$ |
| $k$ | reaction rate (s$^{-1}$) |
| $K(\cdot)$ | complete elliptic integral of the first kind |
| $K\{T\}$ | Kirchoff transform |
| $Kn$ | Knudsen number $(\equiv \Lambda / \ell)$ |
| $\ell$ | arbitrary depth (m) |
| $\mathcal{L}$ | arbitrary length scale (m) |

| | |
|---|---|
| $L, L_1, L_2$ | length (m) |
| $L_f$ | latent heat of fusion (J/kg) |
| $m, n$ | indices for summations |
| $m, n$ | Hertz parameters |
| $m, m_1, m_2$ | surface slope |
| $M$ | rarefaction parameter (m) |
| $M^*$ | dimensionless rarefaction parameter ($\equiv M/d$) |
| $n$ | normal direction (m) |
| $n$ | hyper-ellipse shape parameter |
| $n$ | contact spot density $N/A_a$ (m$^{-2}$) |
| $N$ | number of contact spots |
| $N$ | number of heat sources |
| $N$ | number of sides of a polygon |
| $P$ | contact pressure (MPa) |
| $P_g$ | gap gas pressure (Pa) |
| $Pe$ | Peclet number ($V\mathcal{L}/\alpha$) |
| $p$ | sink plane conductance distribution parameter |
| $p, s$ | thermal conductivity coefficients |
| $q_0$ | constant uniform heat flux (W/m$^2$) |
| $q$ | heat flux (W/m$^2$) |
| $q^\star$ | dimensionless heat flux ($q\sqrt{A}/k\Delta T$) |
| $Q$ | heat flow rate (W) |
| $Q'$ | heat flow rate per unit depth (W/m) |
| $r$ | cylindrical or spherical radial coordinate (m) |
| $r_i$ | inscribed radius of a polygon |
| $R$ | thermal resistance (K/W) |
| $R_g$ | gap thermal resistance (K/W) |
| $R_r$ | radiation thermal resistance (K/W) |
| $R_c$ | contact thermal resistance (K/W) |
| $R_j$ | joint thermal resistance (K/W) |
| $R_s$ | thermal spreading resistance (K/W) |
| $R_{total}$ | total thermal resistance (K/W) |
| $R_{1D}$ | bulk resistance thermal resistance (K/W) |
| $R^\star$ | dimensionless thermal spreading resistance |
| $R_c''$ | specific thermal resistance ($= 1/h_c$ [m$^2$ K/W]) |
| $S$ | shape factor (m) |
| $S_f$ | material flow stress (GPa) |
| $Ste$ | Stefan number ($c_p\Delta T/L_f$) |
| $t$ | time (s) |
| $t$ | integration variable |
| $t, t_1, t_2$ | total and layer thicknesses (m) |
| $T$ | temperature (K) |
| $T_b$ | bulk material temperature (K) |
| $T_c$ | contact temperature (K) |
| $T_j$ | joint temperature (K) |
| $T_s$ | source temperature (K) |
| $T_{cp}$ | contact plane surface temperature (K) |

| | |
|---|---|
| $T_f, T_\infty$ | sink temperature (K) |
| $\bar{u}$ | velocity (m/s) |
| $u_z$ | elastic displacement (m) |
| $U$ | velocity (m/s) |
| $U$ | Kirchoff transform variable |
| $V$ | velocity of heat sliding heat source (m/s) |
| $x$ | Cartesian coordinate (m) |
| $X_c, Y_c$ | heat source centroid (m) |
| $Y$ | mean plane separation (m) |
| $y$ | Cartesian coordinate (m) |
| $Y_0(\cdot)$ | Bessel function of the second kind of order zero |
| $z$ | Cartesian coordinate (m) |
| $Z_c$ | thermal spreading zone for circle ($\equiv R_c/R_{total}$) |
| $Z_e$ | thermal spreading zone for ellipse ($\equiv R_{se}/R_{total}$) |

## Greek Symbols

| | |
|---|---|
| $\alpha$ | thermal diffusivity ($\equiv k/\rho c_p$) |
| $\alpha$ | dimensionless conductivity ratio |
| $\alpha, \alpha_1, \alpha_2$ | accommodation coefficient |
| $\alpha, \beta$ | semi-axes of an ellipse (m) |
| $\alpha, \beta, \gamma$ | equation coefficients |
| $\beta_n$ | eigenvalues |
| $\beta_{mn}$ | eigenvalues ($\equiv \sqrt{\lambda_m^2 + \delta_n^2}$) |
| $\beta$ | angular measurement |
| $\beta(x, y)$ | Beta function |
| $\gamma$ | orthotropic conductivity variable ($\equiv \sqrt{k_z/k_{xy}}, \sqrt{k_z/k_r}$) |
| $\gamma$ | specific heat ratio |
| $\Gamma(\cdot)$ | Gamma function |
| $\delta_n$ | eigenvalues |
| $\delta$ | perpendicular (m) |
| $\delta$ | local gap thickness (m) |
| $\delta$ | penetration depth (m) |
| $\epsilon$ | aspect ratio ($\equiv b/a$) |
| $\epsilon$ | relative contact area ($\equiv \sqrt{A_r/A_a}$) |
| $\epsilon_1, \epsilon_2$ | surface emissivity |
| $\epsilon_c^*$ | dimensionless contact strain |
| $\zeta$ | dummy variable |
| $\eta$ | dimensionless length ($\equiv y/b$) |
| $\theta$ | temperature excess ($\equiv T - T_f$ [K]) |
| $\bar{\theta}$ | mean temperature excess (K) |
| $\hat{\theta}$ | centroidal temperature excess (K) |
| $\theta_0$ | constant uniform temperature excess (K) |
| $\theta_s$ | source temperature excess (K) |
| $\kappa$ | complementary modulus ($\equiv \sqrt{1 - \epsilon^2}$) |
| $\kappa$ | relative conductivity ($k_2/k_1$) |

| | |
|---|---|
| $\lambda$ | integration variable |
| $\lambda_m$ | eigenvalues |
| $\lambda$ | relative mean plane separation ($\equiv Y/\sigma$) |
| $\Lambda$ | mean free path (m) |
| $\mu$ | heat flux shape parameter |
| $\mu$ | coefficient of dynamic friction |
| $\mu$ | dynamic viscosity (Pa s) |
| $v, v_1, v_2$ | Poisson's ratio |
| $\xi$ | dimensionless length ($\equiv x/a$) |
| $\xi$ | denotes arbitrary eigenvalue in spreading function |
| $\xi$ | orthotropic coordinate transformation variable |
| $\xi$ | hydrodynamic spreading factor |
| $\rho$ | density (kg/m$^3$) |
| $\rho$ | segment length |
| $\rho$ | relative position in polar coordinates ($\equiv r/a$) |
| $\rho$ | radius of curvature (m) |
| $\rho_1, \rho_1', \rho_2, \rho_2'$ | radii of curvature of contacting bodies (m) |
| $\sigma, \sigma_1, \sigma_2$ | RMS surface roughness (m) |
| $\sigma$ | Stefan–Boltzmann constant ($\sigma = 5.67 \times 10^{-8}$ W/(m$^2$ K$^4$)) |
| $\tau_w$ | wall shear stress (Pa) |
| $\tau^\star$ | dimensionless wall shear ($\tau_w \sqrt{A}/\mu U$) |
| $\tau$ | relative thickness ($\equiv t/\mathcal{L}$) |
| $\phi_s$ | area contact ratio ($A_s/A_t$) |
| $\phi$ | thermal spreading function |
| $\varphi$ | reciprocal of thermal spreading function |
| $\psi$ | thermal constriction (spreading) parameter |
| $\psi$ | angular measurement |
| $\psi^*$ | thermal elastoconstriction parameter |
| $\omega$ | angular measurement |
| $\omega_1, \omega_2$ | thermal conductivity coefficients |
| $\Omega$ | omega function in point source method |

## Subscripts

| | |
|---|---|
| 0 | denotes at centroid or reference value |
| $1, 2, 3, \ldots$ | denotes layer number |
| $\sqrt{A}$ | length scale used to define dimensionless resistance |
| $a$ | apparent |
| $b$ | base |
| $b$ | bulk |
| $c$ | centroid |
| $c$ | contact |
| $cp$ | contact plane |
| $cr$ | critical |
| $e$ | effective |
| $e$ | edge |

| | |
|---|---|
| *eff* | effective value |
| *ei* | equivalent isothermal |
| *g* | gap |
| *H* | Hertz |
| *i,j* | denotes the *i*th and *j*th sources |
| *i* | inner |
| *ip* | in plane |
| *i* | inscribed |
| *j* | joint |
| *m* | moving |
| *ma* | macro |
| *mi* | micro |
| *mac* | macro |
| *mic* | micro |
| *o* | outer |
| *p* | polymer |
| *r* | *r*-direction |
| *r* | radiation |
| *r* | real |
| *source* | denotes source |
| *sink* | denotes sink |
| *s* | stationary |
| *s* | denotes source |
| *s* | denotes spreading |
| *total* | denotes total |
| *t* | thermal |
| *t* | denotes flux tube |
| *tp* | through plane |
| *υ* | viscous |
| *x* | associated with *x*-direction |
| *xy* | *xy*-plane |
| *y* | associated with *y*-direction |
| *z* | associated with *z*-direction |

## Superscripts

| | |
|---|---|
| * | denotes dimensionless variable |
| ′ | denotes derivative of function where specified |
| ′ | denotes effective variables where specified |
| $\overline{(\cdot)}$ | mean value |
| $\widehat{(\cdot)}$ | centroid value |
| *i,j* | denotes the *i*th and *j*th sources |
| *q* | isoflux |
| *T* | isothermal |

# 1

# Fundamental Principles of Thermal Spreading Resistance

Thermal spreading (or constriction) resistance results when heat is conducted away from a finite region into a larger region (spreading resistance) or from a larger region through a smaller region (constriction resistance). Spreading or constriction resistance can be determined by solving transient or steady-state heat conduction in a solid region with appropriately applied boundary and initial conditions. When the contact areas result in imperfect contact at the microscopic level, we find an array of discrete points of contacts results. Heat transfer across the apparent contact region occurs through conduction at these discrete points of contact and across the gap if there is a gas or liquid present; otherwise, heat transfer through radiation may occur in a vacuum. In these instances, we have contact resistance. In this text, we strive to present the general theory and applications of thermal spreading and contact resistance.

The earliest known thermal spreading solution is the classic problem of an isopotential (or electrified) disk on a semi-infinite region (see Figure 1.1). This problem was first examined by Weber (1873) (see also Gray and Mathews (1966)). Since that time, countless other solutions have been found for semi-infinite (half-space) and finite regions using several techniques. Kennedy (1960) obtained solutions for circular disk spreaders in semiconductor applications. These solutions represent the earliest known results used by engineers to predict heat flow from discrete heat sources. Additional results for a rectangular heat source also appeared in Carslaw and Jaeger (1959). Throughout the last six decades, many useful solutions have been obtained by researchers to aid in calculating heat flow or source temperature in a wide array of applications.

A significant number of these thermal spreading solutions are summarized in the two heat transfer handbook chapters by one of the authors: Yovanovich (1998) and Yovanovich and Marotta (2003). More recently, Razavi et al. (2016) provided a comprehensive review of the literature between 2003 and 2016. While these compendiums are useful, the limitations of space leave most of the details of solutions for the reader in the archival literature. For this reason, we have chosen to provide a comprehensive text on this important branch of heat transfer. We will develop the solutions to the most fundamental problems in great detail, while for others provide only the essential details to assist the reader in understanding the solution process. This approach is necessary, since many problems have rather direct easy-to-follow solutions, while others involve many additional steps and often require some numerical analysis.

In this chapter, we will introduce the reader to the most common geometries, governing equations, boundary conditions, solution methods, and applications of thermal spreading resistance. In Chapters 2–9, we go deeper into the details of many types of thermal spreading problems and their solutions. Finally, in Chapters 10–12, we deal specifically with thermal contact resistance principles and models.

*Thermal Spreading and Contact Resistance: Fundamentals and Applications*, First Edition.
Yuri S. Muzychka and M. Michael Yovanovich.
© 2023 John Wiley & Sons, Inc. Published 2023 by John Wiley & Sons, Inc.

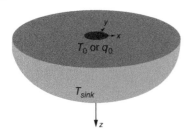

**Figure 1.1** Thermal spreading (constriction resistance) from an isothermal – $T_0$ source or an isoflux – $q_0$ source on a semi-infinite region (half-space).

## 1.1 Applications

Applications of thermal spreading resistance are numerous and include but are not limited to: thermal management in electronics devices and systems, thermal contact (interface) resistance, tribology applications such as roller bearings, phase change phenomena for droplets on surfaces, and thermal-fluid transport on super-hydrophobic surfaces and channels. Several examples are provided in Figure 1.2.

One of the earliest applications of thermal spreading/constriction resistance theory is in the area of thermal contact resistance [Yovanovich (1998)]. When two surfaces are brought into contact, one or more discrete points of contact result due to the geometry and/or inherent roughness of surfaces. The real contact area is much smaller than the nominal contact area and heat flow is disrupted while passing from one surface into the other surface, leading to an interface or contact resistance. This resistance is typically modeled using a contact conductance $h_c$. This contact conductance depends on the mechanical loading of the surfaces (contact pressure), the surface properties (roughness), mechanical properties (hardness, Young's modulus), and thermal properties of the materials in contact (thermal conductivity).

The microelectronics cooling area is rich in applications involving discrete heat sources at package scale which requires thermal management using heat sinks. The heat sink baseplate acts as a thermal spreader for one or more heat sources, and usually has fins which are convectively cooled. These fins may be reduced to a single value of conductance $h_s$ and an appropriate thermal spreading model used to predict the average (or maximum) source temperature or the maximum heat transfer rate for a temperature-constrained application.

At the microscale, device miniaturization produces thumbnail-sized electronic chips with many distributed discrete heat sources. Some examples are high electron mobility transistor (HEMT) devices and LED lighting. Analysis of these devices using analytical methods has been quite fruitful, and many useful solutions are available for a wide range of applications.

Other heat transfer applications arise in condensation, evaporation, or solidification problems involving droplets on surfaces. In these applications, the substrate provides a conduction path for heat and controls the phase change process. These surfaces may be treated or coated with a thin resistive layer, i.e. paints or other protective or even thermally enhancing coatings. Other interesting applications involve superhydrophobic surfaces and channels which have precisely engineered microstructural surface features such as regularly dispersed pillars or transverse or longitudinal ridges. Thermal transport from these complex surfaces relies upon thermal spreading resistance solutions to predict the heat transfer characteristics across these complex boundaries into the fluid in contact.

In tribology applications where sliding or rolling contact occurs between surfaces, thermal spreading resistance plays an important role in predicting the microcontact temperatures. In these applications, one surface is often modeled as rough and flat and the other surface as smooth and

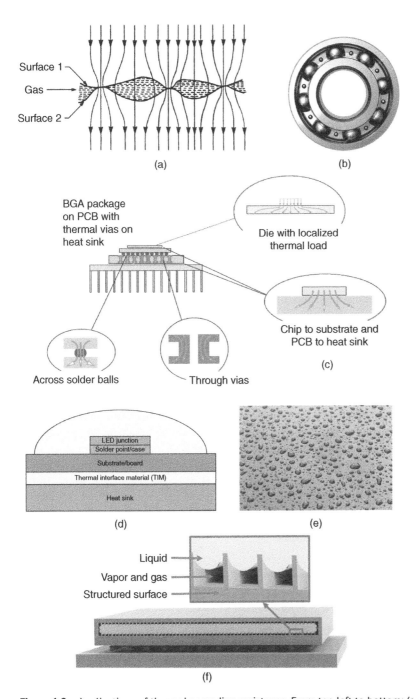

**Figure 1.2** Applications of thermal spreading resistance. From top left to bottom: (a) thermal contact resistance. Source: Hegazy (1985), (b) ball bearings. Source: Scanrail/Adobe Stock, (c) device cooling on heat sinks, Lee et al. (1995). Used with the permission of the American Society of Mechanical Engineers (ASME). (d) LED lighting, (e) phase change with droplets on surfaces. Source: Monsterkoi/Pixabay, and (f) super-hydrophobic channels, Karamanis et al. (2018). Used with the permission of the American Society of Mechanical Engineers (ASME).

flat. Heat generated through sliding or rolling friction is conducted into each surface, one of which is assumed to be in motion relative to the other surface. In this instance, a sliding thermal spreading resistance and stationary thermal spreading resistance are required to predict the average surface contact temperature. In applications involving roller bearings, thermal constriction zones are created between the ball and race.

Additional applications can be found in other fields and will be dealt with as we progress through the text. One unique application is in mass diffusion from discrete regions containing a chemical species such as in pharmaceutical patches. With some modification, thermal spreading solutions can be easily used to model analogous mass diffusion applications.

## 1.2 Semi-Infinite Regions, Flux Tubes, Flux Channels, and Finite Spreaders

Since thermal spreading occurs by means of conduction into a solid region, three distinct classes of geometry have been widely analyzed. These are the (i) the semi-infinite domain or half-space, (ii) the semiinfinite flux tube or channel, and (iii) the finite disk or rectangular channel. Figure 1.3 illustrates the relationship among these three types of geometry. With appropriate care, solutions for one system can yield solutions for other systems, though they may be more complex computationally. But with today's computing power, this has become less of an issue.

In the semiinfinite domain, heat is conducted in all directions away from a finite source and there are no finite boundaries other than the heat source plane. In the source plane, the region outside of the heat source area is assumed to be adiabatic, i.e. all heat flows into the infinite sink. The semiinfinite flux tube or channel configuration results when heat flows from a finite source into a larger but finite specified region which is bounded laterally by adiabatic boundaries for which the heat flow is then directed into a semiinfinite direction $z > 0$. If the flux tube or channel is sufficiently thick, the flux lines become parallel to the lateral boundaries. The finite disk/channel geometry has boundaries specified laterally and transversely. Heat is conducted away from a finite

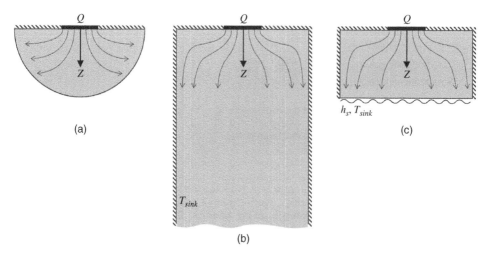

**Figure 1.3** Thermal spreading problems with adiabatic boundaries. (a) Half-space, (b) flux tube/channel, and (c) finite disk.

source area in the source plane and flows toward a sink plane located at some fixed distance $z = \ell$, which is usually maintained at constant temperature $T_{sink}$ or with a uniform conductance $h_s$ prescribed over the sink plane.

Many adaptations to these fundamental configurations have also considered nonadiabatic boundaries with finite conductance $h_e$ specified uniformly along edges as shown in Figure 1.4. In thin disk spreaders edge losses are usually negligible, but in thicker spreaders edge losses contribute to reducing the thermal spreading resistance. While seemingly more complex, these problems are easily solved using an assortment of mathematical techniques. The vast majority of solved problems are dealt with using either circular cylindrical coordinates or rectangular co-ordinates. Solutions are usually found using classical methods of heat conduction analysis such as separation of variables or integral transforms.

In all cases, the geometry being analyzed may be either isotropic (single value of thermal conductivity), orthotropic (different values of thermal conductivity in each of the principle orthogonal co-ordinate directions)), or compound or multilayered (with isotropic or orthotropic layers), see Figure 1.5. In problems with more than one layer, the complexity of the solution increases, but many systems have been solved analytically. In Chapters 3, 4, 5, and 6, we delve into these problems which are primarily motivated by microelectronics cooling applications.

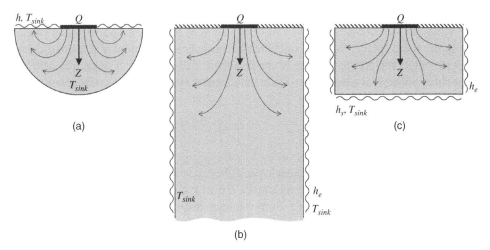

**Figure 1.4** Thermal spreading problems with edge conductance. (a) Half-space, (b) flux tube/channel, and (c) finite disk.

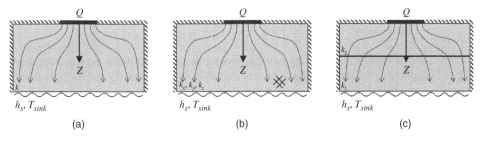

**Figure 1.5** Thermal spreading problems in isotropic (a), orthotropic (b), and compound (c) systems.

## 1.3   Governing Equations and Boundary Conditions

The fundamental equation which is solved in thermal spreading resistance applications is Laplace's equation for steady-state problems:

$$\nabla^2 T = 0 \tag{1.1}$$

or the heat diffusion equation:

$$\frac{1}{\alpha}\frac{\partial T}{\partial t} = \nabla^2 T \tag{1.2}$$

for transient problems.

For most steady problems, we are concerned with solving Laplace's equation in either 2D or 3D Cartesian coordinates or 2D circular cylindrical coordinates:

$$\frac{\partial^2 T}{\partial x^2} + \frac{\partial^2 T}{\partial y^2} + \frac{\partial^2 T}{\partial z^2} = 0 \tag{1.3}$$

or

$$\frac{\partial^2 T}{\partial r^2} + \frac{1}{r}\frac{\partial T}{\partial r} + \frac{\partial^2 T}{\partial z^2} = 0 \tag{1.4}$$

with appropriately specified boundary conditions such as: adiabatic edges or convectively cooled boundaries, symmetry conditions along the axis at the centroid of the heat source, and a single heat source in the source plane. Other co-ordinate systems may also be employed for particular problems. For example the classic problem of the isopotential disk on a half space may also be solved using oblate spheroidal coordinates [Moon and Spencer (1961)] or ellipsoidal coordinates for the elliptic disk.

### 1.3.1   Source Plane Conditions

The heat source region is conventionally modeled as either constant temperature $T_0$ or constant heat flux $q_0$. In other problems, a finite conductance $h$ over the source area or over the region outside of the source area may be considered. However, more often than not, a uniform flux distribution is prescribed when the outer region of the source plane is considered adiabatic. This avoids having to solve the more difficult mixed potential boundary condition, e.g. having temperature specified in one region and (zero) heat flux specified in another region of the same boundary. Most often, the following condition is applied:

$$\left.\begin{array}{rcl} \left.\dfrac{\partial T}{\partial z}\right|_{z=0} &=& -\dfrac{q_0}{k}, \text{ Within Source Region} \\[2mm] \left.\dfrac{\partial T}{\partial z}\right|_{z=0} &=& 0, \quad \text{Outside Source Region} \end{array}\right\} \tag{1.5}$$

Additionally, when a heat flux boundary condition is employed, a nonuniform flux may also be defined over the source region. Typical heat flux distributions include: (i) inverted parabolic heat flux, (ii) uniform heat flux, or (iii) a parabolic heat flux. In cylindrical co-ordinates, we may conveniently solve many problems with the following heat flux distribution:

$$q(r) = q_0(1+\mu)\left(1 - \frac{r^2}{a^2}\right)^{\mu}, \quad 0 \le r \le a \tag{1.6}$$

or for a two-dimensional strip region:

$$q(x) = \frac{Q'\Gamma(\mu + 3/2)}{\sqrt{\pi}\, a\, \Gamma(\mu + 1)} \left(1 - \frac{x^2}{a^2}\right)^{\mu}, \quad 0 \leq x \leq a \tag{1.7}$$

where $Q' = Q/L$, the heat input per unit depth.

The above can be easily applied for cases when $\mu = -1/2$, $\mu = 0$, and $\mu = 1/2$. The first case represents a uniform temperature source on a half-space or an approximately isothermal source on a flux tube or channel for small aspect ratios. For circular heat source on a flux tube, we often develop solutions for the following three special cases:

$$q(r) = \begin{cases} \dfrac{Q}{2\pi a^2} \dfrac{1}{\sqrt{1 - r^2/a^2}} & \text{for} \quad \mu = -1/2 \\[2ex] \dfrac{Q}{\pi a^2} & \text{for} \quad \mu = 0 \\[2ex] \dfrac{3Q}{2\pi a^2} \sqrt{1 - r^2/a^2} & \text{for} \quad \mu = 1/2 \end{cases} \tag{1.8}$$

while for a simple strip source on a flux channel, we often use:

$$q(x) = \begin{cases} \dfrac{Q'}{\pi a} \dfrac{1}{\sqrt{1 - x^2/a^2}} & \text{for} \quad \mu = -1/2 \\[2ex] \dfrac{Q'}{2a} & \text{for} \quad \mu = 0 \\[2ex] \dfrac{2Q'}{\pi a} \sqrt{1 - x^2/a^2} & \text{for} \quad \mu = 1/2 \end{cases} \tag{1.9}$$

These flux distributions also provide a convenient means to bound the thermal resistance, since in many applications, the actual flux distribution is not known.

### 1.3.2 Sink Plane Conditions

The sink plane may be modeled as semi-infinite for flux tubes and channels or with a constant or distributed conductance for finite disks and rectangular channels. In the case of semi-infinite flux tubes/channels, we often prescribe a uniform flow:

$$-k\frac{\partial T}{\partial z} = Q/A_t, \quad z \to \infty \tag{1.10}$$

in a region with cross-sectional area $A_t$ located at a significant distance from the contact plane.

In the case of finite heat spreaders, a conductance may be a result of applied convective heat transfer or a combination of convection/conduction using fins. In most cases, $h_s$ is assumed to be constant over the surface such that:

$$-k\frac{\partial T}{\partial z}\bigg|_{z=\ell} = h_s[T_{z=\ell} - T_{sink}] \tag{1.11}$$

In many microelectronics cooling applications involving heat sinks, the fins may not be uniform in height and/or spacing, or may be impingement cooled with fans. These situations create a nonuniform conductance along the sink plane such that $h_s$ is symmetrically or asymmetrically distributed over the surface as $h(r)$, $h(x)$, or $h(x, y)$. In Chapter 8, we will consider nonuniform conductance applications which can occur in many heat sink applications with fan cooling and variable fin height.

### 1.3.3 Interface Conditions

Finally, in compound (or multilayered systems), we must also include additional boundary conditions at the interfaces of each material pair. These are usually equality of temperature and heat flux,

$$
\left.
\begin{aligned}
T_1 &= T_2 \\
k_1 \frac{\partial T_1}{\partial z_1}\bigg|_{z_1=t_1} &= k_2 \frac{\partial T_2}{\partial z_2}\bigg|_{z_2=0}
\end{aligned}
\right\}
\tag{1.12}
$$

or equality of heat flux and a mixed boundary condition with interfacial conductance, $h_c$:

$$
\left.
\begin{aligned}
-k_1 \frac{\partial T_1}{\partial z_1}\bigg|_{z_1=t_1} &= h_c[T_1 - T_2] \\
k_1 \frac{\partial T_1}{\partial z_1}\bigg|_{z_1=t_1} &= k_2 \frac{\partial T_2}{\partial z_2}\bigg|_{z_2=0}
\end{aligned}
\right\}
\tag{1.13}
$$

For multilayered systems, the addition of each pair of interface conditions makes the problem statement significantly more complex, but a simple recursive method for accounting for each new layer can be developed.

## 1.4 Thermal Spreading Resistance

Thermal spreading resistance arises in multidimensional applications where heat enters a solid region through a portion of a surface, i.e. a finite area within the system boundary. The solution of such problems involves finding the temperature distribution which satisfies Laplace's equation in two or three dimensions along with the appropriately prescribed boundary conditions as outlined earlier. When a temperature field solution is found, thermal resistance may be defined using a number of different definitions. The definition used to determine the thermal spreading resistance depends upon the type of problem.

### 1.4.1 Half-Space Regions

Heat may enter or leave an isotropic half-space (a region whose dimensions are much larger than the characteristic length of the heat source area) through planar singly or doubly connected areas (e.g. circular or annular area). The free surface of the half-space is adiabatic except for the source area. If heat enters the half-space, the flux lines "spread" apart as the heat is conducted away from the small source area (Figure 1.3), then the thermal resistance is called *spreading* resistance. If the heat leaves the half-space through a small area, then the flux lines are "constricted" and the thermal resistance is called *constriction* resistance. The heat transfer may be steady or transient. The temperature field $T$ in the half-space is, in general, three-dimensional, and steady or transient. The temperature in the source area may be two-dimensional, and steady or transient.

If heat transfer is into the half-space, then the spreading resistance is defined as [Carslaw and Jaeger (1959), Yovanovich (1976), Madhusudana (1996), Yovanovich and Antonetti (1988)]:

$$
R_s = \frac{T_{source} - T_{sink}}{Q}
\tag{1.14}
$$

where $T_{source}$ is a convenient source temperature and $T_{sink}$ is a convenient thermal sink temperature; and where $Q$ is the steady or transient heat transfer rate:

$$Q = \iint_A q_n \, dA = \iint_A -k\frac{\partial T}{\partial n} dA \tag{1.15}$$

where $q_n$ is the heat flux component normal to the area and $\partial T/\partial n$ is the temperature gradient normal to the area. If the heat flux distribution is uniform over the area, then $Q = qA$. For singly and doubly connected source areas, three source temperatures have been used in the definition: (i) the maximum temperature, (ii) the centroid temperature, and (iii) the area-average temperature which is defined according to Yovanovich (1976) as

$$\overline{T}_{source} = \frac{1}{A} \iint_A T_{source} \, dA \tag{1.16}$$

where $A$ is the source area. Since the sink area is much larger than the source area, it is, by convention, assumed to be isothermal, i.e. $T_{sink} = T_\infty$. The maximum and centroid temperatures are identical for singly connected, axisymmetric source areas; otherwise, they are different [Yovanovich (1976), Yovanovich and Burde (1977), Yovanovich et al. (1977)]. For doubly connected source areas (e.g. circular annulus), the area-average source temperature is used [Yovanovich and Schneider (1977)]. If the source area is assumed to be isothermal, then $\overline{T}_{source} = T_0$.

The general definition of spreading (or constriction) resistance leads to the following relationship for the dimensionless spreading resistance:

$$k\mathcal{L}R_s = \frac{\mathcal{L}}{A} \frac{\iint_A \theta \, dA}{\iint_A -\left(\frac{\partial \theta}{\partial z}\right)_{z=0} dA} \tag{1.17}$$

where $\theta = T(x, y, 0) - T_\infty$, the rise of the source temperature above the sink temperature. The arbitrary characteristic length scale of the source area is denoted as $\mathcal{L}$.

The spreading resistance definition holds for transient conduction into or out of the half-space. If the heat flux is uniform over the source area, the temperature is nonuniform; and if the temperature of the source area is uniform, the heat flux in nonuniform [Carslaw and Jaeger (1959), Yovanovich (1976)]. For convenience, the dimensionless spreading resistance, denoted as $\psi = k\mathcal{L}R_s$ [Yovanovich (1976), Yovanovich and Antonetti (1988)], is called the spreading (or constriction) resistance parameter. This parameter depends on the heat flux distribution over the source area and the shape and aspect ratio of the singly or doubly connected source area. For simple contacts like the circular disk, $\mathcal{L} = a$ is chosen. For other geometries such as rectangular or elliptical contacts the choice is less obvious, but frequently $\mathcal{L} = \sqrt{A_s}$ is chosen, as it has been shown to minimize the effect of geometry on dimensionless results for sources of similar aspect ratio. We usually define three specific cases for the spreading parameter $\psi$ for comparison purposes. These are:

$$k\mathcal{L}R_s = \begin{cases} \psi^T & -\text{Isothermal} \\ \psi^q & -\text{Isoflux} \\ \psi_{ei}^T & -\text{Equivalent Isothermal} \end{cases} \tag{1.18}$$

Since many solutions for the constriction parameter are only for uniform flux, in these cases, we drop the superscript as it is implied that it is for an isoflux heat source. We only make the distinction when multiple boundary conditions are considered.

The relationship for the dimensionless spreading resistance is mathematically identical to the dimensionless constriction resistance for identical boundary conditions on the source area. For a nonisothermal singly connected area, the spreading resistance can also be defined with respect to its maximum temperature or the temperature at its centroid [Carslaw and Jaeger (1959), Yovanovich (1976), Yovanovich and Burde (1977)]. These temperatures, in general, are not identical, and they are greater than the area-average temperature [Yovanovich and Burde (1977)].

The definition of spreading resistance for the isotropic half-space is applicable for single and multiple isotropic layers which are placed in perfect thermal contact with the half-space, and the heat which leaves the source area is conducted through the layer or layers before entering into the half-space.

### 1.4.2 Semi-Infinite Flux Tubes and Channels

If a circular heat source of area $A_s$ is in contact with a very long circular flux tube or channel of cross-sectional area $A_t$ (Figure 1.3), the flux lines are constrained by the adiabatic sides to "bend," and then become parallel to the axis of the flux tube or channel at some distance $z = \ell$ from the contact plane at $z = 0$.

The isotherms, shown as dashed lines, are everywhere orthogonal to the flux lines. The temperature in planes $z = \ell \gg \sqrt{A_t}$ "far" from the contact plane $z = 0$ becomes isothermal, while the temperature in planes near $z = 0$ is two- or three-dimensional. The thermal conductivity of the flux tube or channel is generally assumed to be constant.

The total thermal resistance $R_{total}$ for steady conduction from the heat source area in $z = 0$ to the arbitrary plane $z = \ell$ is given by the relationship:

$$QR_{total} = \overline{T}_s - \overline{T}_{z=\ell} \tag{1.19}$$

where $\overline{T}_s$ is the mean source temperature and $\overline{T}_{z=\ell}$ is the mean temperature of the arbitrary plane. The one-dimensional resistance of the region bounded by $z = 0$ and $z = \ell$ is given by the relation:

$$QR_{1D} = \overline{T}_{z=0} - \overline{T}_{z=\ell} \tag{1.20}$$

The total resistance is equal to the sum of the one-dimensional resistance and the spreading resistance:

$$R_{total} = R_{1D} + R_s \quad \text{or} \quad R_{total} - R_{1D} = R_s \tag{1.21}$$

By subtraction, the relationship for the spreading resistance proposed by Mikic and Rohsenow (1966) is:

$$R_s = \left( \frac{\overline{T}_s - \overline{T}_\ell}{Q} - \frac{\overline{T}_{z=0} - \overline{T}_\ell}{Q} \right) = \frac{\overline{T}_s - \overline{T}_{z=0}}{Q} \tag{1.22}$$

where $Q$ is the total heat transfer rate from the source area into the flux tube. It is given by

$$Q = \iint_{A_s} -k \frac{\partial T}{\partial z} \bigg|_{z=0} dA_s \tag{1.23}$$

The dimensionless spreading resistance parameter $\psi = k\mathcal{L}R_s$ is also introduced for convenience. The arbitrary length scale $\mathcal{L}$ is related to some dimension of the source area. In general, $\psi$ depends on the shape and aspect ratio of the source area, the shape and aspect ratio of the flux tube cross-section, the relative size of the source area to flux tube area, the orientation of the source area relative to the cross-section of the flux tube, the boundary condition on the source area, and the temperature basis for definition of the spreading resistance.

### 1.4.3 Finite Disks and Channels

In applications involving finite heat spreaders, the system is idealized as having a central heat source placed on one of the heat spreader surfaces, while the lower surface is cooled with a constant conductance which may represent a heat sink, contact conductance, or convective heat transfer coefficient, see Figure 1.3. All edges are usually assumed to be adiabatic. As well, the region outside the heat source in the source plane is also frequently modeled as adiabatic.

In this idealized system, the total thermal resistance of the system is defined as:

$$R_{total} = \frac{\overline{T}_s - T_{sink}}{Q} = \frac{\overline{\theta}_s}{Q} = R_{1D} + R_s \tag{1.24}$$

where $\overline{\theta}_s$ is the mean source temperature excess and $Q$ is the total heat input of the device. In these instances, the total resistance is the sum of the one-dimensional bulk resistance and the spreading resistance. If there is edge conductance (see Figure 1.4), then the resistance cannot be easily split into these two components, and we merely define the total resistance.

We will examine a number of fundamental problems containing a thermal spreading or constriction resistance, beginning first with the simplest problem of a finite heat source on an adiabatic isotropic half-space, and then proceeding to problems involving flux tubes/channels, and finite spreaders.

## 1.5 Solution Methods

Various analytical and numerical approaches have been applied to thermal spreading problems. The nature of the solution method usually depends upon the type of problem being considered. For example, for all problems involving flux tube/channel or disk/channel type of geometry, the separation of variables method is usually applied with great success depending upon the problem complexity. In many problems, a full analytical solution is obtained. Other problems have utilized integral transform techniques such as the Laplace transform, Fourier transform, and Hankel transform, which are used in time-dependent solutions or those specified in circular cylindrical coordinates. By and large, these methods are the most successful for obtaining solutions.

In other problems, such as those with mixed potential boundary conditions specified in the source plane, or problems with nonuniform conductance in the sink plane, an approximate solution is usually found using the method of least squares for approximating the usual Fourier series expansions that result from application of the final boundary condition in the source plane. In many of the half-space problems, point source integration is used for regular- and irregular-shaped constant flux heat sources. The technique, while mathematically elegant, can lead to the need for numerical integration, though many applications of the method in Chapter 2 lead to closed-form solutions.

In potential theory applications, conformal mapping methods have also been used to obtain a number of solutions. Several solutions found in electric field theory applications have been obtained using this approach [Smythe (1968)] and are presented in Chapters 2–4. More recently, conformal mapping techniques are being utilized in solutions being developed for super-hydrophobic surfaces, whereby complex boundaries involving the curved meniscus of a liquid (see Figure 1.2f) are being mapped into more appropriate geometry prior to solution.

Finally, full numerical solutions using finite difference, finite volume, or finite element-based schemes have also been regularly employed. Computational approaches using discretized methods can be time-consuming and it is tempting to simply rely on these approaches over analytical methods. Often, each computational run requires an adequate grid, and as such, simple changes in boundary condition in the source plane, i.e. relocating the source or adding another source, can often add significant extra effort in meshing and subsequent computation time. Thus, being able to avail of a simpler analytic solution may often result in significant time savings. Each of these approaches will be discussed in more detail in Chapters 6 and 8, where we compare analytical results to finite element solutions. Since we are advocating the use of analytical modeling wherever possible, we will not delve into the specific details of these methods using modern codes.

## 1.6 Summary

We reviewed the fundamentals of thermal spreading resistance. In Chapters 2–8, we will discuss fundamental solutions for many thermal spreading problems in cylindrical and rectangular coordinate systems. These solutions have a wide range of potential applications, many of which are discussed in Chapter 9. In Chapter 10, we introduce thermal contact resistance and discuss the required concepts of joint conductance, and what is required to predict it. We conclude the book with two detailed chapters (Chapters 11 and 12) on joint conductance models for nonconforming smooth solids and conforming rough surfaces.

## References

Carslaw, H., and Jaeger, J.C., *Conduction of Heat in Solids*, Oxford Press, Oxford, UK, pp. 215–216, 1959.

Gray, A., and Mathews, G.B., *A Treatise on Bessel Functions and Their Applications*, Dover Publications, New York, p. 141, 1966.

Hegazy, A., *Thermal Joint Conductance of Conforming Rough Surfaces: Effect of Surface Micro-Hardness Variation*, Phd Thesis, University of Waterloo, 1985.

Karamanis, G., Hodes, M., Kirk, T., and Papageordgiou, D.T., "Solution of the Extended Graetz-Nusselt Problem for Liquid Flow Over Isothermal Parallel Ridges", *Journal of Heat Transfer*, Vol. 140, 061703-1–061703-15, 2018.

Kennedy, D.P., "Spreading Resistance in Cylindrical Semi-conductor Devices", *Journal of Applied Physics*, Vol. 31, no. 8, pp. 1490–1497, 1960.

Lee, S., Song, S., Au, V., and Moran, K.P., "Constriction/Spreading Resistance Model for Electronics Packaging", *Proceedings of the ASME/JSME Thermal Engineering Conference*, Vol. 4, pp. 199–206, 1995.

Madhusudana, C.V., *Thermal Contact Conductance*, Springer-Verlag, New York, 1996.

Mikic, B., and Rohsenow, W.M., "Thermal Contact Conductance", MIT Report, DSR 74542-41, 1966.

Moon, P., and Spencer, D.E., *Field Theory for Engineers*, D. Van Nostrand Company, Princeton, NJ, p. 275, 1961.

Razavi, M., Muzychka, Y.S., and Kocabiyik, S., "Review of Advances in Thermal Spreading Resistance Problems", *AIAA Journal of Thermophysics and Heat Transfer*, Vol. 30, no. 4, pp. 863–879, 2016.

Smythe, W.R., *Static and Dynamic Electricity*, 3rd ed., McGraw-Hill, New York, 1968.

Weber, H., "Ueber die Besselschen Functionen und ihre Anwendung auf die Theorie der Elektrischen Ströme", *Journal für die Reine und Angewandte Mathematik*, Vol. 1873, no. 75, pp. 75–105, 1873.

Yovanovich, M.M., "Thermal Constriction Resistance of Contacts on a Half-Space: Integral Formulation", in *Progress in Astronautics and Aeronautics: Radiative Transfer and Thermal Control*, Vol. 49, AIAA, New York, pp. 397–418, 1976.

Yovanovich, M.M., "Chapter 3. Conduction and Thermal Contact Resistances (Conductances)", in *Handbook of Heat Transfer*, eds. W.M. Rohsenow, J.P. Hartnett, and Y.I. Cho, McGraw-Hill, New York, 1998.

Yovanovich, M.M., and Antonetti, V.W., "Application of Thermal Contact Resistance Theory to Electronic Packages, in *Advances in Thermal Modeling of Electronic Components and Systems*, Vol. 1, eds. A. Bar-Cohen and A.D. Kraus, Hemisphere Publishing, New York, pp. 79–128, 1988.

Yovanovich, M.M., and Burde, S.S., "Centroidal and Area Average Resistances of Non-symmetric, Singly Connected Contacts", *AIAA Journal*, Vol. 15, no.10, pp. 1523–1525, 1977.

Yovanovich, M.M., and Marotta, E., "Chapter 4. Thermal Spreading and Contact Resistances", in *Heat Transfer Handbook*, eds. A. Bejan and A.D. Kraus, Wiley, New York, 2003.

Yovanovich, M.M., and Schneider, G.E., "Thermal Constriction Resistance Due to a Circular Annular Contact", in *Progress in Astronautics and Aeronautics*, Vol. 56, AIAA, New York, pp. 141–154, 1977.

Yovanovich, M.M., Burde, S.S., and Thompson, J.C., "Thermal Constriction Resistance of Arbitrary Planar Contacts with Constant Flux", in *Progress in Astronautics and Aeronautics: Thermophysics of Spacecraft and Outer Planet Entry Probes*, Vol. 56, AIAA, New York, pp. 127–139, 1977.

# 2

# Thermal Spreading in Isotropic Half-Space Regions

Steady or transient heat transfer occurs in a semi-infinite region or half-space $z > 0$, which may be isotropic or it may consist of one or more thin isotropic layers bonded to the isotropic half-space. The heat source is some planar singly or doubly connected area such as a circular annulus located in the "free" surface $z = 0$ of the half-space. The dimensions of the half-space are much larger than the largest dimension of the source area. The "free" surface $z = 0$ of the half-space outside the source area is traditionally assumed to be adiabatic.

If the source area is isothermal, then the heat flux over the source area is nonuniform. Otherwise, if the source is subjected to a uniform heat flux, then the source area is nonisothermal. In other cases, the source may have a distributed heat flux such as parabolic or Hertz flux distribution. In this case, the heat flux is zero at the edge of the source and maximum at the centroid. The majority of solutions for discrete sources on a half-space have been obtained for the isothermal or isoflux heat sources. In a select few studies, solutions for a source with finite conductance have been found. Additionally, in other cases, several problems for the circular source having conductance in the region outside of the source area have also been considered.

Solutions to thermal spreading problems in half-spaces are either analytical or numerical in nature. Generally speaking, only the circular, elliptical, rectangular, and annular ring sources have closed-form solutions. Other solutions are obtained in terms of integrals which often require numerical integration to complete the solution. A method involving the superposition of point sources has also been used with great success in finding thermal resistance solutions. We outline the method later and provide a summary of useful results obtained using the method.

## 2.1 Circular Area on a Half-Space

There are two classical steady-state solutions available for the circular source area of radius $a$ on the surface of a half-space of thermal conductivity $k$. The solutions are for the isothermal and isoflux source areas, see Figure 2.1. In both problems, the temperature field is two-dimensional in circular-cylinder coordinates, i.e. $\theta(r, z) = T - T_{sink}$. There is also a solution for the special case of a distributed parabolic heat flux. The important results are presented here with details of the solution methods for each.

*Thermal Spreading and Contact Resistance: Fundamentals and Applications*, First Edition.
Yuri S. Muzychka and M. Michael Yovanovich.
© 2023 John Wiley & Sons, Inc. Published 2023 by John Wiley & Sons, Inc.

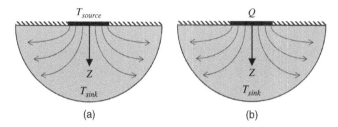

**Figure 2.1** Thermal spreading from a circular heat source on a half-space. (a) Isothermal. (b) Isoflux.

### 2.1.1 Isothermal Circular Source

In the case of a circular isothermal heat source on a half-space, the governing equation for this problem is Laplace's equation:

$$\frac{\partial^2 \theta}{\partial r^2} + \frac{1}{r}\frac{\partial \theta}{\partial r} + \frac{\partial^2 \theta}{\partial z^2} = 0 \tag{2.1}$$

In this problem the mixed-boundary conditions [Sneddon (1966)] on the free surface are:

$$\begin{aligned} z = 0 \quad 0 \leq r < a \quad &\theta = \theta_0 \\ z = 0 \quad r > a \quad &\frac{\partial \theta}{\partial z} = 0 \end{aligned} \tag{2.2}$$

and the condition at remote points is: as $\sqrt{r^2 + z^2} \to \infty$, then $\theta \to 0$.

The temperature distribution throughout the half-space $z \geq 0$ is given by the infinite integral [Carslaw and Jaeger (1959)]:

$$\theta = \frac{2}{\pi}\theta_0 \int_0^\infty e^{-\lambda z} J_0(\lambda r) \sin(\lambda a)\frac{d\lambda}{\lambda} \tag{2.3}$$

where $J_0(x)$ is the Bessel function of the first kind of order zero [Abramowitz and Stegun (1965)], and $\lambda$ is a dummy variable. The solution can be written in the following alternative form according to Carslaw and Jaeger (1959):

$$\theta = \frac{2}{\pi}\theta_0 \sin^{-1}\left[\frac{2a}{\sqrt{(r-a)^2 + z^2} + \sqrt{(r+a)^2 + z^2}}\right] \tag{2.4}$$

The heat flow rate from the isothermal circular source into the half-space is found from:

$$\begin{aligned} Q &= \int_0^a -k\left[\frac{\partial \theta}{\partial z}\right]_{z=0} 2\pi r \, dr \\ &= 4\,k\,a\,\theta_0 \int_0^\infty J_1(\lambda a) \sin(\lambda a)\frac{d\lambda}{\lambda} \\ &= 4\,k\,a\,\theta_0 \end{aligned} \tag{2.5}$$

From the definition of thermal resistance, one finds the relationship for the spreading resistance [Carslaw and Jaeger (1959)]:

$$R_s = \frac{\theta_0}{Q} = \frac{1}{4ka} \tag{2.6}$$

The heat flux distribution over the isothermal heat source area is axisymmetric [Carslaw and Jaeger (1959)]:

$$q(r) = \frac{Q}{2\pi a^2}\frac{1}{\sqrt{1 - (r/a)^2}}, \quad 0 \leq r < a \tag{2.7}$$

This flux distribution is minimum at the centroid $r = 0$ and becomes unbounded at the edge $r = a$.

Equation (2.6) may be used to define a dimensionless constriction or thermal spreading parameter denoted $\psi$ as:

$$\psi = 4kaR_s \tag{2.8}$$

This leads to $\psi = 1$ for an isothermal circular source. Future results may be defined in this manner to assess the effect of heat source shape and boundary condition on the dimensionless thermal resistance. Further variations on the definition of $\psi$ will be considered in the context of source shape effect using a novel length scale $\sqrt{A}$ in place of the source radius $a$, where $A$ is the source area. In these cases, we will define a dimensionless spreading resistance $R^\star$ as:

$$R^\star = k\sqrt{A}R_s \tag{2.9}$$

The above results may also be obtained by considering the problem statement in oblate spheroidal coordinates [Moon and Spencer (1961)]. The oblate spheroidal coordinate system allows for the mixed boundary condition in the source plane to be easily satisfied since the regions within and outside the source represent two distinct boundary conditions as compared to the same condition specified piecewise above.

### 2.1.2 Isoflux Circular Source

We now consider the problem for the isoflux heat source. This requires modification to the source plane boundary conditions. In this problem, the boundary conditions on the free surface are:

$$
\begin{aligned}
z = 0 \quad 0 \leq r < a \quad &\frac{\partial \theta}{\partial z} = -\frac{q_0}{k} \\
z = 0 \quad r > a \quad &\frac{\partial \theta}{\partial z} = 0
\end{aligned}
\tag{2.10}
$$

where $q_0 = Q/\pi a^2$ is the uniform heat flux. The condition at remote points is identical. The temperature distribution throughout the half-space $z \geq 0$ is given by the infinite integral [Carslaw and Jaeger (1959)]:

$$\theta = \frac{q_0 a}{k} \int_0^\infty e^{-\lambda z} J_0(\lambda r) J_1(\lambda a) \frac{d\lambda}{\lambda} \tag{2.11}$$

where $J_1(x)$ is the Bessel function of the first kind of order one [Abramowitz and Stegun (1965)], and $\lambda$ is a dummy variable. The temperature rise in the source area $0 \leq r \leq a$ is axisymmetric and it is given by [Carslaw and Jaeger (1959)]:

$$\theta(r) = \frac{q_0 a}{k} \int_0^\infty J_0(\lambda r) J_1(\lambda a) \frac{d\lambda}{\lambda} \tag{2.12}$$

The alternative form of the solution according to Yovanovich (1976c) is

$$\theta(r) = \frac{2}{\pi} \frac{q_0 a}{k} E\left(\frac{r}{a}\right) \qquad 0 \leq r \leq a \tag{2.13}$$

where $E(r/a)$ is the complete elliptic integral of the second kind of modulus $r/a$ [Byrd and Friedman (1971)] which is tabulated, and it can be calculated by means of Computer Algebra Systems. The temperatures at the centroid $r = 0$ and the edge $r = a$ of the source area are, respectively:

$$\theta(0) = \frac{q_0 a}{k} \qquad \text{and} \qquad \theta(a) = \frac{2}{\pi} \frac{q_0 a}{k} \tag{2.14}$$

**Table 2.1** Dimensionless source temperature.

| $r/a$ | $k\theta(r/a)/(q_0a)$ | $r/a$ | $k\theta(r/a)/(q_0a)$ |
|-----|-----|-----|-----|
| 0.0 | 1.000 | 0.6 | 0.9028 |
| 0.1 | 0.9975 | 0.7 | 0.8630 |
| 0.2 | 0.9899 | 0.8 | 0.8126 |
| 0.3 | 0.9771 | 0.9 | 0.7459 |
| 0.4 | 0.9587 | 1.0 | 0.6366 |
| 0.5 | 0.9342 | | |

The centroid temperature rise relative to the temperature rise at the edge is greater by approximately 57%. The values of the dimensionless temperature rise defined as $k\theta(r/a)/(q_0a)$ are presented in Table 2.1.

The area-average source temperature is

$$\bar{\theta} = \frac{1}{\pi a^2}\frac{q_0a}{k}\int_0^a \left[\int_0^\infty J_0(\lambda r)J_1(\lambda a)\frac{d\lambda}{\lambda}\right] 2\pi r\,dr \tag{2.15}$$

The integrals can be interchanged giving the result [Carslaw and Jaeger (1959)]:

$$\bar{\theta} = \frac{2q_0}{k}\int_0^\infty J_1^2(\lambda a)\frac{d\lambda}{\lambda^2} = \frac{8}{3\pi}\cdot\frac{q_0a}{k} \tag{2.16}$$

According to the definition of spreading resistance, one obtains for the isoflux circular source the relation [Carslaw and Jaeger (1959)]:

$$\bar{R}_s = \frac{\bar{\theta}}{Q} = \frac{8}{3\pi^2}\cdot\frac{1}{ka} \tag{2.17}$$

or

$$\psi = 4ka\bar{R}_s = \frac{32}{3\pi^2} \tag{2.18}$$

We may also define a spreading resistance based on the maximum centroidal temperature:

$$\hat{R}_s = \frac{\theta(0)}{Q} = \frac{1}{\pi ak} \tag{2.19}$$

or

$$\psi = 4ka\hat{R}_s = \frac{4}{\pi} \tag{2.20}$$

The spreading resistance for the isoflux source area based on the area-average temperature rise is greater than the value for the isothermal source by the factor:

$$\frac{(\bar{R}_s)_{isoflux}}{(R_s)_{isothermal}} = \frac{32}{3\pi^2} = 1.08076 \tag{2.21}$$

Thus, the thermal spreading resistance due to an isoflux heat source versus an isothermal heat source is approximately 8% greater.

### 2.1.3  Parabolic Flux Circular Source

The solution for a parabolic (or Hertz profile) is also desirable. The heat flux distribution over the heat source area is axisymmetric and given by:

$$q(r) = \frac{3Q}{2\pi a^2}\sqrt{1 - (r/a)^2} \quad 0 \leq r < a \tag{2.22}$$

The solution is easily reproduced from the analogous elastic contact problem [Johnson (1985)]. Elastic contact problems also utilize a method of analysis similar to the superposition of point sources (in this case point loads) to obtain solutions to displacement. We will outline the method later for irregular heat sources. In the present case, we will draw upon the results in Johnson (1985) to complete the analysis here.

In elastic point loading problems, the transformation to an equivalent thermal problem requires the following definitions be used: the effective elastic modulus $E/2(1 - v^2) = k$, the displacement in the plane of loading $u_z = \theta$, the total load $P = Q$, and the pressure $p = q$. Using the result for the displacement under a Hertz pressure distribution, the following result is obtained for the temperature distribution within the source area in terms of the total heat input $Q$ distributed according to Eq. (2.22):

$$\theta(r) = \frac{3Q}{16ka^3}(2a^2 - r^2) \tag{2.23}$$

With the temperature distribution known, we may integrate the above equation to obtain a mean source temperature:

$$\bar{\theta} = \frac{1}{\pi a^2}\int_0^a \theta(r)2\pi r \, dr = \frac{9Q}{32ka} \tag{2.24}$$

Defining a thermal spreading resistance as before yields:

$$\bar{R}_s = \frac{\bar{\theta}}{Q} = \frac{9}{32ka} \tag{2.25}$$

or

$$\psi = 4ka\bar{R}_s = \frac{9}{8} \tag{2.26}$$

We may also use the maximum centroidal temperature $\theta(0)$ to obtain:

$$\hat{R}_s = \frac{\theta(0)}{Q} = \frac{3}{8ka} \tag{2.27}$$

or

$$\psi = 4ka\hat{R}_s = \frac{3}{2} \tag{2.28}$$

The spreading resistance for the parabolic source area based on the area-average temperature rise is greater than the value for the isothermal source by the factor:

$$\frac{(\bar{R}_s)_{parabolic}}{(R_s)_{isothermal}} = \frac{9}{8} = 1.125 \tag{2.29}$$

Thus, the thermal spreading resistance due to a parabolic heat flux source versus an isothermal heat source is approximately 12.5% greater.

**Table 2.2**   Effect of boundary conditions on stationary circular heat source.

| Case | Flux distribution | $\bar{R}_s\,k\,a$ | $\hat{R}_s\,k\,a$ |
|---|---|---|---|
| A – Isothermal | $\dfrac{1}{2}q_0\dfrac{1}{\sqrt{1-(r/a)^2}}$ | $\dfrac{1}{4}=0.250$ | $\dfrac{1}{4}=0.250$ |
| B – Isoflux | $q_0$ | $\dfrac{8}{3\pi^2}\approx 0.270$ | $\dfrac{1}{\pi}\approx 0.318$ |
| C – Parabolic flux | $\dfrac{3}{2}q_0\sqrt{1-(r/a)^2}$ | $\dfrac{9}{32}\approx 0.281$ | $\dfrac{3}{8}\approx 0.375$ |

### 2.1.4   Summary of Circular Source Thermal Spreading Resistance

The three solutions for a circular source are summarized in Table 2.2 which provides the flux distribution and the thermal spreading resistance based upon the mean source temperature and centroidal source temperature. In each case, the nominal mean flux over the surface is taken as $q_0 = Q/\pi a^2$. If exact flux distributions are unknown, these special cases can be used to bound the thermal spreading resistance in these applications.

## 2.2   Elliptical Area on a Half-Space

The natural extension of the circular source on a half-space is to consider an elliptic source on a half-space. In this case, the source is no longer circular, but has major and minor axes of $2a$ and $2b$, respectively. The general solutions for the elliptic source contain the special case of the circular source when the major and minor axes are equal. Results for three cases are available: isothermal source, isoflux source, and parabolic heat flux. Though, only for the case of the isothermal result is the solution shown to be fully analytical.

### 2.2.1   Isothermal Elliptical Source

The spreading resistance for an isothermal elliptical source area with semiaxes $a \geq b$ is available in closed form. The results are obtained from a solution which follows the classical solution presented for finding the capacitance of a charged elliptical disk placed in free space as given by Jeans (1963), Smythe (1968), and Stratton (1941). Holm (1967) gave the solution for the electrical resistance for current flow from an isopotential elliptical disk. The thermal solution presented next will follow the analysis of Yovanovich (1971).

   The elliptical contact area defined by $x^2/a^2 + y^2/b^2 = 1$ produces a three-dimensional temperature field where the isotherms are ellipsoids which are described by the relationship:

$$\frac{x^2}{a^2+\lambda}+\frac{y^2}{b^2+\lambda}+\frac{z^2}{\lambda}=1 \tag{2.30}$$

The three-dimensional Laplace equation in Cartesian coordinates can be transformed into the one-dimensional Laplace equation in ellipsoidal coordinates:

$$\nabla^2\theta = \frac{\partial}{\partial\lambda}\left(\sqrt{f(\lambda)}\,\frac{\partial\theta}{\partial\lambda}\right)=0 \tag{2.31}$$

where $\lambda$ is the ellipsoidal coordinate for the ellipsoidal temperature rise $\theta(\lambda)$, and where

$$\sqrt{f(\lambda)} = \sqrt{(a^2+\lambda)(b^2+\lambda)\lambda} \tag{2.32}$$

The solution of the differential equation according to Yovanovich (1971) is:

$$\theta = C_2 - C_1 \int_{\lambda}^{\infty} \frac{d\lambda}{\sqrt{f(\lambda)}} \qquad (2.33)$$

The boundary conditions are specified in the contact plane ($z = 0$) where

$$\theta = \theta_0 \qquad \text{within} \qquad \frac{x^2}{a^2} + \frac{y^2}{b^2} = 1$$

$$\frac{\partial \theta}{\partial z} = 0 \qquad \text{outside} \qquad \frac{x^2}{a^2} + \frac{y^2}{b^2} = 1 \qquad (2.34)$$

The regular condition at points remote to the elliptical area is $\theta \to 0$ as $\lambda \to \infty$. This condition is satisfied by $C_2 = 0$, and the condition in the contact plane is satisfied by $C_1 = -Q/(4\pi k)$, where $Q$ is the total heat flow rate from the isothermal elliptical area. The solution is therefore according to Yovanovich (1971):

$$\theta = \frac{Q}{4\pi k} \int_{\lambda}^{\infty} \frac{d\lambda}{\sqrt{(a^2 + \lambda)(b^2 + \lambda)\lambda}} \qquad (2.35)$$

When $\lambda = 0, \theta = \theta_0$, constant for all points within the elliptical area, and when $\lambda \to \infty, \theta \to 0$ for all points far from the elliptical area. According to the definition of spreading resistance for an isothermal contact area, we find

$$R_s = \frac{\theta_0}{Q} = \frac{1}{4\pi k} \int_{0}^{\infty} \frac{d\lambda}{\sqrt{(a^2 + \lambda)(b^2 + \lambda)\lambda}} \qquad (2.36)$$

The last equation can be transformed into a standard form by setting $\sin t = a/\sqrt{a^2 + \lambda}$. The alternative form for the spreading resistance is

$$R_s = \frac{1}{2\pi k a} \int_{0}^{\pi/2} \frac{dt}{\left[1 - \left(\frac{a^2 - b^2}{a^2}\right) \sin^2 t\right]^{1/2}} \qquad (2.37)$$

The spreading resistance depends on the thermal conductivity of the half-space, the semimajor axis $a$, and the aspect ratio of the elliptical area $b/a \leq 1$. It is clear that when the axes are equal, i.e. $b = a$, the elliptical area becomes a circular area and the spreading resistance is $R_s = 1/(4ka)$. The integral is the complete elliptic integral of the first kind $K(\kappa)$ of modulus $\kappa = \sqrt{(a^2 - b^2)/a^2}$ [Byrd and Friedman (1971), Gradshteyn and Ryzhik (1965)]. The spreading resistance for the isothermal elliptical source area can be written as:

$$R_s = \frac{1}{2\pi k a} K(\kappa) \qquad (2.38)$$

The complete elliptic integral is tabulated [Abramowitz and Stegun (1965), Byrd and Friedman (1971)]. It can also be computed efficiently and very accurately by Computer Algebra Systems.

To compare the spreading resistances of the elliptical area and the circular area, it is necessary to nondimensionalize the two results. For the circle, the radius appears as the length scale, and for the ellipse, the semimajor axis appears as the length scale. For proper comparison of the two geometries, it is important to select a length scale that best characterizes the two geometries. The proposed length scale is based on the square root of the "active" area of each geometry, i.e. $\mathcal{L} = \sqrt{A}$ [Yovanovich (1976c), Yovanovich and Burde (1977), Yovanovich et al. (1977)]. Therefore, the dimensionless spreading resistances for the circle and ellipse are:

$$\left(k\sqrt{A}\, R_s\right)_{circle} = \frac{\sqrt{\pi}}{4}$$

**Table 2.3** Dimensionless spreading resistance of isothermal ellipse.

| $a/b$ | $k\sqrt{A}\,R_s$ | $a/b$ | $k\sqrt{A}\,R_s$ |
|-------|---------|-------|---------|
| 1 | 0.4431 | 6 | 0.3678 |
| 2 | 0.4302 | 7 | 0.3566 |
| 3 | 0.4118 | 8 | 0.3466 |
| 4 | 0.3951 | 9 | 0.3377 |
| 5 | 0.3805 | 10 | 0.3297 |

$$\left(k\sqrt{A}\,R_s\right)_{ellipse} = \frac{1}{2\sqrt{\pi}}\sqrt{\frac{b}{a}}\,K(\kappa)$$

where $\kappa = \sqrt{1-(b/a)^2}$. The dimensionless spreading resistance values for an isothermal elliptical area are presented in Table 2.3 for a range of the semiaxes ratio $a/b$.

The tabulated values of the dimensionless spreading resistance reveal an interesting trend beginning with the first entry which corresponds to the circle. The dimensionless resistance values decrease with increasing values of $a/b$. Ellipses with larger values of $a/b$ have smaller spreading resistances than the circle; however, the decrease has a relatively weak dependence on $a/b$. For the same area, the spreading resistance of the ellipse with $a/b = 10$ is approximately 74% of the spreading resistance for the circle.

Two approximations are presented for quick calculator estimations of the dimensionless spreading resistance for isothermal elliptical areas:

$$k\sqrt{A}\,R_s = \begin{cases} \dfrac{\sqrt{\pi\alpha}}{\left(\sqrt{\alpha}+1\right)^2} & \text{for} \quad 1 \le \alpha \le 5 \\[3mm] \dfrac{1}{2\sqrt{\pi\alpha}}\ln(4\alpha) & \text{for} \quad 5 \le \alpha < \infty \end{cases} \tag{2.39}$$

where $\alpha = a/b \ge 1$. Although both approximations can be used at $\alpha = 5$, the second approximation is slightly more accurate, and therefore, it is recommended.

The heat flux distribution over the elliptical area is given by [Yovanovich (1971)]:

$$q(x,y) = \frac{Q}{2\pi ab}\left[1-\left(\frac{x}{a}\right)^2-\left(\frac{y}{b}\right)^2\right]^{-1/2} \tag{2.40}$$

The heat flux is minimum at the centroid, where its magnitude is $q(0,0) = Q/(2\pi ab)$, and it is "unbounded" on the perimeter of the ellipse.

### 2.2.2 Isoflux Elliptical Source

No simple analytical solution for an isoflux elliptic solution exists. Even the analogous elastic contact problem has no simple solution reported [Johnson (1985)]. However, an approximate model which is found to be quite accurate can be developed and validated with the data of Yovanovich et al. (1977).

If the solution for the isothermal elliptic heat source is normalized with the limiting case of the circular heat source, a general "shape" function arises which accounts for the effect of aspect ratio for an isothermal elliptic heat source:

$$\frac{R_{s,elliptic}}{R_{s,circular}} = \frac{2}{\pi}K(\kappa) \tag{2.41}$$

If this "shape" function is now multiplied by the limiting cases for the isoflux circular heat source, one obtains the following expressions for the spreading resistance based on the mean and centroidal temperature:

$$\overline{R}_s\,k\,a = \frac{16}{3\pi^3}K(\kappa) \tag{2.42}$$

and

$$\widehat{R}_s\,k\,a = \frac{2}{\pi^2}K(\kappa) \tag{2.43}$$

Equations (2.42) and (2.43) accurately predict the numerical results presented in Yovanovich et al. (1977) for the isoflux elliptic contact which were obtained using the method of superposition of point heat sources. A plot of the numerical data points with the above expressions is given in Figure 2.2.

### 2.2.3 Parabolic Flux Elliptical Source

A solution for the parabolic heat flux distribution applied over an elliptical region may be obtained by analogy with the elastic contact problem [Johnson (1985)] as we did in Section 2.1.3, or using an approach similar to the one applied for the isoflux elliptic contact. Following Johnson (1985), the heat flux distribution having a Hertz profile is given by:

$$q(x,y) = \frac{3Q}{2\pi ab}\left[1 - \left(\frac{x}{a}\right)^2 - \left(\frac{y}{b}\right)^2\right]^{1/2} \tag{2.44}$$

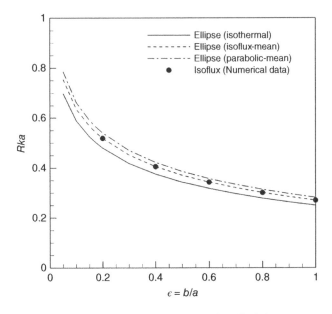

**Figure 2.2** Thermal spreading resistance for elliptic heat sources.

The general solution for the surface ($z = 0$) temperature field may be obtained from the elastic contact problem [Johnson (1985)]. Using the appropriate substitutions, we may write the temperature field as:

$$\theta(x, y) = \frac{1}{2\pi k}(L - Mx^2 - Ny^2) \tag{2.45}$$

where

$$L = \frac{3Q}{2a}K(\kappa) \tag{2.46}$$

$$M = \frac{3Q}{2\kappa^2 a^3}[K(\kappa) - E(\kappa)] \tag{2.47}$$

$$N = \frac{3Q}{2\kappa^2 a^3}\left[\frac{a^2}{b^2}E(\kappa) - K(\kappa)\right] \tag{2.48}$$

where $E(\kappa)$ and $K(\kappa)$ are elliptic integrals of modulus $\kappa = \sqrt{1 - \epsilon^2}$, and $\epsilon = b/a$ is the source aspect ratio.

The maximum or centroidal temperature is found at the origin and is :

$$\theta(0, 0) = \frac{1}{2\pi k}L = \frac{3Q}{4\pi ka}K(\kappa) \tag{2.49}$$

Defining the spreading resistance as before:

$$\widehat{R}_s = \frac{\theta(0, 0)}{Q} = \frac{3}{4\pi ka}K(\kappa) \tag{2.50}$$

or

$$\widehat{R}_s ka = \frac{3}{4\pi}K(\kappa) \tag{2.51}$$

The above equation reduces to the limiting case in Table 2.2 when $\epsilon = b/a = 1$ or $\kappa = 0$.

The effect of aspect ratio on the elliptic heat source for this case is:

$$f(\epsilon) = \frac{2}{\pi}K(\kappa) \tag{2.52}$$

which is the same as for the isothermal source case.

Thus, the solution for the parabolic heat flux distribution is the function $f(\epsilon)$ multiplied by the values for the resistance for Case C in Table 2.2.

The temperature distribution, Eq. (2.45), can be integrated to obtain the mean source temperature:

$$\overline{\theta} = \frac{1}{\pi ab}\iint_A \theta(x, y)dx\, dy \tag{2.53}$$

Equation (2.53) may be integrated by means of the Dirichlet integrals to yield:

$$\overline{\theta} = \frac{2}{\pi k}\left[\frac{L}{4} - \frac{Ma^2}{16} - \frac{Nb^2}{16}\right] \tag{2.54}$$

After simplifying, the following result is obtained for the mean source temperature:

$$\overline{\theta} = \frac{9Q}{16\pi ka}K(\kappa) \tag{2.55}$$

Defining the spreading resistance yields:

$$\overline{R}_s = \frac{\overline{\theta}}{Q} = \frac{9}{16\pi ka}K(\kappa) \tag{2.56}$$

or

$$\overline{R}_s \, k \, a = \frac{9}{16\pi} K(\kappa) \tag{2.57}$$

It is now clear the above results could also have been obtained by multiplying the "shape" function by the limiting cases for the circular source in Table 2.2. In other words, we can define all the results for the elliptic heat source using the results for the circular source in Table 2.2 multiplied by the universal shape function:

$$R_s = R_{s,circle} \cdot f(\epsilon) = R_{s,circle} \left[ \frac{2}{\pi} K(\kappa) \right] \tag{2.58}$$

for either the centroidal or mean source temperature for any of the three cases: isothermal, isoflux, or parabolic flux.

## 2.3 Method of Superposition of Point Sources

Extensive analysis of heat conduction from finite heat sources on a half-space has been performed by a number of researchers [Carslaw and Jaeger (1959), Holm (1967), Yovanovich (1976c, 1986), Yovanovich et al. (1977)]. Solutions are generally obtained using a superposition of point heat sources or point loads [Lur'e (1964)]. We outline the method below.

Consider a point $P(x,y,z)$ within the half-space region and a point on the surface $M(x_0,y_0,0)$ inside the contact region which has a finite source strength $q(x_0,y_0)dA$; refer to Figure 2.3. The effect of the point heat source felt at point $P$ can be found from Fourier's law:

$$q(x_0,y_0)dA = -k(2\pi r^2)\frac{dT}{dr} \tag{2.59}$$

We may do this since the field produced by the point source is hemispherical. Rearranging and integrating Eq. (2.59) keeping in mind that $q(x_0,y_0)dA$ is constant, we find that:

$$-\int_{r_1}^{r_2} \frac{[q(x_0,y_0)dA]dr}{2\pi k r^2} = \int_{T_1}^{T_2} dT \tag{2.60}$$

which simplifies to give:

$$T_1 - T_2 = \frac{q(x_0,y_0)dA}{2\pi k}\left(\frac{1}{r_1} - \frac{1}{r_2}\right) \tag{2.61}$$

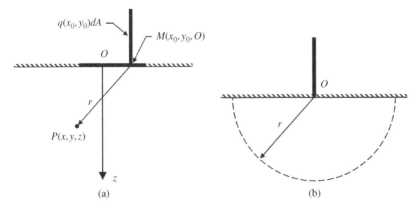

**Figure 2.3** Point heat source on a half space. (a) Point source within the heat source region. (b) Point source on a half space. Source: Yovanovich (1976c)/University of Waterloo. [Adapted from Yovanovich (1976c)].

For convenience, we can take $T_1$ as $T(x, y, z)$, the temperature at the arbitrary point, and $T_2 = 0$ the reference half-space temperature when $r \rightarrow \infty$. Thus, we arrive at the following fundamental result for the temperature at $P(x, y, z)$ due to a point heat source at $M(x_0, y_0, 0)$ on the surface:

$$T(x, y, z) = \frac{q(x_0, y_0)dA}{2\pi k r} \tag{2.62}$$

If we now consider the entire effect of the distributed heat source in the contact region by integrating over the contact region, we obtain:

$$T(x, y, z) = \frac{1}{2\pi k} \iint_A \frac{q(x_0, y_0)dA}{r} \tag{2.63}$$

The above integral is expanded to include the radial distance between the point $P$ and the point $M$ at the surface and becomes:

$$T(x, y, z) = \frac{1}{2\pi k} \iint_A \frac{q(x_0, y_0)dx_0 \, dy_0}{\sqrt{(x_0 - x)^2 + (y_0 - y)^2 + z^2}} \tag{2.64}$$

We are mainly interested in the temperature at the surface and within the contact area due to a constant uniform heat flux $q$; thus, we may use the simpler form:

$$T(x, y, 0) = \frac{q}{2\pi k} \iint_A \frac{dx_0 \, dy_0}{\sqrt{(x_0 - x)^2 + (y_0 - y)^2}} \tag{2.65}$$

Once the temperature field is obtained, the spreading resistance is defined as:

$$R_s = \frac{\overline{T}}{Q} = \frac{\frac{1}{A} \iint_A T(x, y, 0)dA}{\iint_A q dA} \tag{2.66}$$

or, assuming a constant heat flux:

$$R_s = \frac{1}{2\pi k A^2} \iint_A \left[ \iint_A \frac{dx_0 dy_0}{\sqrt{(x_0 - x)^2 + (y_0 - y)^2}} \right] dA \tag{2.67}$$

Equation (2.67) looks quite formidable, but it has been solved for many applications, theoretically and numerically. It clearly only depends on geometry as the inner integral is only a function of position. We will review some solutions obtained using the above formulation. For more details, the reader is directed to the paper by Yovanovich (1976c).

### 2.3.1 Application to a Circular Source

The simplest contact geometry is the circular contact. The solutions have already been obtained for three heat flux distributions: the uniform heat flux, parabolic heat flux, and the inverse parabolic heat flux. We will review the process outlined above for the case of a uniform flux contact as it will be used subsequently for obtaining additional results for noncircular uniform flux heat sources.

Beginning first with Eq. (2.67), we may introduce the following coordinate transformations in the contact plane inside the contact area:

$$x_0 - x = \rho \cos \omega \tag{2.68}$$

and

$$y_0 - y = \rho \sin \omega \tag{2.69}$$

This leads to Eq. (2.67) becoming:

$$T(r, 0) = \frac{q}{2\pi k} \iint_A d\omega \, d\rho = \frac{q}{2\pi k} \int_0^{2\pi} \rho(\omega) d\omega \tag{2.70}$$

where $dx_0 dy_0 = \rho d\rho d\omega$.

The function $\rho(\omega)$ is the length of a segment passing through the point $P(x, y)$ in the contact region and extending to the boundary of the contact. We may define:

$$\rho(\omega) = \rho_1(\omega) + \rho_2(\omega) \tag{2.71}$$

and evaluate:

$$T(r, 0) = \frac{q}{2\pi k} \int_0^{\pi/2} 2[\rho_1(\omega) + \rho_2(\omega)] d\omega \tag{2.72}$$

We need only rotate the segment through $\pi/2$ and use symmetry to account for the fact that the area swept out in the interval $[0, \pi/2]$ equals one half of the total area. Thus, we must now determine the length of each portion of the segment. Simple analysis of the circle shows that a perpendicular extending from the origin to the segment bisects the segment equally. It can be shown that:

$$\rho_1 + \rho_2 = 2\sqrt{a^2 - r^2 \sin^2 \omega} \tag{2.73}$$

which leads to the following integral:

$$T(r, 0) = \frac{2}{\pi} \frac{qa}{k} \int_0^{\pi/2} \sqrt{1 - (r/a)^2 \sin^2 \omega} \, d\omega \tag{2.74}$$

The integral given above is an elliptic integral. Thus, the temperature distribution in the contact region is now found to be:

$$T(r) = \frac{2}{\pi} \frac{qa}{k} E(r/a) \tag{2.75}$$

where $E(r/a)$ is a complete elliptic integral of the second kind of modulus $\eta = r/a$.

The thermal resistance may be defined with respect to the average contact temperature such that

$$\overline{R}_s = \frac{\overline{T}}{Q} \tag{2.76}$$

or with respect to the maximum contact temperature such that

$$\hat{R}_s = \frac{\hat{T}}{Q} \tag{2.77}$$

Using the solution for the temperature field, we find:

$$\hat{T} = T(0) = \frac{qa}{k} \tag{2.78}$$

and

$$\overline{T} = \frac{1}{A} \iint_A T dA = \frac{1}{\pi a^2} \int_0^a \frac{2}{\pi} \frac{qa}{k} E(r/a) 2\pi r \, dr \tag{2.79}$$

or

$$\overline{T} = \frac{4qa}{\pi k} \int_0^1 E(\eta) \eta \, d\eta = \frac{8}{3\pi} \frac{qa}{k} \tag{2.80}$$

where $\eta = r/a$ has been introduced for convenience. The above results are the same as those presented earlier.

The method of superposition of point sources has been used to obtain a number of useful results for other source shapes. Its application will be discussed in Sections 2.5–2.7 in more detail for specific source shapes.

### 2.3.2 Application to Triangular Source Areas

The method will now be applied to an arbitrary triangular region with uniform heat flux. This solution will provide the basic building block for the method to be applied to other source shapes such as triangles, rectangles, and polygons.

Consider the basic triangular areas shown in Figure 2.4a–c. The temperature at the vertex $P$ or point $P(x,y)$ due to a uniform heat flux over the entire area can be determined. In Figure 2.4a, a perpendicular from the vertex at $P$ to the opposite side $AB$ intersects at $C$ and divides the area into two right angle triangles $PAC$ and $PBC$ with angles $\omega_1$ and $\omega_2$ subtended at vertex $P$. The length $PC$ will be denoted $\delta$ for convenience.

Consider the triangle $PAC$ alone as shown in Figure 2.4c. The effect of a uniform heat flux distributed over the shaded elemental area is:

$$T(x,y) = \frac{q}{2\pi k} \int_0^{\omega_1} \frac{\delta d\omega}{\cos\omega} = \frac{q\delta}{2\pi k} \int_0^{\omega_1} \frac{d\omega}{\sqrt{1-\sin^2\omega}} \tag{2.81}$$

where $\omega_1 = \tan^{-1}(AC/\delta)$. The above equation is easily integrated and gives:

$$T(x,y) = \frac{q\delta}{2\pi k} \ln\left[\tan\left(\frac{\pi}{4}+\frac{\omega_1}{2}\right)\right] = \frac{q\delta}{2\pi k}\Omega(\omega_1) \tag{2.82}$$

For convenience, we introduce the function:

$$\Omega(\omega) = \ln\left[\tan\left(\frac{\pi}{4}+\frac{\omega}{2}\right)\right] = \frac{1}{2}\ln\left[\frac{1+\sin\omega}{1-\sin\omega}\right] \tag{2.83}$$

We may also recognize the second term in Eq. (2.81) as an incomplete elliptic integral of the first kind of $\omega_1$ and modulus unity $k=1$, such that we may write:

$$\Omega(\omega_1) = F(\omega_1, 1) \tag{2.84}$$

This allows values of $\Omega(\omega_1)$ to be easily read from tables of elliptic integrals or evaluated using mathematical software tools.

In a similar manner, the effect of a uniform heat flux distributed over the right triangle $PBC$ can now be written as:

$$T(x,y) = \frac{q\delta}{2\pi k} \ln\left[\tan\left(\frac{\pi}{4}+\frac{\omega_2}{2}\right)\right] \tag{2.85}$$

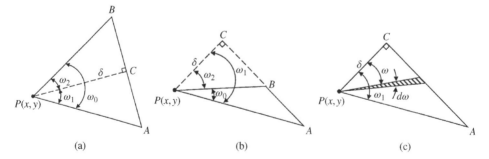

**Figure 2.4** Triangular contacts. Temperature rise at point P for a triangular element: (a) scalene, (b) equivalent, and (c) right isoflux triangles. [Adapted from Yovanovich (1976c)].

where $\omega_2 = \tan^{-1}(BC/\delta)$. Thus,

$$T(x,y) = \frac{q\delta}{2\pi k}\left[\Omega(\omega_1) + \Omega(\omega_2)\right] \tag{2.86}$$

Finally, if we consider the triangle *PAB* as shown in Figure 2.4b, by virtue of superposition, we may write:

$$T(x,y) = \frac{q\delta}{2\pi k}\left[\Omega(\omega_1) - \Omega(\omega_2)\right] \tag{2.87}$$

where

$$\Omega(\omega_1) = \ln\left[\tan\left(\frac{\pi}{4} + \frac{\tan^{-1}(AC/\delta)}{2}\right)\right] \tag{2.88}$$

$$\Omega(\omega_2) = \ln\left[\tan\left(\frac{\pi}{4} + \frac{\tan^{-1}(BC/\delta)}{2}\right)\right] \tag{2.89}$$

Equation (2.83) will prove useful in subsequent sections for calculating thermal spreading resistance from many heat source shapes that can be reduced to a superposition of the basic triangular element. This approach readily provides the maximum temperature at the centroid, but depending on the shape, finding the mean source temperature may be a bit more involved.

## 2.4 Rectangular Area on a Half-Space

One of the more useful solutions for thermal spreading resistance on a half space is that for the rectangular heat source of dimension $2a$ by $2b$ such that $a > b$. The practical uses for these solutions include electrical devices on thick substrates or on a larger scale for modeling heat loss from the foundations of buildings into the ground. Solutions for both isothermal and isoflux sources have been obtained.

### 2.4.1 Isothermal Rectangular Area

Though the problem for an isothermal rectangular source may be easily stated, no analytical approach is possible for this type of heat source on a half-space. However, Schneider (1978) obtained numerical values for the special case of an isothermal rectangular source. A correlation of those values for the dimensionless spreading resistance of an isothermal rectangle for the aspect ratio range: $1 \leq a/b \leq 4$ was also developed. This correlation equation is

$$k\sqrt{A}\,R_s = \sqrt{\frac{a}{b}}\left[0.06588 - 0.00232\left(\frac{a}{b}\right) + \frac{0.6786}{a/b + 0.8145}\right] \tag{2.90}$$

where $\sqrt{A} = \sqrt{4ab}$ is chosen for convenience as the length scale used to nondimensionalize the results. The numerical values are also given in Table 2.4.

A comparison of the values for the isothermal rectangular area and the isothermal elliptical area reveals a very close relationship; see Figure 2.5. The maximum difference of approximately $-0.7\%$ is found at $a/b = 4$. It is expected that the close agreement observed for the four aspect ratios will hold for higher aspect ratios because the dimensionless spreading resistance is a weak function of the shape if the areas are geometrically similar. In fact, the correlation values for the rectangle and the analytical values for the ellipse are within $\pm 1.5\%$ over the wider range, $1 \leq a/b \leq 13$.

**Table 2.4** Dimensionless spreading resistance of isothermal rectangular area.

| $a/b$ | 1 | 2 | 3 | 4 |
|---|---|---|---|---|
| $k\sqrt{A}\,R_s$ | 0.4412 | 0.4282 | 0.4114 | 0.3980 |

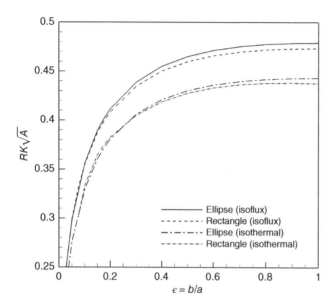

**Figure 2.5** Dimensionless thermal spreading resistance for isothermal and isoflux rectangular and elliptical sources.

### 2.4.2 Isoflux Rectangular Source

In the case of a rectangular isoflux heat source, a solution using the superposition method can be obtained. Carslaw and Jaeger (1959) report the general results for mean and maximum source temperature. We outline the method of solution here using the results from Section 2.3.2.

The temperature of an internal point $P(x,y)$ of a rectangular contact area of $2a$ by $2b$ can be determined from the superposition of eight right triangle areas as shown in Figure 2.6. With the origin located at the center of the rectangle and the $x$ and $y$ axes running parallel to the sides of length $2a$ and $2b$, respectively. The four perpendiculars from $P(x,y)$ to the four sides of the rectangular area are:

$$\delta_1 = a - x = a(1 - \xi) \tag{2.91}$$

$$\delta_2 = b - y = b(1 - \eta) \tag{2.92}$$

$$\delta_3 = a + x = a(1 + \xi) \tag{2.93}$$

$$\delta_4 = b + y = b(1 + \eta) \tag{2.94}$$

where $\xi = x/a$ and $\eta = y/b$.

**Figure 2.6** Rectangular contact. Source:
Yovanovich (1976c)/University of Waterloo.

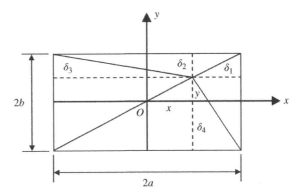

There are eight triangles whose vertices have the common point $P(x,y)$. The eight angles subtended at the point $P$ are:

$$\omega_1 = \tan^{-1}\frac{b-y}{\delta_1} = \tan^{-1}\frac{b(1-\eta)}{a(1-\xi)} \tag{2.95}$$

$$\omega_2 = \tan^{-1}\frac{a-x}{\delta_2} = \tan^{-1}\frac{a(1-\xi)}{b(1-\eta)} \tag{2.96}$$

$$\omega_3 = \tan^{-1}\frac{a+x}{\delta_2} = \tan^{-1}\frac{a(1+\xi)}{b(1-\eta)} \tag{2.97}$$

$$\omega_4 = \tan^{-1}\frac{b-y}{\delta_3} = \tan^{-1}\frac{b(1-\eta)}{a(1+\xi)} \tag{2.98}$$

$$\omega_5 = \tan^{-1}\frac{b+y}{\delta_3} = \tan^{-1}\frac{b(1+\eta)}{a(1+\xi)} \tag{2.99}$$

$$\omega_6 = \tan^{-1}\frac{a+x}{\delta_4} = \tan^{-1}\frac{a(1+\xi)}{b(1+\eta)} \tag{2.100}$$

$$\omega_7 = \tan^{-1}\frac{a-x}{\delta_4} = \tan^{-1}\frac{a(1-\xi)}{b(1+\eta)} \tag{2.101}$$

$$\omega_8 = \tan^{-1}\frac{b+y}{\delta_1} = \tan^{-1}\frac{b(1+\eta)}{a(1-\xi)} \tag{2.102}$$

The solution for the general temperature field within the rectangular area is the sum:

$$T(x,y) = \frac{q}{2\pi k}\sum_{i=1}^{8}\delta_i\Omega(\omega_i) \tag{2.103}$$

where $\Omega(\omega_i)$ is determined from Eq. (2.83). Note that $\delta_1$ is common to triangles (1) and (8), while $\delta_2$ is common to triangles (2) and (3), and so on, as we move counter-clockwise around the rectangle. After expanding, we obtain:

$$\begin{aligned}T(x,y) = \frac{q}{2\pi k}[&\delta_1\Omega(\omega_1) + \delta_2\Omega(\omega_2) + \delta_2\Omega(\omega_3) + \delta_3\Omega(\omega_4)\\ &+\delta_3\Omega(\omega_5) + \delta_4\Omega(\omega_6) + \delta_4\Omega(\omega_7) + \delta_1\Omega(\omega_8)]\end{aligned} \tag{2.104}$$

The maximum temperature will occur at the centroid (or origin) of the rectangle and can be evaluated easily using symmetry, given there are only two sets of triangles which are identical. This leads to:

$$T(0,0) = \frac{q}{2\pi k}4[\delta_1\Omega(\omega_1) + \delta_2\Omega(\omega_2)] \tag{2.105}$$

Finally, using $\delta_1 = a$ and $\delta_2 = b$ along with $\omega_1 = \tan^{-1}(b/a)$ and $\omega_2 = \tan^{-1}(a/b)$ we obtain:

$$T(0,0) = \frac{2q}{\pi k} \left( a \ln \left[ \tan \left( \frac{\pi}{4} + \frac{\tan^{-1}(b/a)}{2} \right) \right] + b \ln \left[ \tan \left( \frac{\pi}{4} + \frac{\tan^{-1}(a/b)}{2} \right) \right] \right) \quad (2.106)$$

The dimensionless spreading resistances of the rectangular source area $-a \leq x \leq a, -b \leq y \leq b$ with aspect ratio $a/b \geq 1$ are found by means of the integral method [Yovanovich (1971)]. Employing the definition of the spreading resistance based on the area-average temperature rise with $Q = q(4ab)$ gives the following dimensionless relationship [Yovanovich (1976c), Carslaw and Jaeger (1959)]:

$$k\sqrt{A}\, \overline{R}_s = \frac{\sqrt{\epsilon}}{\pi} \left\{ \sinh^{-1}\frac{1}{\epsilon} + \frac{1}{\epsilon}\sinh^{-1}\epsilon + \frac{\epsilon}{3}\left[ 1 + \frac{1}{\epsilon^3} - \left( 1 + \frac{1}{\epsilon^2} \right)^{3/2} \right] \right\} \quad (2.107)$$

where $\epsilon = a/b \geq 1$. Employing the definition based on the centroid temperature rise, the dimensionless spreading resistance is obtained from the relationship [Carslaw and Jaeger (1959)]:

$$k\sqrt{A}\, \widehat{R}_s = \frac{\sqrt{\epsilon}}{\pi} \left[ \frac{1}{\epsilon}\sinh^{-1}\epsilon + \sinh^{-1}\left( \frac{1}{\epsilon} \right) \right] \quad (2.108)$$

Typical values of the dimensionless spreading resistance for the isoflux rectangle based on the area-average temperature rise for $1 \leq a/b \leq 10$ are given in Table 2.5.

The expressions for the dimensionless thermal spreading resistance for the elliptic source and rectangular source based upon the square root of source area are provided in Figure 2.5. It is clear that the use of $\sqrt{A}$ as a length scale has merit. Later results will also show this to be the most appropriate length scale for presenting nondimensional results of other source shapes.

**Example 2.1**  Calculate and compare the spreading resistances of isoflux circular and square micro-contact areas when $k = 160\,\text{W}/(\text{m K})$, and the areas of the two heat sources are identical. The radius of the circular micro-contact is $a = 10\,\mu\text{m}$. The spreading resistances should be based on the average temperature of the source.

The area of the circular micro heat source is $A = \pi a^2 = 3.14159 \times 10^{-10}\,\text{m}^2$. The resistance for an isoflux source from Table 2.2 is $\overline{R}_s ka = 8/3\pi^2$. This gives:

$$\overline{R}_s = \frac{8}{3\pi^2}\frac{1}{ka} = \frac{8}{3\pi^2}\left( \frac{1}{160 \cdot 10 \times 10^{-6}} \right) = 168.86\,\text{K/W}$$

For the square heat source of the same area, we have from Table 2.5, $k\sqrt{A}\overline{R}_s = 0.4732$. This gives:

$$\overline{R}_s = 0.4732\frac{1}{k\sqrt{A}} = 0.4732\left( \frac{1}{160 \cdot 1.772 \times 10^{-5}} \right) = 166.85\,\text{K/W}$$

**Table 2.5**  Dimensionless spreading resistance for isoflux rectangular area.

| $a/b$ | $k\sqrt{A}\,\overline{R}_s$ | $a/b$ | $k\sqrt{A}\,\overline{R}_s$ |
| --- | --- | --- | --- |
| 1 | 0.4732 | 6 | 0.3950 |
| 2 | 0.4598 | 7 | 0.3833 |
| 3 | 0.4407 | 8 | 0.3729 |
| 4 | 0.4234 | 9 | 0.3636 |
| 5 | 0.4082 | 10 | 0.3552 |

The two results are almost equal for sources of the same area. The effect of source shape is quite small in this case, being on the order of 1% difference.

## 2.5 Spreading Resistance of Symmetric Singly Connected Areas: The Hyperellipse

The spreading resistances for isoflux, singly connected source areas were obtained by means of numerical methods applied to the integral formulation of the spreading resistance. The source areas examined were isosceles triangles having a range of aspect ratios, the semicircle, L-shaped source areas (squares with corners removed), and the hyperellipse area defined by

$$\left(\frac{x}{a}\right)^n + \left(\frac{y}{b}\right)^n = 1 \tag{2.109}$$

where $a$ and $b$ are the semiaxes along the $x$-axis and $y$-axis, respectively. The shape parameter $n$ lies in the range: $0 < n < \infty$. Many interesting geometries can be generated by the parameters $a, b, n$. The area of the hyperellipse is given by the relationship:

$$A = \frac{4ab}{n}\frac{\Gamma(1+1/n)\Gamma(1/n)}{\Gamma(1+2/n)} \tag{2.110}$$

where $\Gamma(x)$ is the gamma function which is tabulated [Abramowitz and Stegun (1965)], and it can be computed accurately by means of Computer Algebra Systems. The dimensionless spreading resistance was found to be a weak function of the shape of the source area for a wide range of values of $n$. Typical values are given in Table 2.6 for the case when $a = b$.

Yovanovich et al. (1977) obtained the following result for the thermal spreading resistance based on the centroidal temperature:

$$k\sqrt{A}\hat{R}_s = \frac{1}{\pi}\left[\frac{\epsilon n}{B(\frac{n+1}{n}, \frac{1}{n})}\right]^{1/2}\int_0^{\pi/2}\frac{d\theta}{[\sin^n\theta + \epsilon^n\cos^n\theta]^{1/n}} \tag{2.111}$$

where $B(\cdot)$ is the Beta function. The above expression is easily evaluated numerically. Results for $\epsilon = 1$ and different values of $n$ are reported in Tables 2.6 and 2.7, for varying aspect ratios and $n$. Results for the thermal spreading resistance based on mean source temperature are more complicated to obtain. Details are provided in Yovanovich et al. (1977). Values for the dimensionless thermal spreading resistance are provided in Table 2.7 for several special cases.

The dimensionless spreading resistances were based on the centroid temperature rise denoted as $\hat{R}_s$ and the area-average temperature rise denoted as $\overline{R}_s$. The dimensionless spreading resistance was based on the length scale $\mathcal{L} = \sqrt{A}$. All numerical results were found to lie in narrow

**Table 2.6** Effect of $n$ on dimensionless spreading resistances for $b/a = 1$.

| $n$ | $k\sqrt{A}\,\overline{R}_s$ | $k\sqrt{A}\,\hat{R}_s$ |
| --- | --- | --- |
| 0.5 | 0.4440 | 0.5468 |
| 1 | 0.4728 | 0.5611 |
| 2 | 0.4787 | 0.5642 |
| 4 | 0.4770 | 0.5631 |
| $\infty$ | 0.4732 | 0.5611 |

**Table 2.7** Dimensionless resistance for isoflux hyperellipse contacts.

| $\epsilon$ | $n = 1/2$ | | $n = 1$ | |
| --- | --- | --- | --- | --- |
| | $k\sqrt{A}\,\overline{R}_s$ | $k\sqrt{A}\,\widehat{R}_s$ | $k\sqrt{A}\,\overline{R}_s$ | $k\sqrt{A}\,\widehat{R}_s$ |
| 1.0 | 0.4440 | 0.5468 | 0.4728 | 0.5611 |
| 0.8 | 0.4428 | 0.5458 | 0.4713 | 0.5597 |
| 0.6 | 0.4376 | 0.5420 | 0.4651 | 0.5540 |
| 0.4 | 0.4237 | 0.5310 | 0.4487 | 0.5385 |
| 0.2 | 0.3860 | 0.5005 | 0.4052 | 0.4957 |

| $\epsilon$ | $n = 2$ | | $n = \infty$ | |
| --- | --- | --- | --- | --- |
| | $k\sqrt{A}\,\overline{R}_s$ | $k\sqrt{A}\,\widehat{R}_s$ | $k\sqrt{A}\,\overline{R}_s$ | $k\sqrt{A}\,\widehat{R}_s$ |
| 1.0 | 0.4787 | 0.5642 | 0.4732 | 0.5611 |
| 0.8 | 0.4772 | 0.5624 | 0.4718 | 0.5590 |
| 0.6 | 0.4711 | 0.5551 | 0.4658 | 0.5503 |
| 0.4 | 0.4548 | 0.5360 | 0.4502 | 0.5279 |
| 0.2 | 0.4112 | 0.4845 | 0.4082 | 0.4706 |

ranges: $0.4424 \leq k\sqrt{A}\,\overline{R}_s \leq 0.4733$ and $0.5197 \leq k\sqrt{A}\,\widehat{R}_s \leq 0.5614$. The corresponding values for the equilateral triangle are $k\sqrt{A}\,\overline{R}_s = 0.4600$ and $k\sqrt{A}\,\widehat{R}_s = 0.5616$, and for the semicircle, they are $k\sqrt{A}\,\overline{R}_s = 0.4610$ and $k\sqrt{A}\,\widehat{R}_s = 0.5456$.

The following approximations were recommended by Yovanovich and Burde (1977) for quick approximate calculations: $k\sqrt{A}\,\widehat{R}_s = 5/9$ and $k\sqrt{A}\,\overline{R} = 0.84\,k\sqrt{A}\,\widehat{R}_s$.

The ratios of the area-average and centroid temperature rises for all geometries examined were found to be closely related such that $\overline{\theta}/\theta_0 = 0.84 \pm 1.7\%$.

Table 2.7 presents the dimensionless resistance based upon the average and centroidal values of temperature for different values of the parameter $n$, Yovanovich et al. (1977) for $0 < \epsilon = b/a < 1$. It is clearly seen that the dimensionless resistance varies very little with aspect ratio $\epsilon = b/a$ and shape parameter $n$. All the data in Table 2.7 are plotted in Figure 2.7 along with the simple expressions for the elliptic source.

## 2.6  Regular Polygonal Isoflux Sources

Thermal spreading solutions from polygonal sources on a half space have also received some attention. The regular polygons include the family of $N$ sided shapes beginning with the equilateral triangle ($N = 3$) and ending with a circle ($N \to \infty$). Solutions for isoflux sources use the integration of point sources to obtain results. No solutions for isothermal sources exist except for the special cases of the square and circular sources discussed earlier.

The spreading resistances of isoflux regular polygonal areas have been examined extensively. The regular polygonal areas are characterized by the number of sides $N \geq 3$, the side dimension $s$, and the radius of the inscribed circle denoted as $r_i$. The perimeter is $P = Ns$, the relationship between

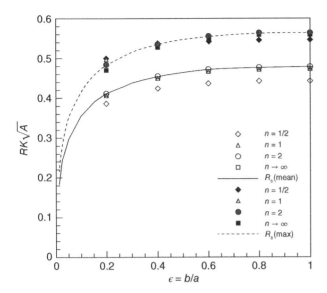

**Figure 2.7** Data for selected source shapes and aspect ratios of the hyperellipse.

**Table 2.8** Dimensionless spreading resistance for polygons.

| N | $k\sqrt{A}\,\hat{R}_s$ |
| --- | --- |
| 3 | 0.5517 |
| 4 | 0.5611 |
| 5 | 0.5630 |
| 6 | 0.5637 |
| 7 | 0.5639 |
| 8 | 0.5640 |
| $N \to \infty$ | 0.5642 |

the inscribed radius and the side dimension is $s/r_i = 2\tan(\pi/N)$. The area of the regular polygon is $A = Nr_i^2\tan(\pi/N)$. The temperature rises from the minimum values located on the edges to a maximum value at the centroid. It can be found easily by means of integral methods based on the superposition of point sources. Results are given in Table 2.8.

A regular polygon of $N$ sides of length $s$ can be divided into $N$ basic triangular regions with vertices at the centroid of the polygon. These triangles can be further divided into $2N$ right triangles with a subtended angle of $\pi/N$. Using Eq. (2.83), we may write:

$$T(0,0) = 2N\frac{qr_i}{2\pi k}\Omega(\pi/N) \tag{2.112}$$

where $\delta = r_i$ and

$$\Omega(\pi/N) = \frac{1}{2}\ln\left[\frac{1+\sin(\pi/N)}{1-\sin(\pi/N)}\right] \tag{2.113}$$

The general relationship for the spreading resistance based on the *centroid* temperature rise is found to be:

$$k\sqrt{A}\,\widehat{R}_s = \frac{1}{2\pi}\sqrt{\frac{N}{\tan(\pi/N)}}\,\ln\left[\frac{1+\sin(\pi/N)}{1-\sin(\pi/N)}\right], \quad N \geq 3 \tag{2.114}$$

The above expression gives $k\sqrt{A}\widehat{R}_s = 0.5517$ for the equilateral triangle $N = 3$ which is approximately 2.3% smaller than the value for the circle, where $N \to \infty$. Numerical methods are required to find the dimensionless spreading resistance for regular polygons subjected to a uniform heat flux. The corresponding value for the area-average basis was reported by Yovanovich and Burde (1977) to be $k\sqrt{A}\overline{R}_s = 0.4600$ for the equilateral triangle which is approximately 4% smaller than the value for the circle.

## 2.7 Additional Results for Other Source Shapes

Additional results for the maximum or centroid-based thermal spreading resistance can be found using the superposition method. Yovanovich (1976a,b,c) obtained a solution for the right triangle and Sadeghi et al. (2010) obtained solutions for the equilateral triangle, rhombus, trapezoid, rectangle with rounded ends, rectangle with semicircular ends, circular sector, and circular segment. We provide several here for completeness without details of their solution.

### 2.7.1 Triangular Source

The solution for an isosceles triangle was found by Sadeghi et al. (2010) using the method of Yovanovich (1976a,b,c). The thermal spreading resistance based on the centroidal temperature was found to be:

$$\begin{aligned}
k\widehat{R}_s\sqrt{A} = \frac{\sqrt{2\beta}}{3\pi}\bigg[&\ln\left[\tan\left(\frac{\pi}{4}+\frac{\omega_1}{2}\right)\right] \\
&+2\sin(\cot^{-1}2\beta)\ln\left[\tan\left(\frac{\pi}{4}+\frac{\omega_2}{2}\right)\tan\left(\frac{\pi}{4}+\frac{\omega_3}{2}\right)\right]\bigg]
\end{aligned} \tag{2.115}$$

where

$$\omega_1 = \tan^{-1}(3/2\beta)$$

$$\omega_2 = \pi/2 - \cot^{-1}(2\beta)$$

$$\omega_3 = \pi - \omega_1 - \omega_2$$

and $\beta = b/a$ is the ratio of the height $2b$ to base $2a$ of the equilateral triangle.

### 2.7.2 Rhombic Source

The rhombic source is a special case of the hyperellipse with $n = 1$. In the case of the thermal resistance based on the centroidal temperature, an alternative expression was proposed by Sadeghi et al. (2010):

$$k\widehat{R}_s\sqrt{A} = \frac{\sqrt{2}\sin(\omega_1)}{\pi\sqrt{\epsilon}}\ln\left[\tan\left(\frac{\pi}{4}+\frac{\omega_1}{2}\right)\tan\left(\frac{\pi}{4}+\frac{\omega_2}{2}\right)\right] \tag{2.116}$$

where

$$\omega_1 = \tan^{-1}(\epsilon)$$

$$\omega_2 = \pi/2 - \omega_1$$

and $\epsilon = b/a$. The area of the rhombus is given by $A = 2ab$, where $a$ and $b$ are the semiaxis lengths.

### 2.7.3 Rectangular Source with Rounded Ends

Sadeghi et al. (2010) applied the superposition method to obtain the solution for a rectangular source $2a \times 2b$ with rounded circular segment ends, where the radius of the circular segment ends is measured from the centroid of the rectangle. The solution for the thermal spreading resistance based on the centroid temperature is:

$$k\hat{R}_s \sqrt{A} = \frac{\sqrt{2}}{\pi} \frac{\beta \ln\left[\tan\left(\frac{\pi}{4} + \frac{\omega_1}{2}\right)\right] + \sqrt{1+\beta^2}\tan^{-1}\beta}{\sqrt{(1+\beta^2)\tan^{-1}\beta + \beta}} \tag{2.117}$$

where

$$\omega_1 = \pi/2 - \tan^{-1}(\beta)$$

and $\beta = b/a$ and $A = 2a^2[(1+\beta^2)\tan^{-1}\beta + \beta]$. The lengths $2a$ and $2b$ are the major and minor axes of the rectangular region, respectively. Note that the parameter $\beta = b/a$ is not the aspect ratio of the source, but rather the aspect ratio of the central rectangle. Later when we consider a general model, the effective aspect ratio for the source will be defined as $\epsilon = \beta/\sqrt{1+\beta^2}$. As $\beta$ becomes large $a \ll b$, the source becomes a circle, since the radius of curvature of the end cap approaches the semiaxis $b$, since $a$, $b$, and $r$ are not independent of each other, see Figure 2.9.

### 2.7.4 Rectangular Source with Semicircular Ends

Sadeghi et al. (2010) applied the superposition method to obtain the solution for a rectangular source $2a \times 2b$ with semicircular ends. The radius of the ends is taken to be equal to the semiminor axis of the rectangular region, $b$. The solution for the thermal spreading resistance based on the centroid temperature is :

$$k\hat{R}_s \sqrt{A} = \frac{2}{\pi\sqrt{4\beta + \pi\beta^2}}\left[\beta \ln\left[\frac{\pi}{4} + \frac{\omega_1}{2}\right] + \int_0^\alpha \left(\cos\omega + \sqrt{\beta^2 - \sin^2\omega}\right)d\omega\right] \tag{2.118}$$

where

$$\alpha = \tan^{-1}\beta$$

$$\omega_1 = \pi/2 - \tan^{-1}\beta$$

and $\beta = b/a$ and $A = a^2[4\beta + \pi\beta^2]$. The lengths $2a$ and $2b$ are the major and minor axes of the rectangular region, respectively. Once again, the parameter $\beta = b/a$ is not the aspect ratio of the source, but rather the aspect ratio of the central rectangle. Later when we consider a general model, the effective aspect ratio for this source will be defined as $\epsilon = \beta/(1+\beta)$. As $\beta$ becomes large, $a \ll b$, the source becomes a circle.

## 2.8   Model for an Arbitrary Singly Connected Heat Source on a Half-Space

Having considered so many unique solutions to thermal spreading in a half-space, it is now desirable to provide a means to easily predict the thermal spreading resistance in an approximate manner. We have seen thus far that choosing the appropriate length scale to nondimensional thermal resistance shows great promise in unifying results. Figure 2.5 illustrates that only a small difference in dimensionless thermal resistance is experienced when the length scale is chosen as the square root of the heat source area. This was further extended to the shapes defined by the hyperellipse for varying values of $n$. Therefore, the appropriate length scale to use for nondimensionalizing the thermal resistance results is $\mathcal{L} = \sqrt{A}$.

Appropriately defining a heat source aspect ratio $\epsilon$ is also a critical factor in predicting the dimensionless thermal resistance. In the previous comparison of Figures 2.5 and 2.7, the source aspect ratio is an easy choice given the symmetry of the heat source. For irregular or nonsymmetric-shaped heat sources a simple definition of source aspect ratio may be taken as the ratio of the smallest and largest perpendicular lengths of a particular heat source. For many cases, this can be easily related to the geometry of interest and its characteristic dimensions as shown in Figure 2.8. By means of the appropriate or effective aspect ratio, the results for many useful shapes can be predicted within 5% using the expressions for the elliptic heat source.

Sadeghi et al. (2010) showed that all of the results for isoflux heat sources fall close to the solution for an isoflux elliptic heat source if an appropriate aspect is defined. This result developed earlier in section is recast using the square root of source area as a characteristic length scale.

For an isoflux heat source using the mean source temperature, we have:

$$\overline{R}_s \, k \, \sqrt{A} = \frac{16\sqrt{\epsilon}}{3\pi^{5/2}} K(\kappa) \quad \text{Isoflux Mean Model} \tag{2.119}$$

or if using the maximum source temperature, we may define:

$$\hat{R}_s \, k \, \sqrt{A} = \frac{2\sqrt{\epsilon}}{\pi^{3/2}} K(\kappa) \quad \text{Isoflux Centroid Model} \tag{2.120}$$

The ratio of the thermal spreading resistance based on the mean temperature and that based on the centroidal temperature is :

$$\frac{\overline{R}}{\hat{R}} \approx 0.8488 \tag{2.121}$$

and is independent of aspect ratio for the elliptic heat source. Yovanovich et al. (1977) showed that the ratio for all cases of the hyperellipses that were considered provided a slightly lower value of 0.84. Note, in Sadeghi et al. (2010), an error is present in the results given for the isoflux heat source. The correct results are given by Eqs. (2.119) and (2.120). Results for 12 heat source shapes are provided in Figure 2.9 for the case of dimensionless thermal resistance based on the maximum heat source temperature, Eq. (2.120). The centroidal temperature reference is used due to fact that the superposition method for irregular heat sources only provides the result based on the maximum temperature. It is seen that excellent agreement is obtainable using the elliptic source as a universal model.

It is reasonable to expect that for an isothermal heat source, the dimensionless thermal resistance can be determined from:

**Figure 2.8** Definitions of effective aspect ratio. $\beta = b/a$ is the nominal aspect ratio of the geometry. Source: Adapted from Sadeghi et al. (2010).

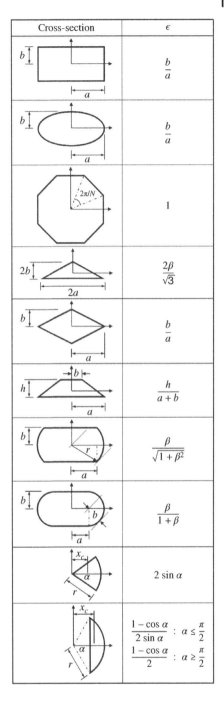

| Cross-section | $\epsilon$ |
|---|---|
| | $\dfrac{b}{a}$ |
| | $\dfrac{b}{a}$ |
| | $1$ |
| | $\dfrac{2\beta}{\sqrt{3}}$ |
| | $\dfrac{b}{a}$ |
| | $\dfrac{h}{a+b}$ |
| | $\dfrac{\beta}{\sqrt{1+\beta^2}}$ |
| | $\dfrac{\beta}{1+\beta}$ |
| | $2\sin\alpha$ |
| | $\dfrac{1-\cos\alpha}{2\sin\alpha}\;:\;\alpha \le \dfrac{\pi}{2}$ <br> $\dfrac{1-\cos\alpha}{2}\;:\;\alpha \ge \dfrac{\pi}{2}$ |

$$R_s\, k\, \sqrt{A} = \frac{\sqrt{\epsilon}}{2\sqrt{\pi}} K(\kappa) \quad \text{Isothermal Model} \tag{2.122}$$

as evidenced by the results shown in Figure 2.3. The usefulness of the square root of area for collapsing dimensionless thermal resistance will also be taken advantage of in Chapters 3 and 4 when dealing with flux channels and flux tubes.

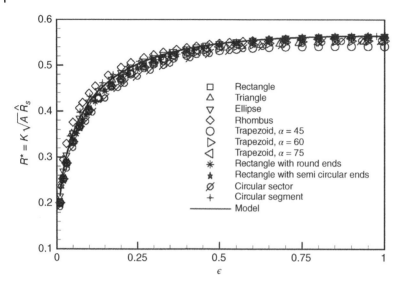

**Figure 2.9** Results for isoflux heat sources based on the maximum or centroidal temperature. Source: After Sadeghi et al. (2010).

## 2.9 Circular Annular Area on a Half-Space

Analytical methods have been used to obtain the spreading resistance for the isoflux and isothermal circular annulus of radii $a, b$ on the surface of an isotropic half-space such that $a < b$ and having thermal conductivity $k$.

### 2.9.1 Isothermal Circular Annular Ring Source

The spreading resistance for the isothermal circular annulus cannot be obtained directly by the integral method. Mathematically, this is a mixed boundary-value problem which requires special solution methods which are discussed by Sneddon (1966). Smythe (1951) reported the solution for the capacitance of a charged annulus. Yovanovich and Schneider (1977) used the two results of Smythe to determine the spreading resistance. Yovanovich and Schneider (1977) reported the following relationships for the spreading resistance of an isothermal circular annular contact area:

$$kbR_s = \frac{1}{\pi^2} \left[ \frac{\ln 16 + \ln[(1 + \epsilon)/(1 - \epsilon)]}{(1 + \epsilon)} \right] \tag{2.123}$$

for $1.000 < 1/\epsilon < 1.10$, and

$$kbR_s = \frac{\pi/8}{\left[ \cos^{-1}\epsilon + \sqrt{1 - \epsilon^2}\tanh^{-1}\epsilon \right] \left[ 1 + 0.0143\epsilon^{-1}\tan^3(1.28\epsilon) \right]} \tag{2.124}$$

for $1.1 < 1/\epsilon < \infty$. When $\epsilon = 0$, the annulus becomes a circle, and the spreading resistance gives $R_s = 1/(4kb)$ in agreement with the result obtained for the isothermal circular area.

### 2.9.2 Isoflux Circular Annular Ring Source

The temperature rise of points in the annular area $a \le r \le b$ was reported by Yovanovich and Schneider (1977) to have the distribution:

$$\theta(r) = \frac{2}{\pi} \frac{qb}{k} \left\{ E\left(\frac{r}{b}\right) - \left(\frac{r}{b}\right) E\left(\frac{a}{r}\right) + \left(\frac{r}{b}\right) \left[1 - \left(\frac{a}{r}\right)^2\right] K\left(\frac{a}{r}\right) \right\} \quad (2.125)$$

where the special functions $K(x)$ and $E(x)$ are the complete elliptic integrals of the first and second kinds, respectively, of arbitrary modulus $x$ [Abramowitz and Stegun (1965), Byrd and Friedman (1971)]. The dimensionless spreading resistance, based on the area-average temperature rise, of the isoflux circular annulus was reported by Yovanovich and Schneider (1977) to have the form:

$$kbR_s = \frac{8}{3\pi^2} \frac{1}{(1 - \epsilon^2)^2} \left[1 + \epsilon^3 - (1 + \epsilon^2)E(\epsilon) + (1 - \epsilon^2)K(\epsilon)\right] \quad (2.126)$$

where the modulus is $\epsilon = a/b < 1$. When $\epsilon = 0$, the annulus becomes a circle of radius $b$, and the above relationship gives $kbR_s = 8/(3\pi^3)$, which is in agreement with the result obtained for the isoflux circular area.

## 2.10 Other Doubly Connected Areas on a Half-Space

The numerical data of spreading resistance from Martin et al. (1984) for three doubly connected regular polygons: equilateral triangle, square, and circle were nondimensionalized as $k\sqrt{A_c}R_s$. The dimensionless spreading resistance is a function of $\epsilon = \sqrt{A_i/A_o}$, where $A_i$ and $A_o$ are the inner and outer projected areas of the polygons. The "active" area is $A_c = A_o - A_i$.

Accurate correlation equations with a maximum relative error of 0.6% were given. For the range: $0 \le \epsilon \le 0.995$,

$$k\sqrt{A_c}R_s = a_0 \left[1 - \left(\frac{\epsilon}{a_1}\right)^{a_2}\right]^{a_3} \quad (2.127)$$

and for the range: $0.995 \le \epsilon \le 0.9999$,

$$kP_oR_s = a_5 \ln\left[\frac{a_4}{\frac{1}{\epsilon} - 1}\right] \quad (2.128)$$

where $P_o$ is the outer perimeter of the polygons, and the correlation coefficients: $a_0$ through $a_5$ are given in Table 2.9.

The correlation coefficient $a_0$ represents the dimensionless spreading resistance of the full contact area in agreement with results presented above. Since the results for the square and the circle are very close for all values of the parameter $\epsilon$, the correlation equations for the square or the circle may be used for other doubly connected regular polygons such as pentagons and hexagons.

**Table 2.9** Correlation coefficients for doubly-connected polygons.

|        | Circle  | Square  | Triangle |
|--------|---------|---------|----------|
| $a_0$  | 0.4789  | 0.4732  | 0.4602   |
| $a_1$  | 0.99957 | 0.99980 | 1.00010  |
| $a_2$  | 1.5056  | 1.5150  | 1.5101   |
| $a_3$  | 0.35931 | 0.37302 | 0.38637  |
| $a_4$  | 39.66   | 68.59   | 115.91   |
| $a_5$  | 0.31604 | 0.31538 | 0.31529  |

## 2.11   Problems with Source Plane Conductance

Contact conductance in the source plane has been considered for two specific problems. The first involves relaxing the adiabatic region outside of the source area by applying a mixed-boundary condition with constant uniform conductance. The other involves applying a constant uniform conductance over the source region while prescribing the adiabatic condition in the region outside of the source area. We will present some useful results for both types of problems.

### 2.11.1   Isoflux Heat Source on a Convectively Cooled Half-Space

We now conclude with thermal spreading into a half-space from a circular heat source with conductance in the region outside of the source area, see Figure 2.10. The region outside the heat source is convectively cooled. Heat enters the half-space through the circular source and ultimately leaves through the cooled surface or the remainder goes into the half-space at $\sqrt{r^2 + z^2} \to \infty$.

The governing equation for this problem is Laplace's equation:

$$\frac{\partial^2 \theta}{\partial r^2} + \frac{1}{r}\frac{\partial \theta}{\partial r} + \frac{\partial^2 \theta}{\partial z^2} = 0 \tag{2.129}$$

where $\theta = T - T_\infty$. The following boundary conditions are required:

$$\begin{aligned} r = 0 \quad & z > 0 \quad && \theta \neq \infty \\ r = 0 \quad & z \to \infty \quad && \theta \to 0 \end{aligned} \tag{2.130}$$

In the source plane we specify:

$$\left. \begin{aligned} \frac{\partial \theta}{\partial z}\bigg|_{z=0} &= -\frac{q}{k}, \quad && 0 < r < a \\ \frac{\partial \theta}{\partial z}\bigg|_{z=0} &= +\frac{h}{k}\theta, \quad && a < r < \infty \end{aligned} \right\} \tag{2.131}$$

where $q = Q/\pi a^2$ is the constant heat flux, $k$ is the thermal conductivity of the half-space, and $h$ is the conductance in the region outside of the source area. This problem, and a number of other variations, was examined by Lemczyk (1986) with several fundamental results summarized in Lemczyk and Yovanovich (1988a,b).

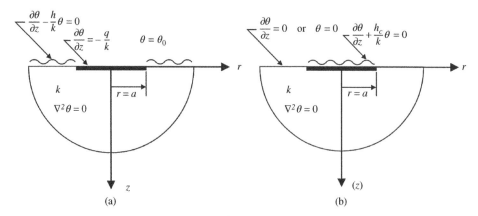

**Figure 2.10**   Thermal spreading from a circular source with conductance: (a) source plane conductance and (b) source conductance. Source: Adapted from Lemczyk (1986).

Lemczyk (1986) solved the above problem using Hankel transforms while considering a superposition of two solutions. The solution process is as follows. The general solution is obtained assuming:

$$\theta = \theta_1 + \theta_2 \tag{2.132}$$

The boundary condition in the source plane for $\theta_1$ is:

$$\left.\frac{\partial \theta_1}{\partial z}\right|_{z=0} = -\frac{q}{k}, \quad 0 < r < a \\ \theta_1 = 0, \quad a < r < \infty \tag{2.133}$$

While the boundary condition in the source plane for $\theta_2$ is:

$$\left.\frac{\partial \theta_2}{\partial z}\right|_{z=0} = 0, \quad 0 < r < a \\ \left.\frac{\partial \theta_2}{\partial z}\right|_{z=0} - \frac{h}{k}\theta_2, = -\frac{\partial \theta_1}{\partial z}, \quad a < r < \infty \tag{2.134}$$

The solution for $\theta_1$ is easily found using the Hankel transform method and provides the following:

$$\theta_1(r,0) = \frac{2qa}{\pi k}[1 - (r/a)^2]^{1/2}, \quad r < a \tag{2.135}$$

while

$$\frac{\partial \theta_1}{\partial z} = \frac{2q}{\pi k}\left[\sin^{-1}(a/r) - \{r^2/a^2 - 1\}^{-1/2}\right], \quad r > a \tag{2.136}$$

The solution for $\theta_2$ is somewhat more involved and as such will not be presented here. The full solution has two special limits when $Bi \to 0$ and $Bi \to \infty$:

$$Bi \to 0, \quad \psi \to \frac{32}{3\pi^2} \tag{2.137}$$

and

$$Bi \to \infty, \quad \psi \to \frac{16}{3\pi^2} \tag{2.138}$$

The $Bi \to 0$ is the solution for perfect insulation in the region outside of the heat source, while for the latter case when $Bi \to \infty$, it is for the special case when $\theta = 0$ in the region outside of the heat source. For this case, the solution is determined from the temperature profile given by $\theta_1$. The full solution smoothly transitions between these two limits in a sigmoidal manner as seen in Figure 2.11. Lemczyk (1986) tabulated results and presented them graphically.

The solution for the heat source having constant temperature $\theta_0$ within the source region and uniform conductance outside of the source region is even more involved and requires the superposition of three Hankel transform solutions [Lemczyk (1986)]. As such, we will not review the details, but rather only provide a simple useful equation for calculating the constriction parameter. Due to the complexity of the exact mathematical solutions in both cases, Lemczyk (1986) developed correlations of the form:

$$\psi = c_1 - c_2 \tanh(c_3 \ln Bi - c_4) \tag{2.139}$$

for both cases.

The coefficients $c_1$–$c_4$ are tabulated in Table 2.10 for both the isoflux and isothermal source cases using the Biot number defined as $Bi = ha/k$. The exact results of Lemczyk (1986) along with the results of the above correlation are provided in Figure 2.11.

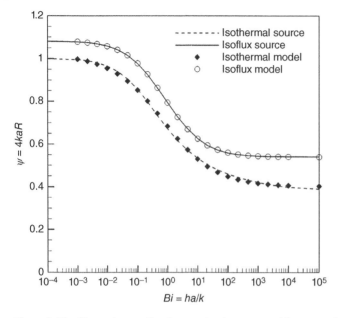

**Figure 2.11** Thermal spreading from a circular source with source plane conductance.

**Table 2.10** Coefficients for isoflux and isothermal heat sources.

|       | Isoflux       | Isothermal    |
| ----- | ------------- | ------------- |
| $c_1$ | 0.81179902    | 0.71006139    |
| $c_2$ | 0.27273615    | 0.30889793    |
| $c_3$ | 0.33492137    | 0.25316704    |
| $c_4$ | −0.065961153  | −0.08602627   |

### 2.11.2 Effect of Source Contact Conductance on Spreading Resistance

Martin et al. (1984) used a novel numerical technique to determine the effect of a uniform contact conductance $h$ on the spreading resistance of square and circular contact areas. Lemczyk and Yovanovich (1988a,b) further examined the issue of source conductance in more detail for both uncoated and coated half-spaces.

Martin et al. (1984) proposed that the dimensionless spreading resistance values they calculated could be correlated with an accuracy of 0.1% by the following relationship:

$$k\sqrt{A}\,R_s = c_1 - c_2\,\tanh\left(c_3 \ln Bi - c_4\right), \qquad 0 \le Bi < \infty \tag{2.140}$$

with $Bi = h\sqrt{A}/k$. The correlation coefficients $c_1$ through $c_4$ are given in Table 2.11.

When $Bi \le 0.1$, the predicted values approach the values corresponding to the isoflux boundary condition, and when $Bi \ge 100$, the predicted values are within 0.1% of the values obtained for the isothermal boundary condition. The transition from the isoflux values to the isothermal values occurs in the range: $0.1 \le Bi \le 100$.

**Table 2.11** Coefficients for square and circle.

|       | Circle   | Square   |
|-------|----------|----------|
| $c_1$ | 0.46159  | 0.45733  |
| $c_2$ | 0.017499 | 0.016463 |
| $c_3$ | 0.43900  | 0.47035  |
| $c_4$ | 1.1624   | 1.1311   |

Lemczyk (1986) also considered the analytical solution of the above problem for the case of a circular source. Additionally, he also considered a number of extensions of the problem for the case of a layered half-space to be discussed in Section 2.12.

## 2.12 Circular Area on Single Layer (Coating) on Half-Space

Integral solutions are available for the spreading resistance for a circular source of radius $a$ in contact with an isotropic layer of thickness $t_1$ and thermal conductivity $k_1$ which is in perfect thermal contact with an isotropic half-space of thermal conductivity $k_2$, [Dryden (1983), Dryden et al. (1985)]. The solutions obtained for two heat flux distributions corresponding to the flux parameter values are $\mu = -1/2$ and $\mu = 0$.

### 2.12.1 Equivalent Isothermal Circular Contact

Dryden (1983) obtained the solution for the equivalent isothermal circular contact flux distribution:

$$q(r) = \frac{Q}{2\pi a^2 \sqrt{1 - r^2/a^2}} \qquad 0 \le r \le a \tag{2.141}$$

The problem is depicted in Figure 2.12.

The dimensionless spreading resistance, based on the area-average temperature, is obtained from the integral [Dryden (1983)]:

$$\psi = 4k_2 a R_s = \frac{4}{\pi} \frac{k_2}{k_1} \int_0^\infty \left[ \frac{\lambda_2 \exp\left(\zeta t_1/a\right) + \lambda_1 \exp\left(-\zeta t_1/a\right)}{\lambda_2 \exp\left(\zeta t_1/a\right) - \lambda_1 \exp\left(-\zeta t_1/a\right)} \right] \frac{J_1\left(\zeta\right) \sin \zeta}{\zeta^2} d\zeta \tag{2.142}$$

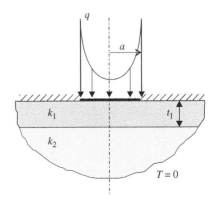

**Figure 2.12** Coated half-space with equivalent isothermal flux distribution.

with $\lambda_1 = (1 - k_2/k_1)/2$ and $\lambda_2 = (1 + k_2/k_1)/2$. The parameter $\zeta$ is a dummy variable of integration. The constriction resistance depends on the thermal conductivity ratio $k_1/k_2$ and the relative layer thickness $t_1/a$. Dryden (1983) presented simple asymptotes for thermal spreading in thin layers: $t_1/a \leq 0.1$ and in thick layers: $t_1/a \geq 10$. These asymptotes were also presented as dimensionless spreading resistances defined as $4k_2aR_s$. They are:

**Thin Layer Asymptote**

$$(4k_2aR_s)_{\text{thin}} = 1 + \left(\frac{4}{\pi}\right)\left(\frac{t_1}{a}\right)\left[\frac{k_2}{k_1} - \frac{k_1}{k_2}\right] \tag{2.143}$$

**Thick Layer Asymptote**

$$(4k_2aR_s)_{\text{thick}} = \frac{k_2}{k_1} - \left(\frac{2}{\pi}\right)\left(\frac{a}{t_1}\right)\left(\frac{k_2}{k_1}\right)\ln\left(\frac{2}{1 + k_1/k_2}\right) \tag{2.144}$$

These asymptotes provide results which are within 1% of the full solution for relative layer thickness: $t_1/a < 0.5$ and $t_1/a > 2$.

The dimensionless spreading resistance is based on the substrate thermal conductivity $k_2$. The above general solution is valid for conductive layers, where $k_1/k_2 > 1$ as well as resistive layers, where $k_1/k_2 < 1$. The infinite integral can be evaluated numerically by means of Computer Algebra Systems which provide accurate results.

**Example 2.2**  An isothermal circular micro-contact spot of radius $a = 3\,\mu m$ is in perfect thermal contact with a thin, uniformly thick, silver layer which was diffusion bonded to a Ni 200 substrate whose thermal conductivity is $k_2 = 64.5\,W/(m\ K)$. The thickness of the silver layer is $t_1 = 2\,\mu m$ and its thermal conductivity is $k_1 = 400\,W/(m\ K)$. Calculate the value of the spreading resistance for this system. If the thickness of the silver layer is increased to $t_1 = 10\,\mu m$ and the radius is unchanged, calculate the new value of the spreading resistance. Use the full equivalent isothermal integral solution for the calculations. Finally, calculate the minimum and maximum values of the spreading resistance.

Using Eq. (2.142), we numerically integrate and find $\psi = 4k_2aR_s = 0.3329$. This gives:

$$R_s = \frac{\psi}{4k_2a} = \frac{0.3329}{4 \cdot 64.5 \cdot 3 \times 10^{-6}} = 430.1\ \text{K/W}$$

If the thickness of the silver layer is increased to $t = 10\,\mu m$, we find the new value of $\psi = 4k_2aR_s = 0.2004$, and the resistance is now:

$$R_s = \frac{\psi}{4k_2a} = \frac{0.2004}{4 \cdot 64.5 \cdot 3 \times 10^{-6}} = 258.9\ \text{K/W}$$

Finally, the limits of the thermal spreading resistance are found by considering the cases $t_1 \to 0$ and $t_1 \to \infty$ using the resistance $R_s = 1/4ka$. This gives the bounds on $R_s$ as :

$$\frac{1}{4k_1a} < R_s < \frac{1}{4k_2a}$$

or

$$208.33\ \text{K/W} < R_s < 1291.98\ \text{K/W}$$

## 2.12.2 Isoflux Circular Contact

Hui and Tan (1994) presented an integral solution for the isoflux circular source. The dimensionless spreading resistance is

$$4k_2 a R_s = \frac{32}{3\pi^2}\left(\frac{k_2}{k_1}\right)^2 + \frac{8}{\pi}\left[1 - \left(\frac{k_2}{k_1}\right)^2\right]\int_0^\infty \frac{J_1^2(\zeta)\,d\zeta}{\left[1 + \frac{k_1}{k_2}\tanh\left(\zeta t_1/a\right)\right]\zeta^2} \tag{2.145}$$

which depends on the thermal conductivity ratio $k_1/k_2$ and the relative layer thickness $t_1/a$. The dimensionless spreading resistance is based on the substrate thermal conductivity $k_2$. The above general solution is valid for conductive layers, where $k_1/k_2 > 1$ as well as resistive layers, where $k_1/k_2 < 1$.

## 2.12.3 Isoflux, Equivalent Isothermal, and Isothermal Solutions

Negus et al. (1985) obtained solutions by application of the Hankel transform method for flux-specified boundary conditions and with a novel technique of linear superposition for the mixed boundary condition (isothermal contact area and zero flux outside the contact area). They reported results for three flux distributions: (i) isoflux, (ii) equivalent isothermal flux, and (iii) true isothermal source. Their results are presented below.

### 2.12.3.1 Isoflux Contact Area
For the isoflux boundary condition, they reported the result for $\psi^q = 4k_1 a R_s$

$$\psi^q = \frac{32}{3\pi^2} + \frac{8}{\pi^2}\sum_{n=1}^\infty (-1)^n \alpha^n I_q \tag{2.146}$$

The first term is the dimensionless isoflux spreading resistance of an isotropic half-space of thermal conductivity $k_1$ and the second term accounts for the effect of the layer relative thickness and relative thermal conductivity. The thermal conductivity parameter $\alpha$ is defined as

$$\alpha = \frac{1 - \kappa}{1 + \kappa}$$

with $\kappa = k_1/k_2$. The layer thickness-conductivity parameter $I_q$ is defined as

$$I_q = \frac{1}{2\pi}\left\{2\sqrt{2(\gamma+1)}\,E\left(\sqrt{2/(\gamma+1)}\right) - \frac{\pi}{2\sqrt{2\gamma}}I_\gamma - 2\pi n\tau_1\right\}$$

with

$$I_\gamma = \left(1 + \frac{0.09375}{\gamma^2} + \frac{0.0341797}{\gamma^4} + \frac{0.00320435}{\gamma^6}\right)$$

The relative layer thickness is $\tau_1 = t/a$ and the relative thickness parameter is

$$\gamma = 2n^2\tau_1^2 + 1$$

The special function $E(\cdot)$ is the complete elliptic integral of the second kind [Abramowitz and Stegun (1965)].

### 2.12.3.2 Equivalent Isothermal Contact Area

For the equivalent isothermal flux boundary condition, they reported the result for $\psi_{ei}^T = 4k_1 a R_s$

$$\psi_{ei}^T = 1 + \frac{8}{\pi} \sum_{n=1}^{\infty} (-1)^n \alpha^n I_{ei} \tag{2.147}$$

where as discussed above the first term represents the dimensionless spreading resistance of an isothermal contact area on an isotropic half-space of thermal conductivity $k_1$ and the second term accounts for the effect of the layer relative thickness and the relative thermal conductivity. The thermal conductivity parameter $\alpha$ is defined above. The relative layer thickness parameter $I_{ei}$ is defined as

$$I_{ei} = \left[ \sqrt{1 - \beta^{-2}} \left( \beta - \beta^{-1} \right) + \frac{1}{2} \sin^{-1} \left( \beta^{-1} \right) - 2n\tau_1 \right]$$

with $\tau_1 = t/a$ and

$$\beta = n\tau_1 + \sqrt{n^2 \tau_1^2 + 1}$$

### 2.12.3.3 Isothermal Contact Area

For the isothermal contact area, Negus et al. (1985) reported a correlation equation for their numerical results. They reported $\psi^T = 4k_1 a R_s$ in the form

$$\psi^T = F_1 \tanh F_2 + F_3 \tag{2.148}$$

where

$$F_1 = 0.49472 - 0.49236\kappa - 0.00340\kappa^2$$

and

$$F_2 = 2.8479 + 1.3337\tau + 0.06864\tau^2 \qquad \text{with} \qquad \tau = \log_{10} \tau_1$$

and

$$F_3 = 0.49300 + 0.57312\kappa - 0.06628\kappa^2$$

where $\kappa = k_1/k_2$. The correlation equation was developed for resistive layers: $0.01 \le \kappa \le 1$ over a wide range of the relative thicknesses: $0.01 \le \tau_1 \le 100$. The maximum relative error associated with the correlation equation is approximately 2.6% at $\tau_1 = 0.01$ and $\kappa = 0.2$. Numerical results for $\psi^q, \psi_{ei}^T, \psi^T$ for a range of values of $\tau_1$ and $\kappa$ were presented in tabular form for easy comparison. They found that the values for $\psi^q > \psi_{ei}^T$ and that $\psi_{ei}^T \le \psi^T$. The maximum difference between $\psi^q$ and $\psi^T$ was approximately 8%. The values for $\psi^T > \psi_{ei}^T$ for very thin layers: $\tau_1 \le 0.1$ and for $\kappa \le 0.1$; however, the differences were less than approximately 8%. For most applications, the equivalent isothermal flux solution and the true isothermal solution are similar.

## 2.13 Thermal Spreading Resistance Zone: Elliptical Heat Source

In this section, we consider the thermal spreading resistance zone associated with an isothermal elliptical source with semiaxes $a, b$ with $b < a$ situated in the free surface $z = 0$ of an isotropic half-space of thermal conductivity $k$. The temperature field is three-dimensional $T(x, y, z)$ and the isotherms are ellipsoidal surfaces described by the following equation in terms of the parameter $\lambda$:

$$\frac{x^2}{a^2 + \lambda} + \frac{y^2}{b^2 + \lambda} + \frac{z^2}{\lambda} = 1 \tag{2.149}$$

On the isothermal ellipse $\lambda = 0$ and for points far from the ellipse $\lambda \to \infty$.

The thermal solution given below follows closely the procedure presented by Yovanovich (1971). Extensive solutions for analogous problems are the potential of a uniform ellipsoid in a large domain [Jeans (1963)] and the electromagnetic field of a charged ellipsoid [Mason and Weaver (1929), Smythe (1951), Stratton (1941), Moon and Spencer (1961, 1971)]. Holm (1967) obtained an approximate solution for current flow from an isopotential elliptical electrical source. These books are recommended as general references for obtaining solutions in ellipsoidal coordinates for different physical problems.

The temperature field $T(\lambda)$ is the solution of

$$\nabla^2 T = \frac{d}{d\lambda} \left[ f(\lambda) \frac{dT}{d\lambda} \right] = 0 \tag{2.150}$$

where

$$f(\lambda) = \sqrt{(a^2 + \lambda)(b^2 + \lambda)\lambda} \tag{2.151}$$

The temperature gradient with respect to $\lambda$ is

$$\frac{dT}{d\lambda} = \frac{C_1}{f(\lambda)} \tag{2.152}$$

The boundary conditions on the isothermal elliptical heat source ($\lambda = 0$) and the heat sink ($\lambda \to \infty$) are: at $\lambda = 0$, $T = T_0$ and as $\lambda \to \infty$, $T \to T_\infty$.

The general solution is

$$T = C_1 \int_0^\lambda \frac{d\lambda}{f(\lambda)} + C_2 \tag{2.153}$$

The boundary conditions are satisfied if the constants are $C_1 = -(T_0 - T_\infty)/\int_0^\infty d\lambda/f(\lambda)$ and $C_2 = T_0$.

The solution is therefore

$$\frac{T_0 - T(\lambda)}{T_0 - T_\infty} = \frac{\int_0^\lambda \frac{d\lambda}{f(\lambda)}}{\int_0^\infty \frac{d\lambda}{f(\lambda)}} \qquad 0 < \lambda < \infty \tag{2.154}$$

which can be written in the alternative form based on the incomplete elliptic integrals of the first kind:

$$\frac{T_0 - T(\lambda)}{T_0 - T_\infty} = \frac{F(\vartheta_\lambda, \kappa)}{F(\vartheta, \kappa)} \tag{2.155}$$

where the amplitude angle $\vartheta_\lambda$ is given by

$$\sin \vartheta_\lambda = \sqrt{\frac{a^2}{a^2 + \lambda}} \tag{2.156}$$

and the modulus is

$$\kappa = \sqrt{1 - \frac{b^2}{a^2}} \tag{2.157}$$

with $b/a \le 1$. If $\lambda = 0$, $\vartheta_\lambda = \pi/2$, then $F(\pi/2, \kappa) = K(\kappa)$, where $K(\cdot)$ is the complete elliptic integral of the first kind with modulus $\kappa$.

The temperature distribution throughout the half-space can be written as

$$\frac{T_0 - T(\lambda)}{T_0 - T_\infty} = \frac{F(\vartheta_\lambda, \kappa)}{K(\kappa)} \qquad \text{for} \qquad 0 < \lambda < \infty \tag{2.158}$$

The foregoing relationship gives the temperature drop $T_0 - T(\lambda)$ from the elliptical heat source to an arbitrary isothermal surface at $\lambda$ relative to the overall temperature drop $T_0 - T_\infty$. In its present form, it does not provide information regarding thermal spreading resistances within the half-space.

In order to determine the thermal spreading resistances, it's necessary to find the heat flow rate $Q$ within the half-space $0 < \lambda < \infty$.

We consider steady conduction through the hemispherical surface located at points $\lambda \gg a > b$, where $A(r) = 2\pi r^2$ and where $\lambda = r^2$ and $d\lambda = 2rdr$. This shows that $d\lambda/dr = 2r = 2\sqrt{\lambda}$.

The heat flow rate is given by

$$Q = -kA(r)\frac{dT}{dr} = -k2\pi\lambda\frac{dT}{d\lambda}\frac{d\lambda}{dr} \tag{2.159}$$

Hence, the heat flow rate is

$$Q = -k4\pi\lambda^{3/2}\frac{dT}{d\lambda} \tag{2.160}$$

From the first integration of the governing equation, we have

$$\frac{dT}{d\lambda} = \frac{C_1}{f(\lambda)} = \frac{C_1}{\lambda^{3/2}} \tag{2.161}$$

because for $\lambda \gg a > b, f(\lambda) = \lambda^{3/2}$. The first integration constant is

$$C_1 = -\frac{T_0 - T_\infty}{\int_0^\infty \dfrac{d\lambda}{f(\lambda)}} \tag{2.162}$$

Finally, the heat flow rate has the form

$$Q = 4\pi k\frac{T_0 - T_\infty}{\int_0^\infty \dfrac{d\lambda}{f(\lambda)}} \tag{2.163}$$

The factor in the denominator can be written in terms of the incomplete elliptic integral of the first kind such as

$$\int_0^\infty \frac{d\lambda}{f(\lambda)} = \frac{a}{2}F(\pi/2, \kappa) \tag{2.164}$$

with modulus $\kappa = \sqrt{1 - b^2/a^2}$. With $F(\pi/2, \kappa) = K(\kappa)$, where

$$K(\kappa) = \int_0^{\pi/2} \frac{dt}{\sqrt{1 - \kappa^2 \sin(t)^2}} \tag{2.165}$$

The heat flow rate from the isothermal elliptical area is

$$Q = 2\pi ka\frac{T_0 - T_\infty}{K(\kappa)} \tag{2.166}$$

Since the thermal spreading resistance is defined as $R_s = (T_0 - T_\infty)/Q$, the overall spreading resistance is

$$R_s = \frac{K(\kappa)}{2\pi ka} \tag{2.167}$$

for aspect ratios $0 < b/a \le 1$. For the circular area with radius $a$, $b = a, \kappa = 0$ and $K(0) = \pi/2$, which gives the well-known result for the spreading resistance $R_s = 1/4ka$.

The thermal spreading resistance zone can be determined by considering temperature drops within two regions of the half-space $0 < \lambda < \infty$. The overall temperature drop from the elliptical heat source at $\lambda = 0$ to the sink at $\lambda \to \infty$ can be written in terms of the temperature drop from the heat source to an arbitrary isotherm $T(\lambda)$ plus the temperature drop from the arbitrary isotherm to the heat sink. Thus, we have

$$T_0 - T_\infty = \left(T_0 - T(\lambda)\right) + \left(T(\lambda) - T_\infty\right) \tag{2.168}$$

Since each temperature drop is associated with a thermal resistance and the heat flow rate $Q$; after dividing through by $Q$, the overall thermal resistance can be written as

$$R(0 \to \infty) = R(0 \to \lambda) + R(\lambda \to \infty) \tag{2.169}$$

The thermal spreading resistance zone is defined by the resistance $R(0 \to \lambda)$ which is written as

$$R(0 \to \lambda) = R(0 \to \infty) - R(\lambda \to \infty) \tag{2.170}$$

The foregoing relation is normalized with respect to the overall resistance such that

$$\frac{R(0 \to \lambda)}{R(0 \to \infty)} = 1 - \frac{R(\lambda \to \infty)}{R(0 \to \infty)} \tag{2.171}$$

The foregoing relation based on thermal resistances is expressed as

$$\frac{R_{se}}{R_{total}} = 1 - \frac{F(\varphi_\lambda, \kappa)}{F(\pi/2, \kappa)} \tag{2.172}$$

The thermal spreading resistance zone is denoted as $R_{se}$. On the ellipse $\lambda = 0$, the amplitude angle is $\varphi = \pi/2$. The total spreading resistance for the isothermal ellipse is given by

$$R_{total} = \frac{1}{2\pi ka} K(\kappa) \tag{2.173}$$

For the special case of an isothermal circular source, $b = a, \kappa = 0, K(0) = \pi/2$ gives the well-known relation:

$$R_s = \frac{1}{4ka} \tag{2.174}$$

To define the extent of the spreading resistance zone, we can use the semimajor axis of the ellipse such as setting $y = 0$ and $z = 0$ in the equation for an ellipsoid to get

$$x^2 = a^2 + \lambda \tag{2.175}$$

The foregoing relation can be written in terms of the semimajor axis as

$$\frac{x^2}{a^2} = 1 + \frac{\lambda}{a^2} \tag{2.176}$$

The amplitude angle has two forms:

$$\sin \varphi = \frac{a^2}{a^2 + \lambda} \tag{2.177}$$

or

$$\sin \varphi = \frac{a}{x} \tag{2.178}$$

The thermal spreading resistance zone for the isothermal ellipse is defined as

$$Z_e = f\left(\frac{b}{a}, \frac{a}{x}\right) \tag{2.179}$$

which is equal to the ratio

$$Z_e = \frac{R_{s,e}}{R_{total}} \tag{2.180}$$

**Table 2.12** Values of $Z_e$ for several values of $b/a$ and $x/a$.

| $b/a$ | $x/a = 2$ | 4 | 6 | 10 | 20 | 30 |
|---|---|---|---|---|---|---|
| 1 | 0.667 | 0.839 | 0.893 | 0.936 | 0.968 | 0.979 |
| 0.8 | 0.696 | 0.855 | 0.904 | 0.943 | 0.971 | 0.981 |
| 0.6 | 0.730 | 0.873 | 0.916 | 0.950 | 0.975 | 0.984 |
| 0.4 | 0.769 | 0.892 | 0.929 | 0.958 | 0.979 | 0.986 |
| 0.2 | 0.818 | 0.915 | 0.944 | 0.967 | 0.983 | 0.989 |
| 0.1 | 0.851 | 0.931 | 0.955 | 0.973 | 0.987 | 0.991 |

If we let $x = fa$, where $f > 1$, then we can solve for the corresponding $\lambda$ to calculate the amplitude angle and then $F(\varphi, \kappa)$. The extent of the spreading zone can be obtained easily. The calculated values are given in Table 2.12 for several aspect ratios $b/a$ and for six values of $x/a$.

The values in the table reveal interesting features of the spreading resistance associated with the isothermal ellipse. When the extent of the region is $x/a = 2$, then $Z_e > 2/3$ and when $x/a = 10$ over 90% of the total spreading resistance occurs in a relatively thin layer even when the heat source is circular $b/a = 1$.

The thermal spreading resistance zone for an isothermal circular heat source of radius $a$ situated in the surface of an insulated half-space of thermal conductivity $k$ can be found by formulation of the conduction problem using oblate spheroidal coordinates.

As presented in the foregoing solution for the spreading resistance of an elliptical heat source, the temperatures are $T = T_0$ for the circular source and $T \to T_\infty$ for $r \to \infty$ and the distance from the center of the circular source becomes very large. The free surface of the half-space $z = 0, r > a$ is adiabatic.

The temperature field in oblate spheroidal coordinates is one-dimensional, and the solution is relatively straightforward.

The thermal spreading resistance zone defined as $Z_c = R_c/R_{total}$, where $R_{total} = 1/4ka$ is found to be

$$Z_c = \frac{2}{\pi}\cos^{-1}\left(\frac{a}{r}\right) \tag{2.181}$$

for $1 \leq r/a < \infty$. The calculated values for $Z_c$ are in very good agreement with the values given in the table for $b/a = 1$ and for different values of $r/a = x/a$.

## 2.14 Temperature Rise of Multiple Isoflux Sources

In this section, we will develop relationships for the local and area-average steady-state temperature rise of multiple planar isoflux heat sources which lie in the plane $z = 0$ of an isotropic half-space whose thermal conductivity is $k$. The sources can touch, but they cannot intersect [Kellogg (1953), MacMillan (1958), Ramsey (1961)].

### 2.14.1 Two Coplanar Isoflux Circular Sources

We first consider two isoflux circular areas of radii $a_1$ and $a_2$ which have uniform heat fluxes $q_1$ and $q_2$, respectively. The distance between their centroids is denoted as $r_{12}$ with $r_{12} \geq a_1 + a_2$.

The temperature rise of each source due to self-heating is given by the following relationship:

$$\theta_i = \frac{2}{\pi} \frac{q_i a_i}{k} E\left(\frac{r_i}{a_i}\right) \qquad \text{for} \quad i = 1, 2 \tag{2.182}$$

where $r_i$ denotes the local radius with center at the centroid of each source. The temperature rise at the centroid of each source is given by

$$\theta_{0,i} = \frac{q_i a_i}{k} \qquad \text{for} \quad i = 1, 2 \tag{2.183}$$

The temperature rise at the centroid of source 1 due to the heating of source 2 is given by

$$\theta_{1,2} = \frac{2}{\pi} \left(\frac{q_2 a_2}{k}\right) \left\{ \frac{E(a_2/r_{12}) - \left[1 - (a_2/r_{12})^2\right] K(a_2/r_{12})}{a_2/r_{12}} \right\} \tag{2.184}$$

where $K(x)$ and $E(x)$ are the complete elliptic integrals of the first and second kinds of modulus $x$. Similarly, the temperature rise at the centroid of source 2 due to the heating of source 1 is given by

$$\theta_{2,1} = \frac{2}{\pi} \left(\frac{q_1 a_1}{k}\right) \left\{ \frac{E(a_1/r_{12}) - \left[1 - (a_1/r_{12})^2\right] K(a_1/r_{12})}{a_1/r_{12}} \right\} \tag{2.185}$$

The temperature rise at each centroid due to self-heating and due to the heating by the neighboring source is given by the following relationships:

$$\left. \begin{aligned} \theta_1 &= \theta_{0,1} + \theta_{2,1} \\ \theta_2 &= \theta_{0,2} + \theta_{1,2} \end{aligned} \right\} \tag{2.186}$$

If $r_{12} > a_1 + a_2$, the point heat source model can be used with relatively small error. The point heat source relationships are:

$$\left. \begin{aligned} \theta_{1,2} &= \left(\frac{q_2 \pi a_2^2}{2\pi k}\right) \frac{1}{r_{12}} = \frac{1}{2}\left(\frac{q_2 a_2}{k}\right) \frac{a_2}{r_{12}} \\ \theta_{2,1} &= \left(\frac{q_1 \pi a_1^2}{2\pi k}\right) \frac{1}{r_{12}} = \frac{1}{2}\left(\frac{q_1 a_1}{k}\right) \frac{a_1}{r_{12}} \end{aligned} \right\} \tag{2.187}$$

The distance $r_{12}$ may be expressed as

$$r_{12} = a_1 + a_2 + d \qquad \text{with} \quad d \geq 0 \tag{2.188}$$

where $d$ denotes the distance between points of each perimeter that lie on the line that joins the centroids. If the sources touch, then $d = 0$.

**Example 2.3** Two identical coplanar isoflux circular sources supply heat to an isotropic substrate whose thermal conductivity is $k$. The radius of each source is $a$ and the uniform heat flux is $q$. One source is located at the origin $r = 0, z = 0$ and the second source lies in the plane $z = 0$ such that its centroid is at the point $(x = ma, y = 0)$, where $m = 2, 4$, and $6$. The region in the plane outside the two heat sources is adiabatic. Calculate the temperature rise at the centroid and four points on the perimeter of the first source. The five points are located at $P_0(x = 0, y = 0)$, $P_1(x = a, y = 0)$, $P_2(x = 0, y = a)$, $P_3(x = -a, y = 0)$, $P_4(x = 0, y = -a)$.

The temperature rise at $P_0$ due to self-heating is given by

$$\theta_0 = \frac{2}{\pi} \frac{qa}{k} E(0) = \frac{qa}{k}$$

The temperature rise at the points on the perimeter of the first source due to self-heating is given by

$$\theta_{1,i} = \frac{2}{\pi} \frac{qa}{k} E(1) = \frac{2}{\pi} \frac{qa}{k} \qquad \text{for} \quad i = 1, 2, 3, 4$$

The radial distance from the axis of source 2 to a point on the perimeter of source 1 is given by

$$r = a \sqrt{1 + m^2 - 2m \cos \psi} \qquad \text{where} \quad 0 \le \psi \le 2\pi$$

For the points $P_i$, the angle has the values: $\psi_1 = 0, \psi_2 = \pi/2, \psi_3 = \pi, \psi_4 = 3\pi/2$, respectively.

The temperature rise at a point on the perimeter of source 1 due to heating by source 2 is given by

$$\theta_{2,i} = \frac{2}{\pi} \frac{qa}{k} \left\{ \frac{E(a/r) - \left[1 - (a/r)^2\right] K(a/r)}{a/r} \right\} \qquad i = 1, 2, 3, 4$$

The temperature rise at the centroid of source 1 due to self-heating and by heating of source 2 is given by

$$\theta_{centroid} = \frac{qa}{k} + \frac{2}{\pi} \frac{qa}{k} \left\{ \frac{E(1/m) - \left[1 - (1/m)^2\right] K(1/m)}{1/m} \right\}$$

The temperature rise at points on the perimeter of source 1 is given by the following relationship:

$$\theta_{1,i} = \frac{2}{\pi} \frac{qa}{k} + \theta_{2,i}$$

The point heat source model may be used to obtain approximate relationships for the temperature rise at the centroid and the points on the perimeter of source 1. The point source model gives the following relationship for the effect of source 2 on points in source 1:

$$\theta_{ps} = \frac{q\pi a^2}{2\pi k r} = \frac{qa^2}{2kr}$$

For the temperature rise at the centroid of source 1, we have

$$\theta_{centroid} = \frac{qa}{k} + \frac{qa}{2km} = \frac{qa}{k} \left( 1 + \frac{1}{2m} \right)$$

For the points on the perimeter of source 1, we have

$$\theta_{1,i} = \frac{2}{\pi} \frac{qa}{k} + \frac{qa}{2k} \frac{1}{\sqrt{1 + m^2 - 2m \cos \psi}}$$

The dimensionless temperature rise $k\theta/qa$ at the points $P_i$, $i = 1$ to $i = 4$ for three values of $m$ are given below:

|  | $k\theta/qa$ | | |
| --- | --- | --- | --- |
| $m$ | $P_1$ | $P_2 = P_4$ | $P_3$ |
| 2 | 1.2732 | 0.8663 | 0.8057 |
| 4 | 0.8057 | 0.7588 | 0.7371 |
| 6 | 0.7371 | 0.7191 | 0.7083 |

The dimensionless temperature rise $k\theta/qa$ based on the point heat source model at points $P_i$, $i = 1$ to $i = 4$ for three values of $m$ is given below:

|   | | $k\theta/qa$ | |
|---|---|---|---|
| $m$ | $P_1$ | $P_2 = P_4$ | $P_3$ |
| 2 | 1.1366 | 0.8602 | 0.8033 |
| 4 | 0.8033 | 0.7579 | 0.7366 |
| 6 | 0.7366 | 0.7188 | 0.7080 |

The largest difference between the exact values and those based on the point heat source model is about 12%, and it occurs at point $P_1$ when $m = 2$. The differences between the exact and approximate values are less than 1% for all other points, and the differences become negligible for larger values of $m$.

The exact and approximate values of the dimensionless centroid temperature rise for $m = 2, 4, 6$ are listed below.

|   | | $k\theta/qa$ | |
|---|---|---|---|
| $m$ | 2 | 4 | 6 |
| Exact | 1.2587 | 1.1260 | 1.0836 |
| Approximate | 1.2500 | 1.1250 | 1.0833 |

The maximum difference which occurs when $m = 2$ is less than 1%. Therefore, the approximate point heat source model can be used with negligible error.

**Example 2.4**  Five identical circular heat sources of radius $a$ supply heat to an isotropic substrate whose thermal conductivity is $k$. The uniform heat flux on each source is $q$. One source is located at the origin of the Cartesian coordinate system, and the remaining four sources are situated at the corners of a square cluster whose dimensions are $2s$ by $2s$ where $s > a$. The substrate can be modeled as a half-space because its dimensions are much larger than the dimensions of the square cluster.

Use the point heat source model to obtain relationships for the temperature rise at the centroids of the center and corner heat sources.

The temperature rise at the centroid of the center source is due to self-heating and the heating by the four corner sources. Therefore, we can write

$$\theta_0 = \frac{qa}{k} + 4\left(\frac{q\pi a^2}{2\pi k}\right)\frac{1}{\sqrt{2}\,s}$$

which can be expressed as

$$\theta_0 = \frac{qa}{k}\left[1 + \sqrt{2}\left(\frac{a}{s}\right)\right] = \frac{qa}{k}\left[1 + 1.4142\left(\frac{a}{s}\right)\right]$$

The temperature rise at the centroid of the four corner sources is identical. The temperature rise is due to self-heating and the contribution of the other four sources. The distance from the centroid of one corner source to the center source is $r = \sqrt{2}s$, and the distances from the corner centroid to the centroids of the two adjacent corner sources are $r = 2s$, and for the furthest corner source, the distance is $r = \sqrt{2}\,(2s)$. The temperature rise can now be expressed as

$$\theta_1 = \frac{qa}{k} + \frac{q\pi a^2}{2\pi k}\left[\frac{1}{\sqrt{2}s} + 2\left(\frac{1}{2s}\right) + \frac{1}{\sqrt{2}\,(2s)}\right]$$

The relationship can be written as

$$\theta_1 = \frac{qa}{k}\left[1 + \frac{1}{2}\left(\frac{3+2\sqrt{2}}{2\sqrt{2}}\right)\left(\frac{a}{s}\right)\right] = \frac{qa}{k}\left[1 + 1.0303\left(\frac{a}{s}\right)\right]$$

## 2.15 Temperature Rise in an Arbitrary Area

We now consider the arbitrary isoflux area $A^i$ with uniform heat flux $q_i$ and the arbitrary isoflux area $A^j$ with uniform heat flux $q_j$ which lie in the surface of the adiabatic half-space $z > 0$ whose thermal conductivity is $k$. We will derive a relationship for the temperature rise at the arbitrary point $P$ in $A^i$ due to the heating by $A^j$. The differential area $dA^i$ is at the distance $r_i$ from the centroid $0i$ of $A^i$ and the differential area $dA^j$ is at the distance $r_j$ from the centroid $0j$ of $A^j$. The distance between the centroids along the line joining them is denoted as $r_{ij}$. The angle $\psi_i$ is subtended by $0i$ and $r_{ij}$ and the angle $\psi_j$ is subtended by $0j$ and $r_{ij}$.

### 2.15.1 Temperature Rise at Arbitrary Point

The temperature rise at $P$ due to the heating of area $A^j$ is given by

$$\theta_P = \left(\frac{q_j}{2\pi k}\right)\int_{A^j}\frac{dA^j}{r} \tag{2.189}$$

By the cosine law, we have

$$r^2 = r_0^2 + r_j^2 - 2r_j r_0 \cos\psi_j$$

Hence,

$$\theta_P = \left(\frac{q_j}{2\pi k r_0}\right)\int_{A^j}\left(1 + \frac{r_j^2}{r_0^2} - 2\frac{r_j}{r_0}\cos\psi_j\right)^{-1/2}dA^j \tag{2.190}$$

Since $r < r_0$, the binomial theorem may be used to expand the integrand giving,

$$\theta_P = \left(\frac{q_j}{2\pi k r_0}\right)\int_{A^j}\left[1 - \frac{r_j^2}{2r_0^2} + \frac{r_j}{r_0}\cos\psi_j + \frac{3r_j^2\cos^2\psi_j}{2r_0^2} + \cdots\right]dA^j \tag{2.191}$$

The integrations give

$$\theta_P = \left(\frac{q_j}{2\pi k r_0}\right)\left[\int_{A^j}dA^j + \frac{1}{r_0}\int_{A^j}r_j\cos\psi_j\,dA^j \right. $$
$$\left. + \int_{A^j}\left(\frac{2r_j^2 - 3r_j^2\sin^2\psi_j}{2r_0^2}\right)dA^j + \cdots\right] \tag{2.192}$$

The integrals have the following relationships:

$$\int_{A^j}dA^j = A^j \qquad \int_{A^j}r_j\cos\psi_j\,dA^j = 0 \tag{2.193}$$

and

$$\int_{A^j} r_j^2 \, dA^j = I_0^j \qquad\qquad \int_{A^j} r_j^2 \sin^2 \psi_j \, dA^j = I_{RR}^j \tag{2.194}$$

where

$I_0^j$ = polar second moment of *j*th area

$I_{RR}^j$ = radial second moment of *j*th area about the line joining the centroids

The temperature rise at *P* may be conveniently expressed in terms of easily computable geometric quantities,

$$\theta_P = \left(\frac{q_j A^j}{2\pi k r_0}\right)\left[1 + \left(\frac{2I_0^j - 3I_{RR}^j}{2r_0^2 A^j}\right) + \cdots\right] \tag{2.195}$$

This relationship gives the correction to the point heat source model as the *j*th area moves closer to the *i*th area through $r_0$.

### 2.15.2  Average Temperature Rise

The area-average temperature rise of the *i*th area due to the heating of the *j*th area is defined as

$$\overline{\theta}_{ij} = \frac{1}{A^i} \int_{A^i} \theta_P \, dA^i \tag{2.196}$$

Thus,

$$\overline{\theta}_{ij} = \frac{1}{A^i}\left(\frac{q_j}{2\pi k}\right)\int_{A^i}\left[\frac{A^j}{r_0} + \left(\frac{2I_0^j - 3I_{RR}^j}{2r_0^3}\right) + \cdots\right] dA^i \tag{2.197}$$

Since $A^j$ and $I_0^j$ are constants, and in this derivation, it is assumed that $I_{RR}^j$ can be treated as a constant evaluated about the axis joining the centroids of the *i*th and *j*th areas, then the evaluation of $\overline{\theta}_{ij}$ requires the following geometric integrals:

$$\int_{A^i} \frac{dA^i}{r_0} \qquad \text{and} \qquad \int_{A^i} \frac{dA^i}{r_0^3}$$

These integrals are evaluated by making the substitution:

$$r_0^2 = r_{ij}^2 + r_i^2 - 2r_i r_{ij} \cos \psi_i$$

Expanding by the binomial theorem and integrating gives the following results:

$$\int_{A^i} \frac{dA^i}{r_0} = \frac{A^i}{r_{ij}} + \frac{2I_0^i - 3I_{RR}^i}{2r_{ij}^3} \tag{2.198}$$

and

$$\int_{A^i} \frac{dA^i}{r_0^3} = \frac{A^i}{r_{ij}^3} + \frac{12I_0^i - 15I_{RR}^i}{2r_{ij}^5} \tag{2.199}$$

Making the substitutions gives the following relationship for the average temperature rise of the *i*th area due to heating of the isoflux *j*th area in terms of easily computable geometric parameters:

$$\overline{\theta}_{ij} = \left(\frac{q_j}{2\pi k}\right)\left\{\frac{A^j}{r_{ij}} + \frac{2I_0^j - 3I_{RR}^j}{2r_{ij}^3} + \left(\frac{A^j}{A^i}\right)\left(\frac{2I_0^i - 3I_{RR}^i}{2r_{ij}^3}\right)\right.$$
$$\left. + \left(\frac{2I_0^i - 15I_{RR}^i}{2r_{ij}^5}\right)\left(\frac{2I_0^j - 3I_{RR}^j}{2r_{ij}^5}\right) + \cdots \right\} \tag{2.200}$$

For circular heat sources, the second moments of area about the axis passing through the centroid and the diametral axis are related to the radius $a$ as:

$$\left.\begin{aligned} I_0 &= \frac{\pi}{2}a^4 \\ I_{RR} &= \frac{\pi}{4}a^4 \end{aligned}\right\} \tag{2.201}$$

For square heat sources, the second moments of area about the axis passing through the centroid and the diametral axes $x0x$ and $y0y$ are related to the side dimension $s$ as:

$$\left.\begin{aligned} I_0 &= \frac{1}{6}s^4 \\ I_{xx} &= \frac{1}{12}s^4 \end{aligned}\right\} \tag{2.202}$$

where $I_{xx} = I_{yy}$ and $I_0 = 2I_{xx}$. The polar second moments of the circle of area $\pi a^2$ and the square of area $s^2$ have the following properties:

$$\left.\begin{aligned} \frac{I_0}{A^2} &= \frac{1}{2\pi} \qquad \text{for the circle} \\ \frac{I_0}{A^2} &= \frac{1}{6} \qquad \text{for the square} \end{aligned}\right\} \tag{2.203}$$

The difference between the values $1/2\pi$ and $1/6$ is about 4.7%. This interesting result allows one to approximate the heating of a square source by a circular heat source having the same area.

## 2.16 Superposition of Isoflux Circular Heat Sources

In this section, the superposition of several isoflux circular heat sources located in the surface of an adiabatic half-space of constant thermal conductivity $k$ will be considered. The method is general because any number of heat sources having any size, and any heat flux level may be employed. However, without loss of generality, the examples will be limited to heat sources having identical diameters $d = 2a$ and identical heat flux level $q$.

The analytical model is based on solutions that provide the temperature rise at the centroid of an isoflux circular heat source and the temperature rise outside an isoflux heat source.

For the first case, the temperature rise is given by

$$T_0 = \frac{qa}{k} \tag{2.204}$$

where $q$ is the uniform heat flux, $a$ is the radius of the circular heat source, and $k$ is the thermal conductivity of the half-space. The surface outside the heat source is adiabatic.

In the second case, the temperature rise outside an isoflux circular heat source is given by

$$T = \frac{qa}{k}\frac{2}{\pi}\rho\left[E\left(\frac{1}{\rho}\right) - \left(1 - \frac{1}{\rho^2}\right)K\left(\frac{1}{\rho}\right)\right] \tag{2.205}$$

where $K(\cdot)$ and $E(\cdot)$ are complete elliptic integrals of the first and second kinds and the relative position is defined as $\rho = r/a > 1$. For $\rho > 2$, the analytical solution approaches the point heat source solution:

$$T = \frac{qa}{k}\frac{1}{2\rho} \tag{2.206}$$

with errors smaller than 1.4%.

**Figure 2.13** Superposition of seven isoflux circular sources in a strip.

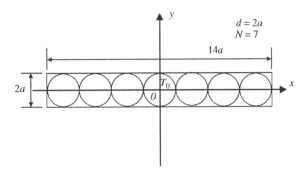

The first example will consist of $N$ aligned and tangent circular heat sources, where $N \geq 3$. One heat source will be located with its centroid coincident with the origin $(0, 0)$ and the remaining heat sources will have centroids located at $x/a = \pm(N - 1)$, where $N = 3, 5, 7, 9, \ldots$. See Figure 2.13 for the case of $N = 7$.

The maximum temperature rise denoted as $T_0$ occurs at the centroid of the central heat source and the temperature rise is due to the self-heating of the central heat source and the mutual effect of the other heat sources.

The total area of all heat sources is $A_s = N\pi a^2$ and the total apparent area of the associated rectangular area is $A_a = N4a^2$. The total apparent area is greater than the total heat source area, $A_a > A_s$.

For $N = 3$ heat sources, the maximum temperature rise is given by the following relation:

$$T_0 = \frac{qa}{k} G \tag{2.207}$$

where

$$G = 1 + 2G_{01} \tag{2.208}$$

and $G_{01} = 0.2586$ is the mutual effect of the heat source at $x/a = 2$ on the central heat source. Similarly, the heat source at $x/a = -2$ produces the same effect on the central heat source.

The thermal spreading resistance of the central heat source is defined as

$$R_0 = \frac{T_0}{Q} = \frac{T_0}{qA_s} \tag{2.209}$$

The dimensionless maximum temperature rise is defined as

$$R_0^{\star} = kL_c R_0 \tag{2.210}$$

where $L_c$ is a convenient characteristic length. Two possible length scales are $L_c = \sqrt{A_s}$ and $L_c = \sqrt{A_a}$. Another option is the arithmetic mean of the two proposed length scales which leads to

$$L_c = \frac{1}{2} \left( \sqrt{A_a} + \sqrt{A_s} \right) \tag{2.211}$$

After substitution and cancellation of parameters $(q, k, a)$ and further simplifications, we obtain the following simple relation:

$$R_0^{\star} = \frac{0.6}{\sqrt{N}} G \qquad N \geq 3 \tag{2.212}$$

applicable for an array of aligned tangent isoflux circular heat sources.

The superposition method is applied to systems consisting of $N = 3, 5, 7, \ldots$ heat sources and one finds the following relations for the parameter $G$:

$$G = 1 + 2(0.2586) \quad \text{for} \quad N = 3 \tag{2.213}$$

and

$$G = 1 + 2(0.2586) + 2(0.1263) \quad \text{for} \quad N = 5 \tag{2.214}$$

and

$$G = 1 + 2(0.2586) + 2(0.1263) + 2(0.08327) \quad \text{for} \quad N = 7 \tag{2.215}$$

and so on. The factor 2 accounts for the mutual effects of heat sources located on the negative $x$-axis. An odd number of heat sources are chosen to maintain a central heat source at the origin and an equal number of heat sources on the positive and negative $x$-axes from symmetry.

The point heat source model is based on elements of the following list which are the distances from the centroid of the central heat source to the centroids of the other heat sources:

$$G_{ij} = \left[ \frac{1}{2}, \frac{1}{4}, \frac{1}{6}, \frac{1}{8}, \frac{1}{10} \right] \tag{2.216}$$

Then the value of $G$ for each case is found from the following:

$$G = 1 + \sum_{1}^{N-1} G_{ij} \tag{2.217}$$

for $N \geq 3$. Calculated values of $G$ are reported in Table 2.14.

The aspect ratio of the apparent rectangular area is defined as the width over the height which is $\epsilon = N2a/2a = N \geq 3$. The analytical value for the dimensionless maximum temperature rise of the isoflux rectangle given by Eq. (2.108) becomes

$$R_0^\star = \frac{1}{\pi \sqrt{N}} \left[ N \sinh^{-1} \frac{1}{N} + \sinh^{-1} N \right] \tag{2.218}$$

for $N \geq 3$ when the aspect ratio is written in terms of the number of isoflux heat sources $\epsilon = N$.

Values for the maximum dimensionless temperature rise at the centroid of an isoflux rectangle with aspect ratios $N \geq 3$ are also reported in Table 2.13. The results of calculations of the models are also reported in Table 2.13.

The first column shows the total number $N$ of aligned heat sources, the second column is the relative spacing of the centroids $\rho = r/a$, the third column shows values of the superposition parameter $G_1$, and the fourth column shows values of $G_2$ which are based on the point source solution, the fifth column shows values for the predicted dimensionless maximum temperature rise of the tangent circular heat sources, and the last column shows the values of the dimensionless maximum temperature rise of an isoflux rectangular heat source. The maximum difference between the model predictions and the analytical values for the isoflux rectangle is about 2.1% when $N = 3$ and the differences are below 1% for all other values of $N$.

The values of $G_1$ and $G_2$ differ by about 1.1% for all values of $N$.

**Table 2.13** Results for $N$ isoflux sources.

| $N$ | $\rho$ | $G_1$ | $G_2$ | $R_{0cir}$ | $R_{0rect}$ |
|-----|--------|-------|-------|------------|-------------|
| 3   | 2      | 1.517 | 1.500 | 0.5255     | 0.5146      |
| 5   | 4      | 1.770 | 1.750 | 0.4749     | 0.4706      |
| 7   | 6      | 1.936 | 1.917 | 0.4390     | 0.4381      |
| 9   | 8      | 2.061 | 2.042 | 0.4122     | 0.4126      |
| 11  | 10     | 2.161 | 2.142 | 0.3911     | 0.3926      |

**Figure 2.14** Superposition of nine isoflux circular sources arranged in a square cluster.

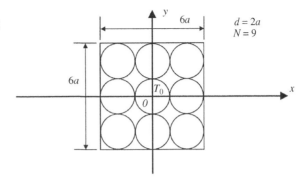

### 2.16.1 Nine Coplanar Circles on Square Cluster

Nine identical circular heat sources (see Figure 2.14) of radius $a$ are situated in the surface of an isotropic half-space of thermal conductivity $k$. The heat flux $q$ on each heat source is uniform and constant. The free surface of the half-space outside the heat sources is adiabatic. The nine circles form a square array with one circle located with its centroid on the origin.

Two tangent circles are on the $x$-axis with centers at $x/a = \pm 2$ and two circles are on the $y$-axis with centers at $y/a = \pm 2$. There are four circles located at the corners of an apparent square area.

The central circle and the tangent circles lie on the $x$-axis and the $y$-axis and perpendicular diagonals of length $d = 6a$ which define an apparent square area having sides of length $6a$. The area of the apparent square is $A_a = 36a^2$, while the real or heat source area of the cluster is $A_s = 9\pi a^2$. The effective area is defined as

$$\sqrt{A_e} = \frac{1}{2}\left[\sqrt{A_a} + \sqrt{A_s}\right] \tag{2.219}$$

The maximum temperature rise of the cluster will occur at the centroid of the central circle and the temperature rise at the centroids of the four tangent circles will be identical and lower than the temperature rise at the origin of the cluster.

The temperature rise at the centroid of the central circle depends on its self-effect and the influence of the four tangent circles. The superposition method is used to obtain the following relation:

$$T_0 = \frac{qa}{k}\left[G_{11} + 4G_{12} + 4G_{13}\right] \tag{2.220}$$

The self-effect coefficient has the value $G_{11} = 1$ and the value of the mutual effect coefficients for the tangent circles are $G_{12} = 0.2586$ and $G_{13} = 0.1797$. Thus, the total effect of the nine heat sources is $G = 2.753$.

The total heat flow rate into the half-space from the nine identical heat sources is $Q = 9\pi a^2 q$. The spreading resistance is defined with respect to the centroid temperature rise and the total heat flow rate $Q$ such as $R_0 = T_0/Q$.

The dimensionless spreading resistance is defined with respect to the total effective surface area of the cluster so that

$$k\sqrt{A_e}R_0 = \frac{1}{2\pi}\left[\frac{2+\sqrt{\pi}}{\sqrt{N}}\right]G \tag{2.221}$$

After substitution for $\sqrt{A_e}$, $G$ and $N$, we get

$$k\sqrt{A_e}R_0 = 0.5511 \tag{2.222}$$

which is 1.8% smaller than the value for the centroid temperature rise for an isoflux square heat source

$$k\sqrt{A_s}R_0 = 0.5611 \tag{2.223}$$

### 2.16.2 Five Coplanar Circles on Square Cluster

Five identical circular heat sources (see Figure 2.15) of radius $a$ are situated in the surface of an isotropic half-space of thermal conductivity $k$. The heat flux $q$ on each heat source is uniform and constant. The free surface of the half-space outside the heat sources is adiabatic. The five circles form a square array with one circle located with its centroid on the origin.

Two tangent circles are on the $x$-axis with centers at $x/a = \pm 2$, and two circles are on the $y$-axis with centers at $y/a = \pm 2$.

The central circle and the tangent circles lie on perpendicular diagonals of length $d = 6a$ which define an apparent square area having sides of length $(2 + 2\sqrt{2})a$. The area of the apparent square is $A_a = 18a^2$, while the heat source area of the cluster is $A_s = 5\pi a^2$. The effective area is defined as

$$\sqrt{A_e} = \frac{1}{2}\left[\sqrt{A_a} + \sqrt{A_s}\right] \tag{2.224}$$

The maximum temperature rise of the cluster will occur at the centroid of the central circle and the temperature rise at the centroids of the four tangent circles will be identical and lower than the temperature rise at the origin of the cluster.

The temperature rise at the centroid of the central circle depends on its self-effect and the influence of the four tangent circles. The superposition method is used to obtain the following relation:

$$T_0 = \frac{qa}{k}\left[G_{11} + 4G_{12}\right] \tag{2.225}$$

The self-effect coefficient has the value $G_{11} = 1$ and the value of the mutual effect coefficient for each tangent circle is $G_{12} = 0.2586$. Thus, $G = 1 + 4(0.2586) = 2.034$.

The total heat flow rate into the half-space from the five identical heat sources is $Q = 5\pi a^2 q$.

The spreading resistance is defined with respect to the centroid temperature rise and the total heat flow rate $Q$ such as $R_0 = T_0/Q$.

The dimensionless spreading resistance is defined with respect to the total effective area of the cluster so that

$$k\sqrt{A_e}R_0 = \frac{(\sqrt{A_a} + \sqrt{A_s})aG}{A_s} \tag{2.226}$$

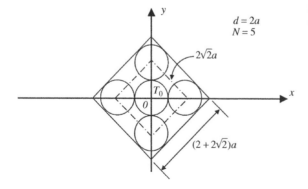

**Figure 2.15** Superposition of five isoflux circular sources arranged in a square cluster.

After substitution for $\sqrt{A_a}$, $\sqrt{A_s}$ and $G$, we get

$$k\sqrt{A_e}R_0 = 0.5693 \tag{2.227}$$

which is 1.8% greater than the value for the centroid temperature rise for an isoflux square heat source

$$k\sqrt{A_s}R_0 = 0.5611 \tag{2.228}$$

### 2.16.3 Four Coplanar Circles on Triangular Cluster

Four identical circular isoflux heat sources (see Figure 2.16) form a triangular cluster. The circles are tangent. The central heat source has its centroid located at the origin while two circles lie on the $x$-axis such that $x/a = \pm 2$. The centroid of the fourth circle is located at $y/a = 2$. The free surface of the half-space is adiabatic. The maximum temperature of the cluster occurs at the centroid of the central circle. The total area of the heat sources is $A_s = 4\pi a^2$.

The centers of the four circles lie on an isosceles triangle of base width $4a$ and height of $2a$, with aspect ratio 0.5.

The four circles lie in an apparent area that has the shape of an isosceles triangle with base $2w = 6a$. The height of the apparent area is $h = 4a$. The total apparent area is $A_a = 1/2wh = 12a^2$.

The dimensionless spreading resistance of the triangular cluster is defined as

$$R_0^\star = k\sqrt{A_e}R_0 = \frac{(\sqrt{A_a} + \sqrt{A_s})aG}{2A_s} \tag{2.229}$$

The self and mutual influence coefficients give $G = 1 + 3(0.2586) = 1.776$. After substitutions for $\sqrt{A_a}$, $\sqrt{A_s}$, and $G$ we find $R_0^\star = 0.4953$.

The analytical solution for an isoflux isosceles triangular heat source gives the dimensionless spreading resistance which is defined as $R_0^\star = k\sqrt{A_s}R_0$ which has values in the range: $0.4336 \leq R_0^\star \leq 0.5516$ for aspect ratios in the range $0.01 \leq \epsilon \leq 1$. The average of the two extreme values is about 0.4926 which is close to the value calculated for the five discrete heat sources.

The foregoing are examples of applications of the superposition method applied to the maximum temperature rise of a set of identical isoflux circular heat sources are located in the surface of an isotropic half-space. The free surface of the half-space outside the heat sources is everywhere adiabatic. The maximum temperature rise occurs at the centroid of the center most heat source.

**Figure 2.16** Superposition of four isoflux circular sources arranged in a triangular cluster.

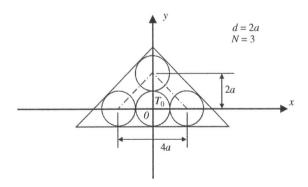

## 2.17 Superposition of Micro- and Macro-Spreading Resistances

The solution methods in this section are based on superposition principles which are applicable for steady heat conduction from multiple isoflux heat sources located in the surface of an isotropic half-space ($z > 0$) of thermal conductivity $k$. In the region outside of the heat sources, the free surface of the half-space is adiabatic.

Figure 2.17a shows a general continuous source area which is subjected to a steady uniform heat flux $q$. The temperature rise $T(x, y)$ of points in the source area varies with the largest values near the centroid and the lowest ones near the perimeter. The temperature rise at any point can be calculated by considering the temperature rise due to the heating of $N$ identical circular heat sources of diameter $2a$. Although other heat sources such as isoflux square sources can be used, the circular sources are chosen because they produce axi-symmetric temperature fields and, therefore, they are easily implemented.

In Figure 2.17b, the continuous source area is replaced by an apparent area consisting of $N$ discrete circular heat source areas of diameter $d = 2a$. The discrete areas are located at equal distances on a square array. Every isoflux heat source $A^i$ heats itself and all others according to the following relation:

$$T_i = \frac{qA^i}{2\pi k r_i} \tag{2.230}$$

There are $N$ identical circular isoflux heat sources of radius $a$ in a regular square array whose dimensions are $W$ by $W$. The discrete heat sources lie in the surface $z = 0$ of an isotropic half-space. Regions of the surface $z = 0$ are adiabatic except, where the heat sources are located.

The heat which diffuses from each heat source is conducted through a spreading resistance zone whose dimensions are related to the radius of the heat source. The isotherms in the respective spreading resistance zones merge to form a single complex isotherm at some distance from the surface $z = 0$. The spreading resistance which is associated with each heat source is called the micro-spreading resistance. The micro-spreading resistances are thermally connected in parallel. The total micro-spreading resistance for $N$ heat sources is

$$R_{mic} = \frac{\psi_{mic}}{Nka} \tag{2.231}$$

when the characteristic length is the heat source radius $L_c = a$ and $\psi_{mic}$ is called the spreading resistance parameter which in general depends on the shape of the heat source, the boundary condition on the heat source, and the spacing between the heat sources.

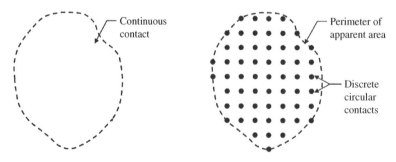

**Figure 2.17** Approximation of continuous arbitrary contact area by discrete contact spots. (a) Continuous contact region. (b) Discrete contact region. Source: After Yovanovich et al. (1983).

If the distance between any two neighboring heat sources is much greater than the radius ($r \geq 40a$), then $\psi_{mic}$ can have different values depending on the definition of the spreading resistance. For an isothermal heat source, $\psi_{mic} = 0.25$. For an isoflux heat source, $\psi_{mic} = 8/3\pi^2 = 0.2702$ if the spreading resistance is based on the mean temperature rise of the heat source, and $\psi_{mic} = 1/\pi = 0.3183$ if the spreading resistance is based on the temperature rise at the centroid of the heat source. In the following work, $\psi_{mic}$ will be based on the mean temperature rise of the heat sources.

To account for the interaction of neighboring heat sources, the spreading resistance parameter is modified in the following manner:

$$\psi_{mic,a} = \frac{1}{\sqrt{\pi}} \psi_{mic,\sqrt{A}} \tag{2.232}$$

where $\psi_{mic,a}$ is based on the heat source radius $a$ and $\psi_{mic,\sqrt{A}}$ is based on the square root of the heat source area. The following general expression can be used for a circle on a circular flux tube, a circle on a square flux tube, or a square source on a square flux tube [Negus et al. (1989)]:

$$\psi_{mic,\sqrt{A}} = 0.475 - 0.62\,\epsilon + 0.13\,\epsilon^3 \tag{2.233}$$

The relative size parameter $\epsilon$ for the $N$ discrete heat sources on a square array is defined as

$$\epsilon = \sqrt{\frac{A_r}{A_a}} = \sqrt{\frac{N\pi a^2}{W^2}} = \frac{a}{W}\sqrt{\pi N} \tag{2.234}$$

The spreading resistance parameter based on $\sqrt{A}$ has the value $\psi_{mic,\sqrt{A}} = 0.475$ as $\epsilon \to 0$. This value can be used for isoflux circular and square heat sources which are far apart.

Conduction from the single complex isotherm to regions which are remote from the heat sources is associated with the macro-spreading resistance $R_{mac}$ which can be expressed as

$$R_{mac} = \frac{\psi_{mac,\sqrt{A}}}{kW} = \frac{0.479}{kW} \tag{2.235}$$

The micro- and macro-spreading resistances are assumed to be independent and thermally connected in series.

The total spreading resistance of an array of $N$ discrete circular heat sources which lie in the apparent area $A_a = W^2$ is based on the simple superposition of the micro- and macro-spreading resistances:

$$R_{total} = R_{mic} + R_{mac} \tag{2.236}$$

The ratio of the total resistance $R_{total}$ divided by the macro-spreading resistance $R_{mac}$ is expressed as

$$\phi_{Model,2} = 1 + \frac{1}{N}\frac{W}{a}\frac{1}{\sqrt{\pi}}\frac{\psi_{mic,\sqrt{A}}}{0.479} \tag{2.237}$$

where the micro-spreading resistance parameter is defined by Eq. (2.233).

If we consider the case where $a = 2$ and $W = 100$, the relative resistance parameter becomes

$$\phi_{Model,2} = 1 + \frac{59.39}{N}\,\psi_{mic,\sqrt{A}} \tag{2.238}$$

which can be written in terms of $N$ and $\epsilon$ as

$$\phi_{Model,2} = 1 + \frac{59.39(0.475 - 0.62\epsilon + 0.13\epsilon^3)}{N} \tag{2.239}$$

with

$$\epsilon = \frac{a}{W}\sqrt{\pi N} \tag{2.240}$$

The accuracy of the method of superposition of micro- and macro-spreading resistances to obtain the total spreading resistance of an array of circular isoflux heat sources will be verified by the application of the superposition method (also called the Surface Element Method) as given by Yovanovich and Martin (1981).

An alternative analytic solution will be presented next. The solution to Laplace's equation $\nabla^2 T = 0$ is written in vector form as

$$T(\vec{r}) = \frac{1}{2\pi k}\iint_A \frac{q(\vec{s})dA}{|\vec{r}-\vec{s}|} \tag{2.241}$$

where $\vec{r}$ is the radius vector to the field point and $\vec{s}$ is the radius vector to the source point and the distance between the field and source points is $|\vec{r}-\vec{s}|$.

In its simplest form, the surface element method (SEM) when applied to spreading resistance problems consists of dividing the source area into a finite number, $N$, of surface elements, $A_j$, over each of which the heat flux $q_j$, is assumed to be uniform. The centroid of a typical surface element is designated by coordinates $(x_i, y_i)$ because $z_i = 0$ in the contact plane. The position vector from the origin is $\vec{r}_i$. The temperature excess (or simply temperature rise if the reference temperature is taken to be zero) at the centroid of the typical surface element $A_i$ is

$$
\begin{aligned}
T_i &= \frac{1}{2\pi k}\iint \frac{q(\vec{s})dA}{|\vec{r}_i-\vec{s}|} \\
&= \sum_{j=1}^{N}\frac{1}{2\pi k}\iint_{A_j} \frac{q_j dA_j}{|\vec{r}_i-\vec{s}|} \\
&= \sum_{j=1}^{N} q_j \left[\frac{1}{2\pi k}\iint_{A_j} \frac{dA_j}{|\vec{r}_i-\vec{s}|}\right]
\end{aligned} \tag{2.242}
$$

The last equation can be conveniently written as

$$T_j = \sum_{j=1}^{N} C_{ij} q_j \tag{2.243}$$

where $C_{ij} q_j$ represents the temperature rise at the field point $(x_j, y_j)$ due to the surface element $A_j$. $C_{ij}$ represents the temperature rise at the centroid of $A_j$ due to heat sources of unit strength distributed over the surface element $A_j$. The influence coefficients $C_{ij}$ are known; they consist of the integrals in Eq. (2.242).

The foregoing equation can be written in more compact form using matrix notation:

$$T = [C](q) \tag{2.244}$$

For boundary conditions of the second kind, Eq. (2.242) can be solved directly for $T_j$ because $q_j$ is known. For boundary conditions of the first kind, $T_j$ are known and Eq. (2.242) must be solved for $q_j$. In matrix form, we have

$$(q) = [C]^{-1} T \tag{2.245}$$

where $[C]^{-1}$ is the inverse matrix of $[C]$.

The thermal spreading resistance is defined as

$$R_c = \frac{\overline{T}}{Q} \tag{2.246}$$

and with the average temperature $\overline{T}$ and total heat flow rate $Q$, we get

$$R_c = \frac{1}{A} \sum_{i=1}^{N} T_i \, dA_i \Big/ \sum_{j=1}^{N} q_j \, dA_j \tag{2.247}$$

For the case of boundary conditions of the second kind $q_j$ are known, and we must use Eq. (2.242) to solve for the unknown $T_i$ to get the average temperature rise and, therefore, the spreading resistance.

When the SEM is employed for the system shown in Figure 2.18, we obtain values of the spreading resistance of the square array of isoflux circular heat sources. For several cases where the total number of heat sources ranges from $N = 9$ to $N = 441$, the density of heat sources increases and more and more of the apparent area is covered by the heat sources. The percentage of the apparent area covered by the discrete heat sources increases from about 1.1% up to about 55.4%. The total spreading resistance defined as $\psi = k\sqrt{A_a}R_c$ becomes $kWR_c = 0.475$ which is the total spreading resistance of an isoflux square heat source of width $W$ when the resistance is based on the average temperature rise. Using Eq. (2.247) along with the $R_c = 0.475/(kW)$, we define $\phi_{Model,1}$ in the same manner as we defined $\phi_{Model,2}$.

When the values of $k\sqrt{A_a}R_c$ for $N < 441$ are normalized with the value for the continuous square heat source, where $N \gg 1$, we get the values shown in the third column of Table 2.14. When $N = 121$ and the discrete heat sources cover 15.2% of the apparent area, the calculated value of the dimensionless spreading resistance is approximately 8.6% greater than the total resistance of the continuous square heat source.

A comparison of numerical results based on the two models is provided in Table 2.14.

The values in the second column are based on the ratio of the total area of the circular heat sources divided by the total apparent area:

$$100\frac{A_s}{A_a} = 100\frac{N\pi a^2}{W^2} \tag{2.248}$$

Values corresponding to $\phi_{Model,1}$ are based on the superposition of identical isoflux circular heat sources on a square array are given in column 3 which depend on the number of heat sources. The values, corresponding to any number of heat sources, are normalized with the value corresponding to an isoflux square heat source.

Values corresponding to $\phi_{Model,2}$ are based on superposition of the micro-spreading resistances of multiple identical circular isoflux heat sources and the macroscopic spreading resistance of a single isoflux square heat source. The micro- and macro-spreading resistances are arranged in series. The values in column 4 are normalized with respect to the macro resistance corresponding to an isolated isoflux square heat source.

**Figure 2.18** Discretization of square contact area. Source: After Yovanovich et al. (1983).

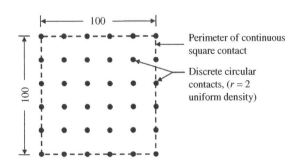

**Table 2.14** Comparison of Model 1 and Model 2.

| $N$ | $A_s/A_a \times$ 100 (%) | $\phi_{Model,1}$ | $\phi_{Model,2}$ | $\phi_{Model,1}/\phi_{Model,2}$ |
|---|---|---|---|---|
| 9 | 1.13 | 3.730 | 3.700 | 1.014 |
| 25 | 3.14 | 1.823 | 1.869 | 0.979 |
| 49 | 6.16 | 1.344 | 1.392 | 0.968 |
| 81 | 10.2 | 1.166 | 1.206 | 0.968 |
| 121 | 15.2 | 1.086 | 1.118 | 0.972 |
| 169 | 21,2 | 1.046 | 1.071 | 0.978 |
| 225 | 28.3 | 1.025 | 1.044 | 0.983 |
| 289 | 36.3 | 1.013 | 1.027 | 0.987 |
| 361 | 45.4 | 1.006 | 1.016 | 0.990 |
| 441 | 55.4 | 1.003 | 1.009 | 0.994 |

The last column of Table 2.14 compares the results of two models. Except for the first value corresponding to nine isoflux circular heat sources located at the four corners, center, and four midpoints of the apparent area, the values of $\phi_{Model,2}$ are lower than the values of $\phi_{Model,1}$. All other values of $\phi_{Model,2}$ are lower than those of $\phi_{Model,1}$ by 1–3%. As the number of heat sources increases ($N \geq 225$), the differences between the two models decrease and fall below 1%. This shows that the model based on the superposition of micro- and macro-spreading resistances becomes very accurate when the number of heat sources is sufficiently large.

The superposition of micro- and macro-spreading resistances is a convenient and easy-to-implement method for obtaining total spreading resistances of arrays of very small isoflux heat sources.

## References

Abramowitz, M., and Stegun, I.A., *Handbook of Mathematical Functions*, Dover Publications, Inc., New York, 1965.

Byrd, P.F., and Friedman, M.D., *Handbook of Elliptic Integrals for Engineers and Scientists*, 2nd Edition, Springer-Verlag, New York, 1971.

Carslaw, H., and Jaeger, J.C., *Conduction of Heat in Solids*, Oxford Press, Oxford, UK, pp. 215–216, 1959.

Dryden, J.R., "The Effect of a Surface Coating on the Constriction Resistance of a Spot on an Infinite Half-Plane", *Journal of Heat Transfer*, Vol. 105, pp. 408–410, 1983.

Dryden, J.R., Yovanovich, M.M., and Deakin, A.S., "The Effect of Coatings on the Steady-State and Short Time Constriction Resistance for an Arbitrary Axisymmetric Flux", *Journal of Heat Transfer*, Vol. 107, pp. 33–38, 1985.

Gradshteyn, I.S., and Ryzhik, I.M., *Table of Integrals, Series, and Products*, Springer-Verlag, New York, 1965.

Holm, R., *Electric Contacts: Theory and Applications*, Springer-Verlag, New York, 1967.

Hui, P., and Tan, H.S., "Temperature Distributions in a Heat Dissipation System Using a Cylindrical Diamond Heat Spreader on a Copper Sink", *Journal of Applied Physics*, Vol. 75, no. 2, pp. 748–757, 1994.

Jeans, J., *The Mathematical Theory of Electricity and Magnetism*, Cambridge University Press, Cambridge, pp. 244–249, 1963.

Johnson, K.L., *Contact Mechanics*, Cambridge University Press, Cambridge, UK, 1985.

Kellogg, O.D., *Foundations of Potential Theory*, Dover Publications, Inc., New York, pp. 188–189, 1953.

Lemczyk, T.F., *Constriction Resistance of Axisymmetric Convective Contacts on Semi-Infinite Domains*, M.A.Sc. Thesis, University of Waterloo, 1986.

Lemczyk, T.F., and Yovanovich, M.M., "Thermal Constriction Resistance with Convection Boundary Conditions: 1. Half-space Contacts", *International Journal of Heat and Mass Transfer*, Vol. 31, no. 9, pp. 1861–1872, 1988a.

Lemczyk, T.F., and Yovanovich, M.M., "Thermal Constriction Resistance with Convection Boundary Conditions: 2. Layered Half-space Contacts", *International Journal of Heat and Mass Transfer*, Vol. 31, no. 9, pp. 1873–1883, 1988b.

Lur'e, A.I., *Three Dimensional Problems of the Theory of Elasticity*, Wiley-Interscience, New York, 1964.

MacMillan, W.D., *The Theory of the Potential*, Dover Publications, Inc., New York, pp. 24–63, 1958.

Martin, K.A., Yovanovich, M.M., and Chow, Y.L., "Method of Moments Formulation of Thermal Constriction Resistance of Arbitrary Contacts", AIAA-84-1745, *AIAA 19th Thermophysics Conference*, Snowmass, CO, USA, June 25–28, 1984.

Mason, M., and Weaver, W., *The Electromagnetic Field*, Dover Publications, Inc., New York, 1929.

Moon, P., and Spencer, D.E., *Field Theory for Engineers*, D. Van Nostrand Company, Princeton, NJ, p. 275, 1961.

Moon, P., and Spencer, D.E., *Field Theory Handbook*, 2nd Edition, Springer-Verlag, New York, 1971.

Negus, K.J., Yovanovich, M.M., and Thompson, J.C., "Thermal Constriction Resistance of Circular Contacts on Coated Surfaces: Effect of Contact Boundary Condition", AIAA-85-1014, *AIAA 20th Thermophysics Conference*, Williamsburg, VA, USA, June 19–21, 1985.

Negus, K.J., Yovanovich, M.M., and Beck, J.V., "On the Non-dimensionalization of Constriction Resistance for Semi-infinite Heat Flux Tubes, *Journal of Heat Transfer*, Vol. 111, pp. 804–807, 1989.

Ramsey, A.S., *Newtonian Attraction*, Cambridge University Press, Cambridge, 1961.

Sadeghi, E., Bahrami, M., and Djilali, N., "Thermal Spreading Resistance of Arbitrary Shape Heat Sources on a Half-space: A Unified Approach", *IEEE Transactions on Components and packaging Technologies*, Vol. 33, no. 2, pp. 267–277, 2010.

Schneider, G.E., "Thermal Resistance Due to Arbitrary Dirichlet Contacts on a Half-Space", *Progress in Astronautics and Aeronautics: Thermophysics and Thermal Control*, 65, pp. 103–109, 1978.

Smythe, W.R., "The Capacitance of a Circular Annulus", *American Journal of Physics*, Vol. 22, no. 8, pp. 1499–1501, 1951.

Smythe, W.R., *Static and Dynamic Electricity*, 3rd Edition, McGraw-Hill Book Co., New York, pp. 122–123, 1968.

Sneddon, I.N., *Mixed Boundary Value Problems in Potential Theory*, North-Holland Publishing, Amsterdam, p. 63, 1966.

Stratton, J.A., *Electromagnetic Theory*, McGraw-Hill Book Co., New York, pp. 207–209, 1941.

Yovanovich, M.M., "Thermal Constriction Resistance Between Contacting Metallic Paraboloids: Application to Instrument Bearings", *Progress in Astronautics and Aeronautics: Fundamentals of Spacecraft Thermal Design*, Vol. 24, pp. 337–358, 1971.

Yovanovich, M.M., "General Thermal Constriction Resistance Parameter for Annular Contacts of Circular Flux Tubes", *AIAA Journal*, Vol. 14, no. 6, pp. 822–824, 1976a.

Yovanovich, M.M., "General Expressions for Constriction Resistances of Arbitrary Flux Distributions", in *Progress in Astronautics and Aeronautics: Radiative Transfer and Thermal Control*, Vol. 49, AIAA, New York, pp. 381–396, 1976b.

Yovanovich, M.M., "Thermal Constriction Resistances of Contacts on a Half-space: Integral Formulation", in *Progress in Astronautics and Aeronautics: Radiative Transfer and Thermal Control*, Vol. 49, AIAA, New York, pp. 397–418, 1976c.

Yovanovich, M.M., "Recent Developments in Thermal Contact, Gap and Joint Conductance Theories and Experiment", *Proceedings of the 8th International Heat Transfer Conference*, San Francisco, CA, USA, Vol. 1, pp. 35–45, 1986.

Yovanovich, M.M., and Burde, S.S., "Centroidal and Area Average Resistances of Non-symmetric, Singly Connected Contacts", *AIAA Journal*, Vol. 15, no. 10, pp. 1523–1525, 1977.

Yovanovich, M.M., and Martin, K.A., "Some Basic Three-Dimensional Influence Coefficients for the Surface Element Method", *Progress in Astronautics and Aeronautics: Heat Transfer and Thermal Control*, Vol. 78, pp. 202–213, 1981.

Yovanovich, M.M., and Schneider, G.E., "Thermal Constriction Resistance Due to a Circular Annular Contact, *Progress in Astronautics and Aeronautics: Thermophysics of Spacecraft and Outer Planet Entry Probes*, Vol. 56, pp. 141–154, 1977.

Yovanovich, M.M., Burde, S.S., and Thompson, J.C., "Thermal Constriction Resistance of Arbitrary Planar Contacts with Constant Flux", *Progress in Astronautics and Aeronautics: Thermophysics of Spacecraft and Outer Planet Entry Probes*, Vol. 56, pp. 127–139, 1977.

Yovanovich, M.M., Thompson, J.C., and Negus, K,J., "Thermal Resistance of Arbitrary Shaped Contacts", *Proceedings of the 3rd International Conference*, Seattle, WA, USA, August 2nd-5th, Edited by R.W. Lewis, J.A. Johnson and W.R. Smith, 1983.

# 3

# Circular Flux Tubes and Disks

The heat transfer literature contains many solutions for thermal spreading resistance in cylindrical co-ordinate systems for single isotropic materials, compound systems, orthotropic systems, and multilayered systems. This chapter focuses primarily on solutions for isotropic and multilayered systems with variable heat flux specified in the source region. Orthotropic systems will be dealt with in Chapter 5.

Single heat source spreading resistance and temperature field for finite disks and flux tubes has been examined in numerous published works [Kennedy (1960), Mikic and Rohsenow (1966), Yovanovich (1975), Yovanovich et al. (1980), Yovanovich et al. (1988), Song et al. (1995)]. Two of the most important solutions are for a circular heat source on a semi-infinite flux tube and for a circular source on a finite disk with constant uniform conductance in the sink plane. These solutions will be presented in detail and various extensions discussed in Sections 3.1 and 3.2 of the chapter.

Additional results can be found for circular annular heat sources on disks and flux tubes [Yip (1969)] and solutions for problems where edge conductance is prescribed [Yovanovich (2003)]. Finally, several useful solutions exist for compound and multilayered disks are shown to be simple extensions of existing solutions [Muzychka et al. (1999), Muzychka et al. (2013), Muzychka (2015)].

Many results are currently summarized in the review chapters by Yovanovich (1998) and Yovanovich and Marotta (2003). Since this time, additional results have been published which are also considered in detail in this chapter.

## 3.1 Semi-Infinite Flux Tube

Having examined thermal spreading resistance from discrete heat sources on a half-space, it is now desirable to consider thermal spreading resistance in a semi-infinite flux tube as shown in Figure 3.1. The flux tube geometry results when an array of contacts results. In this instance, an external boundary is drawn around each contact, and the lateral edges are assumed to be adiabatic. In this way, an array or a bundle of flux tubes results. The problem of finding the thermal spreading resistance in a semi-infinite isotropic circular flux tube has been investigated by many researchers [Roess (1950), Mikic and Rohsenow (1966), Gibson (1976), Yovanovich (1975), Negus and Yovanovich (1984a,b), Negus et al. (1989), Yovanovich et al. (1998)]. We consider the fundamental problem, but generalize it for three common flux distributions.

Starting with an isotropic circular flux tube, we desire to solve:

$$\frac{\partial^2 T}{\partial r^2} + \frac{1}{r}\frac{\partial T}{\partial r} + \frac{\partial^2 T}{\partial z^2} = 0 \tag{3.1}$$

*Thermal Spreading and Contact Resistance: Fundamentals and Applications*, First Edition.
Yuri S. Muzychka and M. Michael Yovanovich.
© 2023 John Wiley & Sons, Inc. Published 2023 by John Wiley & Sons, Inc.

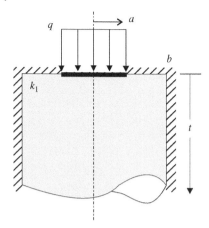

**Figure 3.1**  Semi-infinite flux tube with a uniform heat flux source.

which is subject to

$$\frac{\partial T}{\partial r}\bigg|_{r=0} = 0$$

$$\frac{\partial T}{\partial r}\bigg|_{r=b} = 0 \tag{3.2}$$

at the centroid and edge of the disk, respectively. Over the top surface $z = 0$, we apply

$$\left.\begin{array}{rl} \dfrac{\partial T}{\partial z}\bigg|_{z=0} = & -\dfrac{q(r)}{k}, \quad 0 < r < a \\[3mm] \dfrac{\partial T}{\partial z}\bigg|_{z=0} = & 0, \qquad a < r < b \end{array}\right\} \tag{3.3}$$

where $q(r)$ is a flux distributed over the contact area. Three typical heat flux distributions which are considered are the uniform heat flux, the inverted parabolic heat flux, and the parabolic heat flux. The inverted parabolic flux distribution is a good approximation for the isothermal heat source for smaller aspect ratios, i.e. $a/b < 0.4$. For larger contact ratios, the solution is only approximately isothermal in this case. The flux distribution applied over the source area is assumed to be

$$q(r) = (1 + \mu)\frac{Q}{\pi a^2}\left(1 - \frac{r^2}{a^2}\right)^{\mu} \tag{3.4}$$

Finally, at $z \to \infty$, we prescribe the uniform flow:

$$-k\frac{\partial T}{\partial z} = Q/\pi b^2 \tag{3.5}$$

in a region located at a significant distance from the contact plane.

A solution may be found using separation of variables such that $T(r, z) = R(r) * Z(z)$. Separating variables gives a general solution to Laplace's equation of the form:

$$T(r, z) = A_0 + B_0 z + [A_1 J_0(\lambda r) + B_1 Y_0(\lambda r)][A_2 \exp(-\lambda z) + B_2 \exp(\lambda z)] \tag{3.6}$$

This general solution comprises of a superposition of a uniform one-dimensional temperature field and a two-dimensional spreading field. As the heat source radius approaches the flux tube radius, the flow becomes one-dimensional near the source plane.

The boundary condition along the flux tube axis $r = 0$, precludes the use of the Bessel functions $Y_0(\cdot)$ and hence, $B_1 = 0$. Similarly, as $z \to \infty$, the constant $B_2 = 0$ must also be chosen. Further, to satisfy the $z \to \infty$ condition, the constant $B_0 = -Q/(\pi b^2 k)$. This leads to the simpler form of the general solution:

$$T(r,z) = A_0 - \frac{Q}{\pi b^2 k} z + A_1 J_0(\lambda r) \exp(-\lambda z) \tag{3.7}$$

The boundary condition along $r = b$ requires that

$$\left. \frac{d}{dr} J_0(\lambda r) \right|_{r=b} = J_1(\lambda b) = 0 \tag{3.8}$$

yielding an infinite number of eigenvalues such that

$$\delta_n = \lambda_n b = 3.8317, 7.0156, 10.1735, 13.3237, 16.4706, \dots \tag{3.9}$$

Successive eigenvalues may be approximated by adding $\pi$ to each new value as the above sequence yields $\delta_i - \delta_{i-1} = \pi$ as $i \to \infty$. The solution at this point now becomes

$$T(r,z) = A_0 - \frac{Q}{k\pi b^2} z + \sum_{n=1}^{\infty} A_n J_0(\lambda_n r) \exp(-\lambda_n z) \tag{3.10}$$

In the source plane, we must use:

$$\left. \frac{\partial T}{\partial z} \right|_{z=0} = -\frac{Q}{k\pi b^2} - \sum_{n=1}^{\infty} A_n J_0(\lambda_n r) \lambda_n = \begin{cases} -\dfrac{q(r)}{k} & 0 \le r < a \\ 0 & a < r \le b \end{cases} \tag{3.11}$$

Application of the boundary condition in the source plane requires a Fourier–Bessel series expansion to determine the value of $A_n$. This requires solving the following equation after multiplying each term by $r J_0(\lambda_m r)$ and integrating over the source plane:

$$\int_0^b \frac{Q}{k\pi b^2} r J_0(\lambda_m r) dr + \sum_{n=1}^{\infty} \int_0^b A_n J_0(\lambda_n r) \lambda_n r J_0(\lambda_m r) dr = \int_0^a \frac{q(r)}{k} r J_0(\lambda_m r) dr \tag{3.12}$$

Upon expanding the series and integrating, the orthogonality property of Bessel functions provides nonzero terms in the series only when $m = n$. Further, the first term on the left is zero due to the integral producing $J_1(\lambda_n b)$ which is identically equal to zero from the expression for the eigenvalues. Thus, the equation above simplifies to

$$A_n = \frac{\displaystyle\int_0^a \frac{q(r)}{k} r J_0(\lambda_n r) dr}{\displaystyle\int_0^b r J_0^2(\lambda_n r) \lambda_n \, dr} \tag{3.13}$$

Using the prescribed heat flux distribution, the above integral was evaluated by Yovanovich (1975) and is found to be

$$A_n = \frac{2Q}{\pi k a} \frac{(\mu+1) 2^\mu \Gamma(\mu+1) J_{\mu+1}(\delta_n \epsilon)}{\delta_n^2 J_0^2(\delta_n)(\delta_n \epsilon)^\mu}, \quad \mu \ge -1 \tag{3.14}$$

where $\Gamma(\cdot)$ is the Gamma function, $\epsilon = a/b$ is the contact spot ratio, and $\delta_n$ are the eigenvalues.

For the three special cases under consideration when $\mu = -1/2, 0, 1/2$, the Fourier coefficients simplify. These are

$$A_n = \frac{Q \sin(\delta_n \epsilon)}{\pi k a \delta_n^2 J_0^2(\delta_n)}, \quad \mu = -1/2 \tag{3.15}$$

$$A_n = \frac{2Q J_1(\delta_n \epsilon)}{\pi k a \delta_n^2 J_0^2(\delta_n)}, \quad \mu = 0 \tag{3.16}$$

$$A_n = \frac{3Q \sin(\delta_n \epsilon)}{\pi k a \delta_n^2 J_0^2(\delta_n)} \left( \frac{1}{(\delta_n \epsilon)^2} - \frac{1}{(\delta_n \epsilon) \tan(\delta_n \epsilon)} \right), \quad \mu = 1/2 \tag{3.17}$$

This completes the solution for the temperature distribution.

In order to define the various thermal resistances, we must define three fundamental mean temperatures. These are (i) the mean source temperature, (ii) the mean contact plane temperature, and (iii) a mean sink plane temperature.

The mean temperature of the source is found by integrating the temperature over the source area:

$$\overline{T}_s = \frac{1}{\pi a^2} \int_0^a T(r, 0) 2\pi r \, dr = A_0 + 2 \sum_{n=1}^{\infty} A_n \frac{J_1(\lambda_n a)}{\lambda_n a} \tag{3.18}$$

The mean contact or source plane temperature is found by integrating over the entire source plane area:

$$\overline{T}_{cp} = \frac{1}{\pi b^2} \int_0^b T(r, 0) 2\pi r \, dr = A_0 \tag{3.19}$$

Finally, in an arbitrary sink plane $z = \ell$, we find

$$\overline{T}_{z=\ell} = \frac{1}{\pi b^2} \int_0^b T(r, z = \ell) 2\pi r \, dr = A_0 - \frac{Q\ell}{k\pi b^2} \tag{3.20}$$

In all three cases, the constant $A_0$ is the mean contact plane temperature in the absence of spreading resistance.

Three thermal resistances may be defined using the various mean temperatures defined earlier. These are (i) the total thermal resistance, (ii) the one-dimensional or bulk resistance, and (iii) the spreading (or constriction) resistance.

The total thermal resistance for the flux tube may be defined as

$$R_{total} = \frac{\overline{T}_s - \overline{T}_{z=\ell}}{Q} = \frac{2 \sum_{n=1}^{\infty} A_n \frac{J_1(\lambda_n a)}{\lambda_n a} + \frac{Q\ell}{k\pi b^2}}{Q} = R_s + R_{1D} \tag{3.21}$$

while the one-dimensional resistance for the flux tube is defined as

$$R_{1D} = \frac{\overline{T}_{cp} - \overline{T}_{z=\ell}}{Q} = \frac{\ell}{k\pi b^2} \tag{3.22}$$

Finally, the spreading (or constriction resistance) is defined as

$$R_s = \frac{\overline{T}_s - \overline{T}_{cp}}{Q} = \frac{2 \sum_{n=1}^{\infty} A_n \frac{J_1(\lambda_n a)}{\lambda_n a}}{Q} \tag{3.23}$$

Using the above results for $R_s$ and $A_n$, we can define a dimensionless spreading resistance (or constriction parameter) $\psi = 4kaR_s$ for the general heat flux distribution:

$$\psi = \frac{16}{\pi\epsilon}(\mu+1)2^{\mu}\Gamma(\mu+1)\sum_{n=1}^{\infty}\frac{J_1(\delta_n\epsilon)J_{\mu+1}(\delta_n\epsilon)}{\delta_n^3 J_0^2(\delta_n)(\delta_n\epsilon)^{\mu}}, \quad \mu \geq -1 \tag{3.24}$$

Using the results of $A_n$ for each of the three special flux distributions, we can define a dimensionless spreading resistance $\psi$ for each case. These are

$$\psi = \frac{8}{\pi\epsilon}\sum_{n=1}^{\infty}\frac{J_1(\delta_n\epsilon)\sin(\delta_n\epsilon)}{\delta_n^3 J_0^2(\delta_n)}, \quad \mu = -1/2 \tag{3.25}$$

$$\psi = \frac{16}{\pi\epsilon}\sum_{n=1}^{\infty}\frac{J_1^2(\delta_n\epsilon)}{\delta_n^3 J_0^2(\delta_n)}, \quad \mu = 0 \tag{3.26}$$

$$\psi = \frac{24}{\pi\epsilon}\sum_{n=1}^{\infty}\frac{J_1(\delta_n\epsilon)\sin(\delta_n\epsilon)}{\delta_n^3 J_0^2(\delta_n)}\left(\frac{1}{(\delta_n\epsilon)^2}-\frac{1}{(\delta_n\epsilon)\tan(\delta_n\epsilon)}\right), \quad \mu = 1/2 \tag{3.27}$$

Values for each case over a range of $0 < \epsilon < 1$ are given in Table 3.1. Note that in the case of $\mu = -1/2$, the flux profile is unbounded at the edge of the source, thus as $\epsilon \to 1$ some computational issues leading to roundoff errors lead to erroneous values. It can be seen that the results are approaching a value of $\psi = 0$.

Correlations for the above results were developed by Yovanovich (1975) to make computations for each case easier for small values of $\epsilon$. For the region $\epsilon < 0.1$, Yovanovich (1975) proposed a simple linear correlation given by

$$\psi = \alpha(1 - \beta\epsilon) \tag{3.28}$$

while for somewhat larger values ($\epsilon \leq 0.3$), Yovanovich (1975) proposed:

$$\psi = \alpha(1 - \epsilon)^{\gamma} \tag{3.29}$$

**Table 3.1** Effect of $\mu$ on dimensionless spreading resistance $\psi$.

| $\epsilon$ | $\mu = -\frac{1}{2}$ | $\mu = 0$ | $\mu = \frac{1}{2}$ |
|---|---|---|---|
| 0 | 1.0000 | 1.0807 | 1.1250 |
| 0.01 | 0.9854 | 1.0665 | 1.1109 |
| 0.1 | 0.8580 | 0.9397 | 0.9842 |
| 0.2 | 0.7201 | 0.8008 | 0.8450 |
| 0.3 | 0.5851 | 0.6649 | 0.7085 |
| 0.4 | 0.4556 | 0.5337 | 0.5763 |
| 0.5 | 0.3341 | 0.4092 | 0.4500 |
| 0.6 | 0.2231 | 0.2936 | 0.3316 |
| 0.7 | 0.1262 | 0.1895 | 0.2234 |
| 0.8 | 0.0482 | 0.1008 | 0.1284 |
| 0.9 | — | 0.0331 | 0.0510 |
| 1 | — | 0.0000 | 0.0000 |

**Table 3.2** Coefficients for correlations of $\psi$.

|          | $\mu = -\frac{1}{2}$ | $\mu = 0$ | $\mu = \frac{1}{2}$ |
| -------- | -------------------- | --------- | ------------------- |
| $\alpha$ | 1.0000               | 1.0808    | 1.1252              |
| $\beta$  | 1.4197               | 1.4111    | 1.4098              |
| $\gamma$ | 1.5                  | 1.35      | 1.30                |

where the values of $\alpha$, $\beta$, and $\gamma$ are given in Table 3.2. Equation (3.28) provides accuracy of 0.1%, while Eq. (3.29) provides accuracy better than 1%. For a larger range of $\epsilon$, and only for the special case of $\mu = -1/2$, Roess [see Yovanovich (1975)] developed the following correlation:

$$\psi = 1 - 1.4093\epsilon + 0.2959\epsilon^3 + 0.05254\epsilon^5 + 0.02105\epsilon^7 + 0.01107\epsilon^9 + 0.006312\epsilon^{11} \tag{3.30}$$

Roess did not develop his correlation with values of $\psi$ beyond $0.1 \leq \epsilon \leq 0.6$. But as shown by Yovanovich (1975), his values are in excellent agreement with the Roess correlation. In fact, over the range $0.0 \leq \epsilon \leq 0.8$, the Roess correlation provides an accuracy better than 1%.

The three series solutions given above converge very slowly as $\epsilon \to 0$ which corresponds to the case of a circular contact area on a half-space. The corresponding half-space results which were reported earlier in Chapter 2 are

$$4kaR_s = \begin{cases} 1 & \text{for} \quad \mu = -1/2 \\ 32/3\pi^2 \approx 1.0807 & \text{for} \quad \mu = 0 \\ 36/32 \approx 1.1250 & \text{for} \quad \mu = 1/2 \end{cases} \tag{3.31}$$

Since the three series solutions presented above for the three heat flux distributions $\mu = -1/2$, $0$, $1/2$ converge slowly as $\epsilon \to 0$; therefore, correlation equations for the dimensionless spreading resistance $\psi = 4kaR_s$ for the three flux distributions were developed having the general form:

$$\psi = C_0 + C_1\epsilon + C_3\epsilon^3 + C_5\epsilon^5 + C_7\epsilon^7 \tag{3.32}$$

with the correlation coefficients given in Table 3.3. The correlation equations are applicable for the parameter range $0 \leq \epsilon \leq 0.8$ and provide 4 decimal place accuracy.

### 3.1.1 Isothermal Source on a Flux Tube

Finally, Negus and Yovanovich (1984a) considered the case of a flux tube with a true isothermal contact. They used a superposition method to obtain a solution for the isothermal boundary condition

**Table 3.3** Correlation coefficients for three flux distributions.

| $C_n$   | $\mu = -1/2$ | $\mu = 0$ | $\mu = 1/2$ |
| ------- | ------------ | --------- | ----------- |
| $C_0$   | 1.00000      | 1.08085   | 1.12517     |
| $C_1$   | −1.40981     | −1.41002  | −1.41038    |
| $C_3$   | 0.303641     | 0.259714  | 0.235387    |
| $C_5$   | 0.0218272    | 0.0188631 | 0.0117527   |
| $C_7$   | 0.0644683    | 0.0420278 | 0.0343458   |

and compared it with the case of the approximate isothermal condition determined from the inverse parabolic flux distribution. For $\epsilon > 0.4$, they found their results provided better accuracy over the approximate profile. They recommended the following correlation for the constriction parameter $\psi^T = 4kaR_s$:

$$\psi^T = 1 - 1.40978\epsilon + 0.34406\epsilon^3 + 0.0435\epsilon^5 + 0.02271\epsilon^7 \tag{3.33}$$

The above correlation provides exceptional accuracy when compared with their numerical results and agrees well with the solution of Gibson (1976).

## 3.2 Finite Disk with Sink Plane Conductance

A practical problem in the application of thermal spreaders is to consider a finite disk with conductance in the sink plane, see Figure 3.2. Kennedy (1960) obtained solutions for a disk with an isothermal sink and/or isothermal edges. All of the cases examined by Kennedy (1960) are special cases of results developed in this chapter. This problem can be more easily solved by defining the temperature excess $\theta = T - T_f$, where $T_f$ is the temperature associated with the constant uniform sink plane conductance $h_s$. This leads to the following problem statement:

$$\frac{\partial^2\theta}{\partial r^2} + \frac{1}{r}\frac{\partial\theta}{\partial r} + \frac{\partial^2\theta}{\partial z^2} = 0 \tag{3.34}$$

which is subject to

$$\left.\frac{\partial\theta}{\partial r}\right|_{r=0} = 0$$
$$\left.\frac{\partial\theta}{\partial r}\right|_{r=b} = 0 \tag{3.35}$$

along the axis and edges of the heat spreader, respectively. In the source plane, we specify as before

$$\left.\begin{array}{ll}
\left.\dfrac{\partial\theta}{\partial z}\right|_{z=0} = -\dfrac{q}{k}, & 0 \le r < a \\[2mm]
\left.\dfrac{\partial\theta}{\partial z}\right|_{z=0} = 0, & a < r \le b
\end{array}\right\} \tag{3.36}$$

while in the sink plane, we require the following condition:

$$-k\left.\frac{\partial\theta}{\partial z}\right|_{z=t} = h_s\theta(r,t) \tag{3.37}$$

A general solution may be found using separation of variables such that $\theta(r,z) = R(r) * Z(z)$. Separating variables gives a general solution to Laplace's equation of the form:

$$\theta(r,z) = A_0 + B_0 z + [A_1 J_0(\lambda r) + B_1 Y_0(\lambda r)][A_2 \cosh(\lambda z) + B_2 \sinh(\lambda z)] \tag{3.38}$$

**Figure 3.2**  Finite isotropic disk spreader.

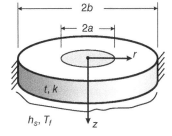

This solution has the general form:

$$\theta(r,z) = \theta_{1D}(z) + \theta_s(r,z) \tag{3.39}$$

The problem can be solved in two parts with the one-dimensional flow solution easily shown to be

$$\theta_{1D}(z) = \frac{Q}{\pi b^2}\left(\frac{t}{k} + \frac{1}{h_s}\right) - \frac{Qz}{k\pi b^2} \tag{3.40}$$

The two-dimensional spreading field can now be solved by applying the boundary conditions to the remaining portion of the problem. The symmetry condition along the axis of the disk $r = 0$ requires that the constant $B_1 = 0$ due to the infinite nature of the Bessel function $Y_0(0)$. The solution becomes

$$\theta_s(r,z) = J_0(\lambda r)[A_2\cosh(\lambda z) + B_2\sinh(\lambda z)] \tag{3.41}$$

with $A_1$ absorbed into the other remaining constants. As before, application of the adiabatic edge condition requires that

$$J_1(\lambda b) = 0 \tag{3.42}$$

yielding the eigenvalues to the above equation such that

$$\delta_n = \lambda_n b = 3.8317, 7.0156, 10.1735, 13.3237, 16.4706, \ldots \tag{3.43}$$

such that the spreading portion of the solution becomes

$$\theta_s(r,z) = \sum_{n=1}^{\infty} J_0(\lambda_n r)[A_2\cosh(\lambda_n z) + B_2\sinh(\lambda_n z)] \tag{3.44}$$

Along the sink plane, we write

$$-k\sum_{n=1}^{\infty} J_0(\lambda_n r)[A_2\sinh(\lambda_n t)\lambda_n + B_2\cosh(\lambda t)\lambda_n]$$

$$= h_s\sum_{n=1}^{\infty} J_0(\lambda_n r)[A_2\cosh(\lambda_n t) + B_2\sinh(\lambda_n t)] \tag{3.45}$$

This leads to the following result

$$B_2 = -A_2\phi \tag{3.46}$$

where $\phi$ may be compactly written as

$$\phi = -\frac{B_2}{A_2} = \frac{\lambda_n\tanh(\lambda_n t) + \dfrac{h_s}{k}}{\lambda_n + \dfrac{h_s}{k}\tanh(\lambda_n t)} \tag{3.47}$$

or more conveniently,

$$\phi = -\frac{B_2}{A_2} = \frac{\delta_n\tanh(\delta_n\tau) + Bi}{\delta_n + Bi\tanh(\delta_n\tau)} \tag{3.48}$$

where $\delta_n = \lambda_n b$, $\tau = t/b$, and $Bi = h_s b/k$. Note that $\phi \to 1$ when $\tau \geq 0.72$ for all eigenvalues $\delta_n$, returning the flux tube limit for thick spreaders.

The above parameter $\phi$ is often referred to as a spreading function. In later problems for compound disks or multilayered disks, it is this function that requires care and attention as other solutions have much in common.

The solution for the spreading portion of the temperature field now becomes

$$\theta_s(r,z) = \sum_{n=1}^{\infty} A_2 J_0(\lambda_n r)[\cosh(\lambda_n z) - \phi \sinh(\lambda_n z)] \tag{3.49}$$

Application of the final piecewise boundary condition in the source plane requires

$$\frac{\partial \theta_s}{\partial z}\bigg|_{z=0} = \sum_{n=1}^{\infty} J_0(\lambda_n r)[-A_2 \phi \lambda_n] = \begin{cases} -q/k & 0 \leq r < a \\ 0 & a < r \leq b \end{cases} \tag{3.50}$$

The final constant is found by taking a Fourier–Bessel expansion such that

$$A_2 = \frac{\dfrac{q}{k} \displaystyle\int_0^a J_0(\lambda_n r) r\, dr}{\lambda_n \phi \displaystyle\int_0^b J_0^2(\lambda_n r) r\, dr} = \frac{2qa}{\phi k} \frac{J_1(\delta_n a/b)}{\delta_n^2 J_0^2(\delta_n)} \tag{3.51}$$

The final solution for the two-dimensional spreading field gives

$$\theta_s(r,z) = \sum_{n=1}^{\infty} \frac{2Q}{\phi \pi a k} \frac{J_1(\delta_n a/b)}{\delta_n^2 J_0^2(\delta_n)} J_0(\delta_n r/b)[\cosh(\delta_n z/b) - \phi \sinh(\delta_n z/b)] \tag{3.52}$$

where we have now introduced $q = Q/\pi a^2$ for consistency.

The full temperature field for the finite disk is now written as

$$\begin{aligned}
\theta(r,z) = {} & \frac{Q}{\pi b^2}\left(\frac{t}{k} + \frac{1}{h_s}\right) - \frac{Qz}{k\pi b^2} \\
& + \sum_{n=1}^{\infty} \frac{2Q}{\phi \pi a k} \frac{J_1(\delta_n a/b)}{\delta_n^2 J_0^2(\delta_n)} J_0(\delta_n r/b)[\cosh(\delta_n z/b) - \phi \sinh(\delta_n z/b)]
\end{aligned} \tag{3.53}$$

The temperature field in the contact plane $\theta(r,0)$ is of greatest interest and found to be

$$\theta(r,0) = \frac{Q}{\pi b^2}\left(\frac{t}{k} + \frac{1}{h_s}\right) + \sum_{n=1}^{\infty} \frac{2Q}{\phi \pi a k} \frac{J_1(\delta_n a/b)}{\delta_n^2 J_0^2(\delta_n)} J_0(\delta_n r/b) \tag{3.54}$$

Integration over the contact region gives

$$\bar{\theta}_s = \frac{1}{\pi a^2} \int_0^a \theta(r,0) 2\pi r\, dr = \frac{Q}{\pi b^2}\left(\frac{t}{k} + \frac{1}{h_s}\right) + \frac{4Qb}{\pi a^2 k} \sum_{n=1}^{\infty} \frac{J_1^2(\delta_n a/b)}{\phi \delta_n^3 J_0^2(\delta_n)} \tag{3.55}$$

The total resistance is now found to be

$$R_{total} = \frac{\bar{\theta}_s}{Q} = \frac{1}{\pi b^2}\left(\frac{t}{k} + \frac{1}{h_s}\right) + \frac{4b}{\pi a^2 k} \sum_{n=1}^{\infty} \frac{J_1^2(\delta_n a/b)}{\phi \delta_n^3 J_0^2(\delta_n)} \tag{3.56}$$

The total thermal resistance of the disk is composed of a one-dimensional bulk resistance and the two-dimensional thermal spreading resistance.

In dimensionless form, the total thermal resistance due to a finite spot on an isotropic disk is given by

$$4ka R_{total} = \frac{4\epsilon}{\pi}\left[\tau + \frac{1}{Bi}\right] + \frac{16}{\pi \epsilon} \sum_{n=1}^{\infty} \frac{J_1^2(\delta_n \epsilon)}{\phi \delta_n^3 J_0^2(\delta_n)} \tag{3.57}$$

From which we obtain the dimensionless thermal spreading resistance component

$$4ka R_s = \frac{16}{\pi \epsilon} \sum_{n=1}^{\infty} \frac{J_1^2(\delta_n \epsilon)}{\phi \delta_n^3 J_0^2(\delta_n)} \tag{3.58}$$

where $\epsilon = a/b$ is the relative size of the heat source to the flux tube. Note that the only difference in the above equation when compared with the semi-infinite flux tube equation (3.26) is the presence of the spreading function $\phi$. This simple observation will allow us to extend the basic solution to many other cases simply by determining the appropriate spreading function.

The above solution has a number of special cases, including the flux tube when $\tau \to \infty$ and the half-space when $\tau \to \infty$ and $\epsilon \to 0$. A variation of the above problem, which we will consider later, involves a circular disk with edge cooling. It has been shown that the spreading function $\phi$ is the only parameter that changes with increasing system complexity, i.e. additional layers and interfacial conductance.

The solution for the uniform isoflux boundary condition was re-examined by Song et al. (1995) and Lee et al. (1995). They nondimensionalized the spreading resistance based on the centroid and area-average temperatures using the square root of the source area [as recommended by Yovanovich (1976), Yovanovich and Burde (1977), Chow and Yovanovich (1982), Negus and Yovanovich (1984a), Yovanovich and Antonetti (1988)] and compared the analytical results against the numerical results reported by Nelson and Sayers (1992) over the full range of the independent parameters: $Bi = hb/k$, $\epsilon = a/b$, and $\tau = t/b$. Nelson and Sayers (1992) also chose the square root of the source area to report their numerical results. The agreement between the analytical and numerical results were reported to be in excellent agreement.

Lee et al. (1995) recommended a simple closed-form expression for the dimensionless constriction resistance based on the area-average and centroid temperatures. They defined the dimensionless spreading resistance parameter as $\psi = k\sqrt{A}R_s$, where $R_s$ is the constriction resistance, and they recommended the following approximations:

**Area Average Temperature**

$$\psi_{ave} = \frac{1}{2}(1 - \epsilon)^{3/2}\varphi_c \tag{3.59}$$

**Centroid Temperature**

$$\psi_{max} = \frac{1}{\sqrt{\pi}}(1 - \epsilon)\varphi_c \tag{3.60}$$

where, for both cases

$$\varphi_c = \frac{Bi\ \tanh\left(\delta_c\tau\right) + \delta_c}{Bi + \delta_c\ \tanh\left(\delta_c\tau\right)} \tag{3.61}$$

and

$$\delta_c = \pi + \frac{1}{\sqrt{\pi}\epsilon} \tag{3.62}$$

The above approximations are within $\pm 10\%$ of the analytical results [Song et al. (1995), Lee et al. (1995)] and the numerical results of Nelson and Sayers (1992). For this reason, they have been widely adopted by thermal packaging engineers for their simplicity and accuracy.

**Example 3.1** Consider a small device heat sink with a square baseplate of dimensions $2w = 5$ cm and a thickness of $t = 3$ mm. The heat source has a diameter $2a = 1$ cm with a strength of 2 W and the sink temperature is $T_f = 20°$C. The thermal spreader has a thermal conductivity of $k = 200$ W/(m K) and an equivalent sink conductance of $h_s = 50$ W/(m² K). Determine the average source temperature using the exact analytical solution and the approximate solution for thermal spreading resistance.

First, we must convert the square baseplate to an equivalent circular baseplate. Equating the area:

$$4w^2 = \pi b^2$$

gives $b = 2.821$ cm. Thus, we have a source aspect ratio of $\epsilon = a/b = 0.1772$. The one-dimensional bulk resistance of the spreader is

$$R_{1D} = \frac{t}{k\pi b^2} + \frac{1}{h_s\pi b^2} = 8.006 \text{ K/W}$$

The spreading resistance found using Eq. (3.58) with $N = 100$ terms is

$$R_s = 0.3824 \text{ K/W}$$

Giving us a total resistance:

$$R_{total} = 8.388 \text{ K/W}$$

Finally, we find that $\overline{\theta}_s = QR_{total} = 16.78°\text{C}$ or $\overline{T}_s = 36.78°\text{C}$. Using the approximate correlations of Lee et al. (1995), we find that

$$R_s = 0.3584 \text{ K/W}$$

and

$$R_{total} = 8.364 \text{ K/W}$$

which leads to $\overline{\theta}_s = QR_{total} = 16.73°\text{C}$ or $\overline{T}_s = 36.73°\text{C}$. In this example, the small error of the approximate result for the spreading resistance is further diminished by the much larger bulk resistance.

### 3.2.1 Distributed Heat Flux over Source Area

The solution given previously assumes a uniform heat flux source. If a variable axi-symmetric heat flux is defined according to

$$q(r) = q_0(1 + \mu)(1 - r^2/a^2)^\mu \tag{3.63}$$

where $q_0 = Q/\pi a^2$ as before. The spreading (or constriction) resistance in the finite disk then accounts for all of the multidimensional effects for a uniform or variable flux heat source. In this case, we have

$$4kaR_s = \frac{16(\mu + 1)}{\pi\epsilon} \sum_{n=1}^{\infty} \frac{D_n}{\phi} \frac{J_1(\delta_n\epsilon)}{\delta_n^3 J_0^2(\delta_n)} \tag{3.64}$$

where $D_n$ is defined according to

$$\mu = -\frac{1}{2}, \quad D_n = \sin(\delta_n\epsilon)$$

$$\mu = 0, \quad D_n = J_1(\delta_n\epsilon) \tag{3.65}$$

$$\mu = +\frac{1}{2}, \quad D_n = \sin(\delta_n\epsilon)\left[\frac{1}{(\delta_n\epsilon)^2} - \frac{1}{(\delta_n\epsilon)\tan(\delta_n\epsilon)}\right]$$

In this manner, when $\mu = -1/2$ an inverse parabolic distribution is obtained which yields an approximately isothermal heat source for $\epsilon < 0.4$, for $\mu = 0$ a uniform heat source is obtained, and when $\mu = 1/2$, a parabolic flux distribution is obtained.

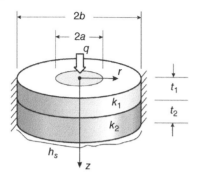

**Figure 3.3** Compound thermal spreader.

## 3.3 Compound Disk

The next system of interest is composed of two layers, which in general, are imperfectly bonded to each other giving rise to an interfacial resistance (Muzychka et al. 2004, Muzychka 2014), see Figure 3.3. Often, a thinner layer with $k_1 > k_2$ is applied to a thermal spreader to promote spreading and reduce the spreading resistance. An interfacial contact conductance $h_c$ is used to model the imperfect contact region. This conductance can also be modeled as an equivalent resistance such that

$$R_i = \frac{t_i}{k_i A_b} = \frac{1}{h_c A_b} \tag{3.66}$$

or

$$h_c = \frac{k_i}{t_i} \tag{3.67}$$

where $k_i$ is the thermal conductivity of the interfacial region or bonding agent, and $t_i$ the nominal thickness of this region when these characteristics are known. This approach is adopted since the interfacial layer is usually thin relative to the adjacent layers, i.e. $t_i \ll t_1, t_2$. It also represents a good approximation to a three-layer system when this criteria is satisfied.

The system is modeled using a local system of coordinates in each layer, such that the heat source has an origin at the center of the disk or channel. The $z$-direction is now measured locally within each layer. This approach is chosen as it leads to a simpler solution for the undetermined coefficients. Further, as before we will define the temperature excess, $\theta = T - T_f$, relative to the sink temperature.

The problem is now stated as follows: the governing equation for each layer in cylindrical co-ordinates is:

$$\frac{\partial^2 \theta_1}{\partial r^2} + \frac{1}{r} \frac{\partial \theta_1}{\partial r} + \frac{\partial^2 \theta_1}{\partial z_1^2} = 0, \quad 0 < z_1 < t_1$$

$$\frac{\partial^2 \theta_2}{\partial r^2} + \frac{1}{r} \frac{\partial^2 \theta_2}{\partial r} + \frac{\partial^2 \theta_2}{\partial z_2^2} = 0, \quad 0 < z_2 < t_2 \tag{3.68}$$

The following boundary conditions are imposed on the system. In the source plane, we specify a variable heat flux distribution such that over the region of the heat source equation (3.63) is specified, while outside of the heat source region, the remainder of the surface is taken as adiabatic. Therefore, the following is assumed for a single heat source:

$$\frac{\partial \theta_1}{\partial z_1} = -q(r)/k_1, \ z_1 = 0 \ \text{ Over Source Region}$$

$$\frac{\partial \theta_1}{\partial z_1} = 0, \qquad z_1 = 0 \ \text{ Outside Source Region}$$

(3.69)

At the interface, we require the following conditions representing the equality of flux and temperature drop due to the imperfect interface:

$$k_1 \frac{\partial \theta_1}{\partial z_1}\bigg|_{z_1=t_1} = k_2 \frac{\partial \theta_2}{\partial z_2}\bigg|_{z_2=0}$$

$$-k_1 \frac{\partial \theta_1}{\partial z_1}\bigg|_{z_1=t_1} = h_c \left[\theta_1(r, t_1) - \theta_2(r, 0)\right]$$

(3.70)

Along the heat sink plane, we specify the following condition:

$$-k_2 \frac{\partial \theta_2}{\partial z_2}\bigg|_{z_2=t_2} = h_s \theta_2(r, t_2)$$

(3.71)

Along the axis and edge of the compound disk (using symmetry), the following conditions are also required:

$$\frac{\partial \theta_i}{\partial r}\bigg|_{r=0} = 0, \ i = 1, 2$$

$$\frac{\partial \theta_i}{\partial r}\bigg|_{r=b} = 0, \ i = 1, 2$$

(3.72)

The solution to the above problem was solved by Yovanovich et al. (1980) without interfacial conductance. We may find a general solution for the thermal spreading resistance using this result and the result for the isotropic disk presented earlier. To extend the solution for the compound disk, we merely need only find the new spreading function $\phi$ through application of the new boundary conditions in the $z$-direction for both problems with and without perfect interfacial conductance. With an additional layer and interfacial conductance, the spreading function $\phi$ in the solution for the isotropic disk can now be defined for two cases: (i) finite interfacial conductance and (ii) perfect interfacial contact along with the corresponding case of infinite sink plane conductance, i.e. a constant uniform temperature surface, Muzychka (2015). Muzychka (2015) showed that with the new spreading function, one could easily extend the solution for a single layer to a system with two layers with little additional effort. In all the expressions below, the eigenvalue $\lambda_n$ is related to the zeros $\delta_n = \lambda_n b$ of the equation $J_1(\delta_n) = 0$, i.e.

$$\delta_n = \lambda_n b = 3.8317, 7.0156, 10.1735, 13.3237, 16.4706, \ldots$$

(3.73)

The new spreading functions for the case of finite interfacial conductance and perfect interfacial contact are given below.

**Finite Interfacial Conductance**

For the general system stated earlier (two isotropic layers with interfacial conductance), the necessary spreading function is:

$$\phi = \frac{C_1 + C_2 \tanh(\lambda_n t_1)}{C_1 \tanh(\lambda_n t_1) + C_2}$$

(3.74)

where

$$C_1 = \left[\lambda_n \tanh(\lambda_n t_2) + h_s/k_2\right] \tag{3.75}$$

and

$$C_2 = k_1/k_2 \left[\lambda_n(1 + h_s/h_c) + (h_s/k_2 + \lambda_n^2 k_2/h_c) \tanh(\lambda_n t_2)\right] \tag{3.76}$$

and

$$R_{1D} = \frac{1}{\pi b^2} \left(\frac{t_1}{k_1} + \frac{1}{h_c} + \frac{t_2}{k_2} + \frac{1}{h_s}\right) \tag{3.77}$$

If the device is in contact with an ideal heat sink, $h_s \rightarrow \infty$, the above expression simplifies to yield:

$$\phi = \frac{1 + \lambda_n k_1/h_c \tanh(\lambda_n t_1) + k_1/k_2 \tanh(\lambda_n t_2) \tanh(\lambda_n t_1)}{\lambda_n k_1/h_c + k_1/k_2 \tanh(\lambda_n t_2) + \tanh(\lambda_n t_1)} \tag{3.78}$$

and

$$R_{1D} = \frac{1}{\pi b^2} \left(\frac{t_1}{k_1} + \frac{1}{h_c} + \frac{t_2}{k_2}\right) \tag{3.79}$$

**Perfect Interfacial Contact**

If there is perfect interfacial contact, then $h_c \rightarrow \infty$, the solution for the spreading function becomes

$$\phi = \frac{\left[\lambda_n \tanh(\lambda_n t_2) + h_s/k_2\right] + k_1/k_2 \left[\lambda_n + h_s/k_2 \tanh(\lambda_n t_2)\right] \tanh(\lambda_n t_1)}{\left[\lambda_n \tanh(\lambda_n t_2) + h_s/k_2\right] \tanh(\lambda_n t_1) + k_1/k_2 \left[\lambda_n + h_s/k_2 \tanh(\lambda_n t_2)\right]} \tag{3.80}$$

and

$$R_{1D} = \frac{1}{\pi b^2} \left(\frac{t_1}{k_1} + \frac{t_2}{k_2} + \frac{1}{h_s}\right) \tag{3.81}$$

Finally, if the device is in contact with an ideal heat sink, $h_s \rightarrow \infty$, the above expression simplifies to yield:

$$\phi = \frac{1 + k_1/k_2 \tanh(\lambda_n t_2) \tanh(\lambda_n t_1)}{\tanh(\lambda_n t_1) + k_1/k_2 \tanh(\lambda_n t_2)} \tag{3.82}$$

and

$$R_{1D} = \frac{1}{\pi b^2} \left(\frac{t_1}{k_1} + \frac{t_2}{k_2}\right) \tag{3.83}$$

In all of the above cases, the thermal spreading resistance is given by Eq. (3.64) defined with respect to the fist layer:

$$4k_1 a R_s = \frac{16(\mu + 1)}{\pi \epsilon} \sum_{n=1}^{\infty} \frac{D_n}{\phi} \frac{J_1(\delta_n \epsilon)}{\delta_n^3 J_0^2(\delta_n)} \tag{3.84}$$

where $D_n$ is the Fourier coefficient defined by Eq. (3.65) for each of the three special cases of heat flux distribution. Note that in each of the spreading functions defined above $\phi$, the eigenvalue $\lambda_n = \delta_n/b$, where the $\delta_n$ are the roots of $J_1(\delta_n) = 0$.

**Example 3.2** Calculate the spreading resistance and total resistance for the isothermal circular source of radius $a = 2.89$ mm which is in perfect thermal contact with a thin coating of thickness $t_1 = 0.289$ mm which is bonded to a circular substrate of radius $b = 12.5$ mm. The thickness of the substrate is $t_2 = 10$ mm, and its lower face is cooled by a fluid through a film coefficient of $h = 500$ W/(m$^2$ K). The thermal conductivities of the coating and substrate are $k_1 = 0.5$ W/(m K)

**Table 3.4** Solutions obtainable from an isoflux source on a compound disk.

| Configuration | Limiting values |
| --- | --- |
| *Circular source on a finite disk* | |
| Finite compound disk | $a, b, t_1, t_2, k_1, k_2, h_s$ |
| Finite isotropic disk | $t_2 = 0, k_2 = k_1$ |
| *Circular source on a flux tube* | |
| Semi-infinite compound flux tube | $t_2 \to \infty$ |
| Semi-infinite isotropic circular flux tube | $t_1 \to \infty$ |
| *Circular source on an semi-infinite plate* | |
| Semi-infinite compound plate | $b \to \infty$ |
| Semi-infinite isotropic plate | $b \to \infty, t_2 = 0, k_2 = k_1$ |
| *Circular source on a half space* | |
| Compound half space | $b \to \infty, t_2 \to \infty$ |
| Isotropic half space | $b \to \infty, t_1 \to \infty$ |

and $k_2 = 50\,\text{W/(m K)}$, respectively. The top surface outside the source area $a < r \leq b$ and the sides $r = b$ are adiabatic.

For this problem, we must use Eq. (3.84) for the thermal spreading resistance with $D_n = \sin(\delta_n \epsilon)$ from Eq. (3.65). The spreading function $\phi$ is defined by Eq. (3.80) and evaluated with $\lambda_n = \delta_n/b$. Finally, the one-dimensional bulk resistance given by Eq. (3.81). The results for the given variables and assuming $N = 500$ in the series for Eq. (3.84) are found to be

$$R_s = 18.78 \text{ K/W}$$

$$R_{1D} = 5.66 \text{ K/W}$$

$$R_{total} = 24.44 \text{ K/W}$$

### 3.3.1 Special Limits in the Compound Disk Solution

With care, the general compound disk solution may be used to obtain several special limiting cases discussed earlier. These include (i) half-space, (ii) layered half-space, (iii) layered flux tube limit, and (iv) thin semi-infinite plate. These special cases have many useful applications. Additionally, the multilayer solution may also be used to calculate the spreading resistance into a multilayered half-space in a similar manner. For convenience, we only tabulate the cases derived from the compound disk solution in Table 3.4.

## 3.4 Multilayered Disks

A simple system is composed of a single isotropic layer with a constant conductance, $h_s$, at the sink plane. In a multilayered system, we have $N$ layers denoted $i = 1 \ldots N$ and $N - 1$ interfaces denoted $j = 1 \ldots N - 1$ as shown in Figure 3.4. These layers are isotropic and may be perfectly or imperfectly bonded to each other giving rise to an interfacial or contact resistance, $h_{c,j}$.

In cylindrical co-ordinates, for $N$ isotropic layers we have

$$\frac{\partial^2 \theta_1}{\partial r^2} + \frac{1}{r}\frac{\partial \theta_1}{\partial r} + \frac{\partial^2 \theta_1}{\partial z_1^2} = 0, \quad 0 < z_1 < t_1$$

$$\frac{\partial^2 \theta_2}{\partial r^2} + \frac{1}{r}\frac{\partial^2 \theta_2}{\partial r} + \frac{\partial^2 \theta_2}{\partial z_2^2} = 0, \quad 0 < z_2 < t_2 \qquad (3.85)$$

$$\vdots \qquad\qquad\qquad \vdots$$

$$\frac{\partial^2 \theta_N}{\partial r^2} + \frac{1}{r}\frac{\partial \theta_N}{\partial r} + \frac{\partial^2 \theta_N}{\partial z_N^2} = 0, \quad 0 < z_N < t_N$$

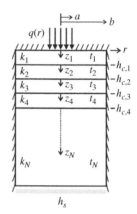

**Figure 3.4** Multilayered thermal spreader.

The following boundary conditions are imposed on the system. In the heat source plane, we specify a discrete heat flux distribution such that over the region of the heat source, a uniform heat flux is specified, while outside of the heat source region, the remainder of the surface is taken as adiabatic. Therefore, the following is assumed for a single heat source:

$$\frac{\partial \theta_1}{\partial z_1} = -q(r)/k_1, \; z_1 = 0 \quad \text{Over Source Region}$$

$$\frac{\partial \theta_1}{\partial z_1} = 0, \qquad\qquad z_1 = 0 \quad \text{Outside Source Region} \qquad (3.86)$$

At the $N-1$ interfaces, we require the following conditions representing the equality of flux and temperature drop due to the imperfect interface:

$$k_i \frac{\partial \theta_i}{\partial z_i}\bigg|_{z_i=t_i} = k_{i+1}\frac{\partial \theta_{i+1}}{\partial z_{i+1}}\bigg|_{z_{i+1}=0}$$

$$-k_i \frac{\partial \theta_i}{\partial z_i}\bigg|_{z_i=t_i} = h_{cj}\left[\theta_i(r,t_i) - \theta_{i+1}(r,0)\right] \qquad (3.87)$$

Finally, along the heat sink plane, we specify the following condition:

$$-k_N \frac{\partial \theta_N}{\partial z_N}\bigg|_{z_N=t_N} = h_s\theta_N(r,t_N) \qquad (3.88)$$

Along the axis and edge of the disk:

$$\frac{\partial \theta_i}{\partial r}\bigg|_{r=0} = 0, \; i = 1\ldots N$$

$$\frac{\partial \theta_i}{\partial r}\bigg|_{r=b} = 0, \; i = 1\ldots N \qquad (3.89)$$

The basic solution for a finite isotropic flux tube can be extended to the more general case of a $N$ layer structure strictly through modification of the definition of the spreading function $\phi$. Recently, Bagnall et al. (2014) presented a recursive procedure for defining a new spreading function in terms of spreading functions for the additional layers. The procedure was developed for a rectangular flux channel, but is equally applicable to a circular flux tube since the spreading function is only associated with the formal solution in the $z$-direction.

In an $N$ layer system, the spreading resistance is now calculated from Eq. (3.64):

$$4k_1 a R_s = \frac{16(\mu+1)}{\pi\epsilon}\sum_{n=1}^{\infty}\frac{D_n}{\phi_1}\frac{J_1(\delta_n\epsilon)}{\delta_n^3 J_0^2(\delta_n)} \qquad (3.90)$$

where $D_n$ is the Fourier coefficient defined by Eq. (3.65) for each of the three special cases of heat flux distribution, $\epsilon = a/b$ is the relative contact size and $\phi_1$ is the spreading function associated with the first layer in the system. The total resistance of the disk considering the uniform flow portion is

$$R_{total} = \sum_{i=1}^{N} \frac{t_i}{k_i \pi b^2} + \sum_{j=1}^{N-1} \frac{1}{h_{c,j} \pi b^2} + \frac{1}{h_s \pi b^2} + R_s \tag{3.91}$$

In an $N$ layer structure, we can define two different spreading functions, one which assumes perfect interfacial contact and one which assumes imperfect contact. Although the former can be modeled from the latter, both are presented for convenience. The spreading function $\phi_1$ in Eq. (3.90) is a recursive function when two or more layers are present in a system and is now denoted as $\phi_1$. The recursive relationship is now given by a general expression for $\phi_i$ such that $\phi_1$ is a function of $\phi_2$ which is a function of $\phi_3$ ..., and so on, until the final spreading function $\phi_N$ is called. Following the work of Bagnall et al. (2014), we may write:

**Perfect Interfacial Contact**

$$\phi_i = \frac{\frac{k_i}{k_{i+1}} \tanh(\delta_n t_i / b) + \phi_{i+1}}{\frac{k_i}{k_{i+1}} + \phi_{i+1} \tanh(\delta_n t_i / b)} \tag{3.92}$$

**Nonperfect Interfacial Contact**

$$\phi_i = \frac{\frac{k_i}{k_{i+1}} \tanh(\delta_n t_i / b) + \phi_{i+1}[1 + \frac{k_i}{h_{c,i} b} \delta_n \tanh(\delta_n t_i / b)]}{\frac{k_i}{k_{i+1}} + \phi_{i+1}[\tanh(\delta_n t_i / b) + \frac{k_i}{h_{c,i} b} \delta_n]} \tag{3.93}$$

In both cases, we denote the $N$th layer spreading function by

$$\phi_N = \left[ \frac{\delta_n \tanh(\delta_n t_N / b) + h_s b / k_N}{\delta_n + h_s b / k_N \tanh(\delta_n t_N / b)} \right] \tag{3.94}$$

Specification of the spreading functions for $N$-layers is now a simple matter. Once defined, it becomes a simple exercise to define sequences of materials in multilayered structures and assess thermal performance. In Muzychka et al. (1999), the authors developed a three-component solution and used it to predict contact conductance between surfaces coated with a diamond-like carbon (DLC). The above model may be used to reduce the analysis to a three-layer system. In the case of the data utilized by Muzychka et al. (1999), the intermediate layer was required to help bind the DLC layer to the aluminum substrates.

**Example 3.3** In this example, we have a heat source of radius $a = 5$ mm, a disk of radius $b = 10$ mm, three layers with $t_1 = 2$ mm and $k_1 = 25$ W/(m K), $t_2 = 2$ mm and $k_2 = 2$ W/(m K), and $t_3 = 5$ mm and $k_3 = 50$ W/(m K). The heat sink plane conductance is $h_s = 1000$ W/(m$^2$ K). All interfaces are considered to be perfect. The results are provided in the table below. Excellent agreement is obtained for all three flux distributions. In this example, we use Eq. (3.90) to find the spreading resistance, Eq. (3.91) to find the total resistance, and Eqs. (3.92) and (3.94) for the spreading functions.

|  | Analytical | | FEM | |
| --- | --- | --- | --- | --- |
|  | $R_{total}$ | $R_s$ | $R_{total}$ | $R_s$ |
| $\mu = -\frac{1}{2}$ | 7.856 | 0.918 | 7.813 | 0.875 |
| $\mu = 0$ | 8.068 | 1.129 | 8.067 | 1.128 |
| $\mu = +\frac{1}{2}$ | 8.187 | 1.248 | 8.189 | 1.250 |

The slight discrepancy in the between the finite element method (FEM) result and the analytical result for $\mu = -1/2$ is due to the approximate isothermal solution not being exactly equal beyond $\epsilon > 0.4$ and also the discontinuity of the heat flux at the edge of the isothermal source.

**Example 3.4** In this example, we consider a heat source of radius $a = 2.5$ mm, a disk of radius $b = 12.5$ mm, five layers with $t_1 = 1$ mm and $k_1 = 400$ W/(m K), $t_2 = 1$ mm and $k_2 = 1$ W/(m K), $t_3 = 2$ mm and $k_3 = 50$ W/(m K), $t_4 = 1$ mm and $k_4 = 1$ W/(m K), and $t_5 = 2$ mm and $k_5 = 25$ W/(m K). The heat sink plane conductance is $h_s = 1000$ W/(m² K). All interfaces are considered to be perfect. The results are provided in the table below. Excellent agreement is obtained for all three flux distributions. As in the previous example, we use Eq. (3.90) to find the spreading resistance, Eq. (3.91) to find the total resistance, and Eqs. (3.92) and (3.94) for the spreading functions.

| | Analytical | | FEM | |
|---|---|---|---|---|
| | $R_{total}$ | $R_s$ | $R_{total}$ | $R_s$ |
| $\mu = -\frac{1}{2}$ | 6.804 | 0.443 | 6.782 | 0.421 |
| $\mu = 0$ | 6.840 | 0.479 | 6.839 | 0.478 |
| $\mu = +\frac{1}{2}$ | 6.860 | 0.499 | 6.864 | 0.503 |

Once again, we see the slight discrepancy between the analytical and FEM results for the approximate isothermal source case $\mu = -1/2$.

## 3.5 Flux Tube with Circular Annular Heat Source

Having examined the thermal spreading resistance from a circular uniform flux source on a semi-infinite flux tube, we can extend the earlier analysis to a flux tube with an annular heat source. The new problem requires the following to be solved [Yip (1969)]:

$$\frac{\partial^2 T}{\partial r^2} + \frac{1}{r}\frac{\partial T}{\partial r} + \frac{\partial^2 T}{\partial z^2} = 0 \tag{3.95}$$

which is subject to:

$$\left.\frac{\partial T}{\partial r}\right|_{r=0} = 0$$
$$\left.\frac{\partial T}{\partial r}\right|_{r=c} = 0 \tag{3.96}$$

along the axis and edge of the flux tube, respectively. Over the top surface in the source plane $z = 0$,

$$\left.\begin{array}{ll}
\left.\dfrac{\partial T}{\partial z}\right|_{z=0} = 0, & 0 < r < a \\[2mm]
\left.\dfrac{\partial T}{\partial z}\right|_{z=0} = -\dfrac{q}{k}, & a < r < b \\[2mm]
\left.\dfrac{\partial T}{\partial z}\right|_{z=0} = 0, & b < r < c
\end{array}\right\} \tag{3.97}$$

where $q$ is a constant flux distributed over the annular contact area.

Finally, at $z \to \infty$, we prescribe the uniform flow:

$$-k\frac{\partial T}{\partial z} = Q/\pi c^2 \tag{3.98}$$

in a region located at a significant distance from the contact plane.

A solution may be found using separation of variables such that $T(r, z) = R(r) * Z(z)$. Separating variables gives a general solution to Laplace's equation of the form:

$$T(r, z) = A_0 + B_0 z + [A_1 J_0(\lambda r) + B_1 Y_0(\lambda r)][A_2 \exp(-\lambda z) + B_2 \exp(\lambda z)] \tag{3.99}$$

This general solution comprises of a superposition of a uniform one-dimensional temperature field and a two-dimensional spreading field.

As before, the boundary condition along the cylinder axis $r = 0$ precludes the use of the Bessel functions $Y_0(\cdot)$, and hence, $B_1 = 0$. Similarly, as $z \to \infty$, the constant $B_2 = 0$ must also be chosen. Further, to satisfy the $z \to \infty$ condition, the constant $B_0 = -Q/(\pi c^2 k)$. This leads to the much simpler form of the general solution:

$$T(r, z) = A_0 - \frac{Q}{\pi c^2 k} z + A_1 J_0(\lambda r) \exp(-\lambda z) \tag{3.100}$$

The boundary condition along $r = c$ requires that

$$\frac{d}{dr} J_0(\lambda r)\bigg|_{r=c} = J_1(\lambda c) = 0 \tag{3.101}$$

yielding an infinite number of eigenvalues such that

$$\delta_n = \lambda_n c = 3.8317, 7.0156, 10.1735, 13.3237, 16.4706, \ldots \tag{3.102}$$

The solution at this point now becomes

$$T(r, z) = A_0 - \frac{Q}{k\pi c^2} z + \sum_{n=1}^{\infty} A_n J_0(\lambda_n r) \exp(-\lambda_n z) \tag{3.103}$$

In the source plane, we must use

$$\frac{\partial T}{\partial z}\bigg|_{z=0} = -\frac{Q}{k\pi c^2} - \sum_{n=1}^{\infty} A_n J_0(\lambda_n r)\lambda_n = \begin{cases} 0 & 0 < r \leq a \\ -\dfrac{q}{k} & a \leq r < b \\ 0 & b < r \leq c \end{cases} \tag{3.104}$$

Application of the boundary condition in the source plane requires a Fourier–Bessel series expansion to determine the value of $A_n$. This requires solving the following equation after multiplying each term by $r J_0(\lambda_m r)$ and integrating over the source plane:

$$\int_0^c \frac{Q}{k\pi b^2} r J_0(\lambda_m r) dr + \sum_{n=1}^{\infty} \int_0^c A_n J_0(\lambda_n r)\lambda_n r J_0(\lambda_m r) dr = \int_a^b \frac{q}{k} r J_0(\lambda_m r) dr \tag{3.105}$$

Upon expanding the series and integrating the orthogonality property of Bessel functions provide nonzero terms in the series only when $m = n$. Further, the first term on the left is zero due to the integral producing $J_1(\lambda_n c)$ which is identically equal to zero from the expression for the eigenvalues. Thus, the equation above simplifies to

$$A_n = \frac{\dfrac{q}{k} \int_a^b r J_0(\lambda_n r) dr}{\int_0^c r J_0^2(\lambda_n r)\lambda_n \, dr} \tag{3.106}$$

or after evaluating the integrals and simplifying, we obtain

$$A_n = \frac{2Q[bJ_1(\lambda_n b) - aJ_1(\lambda_n a)]}{k\pi c^2(b^2 - a^2)\lambda_n^2 J_0^2(\lambda_n c)} \tag{3.107}$$

The mean temperature of the source is found by integrating the temperature over the source area:

$$\overline{T}_s = \frac{1}{\pi(b^2 - a^2)}\int_a^b T(r,0)2\pi r \, dr = A_0 + 2\sum_{n=1}^{\infty} A_n \frac{[bJ_1(\lambda_n b) - aJ_1(\lambda_n a)]}{\lambda_n(b^2 - a^2)} \tag{3.108}$$

while the mean contact or source plane temperature is found by integrating over the entire source plane area:

$$\overline{T}_{cp} = \frac{1}{\pi b^2}\int_0^c T(r,0)2\pi r \, dr = A_0 \tag{3.109}$$

Defining the spreading resistance as before using $R_s = (\overline{T}_s - \overline{T}_{cp})/Q$, we obtain

$$R_s = \frac{4}{\pi(b^2 - a^2)^2}\sum_{n=1}^{\infty} \frac{[bJ_1(\lambda_n b) - aJ_1(\lambda_n a)]^2}{k\lambda_n(\lambda_n c)^2 J_0^2(\lambda_n c)} \tag{3.110}$$

Finally, if we define $\psi = 4kbR_s$ and introduce $\epsilon = b/c$ and $\alpha = (b-a)/b$, we obtain

$$\psi = \frac{16/\pi}{\epsilon\alpha^2(2-\alpha)^2}\sum_{n=1}^{\infty} \frac{[J_1(\delta_n\epsilon) - (1-\alpha)J_1(\delta_n\epsilon(1-\alpha))]^2}{(\delta_n)^3 J_0^2(\delta_n)} \tag{3.111}$$

The above equation reduces to that of the circular spot of radius $b$, when $\alpha = 1$, i.e. $a \to 0$. The above solution may be modified for a finite disk or multilayered disk by introducing the appropriate spreading functions $\phi$ in the denominator of the summation in Eq. (3.111). Yovanovich (1975) developed a general expression for an arbitrarily specified axi-symmetric flux distribution. For the special case of a uniform heat flux, his solution reduces to that obtained by Yip (1969).

## 3.6 Flux Tubes and Disks with Edge Conductance

We are now interested in an application of the solution for a finite disk thermal spreader with edge conductance as shown in Figure 3.5. The addition of edge conductance is quite useful in thermal spreader applications where fins are attached to the edge of a thick spreader which are then air cooled with a fan. This is an extension of the problem covered earlier in Section 3.3. This leads to the following revised problem statement [Yovanovich (2003)]:

$$\frac{\partial^2\theta}{\partial r^2} + \frac{1}{r}\frac{\partial\theta}{\partial r} + \frac{\partial^2\theta}{\partial z^2} = 0 \tag{3.112}$$

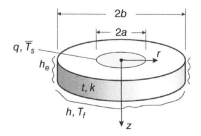

**Figure 3.5** Isotropic spreader with edge conductance.

which is subject to

$$\frac{\partial \theta}{\partial r}\Big|_{r=0} = 0$$

$$-k \frac{\partial \theta}{\partial r}\Big|_{r=b} = h_e \theta(b, z)$$

(3.113)

along the axis and edges of the heat spreader. In the source plane, we specify as before

$$\left.\begin{array}{rl} \frac{\partial \theta}{\partial z}\Big|_{z=0} &= -\frac{q(r)}{k}, \ 0 \leq r < a \\[2ex] \frac{\partial \theta}{\partial z}\Big|_{z=0} &= \quad 0, \quad a < r \leq b \end{array}\right\}$$

(3.114)

where

$$q(r) = q_0(1 + \mu)(1 - r^2/a^2)^\mu$$

(3.115)

Finally, in the sink plane, we require the following condition:

$$-k \frac{\partial \theta}{\partial z}\Big|_{z=t} = h_s \theta(r, t)$$

(3.116)

A general solution may be found using separation of variables such that $\theta(r, z) = R(r) * Z(z)$. Separating variables gives a general solution to Laplace's equation of

$$\theta(r, z) = [A_1 J_0(\lambda r) + B_1 Y_0(\lambda r)][A_2 \cosh(\lambda z) + B_2 \sinh(\lambda z)]$$

(3.117)

The uniform flow solution is absent since the applied edge cooling makes the problem multidimensional. In the case of vanishingly small but nonzero edge conductance, i.e. $h_e \to 0$, but $h_e \neq 0$, the first term of the series solution can be shown to approach the one-dimensional bulk resistance of the system.

The two-dimensional spreading field can now be solved by applying the boundary conditions. The symmetry condition along the axis of the disk $r = 0$ requires that the constant $B_1 = 0$ due to the infinite nature of the Bessel function $Y_0(0)$. The solution becomes

$$\theta(r, z) = J_0(\lambda r)[A_2 \cosh(\lambda z) + B_2 \sinh(\lambda z)]$$

(3.118)

with $A_1$ absorbed into the other remaining constants. Application of the edge condition requires that

$$\delta_n J_1(\delta_n) = \frac{h_e b}{k} J_0(\delta_n)$$

(3.119)

or

$$\delta_n J_1(\delta_n) = Bi_e J_0(\delta_n)$$

(3.120)

where $\delta_n = \lambda_n b$ as before. The above equation may be easily solved for any value of the Biot number $Bi_e = h_e b/k$.

The solution now becomes

$$\theta(r, z) = \sum_{n=1}^{\infty} J_0(\lambda_n r)[A_2 \cosh(\lambda_n z) + B_2 \sinh(\lambda_n z)]$$

(3.121)

Along the sink plane, we write

$$-k\sum_{n=1}^{\infty}J_0(\lambda_n r)[A_2\sinh(\lambda_n t)\lambda_n + B_2\cosh(\lambda t)\lambda_n]$$

$$= h_s\sum_{n=1}^{\infty}J_0(\lambda_n r)[A_2\cosh(\lambda_n t) + B_2\sinh(\lambda_n t)] \tag{3.122}$$

This leads to the following result:

$$B_2 = -A_2\phi \tag{3.123}$$

where $\phi$ may be compactly written as

$$\phi = -\frac{B_2}{A_2} = \frac{\lambda_n\tanh(\lambda_n t) + \dfrac{h_s}{k}}{\lambda_n + \dfrac{h_s}{k}\tanh(\lambda_n t)} \tag{3.124}$$

or more conveniently,

$$\phi = -\frac{B_2}{A_2} = \frac{\delta_n\tanh(\delta_n\tau) + Bi}{\delta_n + Bi\tanh(\delta_n\tau)} \tag{3.125}$$

where $\delta_n = \lambda_n b$, $\tau = t/b$, and $Bi = h_s b/k$.

The solution for the temperature field becomes

$$\theta(r,z) = \sum_{n=1}^{\infty}A_2 J_0(\lambda_n r)[\cosh(\lambda_n z) - \phi\sinh(\lambda_n z)] \tag{3.126}$$

Application of the final piecewise boundary condition in the source plane requires

$$\left.\frac{\partial\theta}{\partial z}\right|_{z=0} = \sum_{n=1}^{\infty}J_0(\lambda_n r)[-A_2\phi\lambda_n] = \begin{cases} -q(r)/k & 0 \le r < a \\ 0 & a < r \le b \end{cases} \tag{3.127}$$

Using the general flux distribution, we can find the Fourier coefficient $A_2$ as before. The resulting expression for the spreading (or constriction) resistance in the finite disk then accounts for all of the multidimensional effects for a uniform or variable flux heat source and the edge conductance. In this case, we have

$$4kaR_{total} = \frac{16(\mu+1)}{\pi\epsilon}\sum_{n=1}^{\infty}\frac{D_n}{\phi}\frac{J_1(\delta_n\epsilon)}{\delta_n^3 J_0^2(\delta_n)} \tag{3.128}$$

where $D_n$ are defined according to Eq. (3.65) with the eigenvalues $\delta_n$ computed from Eq. (3.120).

For the special case of $Bi_e \to 0$, i.e. adiabatic edges, the total resistance is the sum of the one-dimensional bulk resistance and the spreading resistance. It can be shown that the first term in the series of Eq. (3.128) represents the bulk resistance as $Bi_e \to 0$. Hence, the spreading resistance portion becomes

$$4kaR_s = 4kaR_{total} - 4kaR_{1D} \tag{3.129}$$

where

$$4kaR_{1D} = \frac{4\epsilon}{\pi}\left[\tau + \frac{1}{Bi}\right] \tag{3.130}$$

Additional problems with edge conductance have been solved in Section 4.6 for rectangular spreaders. In those cases, the same limiting behavior above is also observed.

## 3.7 Spreading Resistance for an Eccentric Source on a Flux Tube

The problem of an eccentric circular isoflux heat source on a semi-infinite flux tube was considered by Bairi and Laraqi (2004). They applied Fourier Cosine and Hankel transforms to obtain a solution for the spreading resistance. The problem is defined as follows:

$$\frac{\partial^2 T}{\partial r^2} + \frac{1}{r}\frac{\partial T}{\partial r} + \frac{1}{r^2}\frac{\partial^2 T}{\partial \theta^2} + \frac{\partial^2 T}{\partial z^2} = 0 \tag{3.131}$$

which is subject to the following symmetry conditions in the $\theta$-direction:

$$\begin{aligned}
\left.\frac{\partial T}{\partial \theta}\right|_{\theta=0} &= 0 \\
\left.\frac{\partial T}{\partial \theta}\right|_{\theta=\pi} &= 0
\end{aligned} \tag{3.132}$$

while in the $r$-direction, the following are applied:

$$\begin{aligned}
T(0,\theta,z) &\neq \infty \\
\left.\frac{\partial T}{\partial r}\right|_{r=b} &= 0
\end{aligned} \tag{3.133}$$

Over the top surface of the flux tube $z = 0$,

$$\left.\begin{aligned}
\left.\frac{\partial T}{\partial z}\right|_{z=0} &= -\frac{q}{k}, \quad \text{Inside the Source} \\
\left.\frac{\partial T}{\partial z}\right|_{z=0} &= 0, \quad \text{Outside the Source}
\end{aligned}\right\} \tag{3.134}$$

where the source is located eccentrically off the main axis of the flux tube with its centroid displaced from the centroid of the flux tube by a distance $e$.

Finally, as $z \to \infty$, we prescribe $T(r,\theta,z \to \infty) = 0$. We only summarize the final expression for the spreading temperature field and the spreading resistance as the solution is quite involved requiring an additional co-ordinate transformation to complete the required integrations over the source area. Bairi and Laraqi (2004) obtained the following result for the temperature field:

$$\begin{aligned}
T(r,\theta,z) = \frac{2qa}{k}\Bigg[ &\sum_{n=1}^{\infty}\frac{J_0(\beta_n r)J_1(\beta_n a)J_0(\beta_n e)\exp(-\beta_n z)}{(\beta_n b)^2 J_0^2(\beta_n b)} \\
&+ \sum_{m=1}^{\infty}\sum_{n=1}^{\infty}\frac{J_m(\beta_n r)J_1(\beta_n a)J_m(\beta_n e)\cos(m\theta)\exp(-\beta_n z)}{[(\beta_n b)^2 - m^2]J_m^2(\beta_n b)}\Bigg]
\end{aligned} \tag{3.135}$$

In the abovementioned solution, $e$ is the eccentricity (distance of source centroid from flux tube centroid) and $\beta_n$ are the eigenvalues determined from

$$J_m'(\beta_n b) = 0 \tag{3.136}$$

In the abovementioned solution, the eigenvalues in the single summation are determined from the case when $m = 0$, i.e. $J_0'(\beta_n b) = J_1(\beta_n b) = 0$. As such, it can be seen that the solution is computationally challenging when evaluating the double summation as many sets of eigenvalues are required. Bairi and Laraqi (2004) also reported the mean source temperature and the spreading resistance as follows:

$$\overline{T}_s = \frac{4qb}{\pi k}\Bigg[\sum_{n=1}^{\infty}\frac{J_1^2(\beta_n a)J_0^2(\beta_n e)}{(\beta_n b)^3 J_0^2(\beta_n b)} + \sum_{m=1}^{\infty}\sum_{n=1}^{\infty}\frac{J_1^2(\beta_n a)J_m^2(\beta_n e)}{(\beta_n b)[(\beta_n b)^2 - m^2]J_m^2(\beta_n b)}\Bigg] \tag{3.137}$$

$$R_s = \frac{4b}{k\pi^2 a^2} \left[ \sum_{n=1}^{\infty} \frac{J_1^2(\beta_n a)J_0^2(\beta_n e)}{(\beta_n b)^3 J_0^2(\beta_n b)} + \sum_{m=1}^{\infty}\sum_{n=1}^{\infty} \frac{J_1^2(\beta_n a)J_m^2(\beta_n e)}{(\beta_n b)[(\beta_n b)^2 - m^2]J_m^2(\beta_n b)} \right] \tag{3.138}$$

In order to simplify computation, Bairi and Laraqi (2004) developed several simpler correlations using results generated from the solution above. The authors define their constriction parameter as $\psi = k\sqrt{A}R_s$, where $A$ is the area of the heat source. They further define two basic constriction parameters for the case of zero eccentricity and maximum eccentricity. For the case of zero eccentricity, we have

$$\psi_0 = \frac{4b}{\pi^{3/2}a} \sum_{n=1}^{\infty} \frac{J_1^2(\beta_n a)}{(\beta_n b)^3 J_0^2(\beta_n b)} \tag{3.139}$$

while for maximum eccentricity, they provided the following correlation:

$$\frac{\psi_{max}}{\psi_0} = 1.5816\epsilon^{0.0528} \tag{3.140}$$

where $\epsilon = a/b$ is the relative source size. The correlation using Eq. (3.32) may be used to calculate $\psi_0$ for the uniform flux heat source.

Finally, the general constriction parameter $\psi^q$, for any contact size or eccentricity is given by

$$\frac{\psi^q}{\psi_0} = 1 + [1.5816\epsilon^{0.0528} - 1]\left(\frac{e^*}{1-\epsilon}\right)^{1.76}\left(\frac{\epsilon}{1-e^*}\right)^{0.88} \tag{3.141}$$

where $e^* = e/b$ is the dimensionless eccentricity.

For the case of an isothermal eccentric contact, no solution is known. However, Cooper (1969) provided an expression based on electrolytic analogue experiments. The resulting expression can be cast as

$$4kaR_s = \psi^T + \frac{6\epsilon}{2\pi}\left(\frac{e}{b}\right)^2 \tag{3.142}$$

where $\psi^T$ is the spreading parameter for an isothermal coaxial source.

## 3.8 Thermal Spreading with Variable Conductivity Near the Contact Surface

In many applications, where two different metals are in contact, a diffusion zone may form in each material over time, whereby the thermal conductivity of each material in contact varies with position near the contact plane. This diffusion zone may be either resistive or conductive depending upon the materials in contact as shown in Figure 3.6 for a Nickel–Copper interface. In this case, the diffusion layer leads to a complex distribution of thermal conductivity transitioning from one region to the other across the plane of contact. In the Nickel–Copper case shown, the diffusion layer in each region near the interface is a resistive. We may also have a conductive layer in cases where a thin layer of a more conductive material is deposited by means of a diffusion process. Modeling thermal spreading/constriction resistance in these cases can be difficult, but if each region is approximated as having a linearly varying thermal conductivity measured from the plane of contact, a simpler solution can be obtained. Negus et al. (1987) considered the problem shown in Figure 3.7 for both conductive and resistive layers of linearly varying thermal conductivity.

The solution process is similar to that in previously considered problems with some modifications. The domain considers two regions, a layer of linearly varying thermal conductivity, and a

**Figure 3.6** Variation of thermal conductivity for a diffused Nickel–Copper Interface. Source: Adapted from Negus et al. (1987)/American Society of Mechanical Engineers.

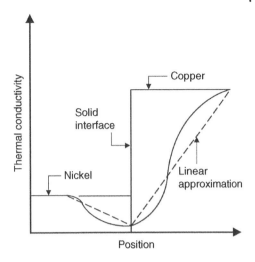

**Figure 3.7** Flux tube with a linearly varying thermal conductivity layer. Source: Adapted from Negus et al. (1987)/American Society of Mechanical Engineers.

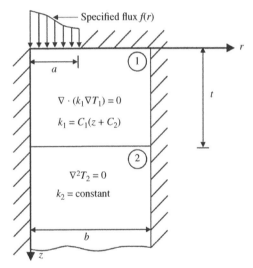

substrate of constant thermal conductivity. The conductivity is defined as

$$k_1 = C_1(z + C_2) \quad 0 < z < t$$
$$k_2 = C_1(t + C_2) \quad t < z < \infty$$

(3.143)

whereby the surface conductivity is denoted as $k_0 = C_1 C_2$ when $z = 0$. The problem requires solving

$$\nabla \cdot (k_1 \nabla T_1) = \frac{\partial^2 T_1}{\partial r^2} + \frac{1}{r}\frac{\partial T_1}{\partial r} + \frac{\partial^2 T_1}{\partial z^2} + \frac{1}{k_1}\frac{\partial k_1}{\partial z}\frac{\partial T_1}{\partial z} = 0 \quad 0 < z < t$$

$$\nabla^2 T_2 = \frac{\partial^2 T_2}{\partial r^2} + \frac{1}{r}\frac{\partial T_2}{\partial r} + \frac{\partial^2 T_2}{\partial z^2} = 0 \quad\quad\quad t < z < \infty$$

(3.144)

which is subject to

$$\left.\frac{\partial T_1}{\partial r}\right|_{r=0} = \left.\frac{\partial T_2}{\partial r}\right|_{r=0} = 0 \quad z \geq 0$$

$$\left.\frac{\partial T_1}{\partial r}\right|_{r=b} = \left.\frac{\partial T_2}{\partial r}\right|_{r=b} = 0 \quad z \geq 0$$

(3.145)

at the centroid and edge of the flux tube, respectively. Over the top surface $z = 0$, we apply:

$$\left.\frac{\partial T_1}{\partial z}\right|_{z=0} = -\frac{q(r)}{k_0}, \quad 0 < r \le a \\ \left.\frac{\partial T_1}{\partial z}\right|_{z=0} = 0, \quad a < r < b \right\}$$

(3.146)

where

$$q(r) = (1 + \mu)\frac{Q}{\pi a^2}\left(1 - \frac{r^2}{a^2}\right)^{\mu}$$

(3.147)

At the interface $z = t$, we apply

$$k_1\left.\frac{\partial T_1}{\partial z}\right|_{z=t} = k_2\left.\frac{\partial T_2}{\partial z}\right|_{z=t}, \quad 0 < r < b \\ T_1(r, t) = T_2(r, t), \quad 0 < r < b \right\}$$

(3.148)

and finally, at $z \to \infty$, we prescribe the uniform flow:

$$-k_2\frac{\partial T_2}{\partial z} = Q/\pi b^2, \quad 0 \le r \le b$$

(3.149)

in a region located at a significant distance from the contact plane. Negus et al. (1987) used the separation of variables method to solve the above problem for both a resistive and conductive layer. In Eq. (3.143), the constant $C_2$ determines whether a resistive or conductive layer is present. If $C_2 > 0$, a resistive layer is present with conductivity increasing with depth to its maximum value in the substrate. For a conductive layer, we must have $C_2 < 0$ and larger than $z$ in magnitude leading to conductivity decreasing with depth to its minimum value in the substrate. This means that the extra term in Laplace's equation (see Eq. (3.144)) for the first layer changes sign depending upon whether it is resistive or conductive, and the solution process will produce different solutions. In the matter of brevity, we will only report the final results for each case. The reader is directed to Negus et al. (1987) for more details pertaining to the actual solution.

The general solution for the constriction parameter $\psi = 4k_0 a R_s$, for the case of $\mu > -1$ is

$$\psi = \frac{16}{\pi \epsilon}(1 + \mu)2^{\mu}\Gamma(\mu + 1)\sum_{n=1}^{\infty}\frac{\theta_n J_1(\delta_n \epsilon)J_{\mu+1}(\delta_n \epsilon)}{\delta_n^3(\delta_n \epsilon)^{\mu}J_0^2(\delta_n)}$$

(3.150)

where, as before, the eigenvalues $\delta_n = \lambda_n b$ are the zeroes of $J_1(\lambda_n b) = 0$.

The following special cases for uniform heat flux ($\mu = 0$) and equivalent isothermal ($\mu = -1/2$) are determined from the general case above

$$\psi^q = \frac{16}{\pi \epsilon}\sum_{n=1}^{\infty}\frac{\theta_n J_1^2(\delta_n \epsilon)}{\delta_n^3 J_0^2(\delta_n)}, \quad \mu = 0$$

(3.151)

and

$$\psi_{ei}^T = \frac{8}{\pi \epsilon}\sum_{n=1}^{\infty}\frac{\theta_n J_1(\delta_n \epsilon)\sin(\delta_n \epsilon)}{\delta_n^3 J_0^2(\delta_n)}, \quad \mu = -1/2$$

(3.152)

In the abovementioned equations, the parameter $\theta_n$ depends on whether a resistive or conductive layer is present in the first layer. Additionally, we define the following dimensionless groups: $\epsilon = a/b$, $\tau = t/a$, and $\kappa = k_0/k_2$. The specific expressions are

**Resistive Layer**

$$\theta_n = \frac{K_0(\lambda_n C_2) + \phi_n I_0(\lambda_n C_2)}{K_1(\lambda_n C_2) - \phi_n I_1(\lambda_n C_2)}$$

(3.153)

where

$$\phi_n = \frac{K_1(\lambda_n(t + C_2)) - K_0(\lambda_n(t + C_2))}{I_1(\lambda_n(t + C_2)) + I_0(\lambda_n(t + C_2))} \tag{3.154}$$

In the abovementioned equations, it can be shown that the following arguments can be presented in terms of nondimensional groups in the following manner:

$$\lambda_n(t + C_2) = \frac{\delta_n \tau \epsilon}{1 - \kappa}$$

$$\lambda_n C_2 = \frac{\delta_n \tau \epsilon \kappa}{1 - \kappa} \tag{3.155}$$

**Conductive Layer**

$$\theta_n = \frac{K_0(-\lambda_n C_2) + \phi_n I_0(-\lambda_n C_2)}{\phi_n I_1(-\lambda_n C_2) - K_1(-\lambda_n C_2)} \tag{3.156}$$

where

$$\phi_n = \frac{K_1(-\lambda_n(t + C_2)) + K_0(-\lambda_n(t + C_2))}{I_1(-\lambda_n(t + C_2)) - I_0(-\lambda_n(t + C_2))} \tag{3.157}$$

In the abovementioned equations, it can be shown that the following arguments can be presented in terms of nondimensional groups in the following manner:

$$-\lambda_n(t + C_2) = \frac{\delta_n \tau \epsilon}{\kappa - 1}$$

$$-\lambda_n C_2 = \frac{\delta_n \tau \epsilon \kappa}{\kappa - 1} \tag{3.158}$$

Finally, in the above expressions, $(I_0, I_1)$ and $(K_0, K_1)$ are the modified Bessel functions of the first and second kind of orders zero and one. The above solutions are useful in thermal contact problems where either dis-similar materials are in contact or where a layer of material is deposited onto a substrate resulting in a diffusion layer of varying thermal conductivity.

## 3.9 Effect of Surface Curvature on Thermal Spreading Resistance in a Flux Tube

The effect of a nonflat contact plane was examined by Mayer et al. (2019) who examined thermal spreading effects in super-hydrophobic channels composed of cylindrical pillars. Further details of these surfaces are provided in Chapter 9. Here, we only consider the extension of the classical cylindrical flux tube for an isothermal heat source. Mayer et al. (2019) analytically solved the problem illustrated in Figure 3.8.

They also calculated numerical results as a validation of their complex solution. The interested reader is referred to original paper for most of the details as the solution is quite complex and rather involved. We only present the final results for illustration and reference. Figures 3.9 and 3.10 provide the dimensionless thermal spreading resistance defined as

$$\tilde{R}''_{sp} = \frac{k}{b}\left(\frac{T_s - \overline{T}_{cp}}{Q}\right) \tag{3.159}$$

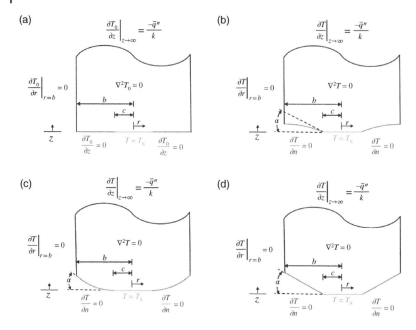

**Figure 3.8** Thermal spreading from a curved boundary: (a) flat contact, (b) nonflat contact with concave surface, (c) nonflat contact with convex surface, and (d) nonflat contact with straight surface. Source: Mayer et al. (2019). Used with the permission of the American Society of Mechanical Engineers (ASME).

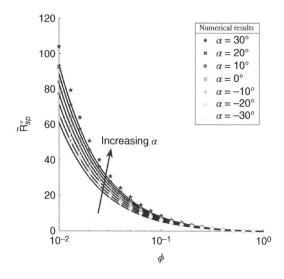

**Figure 3.9** Dimensionless thermal spreading resistance, theory, and numerical results, versus contact ratio for selected angles. Source: After Mayer et al. (2019). Image courtesy of M. Mayer.

which is plotted versus the contact ratio $\phi = A_s/A_t$. Note, in Section 3.1, we use $\epsilon = \sqrt{\phi}$ and we also nondimensionalized the thermal spreading resistance using the source dimension rather than the flux tube dimension. As can be seen in Figures 3.9 and 3.10, significant deviation from the flat surface result is observed for all contact angles of the adiabatic boundary region.

**Figure 3.10** Dimensionless thermal spreading resistance, theory, and numerical results, versus contact ratio for typical contact ratios and angles. Source: After Mayer et al. (2019). Image courtesy of M. Mayer.

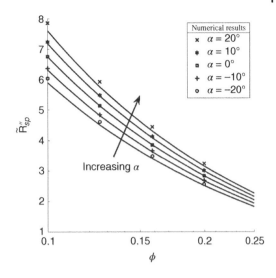

# References

Bagnall, K., Muzychka, Y.S., and Wang, E., "Analytical Solution for Temperature Rise in Complex Multilayer Structures with Discrete Heat Sources", *IEEE Transactions on Components, Packaging, and Manufacturing Technologies*, Vol. 4, no. 5, pp. 817–830, 2014.

Bairi, A., and Laraqi, N., "The Thermal Constriction Resistance for an Eccentric Spot on a Circular Heat Flux Tube", *Journal of Heat Transfer*, Vol. 126, pp. 652–655, 2004.

Chow, Y.L., and Yovanovich, M.M., "The Shape Factor of the Capacitance of a Conductor", *Journal of Applied Physics*, Vol. 53, no. 12, pp. 8470–8475, 1982.

Cooper, M.G., "A Note on Electrolytic Analogue Experiments for Thermal Contact Resistance", *International Journal of Heat and Mass Transfer*, Vol. 12, pp. 1715–1718, 1969.

Gibson, R.D., "The Contact Resistance for a Semi-infinite Cylinder in a Vacuum", *Journal of Applied Energy*, Vol. 2, pp. 57–65, 1976.

Kennedy, D.P., "Spreading Resistance in Cylindrical Semi-conductor Devices", *Journal of Applied Physics*, Vol. 31, no. 8, pp. 1490–1497, 1960.

Lee, S., Song, S., Au, V., and Moran, K.P., "Constriction/Spreading Resistance Model for Electronics Packaging", *Proceedings of the ASME/JSME Thermal Engineering Conference*, Vol. 4, pp. 199–206, 1995.

Mayer, M., Hodes, M., Kirk, T., and Crowdy, D., "Effect of Surface Curvature on Contact Resistance Between Cylinders", *Journal of Heat Transfer*, Vol. 141, pp. 032002-1–032002-12, 2019.

Mikic, B., and Rohsenow, W.M., "Thermal Contact Conductance", MIT Report, DSR 74542-41, 1966.

Muzychka, Y.S., "Spreading Resistance in Compound Orthotropic Flux Tubes and Channels with Interfacial Resistance", *Journal of Thermophysics and Heat Transfer*, doi: http://arc.aiaa .org/doi/abs/10.2514/1.T4203, 2014.

Muzychka, Y.S., "Thermal Spreading Resistance in a Multilayered Orthotropic Circular Disk with Interfacial Resistance and Variable Flux", *Proceedings of InterPack 2015*, San Francisco, CA, USA, July 6–9, 2015.

Muzychka, Y.S., Bagnall, K., and Wang, E., "Thermal Spreading Resistance and Heat Source Temperature in Compound Orthotropic Systems with Interfacial Resistance", *IEEE Transactions on Components, Packaging, and Manufacturing Technologies*, Vol. 3, no. 11, pp. 1826–1841, 2013.

Muzychka, Y.S., Sridhar, M.R., Yovanovich, M.M., and Antonetti, V.W., "Thermal Constriction Resistance in Multilayered Contacts: Applications in Contact Resistance", *Journal of Thermophysics and Heat Transfer*, Vol. 13, pp. 489–494, 1999.

Muzychka, Y.S., Yovanovich, M.M., and Culham, J.R., "Thermal Spreading Resistances in Compound and Orthotropic Systems", *Journal of Thermophysics and Heat Transfer*, Vol. 18, pp. 45–51, 2004.

Negus, K.J., and Yovanovich, M.M., "Constriction Resistance of Circular Flux Tubes with Mixed Boundary Conditions by Linear Superposition of Neumann Solutions", *ASME-84-HT-84*, ASME, New York, 1984a.

Negus, K.J., and Yovanovich, M.M., "Application of the Method of Optimized Images to Steady Three-Dimensional Conduction Problems", *ASME-84-WA/HT-110*, ASME, New York, 1984b.

Negus, K.J., Vanoverbeke, C.A., and Yovanovich, M.M., "Thermal Constriction Reistance with Variable Thermal Conductivity Near the Contact Surface", *24th National Heat Transfer Conference and Exhibition*, Pittsburgh, PA, USA, August 9–12, ASME HTD-Vol. 69, pp. 91–98, 1987.

Negus, K.J., Yovanovich, M.M., and Beck, J.V., "On the Non-dimensionalization of Constriction Resistance for Semi-infinite Heat Flux Tubes", *Journal of Heat Transfer*, Vol. 111, pp. 804–807, 1989.

Nelson, G.J., and Sayers, W.A., "A Comparison of Two-Dimensional Planar, Axi-symmetric and Three-Dimensional Spreading Resistance", *Proceedings of the 8th Annual IEEE Semi-conductor Thermal Measurement and Management Symposium*, pp. 62–68, 1992.

Roess, L.C., "Theory of Spreading Conductance", Appendix A of Report of the Beacon Laboratories of Texas Company, Beacon, NY, 1950 (as given in Yovanovich 1975).

Song, S., Lee, S., and Au, V., "Closed-Form Equation for Thermal Constriction/Spreading Resistances with Variable Resistance Boundary Condition", *Proceedings of the 1994 IEPS Conference*, Atlanta, GA, USA, pp. 111–121, 1995.

Yip, F.C., Thermal Contact-Constriction of Resistance. PhD Thesis, *University of Calgary*, 1969.

Yovanovich, M.M., "General Expression for Circular Constriction Resistances for Arbitrary Heat Flux Distribution", in *Progress in Astronautics and Aeronautics: Radiative Transfer and Thermal Control*, Vol. 49, , AIAA, New York, pp. 301–308, 1975.

Yovanovich, M.M., "General Thermal Constriction Resistance Parameter for Annular Contacts on Circular Flux Tubes", *AIAA Journal*, Vol. 14, pp. 822–824, 1976.

Yovanovich, M.M., "Chapter 8: Theory and Applications of Constriction and Spreading Resistance Concepts for Microelectronic Thermal Management", in *Advances in Cooling Techniques for Computers*, Hemisphere Publishing, 1988.

Yovanovich, M.M., "Chapter 3: Conduction and Thermal Contact Resistance (Conductance)", in *Handbook of Heat Transfer*, eds. W.M. Rohsenow, J.P. Hartnett, and Y.L. Cho, McGraw-Hill, New York, 1998.

Yovanovich, M.M., "Thermal Resistance of Circular Source on Finite Circular Cylinder with Side and End Cooling", *Journal of Electronic Packaging*, Vol. 125, pp. 169–177, 2003.

Yovanovich, M.M., and Antonetti, V.W., "Application of Thermal Contact Resistance Theory to Electronic Packages", *Advances in Thermal Modeling of Electronic Components and Systems*, Vol. 1, eds. A. Bar-Cohen and A.D. Kraus, Hemisphere Publishing, New York, pp. 79–128, 1988.

Yovanovich, M.M., and Burde, S.S., "Centroidal and Area Average Resistances of Non-symmetric, Singly Connected Contacts", *AIAA Journal*, Vol. 15, no. 10, pp. 1523–1525, 1977.

Yovanovich, M.M., and Marotta, E. "Chapter 4: Thermal Spreading and Contact Resistances", *Heat Transfer Handbook*, eds. A. Bejan and A.D. Kraus, Wiley, 2003.

Yovanovich, M.M., Culham, J.R., and Teertstra, P.M., "Analytical Modeling of Spreading Resistance in Flux Tubes, Half Spaces, and Compound Disks", *IEEE Transactions on Components, Packaging, and Manufacturing Technology - Part A*, Vol. 21, pp. 168–176, 1998.

Yovanovich, M.M., Tien, C.H., and Schneider, G.E., "General Solution of Constriction Resistance within a Compound Disk", in *Progress in Astronautics and Aeronautics: Heat Transfer, Thermal Control, and Heat Pipes*, MIT Press, Cambridge, MA, pp. 47–62, 1980.

# 4

# Rectangular Flux Channels

In this chapter, we focus on applications of the rectangular flux channel in two or three dimensions. Significant effort in this area has been achieved in the research literature due the far-reaching applications.

Typical applications include cooling of electronic devices both at the package and board level, and cooling of power semi-conductors using heat sinks. In these applications, heat dissipated by electronic devices is conducted through electronic packages into printed circuit boards or heat sink baseplates which are convectively cooled. In heat sink applications, the convective film resistance is replaced by an effective extended surface conductance which accounts for the thermal resistance of the fins attached to the heat spreader.

Analytical results for a symmetrically located heat source or an eccentrically located heat source on a finite rectangular flux channel have been obtained for isotropic, compound, and multilayered systems. Solutions have also been extended for orthotropic systems but will be discussed in Chapter 5. These solutions may be used to model single or multisource systems by means of superposition. This approach differs from other methods in that the heat source specification is incorporated in the definition of the thermal boundary conditions rather than in the governing partial differential equation [Ellison (1984, 2003)]. This results in analytical expressions which can be easily manipulated in most advanced mathematical modeling packages such as Maple or Matlab.

These fundamental solutions can also be used to model embedded heat sources or sources on a submount connected to a larger thermal spreader using simple modifications of the theory and appropriate energy balances.

We will begin with the simplest cases of the two- and three-dimensional semi-infinite flux channels of Mikic and Rohsenow (1966). We then consider a two-dimensional eccentric strip heat source for a semi-infinite flux channel which was obtained by Veziroglu and Chandra (1969). The solution for a centrally located heat source on a finite isotropic rectangular plate connected to an isothermal sink was obtained by Krane (1991). The authors [Yovanovich et al. (1999)] obtained a solution for a centrally located heat source on a finite compound rectangular flux channel and later obtained a more general solution for an eccentric heat source [Muzychka et al. (2003)] on a compound rectangular flux channel. In the cases where edge conductance is present, Muzychka et al. (2004) obtained a general result for an isotropic rectangular spreader. Finally, Bagnall et al. (2014) obtained a more general solution for a multilayered rectangular spreader. This result allows earlier solutions to be extended more broadly through the use of spreading functions.

In all of the above cases, each solution represents an extension of a previous result, much the same as in Chapter 3 dealing with the cylindrical flux tubes and disks. We will address each carefully highlighting similarities between the flux tube solutions and flux channel solutions.

*Thermal Spreading and Contact Resistance: Fundamentals and Applications*, First Edition.
Yuri S. Muzychka and M. Michael Yovanovich.
© 2023 John Wiley & Sons, Inc. Published 2023 by John Wiley & Sons, Inc.

## 4.1 Two-Dimensional Semi-Infinite Flux Channel

Mikic and Rohsenow (1966) first considered a simple strip heat source in two dimensions. Beginning with the following simple problem statement (later we will extend to three dimensions), we must solve:

$$\nabla^2 T = \frac{\partial^2 T}{\partial x^2} + \frac{\partial^2 T}{\partial z^2} = 0 \tag{4.1}$$

subject to the following boundary conditions:

$$\left.\frac{\partial T}{\partial x}\right|_{x=0} = 0$$
$$\left.\frac{\partial T}{\partial x}\right|_{x=c} = 0 \tag{4.2}$$

at the centroid and edge of the channel, respectively. Over the top surface $z = 0$,

$$\left.\begin{array}{rcl}
-k\left.\dfrac{\partial T}{\partial z}\right|_{z=0} &=& \dfrac{Q'}{2a}, \quad 0 < x < a \\[2mm]
\left.\dfrac{\partial T}{\partial z}\right|_{z=0} &=& 0, \quad a < x < c
\end{array}\right\} \tag{4.3}$$

Finally, at $z \to \infty$, we prescribe the uniform flow:

$$-k\frac{\partial T}{\partial z} = \frac{Q'}{2c}, \quad z \to \infty \tag{4.4}$$

in a region located at a significant distance from the contact plane. $Q' = Q/L$ is the heat input per unit depth into the y-direction. Later, when we consider the three-dimensional spreader $Q' = Q/2d$, where $L = 2d$ is the depth of the channel into the y-direction as shown in Figure 4.1.

The solution can be found using separation of variables assuming that $T(x, z) = X(x) \cdot Z(z)$. The general solution takes the form of

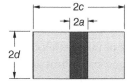

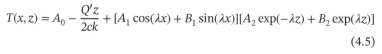

$$T(x, z) = A_0 - \frac{Q'z}{2ck} + [A_1 \cos(\lambda x) + B_1 \sin(\lambda x)][A_2 \exp(-\lambda z) + B_2 \exp(\lambda z)] \tag{4.5}$$

The symmetry condition at $x = 0$ and the requirement that the solution remain finite as $z \to \infty$ requires $B_1 = 0$ and $B_2 = 0$, respectively. This allows us to write the solution more simply as

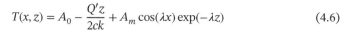

$$T(x, z) = A_0 - \frac{Q'z}{2ck} + A_m \cos(\lambda x) \exp(-\lambda z) \tag{4.6}$$

where $A_m$ is the combined constant of $A_1$ and $A_2$.

Application of the edge boundary condition at $x = c$ gives

$$A_m \lambda \sin(\lambda c) = 0 \tag{4.7}$$

Since neither $\lambda = 0$ nor $A_m = 0$, we choose to define the eigenvalues such that

$$\sin(\lambda_m c) = 0 \tag{4.8}$$

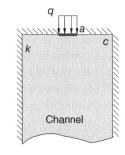

**Figure 4.1** Two dimensional semi-infinite flux channel.

where $\lambda_m c = m\pi$ or $\lambda_m = m\pi/c$. The solution now becomes

$$T(x,z) = A_0 - \frac{Q'z}{2ck} + \sum_{m=1}^{\infty} A_m \cos(m\pi x/c) \exp(-m\pi z/c) \tag{4.9}$$

We may obtain the final constant by means of the boundary condition in the source plane:

$$\left.\frac{\partial T}{\partial z}\right|_{z=0} = -\frac{Q'}{2ck} - \sum_{m=1}^{\infty} A_m \cos(m\pi x/c)(m\pi/c) = \begin{cases} -\dfrac{Q'}{2ak} & 0 \le x < a \\ 0 & a < x \le c \end{cases} \tag{4.10}$$

As before, we take a Fourier expansion by multiplying through by $\cos(n\pi x/c)$ and integrating over the interval. After invoking the orthogonality property for $n = m$, we obtain

$$A_m = \frac{Q' \sin(m\pi a/c)}{k(a/c)(m\pi)^2} \tag{4.11}$$

The solution for the temperature in the two-dimensional spreader now becomes

$$T(x,z) = A_0 - \frac{Q'z}{2ck} + \sum_{m=1}^{\infty} \frac{Q' \sin(m\pi a/c) \cos(m\pi x/c) \exp(-m\pi z/c)}{k(a/c)(m\pi)^2} \tag{4.12}$$

As before we may find the mean temperature of the source by integrating the temperature over the source region:

$$\overline{T}_s = \frac{1}{a} \int_0^a T(x,0)dx = A_0 + \sum_{n=1}^{\infty} A_m \frac{\sin(m\pi a/c)}{(a/c)(m\pi)} \tag{4.13}$$

The mean contact or source plane temperature is found by integrating over the entire source plane area:

$$\overline{T}_{cp} = \frac{1}{c} \int_0^c T(x,0)dx = A_0 \tag{4.14}$$

Finally, in an arbitrary sink plane $z = \ell$, we find

$$\overline{T}_{z=\ell} = \frac{1}{c} \int_0^c T(x, z = \ell)dx = A_0 - \frac{Q'\ell}{2ck} \tag{4.15}$$

In all three cases, the constant $A_0$ is seen to be the mean contact plane temperature. Finally, the spreading (or constriction resistance) using $Q'$ is defined as

$$R_s = \frac{\overline{T}_s - \overline{T}_{cp}}{Q'} = \sum_{n=1}^{\infty} \frac{\sin^2(m\pi a/c)}{k(a/c)^2(m\pi)^3} \tag{4.16}$$

or defining a dimensionless constriction parameter using the depth $L$ of the flux channel, yields $\psi = kLR_s$:

$$\psi = kLR_s = \sum_{n=1}^{\infty} \frac{\sin^2(m\pi\epsilon)}{\epsilon^2(m\pi)^3} \tag{4.17}$$

where $\epsilon = a/c$ is the source aspect ratio.

### 4.1.1  Variable Heat Flux Distributions

If the flux is not uniform of the source region, a general heat flux distribution may be prescribed in the form [Yovanovich and Marotta (2003)]

$$q(x) = \frac{Q'\Gamma(\mu + 3/2)}{\sqrt{\pi}\,a^{1+2\mu}\,\Gamma(\mu + 1)}\left(a^2 - x^2\right)^{\mu}, \quad 0 \le x \le a \tag{4.18}$$

where $Q' = Q/L$ is the total heat transfer rate into the strip per unit depth, and $\Gamma(\cdot)$ is the Gamma function [Abramowitz and Stegun (1965)]. The parameter $\mu$ defines the heat flux distribution on the strip which may have the following values: (i) $\mu = -1/2$ to approximate an isothermal strip provided $a/c < 0.4$, (ii) $\mu = 0$ for an isoflux distribution, and (iii) $\mu = 1/2$ which gives a parabolic flux distribution. The three flux distributions of interest are

$$q(x) = \begin{cases} \dfrac{Q'}{\pi a}\dfrac{1}{\sqrt{1 - x^2/a^2}} & \text{for} \quad \mu = -1/2 \\[3mm] \dfrac{Q'}{2a} & \text{for} \quad \mu = 0 \\[3mm] \dfrac{2Q'}{\pi a}\sqrt{1 - x^2/a^2} & \text{for} \quad \mu = 1/2 \end{cases} \tag{4.19}$$

Using Eq. (4.18) in Eq. (4.10), the dimensionless spreading resistance relationship based on the mean source temperature is

$$kLR_s = \frac{\Gamma(\mu + 3/2)}{\pi^2 \epsilon}\sum_{m=1}^{\infty}\left[\frac{2}{m\pi\epsilon}\right]^{\mu+1/2}\frac{\sin(m\pi\epsilon)}{m^2}J_{\mu+1/2}(m\pi\epsilon) \tag{4.20}$$

The general relationship above gives the following three relationships for the specific flux distributions of interest

$$kLR_s = \begin{cases} \dfrac{1}{\epsilon\pi^2}\displaystyle\sum_{m=1}^{\infty}\dfrac{\sin(m\pi\epsilon)}{m^2}J_0(m\pi\epsilon) & \text{for} \quad \mu = -1/2 \\[4mm] \dfrac{1}{\epsilon^2\pi^3}\displaystyle\sum_{m=1}^{\infty}\dfrac{\sin^2(m\pi\epsilon)}{m^3} & \text{for} \quad \mu = 0 \\[4mm] \dfrac{2}{\epsilon^2\pi^3}\displaystyle\sum_{m=1}^{\infty}\dfrac{\sin(m\pi\epsilon)}{m^3}J_1(m\pi\epsilon) & \text{for} \quad \mu = 1/2 \end{cases} \tag{4.21}$$

The dimensionless spreading resistance for this problem depends on two parameters: (i) the relative size of the strip $\epsilon$, and (ii) the heat flux distribution parameter $\mu$.

Numerical values for $kLR_s$ for three values of $\mu$ are given in Table 4.1. From the tabulated values, it can be seen that the spreading resistance values for the isothermal strip are smaller than the values for the isoflux distribution which are smaller than the values for the parabolic distribution for all values of $\epsilon$. The differences are large as $\epsilon \to 1$; however, the differences become negligibly small as $\epsilon \to 0$. Also, unlike the case for a circular flux tube, there is no single limiting value as $\epsilon \to 0$. The behavior in this limit is logarithmic.

There is a closed-form relationship for the "true" isothermal area on an "infinitely" thick flux channel. According to Sexl and Burkhard (1969), Veziroglu and Chandra (1969) and Yovanovich

**Table 4.1** Dimensionless spreading resistance $kLR_s$ in flux channels.

| $\epsilon$ | $\mu$ | $\mu$ | $\mu$ | Isothermal | Abrupt change |
|------|--------|--------|--------|--------|--------|
|      | $-1/2$ | 0 | 1/2 | | |
| 0.01 | 1.321 | 1.358 | 1.375 | 1.322 | 1.343 |
| 0.1  | 0.5902 | 0.6263 | 0.6430 | 0.5905 | 0.6110 |
| 0.2  | 0.3729 | 0.4083 | 0.4247 | 0.3738 | 0.3936 |
| 0.3  | 0.2494 | 0.2836 | 0.2995 | 0.2514 | 0.2699 |
| 0.4  | 0.1658 | 0.1984 | 0.2134 | 0.1691 | 0.1860 |
| 0.5  | 0.1053 | 0.1357 | 0.1496 | 0.1103 | 0.1249 |
| 0.6  | 0.0607 | 0.0882 | 0.1007 | 0.0675 | 0.0794 |
| 0.7  | 0.0283 | 0.0521 | 0.0628 | 0.0367 | 0.0456 |
| 0.8  | 0.0066 | 0.0255 | 0.0338 | 0.0160 | 0.0214 |

et al. (1999), the relationship is

$$kLR_s = \frac{1}{\pi} \ln \left[ \frac{1}{\sin \left( \frac{\pi \epsilon}{2} \right)} \right] \tag{4.22}$$

Numerical values are given in Table 4.1. A comparison of the values corresponding to $\mu = -1/2$ and those for the true isothermal strip shows close agreement provided $\epsilon < 0.5$. For very narrow strips where $\epsilon < 0.1$, the differences are less than 1%.

Finally, if steady conduction occurs in a two-dimensional channel (as shown in Figure 4.2), whose width decreases from $2a$ to $2b$, there is spreading resistance as heat flows through the common interface. The true boundary condition at the common interface is unknown. The temperature and the heat flux are both nonuniform. Conformal mapping leads to a closed-form solution for the spreading resistance.

The relationship for the spreading resistance is according to Smythe (1968)

$$kLR_s = \frac{1}{2\pi} \left\{ \left( \epsilon + \frac{1}{\epsilon} \right) \ln \left[ \frac{1+\epsilon}{1-\epsilon} \right] + 2 \ln \left[ \frac{1-\epsilon^2}{4\epsilon} \right] \right\} \tag{4.23}$$

where $\epsilon = a/b < 1$. Numerical values are given in the final column of Table 4.1. An examination of the values reveals that they lie between the values for $\mu = -1/2$ and $\mu = 0$. The average value of the first two columns corresponding to $\mu = -1/2$ and $\mu = 0$ are in very close agreement with the values in the last column. The differences are less than 1% for $\epsilon \leq 0.20$, and they become negligible as $\epsilon \rightarrow 0$.

This concludes the discussion on the semi-infinite flux channel. In Sections 4.3–4.5, we will extend the results of the semi-infinite flux channel to allow for a finite flux channel of thickness $t$ with conductance $h_s$, compound, and multilayer systems. In all cases, the above results are simply modified through the appropriate use of the spreading function $\phi$.

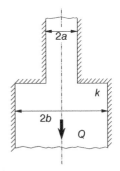

**Figure 4.2**
Semi-infinite flux channel with abrupt change in cross-section.

## 4.2 Three-Dimensional Semi-Infinite Flux Channel

Mikic and Rohsenow (1966) also considered the case of a two-dimensional centrally located heat source on a three-dimensional flux channel as shown in Figure 4.3. In this case, we now must solve Laplace's equation in three dimensions:

$$\nabla^2 T = \frac{\partial^2 T}{\partial x^2} + \frac{\partial^2 T}{\partial y^2} + \frac{\partial^2 T}{\partial z^2} = 0 \tag{4.24}$$

subject to the following boundary conditions (using symmetry):

$$\left.\frac{\partial T}{\partial x}\right|_{x=0} = \left.\frac{\partial T}{\partial y}\right|_{y=0} = 0$$

$$\left.\frac{\partial T}{\partial x}\right|_{x=c} = \left.\frac{\partial T}{\partial y}\right|_{y=d} = 0 \tag{4.25}$$

at the centroid and edges of the channel, respectively. Over the top surface $z = 0$,

$$\left.\begin{aligned}
-k\left.\frac{\partial T}{\partial z}\right|_{z=0} &= \frac{Q}{4ab}, \quad 0 < x < a, \quad 0 < y < b \\
\left.\frac{\partial T}{\partial z}\right|_{z=0} &= 0, \quad a < x < c, \quad b < y < d
\end{aligned}\right\} \tag{4.26}$$

Finally, at $z \to \infty$, we prescribe the uniform flow:

$$-k\frac{\partial T}{\partial z} = \frac{Q}{4cd}, \quad z \to \infty \tag{4.27}$$

in a region located at a significant distance from the contact plane. Note, there is no requirement that the heat source area have side dimensions that scale with the flux channel dimensions, i.e. $a/c$ and $b/d$ having the same ratios. In this case for example, we can find a solution for a square heat source on a rectangular channel.

The solution can be found using separation of variables assuming that $T(x, y, z) = X(x) \cdot Y(y) \cdot Z(z)$. The solution takes the form of

$$T(x, y, z) = A_0 - \frac{Qz}{(4cd)k} + \sum_{m=1}^{\infty} A_m \cos(\lambda_m x) \exp(-\lambda_m z)$$

$$+ \sum_{n=1}^{\infty} A_n \cos(\delta_n y) \exp(-\delta_n z) + \sum_{m=1}^{\infty}\sum_{n=1}^{\infty} A_{mn} \cos(\lambda_m x) \cos(\delta_n y) \exp(-\beta_{mn} z) \tag{4.28}$$

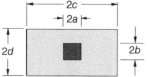

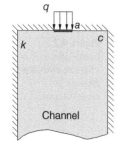

**Figure 4.3** Three dimensional semi-infinite flux channel.

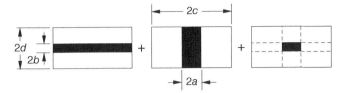

**Figure 4.4** Superposition of solutions.

after considering the boundary conditions in the $x$, $y$ directions and the $z \to \infty$ condition. The eigenvalues are defined as follows:

$$\lambda_m = \frac{m\pi}{c}, \quad \delta_n = \frac{n\pi}{d}, \quad \beta_{mn} = \sqrt{\lambda_m^2 + \delta_n^2} \tag{4.29}$$

The general solution represents a superposition of two strip solutions (single series), a finite rectangle (double series) and a uniform flow solution, see Figure 4.4. Under the special conditions of $b = d$ or $a = c$, the general solution reduces to that of a two-dimensional strip solved earlier.

Application of the boundary condition in the source can be done for each of the two single series and the double series to obtain $A_m$, $A_n$, and $A_{mn}$. This yields the following results:

$$A_m = \frac{Q \sin(m\pi a/c)}{2(m\pi)^2(a/c)dk}$$

$$A_n = \frac{Q \sin(n\pi b/d)}{2(n\pi)^2(b/d)ck} \tag{4.30}$$

$$A_{mn} = \frac{Q \sin(m\pi a/c) \sin(n\pi b/d)}{(m\pi)(n\pi)bak\sqrt{(m\pi/c)^2 + (n\pi/d)^2}}$$

The temperature in the source plane $z = 0$ is

$$T(x, y, 0) = A_0 + \sum_{m=1}^{\infty} A_m \cos(m\pi x/c) + \sum_{n=1}^{\infty} A_n \cos(n\pi y/d)$$

$$+ \sum_{m=1}^{\infty} \sum_{n=1}^{\infty} A_{mn} \cos(m\pi x/c) \cos(n\pi y/d) \tag{4.31}$$

The mean source temperature is found from

$$\overline{T}_{source} = \frac{1}{4ab} \int_{-a}^{a} \int_{-b}^{b} T(x, y, 0) dx \, dy \tag{4.32}$$

or

$$\overline{T}_{source} = A_0 + \sum_{m=1}^{\infty} A_m \frac{\sin(m\pi a/c)}{(a/c)(m\pi)} + \sum_{n=1}^{\infty} A_n \frac{\sin(n\pi b/d)}{(b/d)(n\pi)}$$

$$+ \sum_{m=1}^{\infty} \sum_{n=1}^{\infty} A_{mn} \frac{\sin(m\pi a/c) \sin(n\pi b/d)}{(a/c)(b/d)(m\pi)(n\pi)} \tag{4.33}$$

The mean temperature of the contact plane $z = 0$ is obtained from

$$\overline{T}_{cp} = \frac{1}{4cd} \int_{-c}^{c} \int_{-d}^{d} T(x, y, 0) dx \, dy \tag{4.34}$$

or

$$\overline{T}_{cp} = A_0 \tag{4.35}$$

Finally, defining the spreading resistance defined as before

$$R_s = \frac{\overline{T}_{source} - \overline{T}_{cp}}{Q} \tag{4.36}$$

or after combining with the coefficients, we obtain

$$
\begin{aligned}
kR_s = {} & \sum_{m=1}^{\infty} \frac{\sin^2(m\pi a/c)}{2d(a/c)^2(m\pi)^3} + \sum_{n=1}^{\infty} \frac{\sin^2(n\pi b/d)}{2c(b/d)^2(n\pi)^3} \\
& + \sum_{m=1}^{\infty}\sum_{n=1}^{\infty} \frac{\sin^2(m\pi a/c)\sin^2(n\pi b/d)}{cd(a/c)^2(b/d)^2(m\pi)^2(n\pi)^2\sqrt{(m\pi/c)^2+(n\pi/d)^2}}
\end{aligned} \tag{4.37}
$$

Equation (4.37) while quite complex in appearance is reasonably efficient computationally. One special case that should be highlighted is that of a square source on a square flux channel. In the above equation, if we make $b = a$ and $d = c$, the solution simplifies:

$$
\begin{aligned}
kR_s = {} & 2\sum_{m=1}^{\infty} \frac{\sin^2(m\pi a/c)}{2c(a/c)^2(m\pi)^3} \\
& + \sum_{m=1}^{\infty}\sum_{n=1}^{\infty} \frac{\sin^2(m\pi a/c)\sin^2(n\pi a/c)}{c(a/c)^4(m\pi)^2(n\pi)^2\sqrt{(m\pi)^2+(n\pi)^2}}
\end{aligned} \tag{4.38}
$$

or written in nondimensional form using $\sqrt{A} = 2a$ with $\epsilon = a/c$, we obtain

$$k\sqrt{A_s}R_s = \frac{2}{\pi^3\epsilon}\left[\sum_{m=1}^{\infty}\frac{\sin^2(m\pi\epsilon)}{m^3} + \frac{1}{\pi^2\epsilon^2}\sum_{m=1}^{\infty}\sum_{n=1}^{\infty}\frac{\sin^2(m\pi\epsilon)\sin^2(n\pi\epsilon)}{m^2n^2\sqrt{m^2+n^2}}\right] \tag{4.39}$$

for a semi-infinite square flux channel with a square heat source. This result is quite useful when modeling electronic packages.

### 4.2.1 Correlation Equations for Various Combinations of Source Areas and Boundary Conditions

Solutions are also available for various combinations of source areas and flux tube cross-sectional areas such as circle/circle and circle/square for the uniform flux, true isothermal, and equivalent isothermal boundary conditions [Negus et al. (1989)].

Numerical results were correlated with the polynomial:

$$4kaR_s = C_0 + C_1\epsilon + C_3\epsilon^3 + C_5\epsilon^5 + C_7\epsilon^7 \tag{4.40}$$

The dimensionless spreading (constriction) resistance coefficient $C_0$ is the half-space value, and the correlation coefficients $C_1$ through $C_7$ are given in Table 4.2.

Negus et al. (1989) also show that a single correlation of the form:

$$k\sqrt{A_c}R_s = 0.475 - 0.62\epsilon + 0.13\epsilon^3 \tag{4.41}$$

where $\epsilon = \sqrt{A_s/A_t}$ is used for the source aspect ratio, provides excellent correlation for all combinations of source/tube shapes, including other regular polygons. The success of this correlation further re-iterates the use of the square root of the source $\sqrt{A}$ as a characteristic length scale.

**Table 4.2** Coefficients for correlations of dimensionless spreading resistance $4kaR_s$.

| Flux tube geometry and contact boundary condition | $C_0$ | $C_1$ | $C_3$ | $C_5$ | $C_7$ |
|---|---|---|---|---|---|
| Circle/circle uniform flux | 1.08076 | −1.41042 | 0.26604 | −0.00016 | 0.058266 |
| Circle/circle true isothermal | 1.00000 | −1.40978 | 0.34406 | 0.04305 | 0.02271 |
| Circle/square uniform flux | 1.08076 | −1.24110 | 0.18210 | 0.00825 | 0.038916 |
| Circle/square equiv. isothermal flux | 1.00000 | −1.24142 | 0.20988 | 0.02715 | 0.02768 |
| Square/square uniform flux | 0.9464 | −1.2415 | 0.2396 | 0 | 0 |

## 4.3 Finite Two- and Three-Dimensional Flux Channels

We now consider the case of finite rectangular flux channels in two and three dimensions, see Figure 4.5. The problem statement is the same as that in Sections 4.1 and 4.2 except for the uniform flow condition at $z \to \infty$. Beginning with

$$\nabla^2 T = \frac{\partial^2 \theta}{\partial x^2} + \frac{\partial^2 \theta}{\partial y^2} + \frac{\partial^2 \theta}{\partial z^2} = 0 \tag{4.42}$$

subject to the following boundary conditions:

$$\frac{\partial \theta}{\partial x}\bigg|_{x=0} = \frac{\partial \theta}{\partial y}\bigg|_{y=0} = 0$$

$$\frac{\partial \theta}{\partial x}\bigg|_{x=c} = \frac{\partial \theta}{\partial y}\bigg|_{y=d} = 0 \tag{4.43}$$

at the centroid and edges of the channel, respectively. Over the top surface $z = 0$,

$$\left. \begin{aligned} -k\frac{\partial \theta}{\partial z}\bigg|_{z=0} &= \frac{Q}{4ab}, \quad 0 < x < a, \quad 0 < y < b \\ \frac{\partial \theta}{\partial z}\bigg|_{z=0} &= 0, \quad a < x < c, \quad b < y < d \end{aligned} \right\} \tag{4.44}$$

while the sink plane now has the following boundary condition applied to a finite thickness slab:

$$-k\frac{\partial \theta}{\partial z}\bigg|_{z=t} = h_s \theta(x, y, z = t) \tag{4.45}$$

where $\theta = T - T_\infty$. This modification is essentially the same as that applied for the circular disk in Section 3.2.

This general solution now has the form:

$$\theta(x, y, z) = \theta_{1D}(z) + \theta_s(x, y, z) \tag{4.46}$$

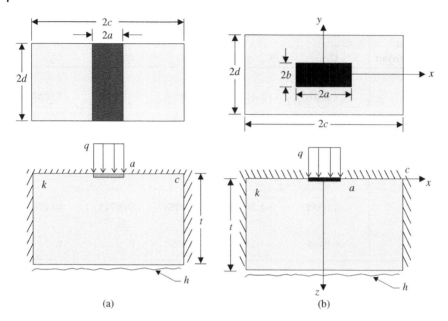

**Figure 4.5** Finite rectangular flux channels: (a) two dimensional finite flux channel and (b) three dimensional finite flux channel.

The problem can be solved in two parts with the one-dimensional flow solution easily shown to be

$$\theta_{1D}(z) = \frac{Q}{4cd}\left(\frac{t}{k} + \frac{1}{h}\right) - \frac{Qz}{4cdk} \tag{4.47}$$

The solution for the multidimensional spreading component which satisfies the $x, y$ boundary conditions is as before

$$\theta_s(x, y, z) = \sum_{m=1}^{\infty} \cos(\lambda_m x)[A_m \cosh(\lambda_m z) + B_m \sinh(\lambda_m z)]$$

$$+ \sum_{n=1}^{\infty} \cos(\delta_n y)[A_n \cosh(\delta_n z) + B_n \sinh(\delta_n z)]$$

$$+ \sum_{m=1}^{\infty}\sum_{n=1}^{\infty} \cos(\lambda_m x)\cos(\delta_n y)[A_{mn} \cosh(\beta_{mn} z) + B_{mn} \sinh(\beta_{mn} z)] \tag{4.48}$$

Application of the sink plane boundary condition yields the following general result for each portion of the solution:

$$B_i = -\phi(\xi)A_i \quad \text{for } i = m, n, \text{ and } mn \tag{4.49}$$

where $\phi$ is the general spreading function found earlier:

$$\phi(\xi) = \frac{\xi \tanh(\xi t) + \dfrac{h_s}{k}}{\xi + \dfrac{h_s}{k}\tanh(\xi t)} \tag{4.50}$$

and $\xi = \lambda_m, \delta_n, \beta_{mn}$ is the eigenvalue in each portion of the solution, i.e. $\lambda_m = m\pi/c$, $\delta_n = n\pi/d$, and $\beta_{mn} = \sqrt{\lambda_m^2 + \delta_n^2}$.

This gives the final general result for the spreading resistance:

$$kR_s = \sum_{m=1}^{\infty} \frac{\sin^2(m\pi a/c)}{2d(a/c)^2(m\pi)^3\phi(\lambda_m)} + \sum_{n=1}^{\infty} \frac{\sin^2(n\pi b/d)}{2c(b/d)^2(n\pi)^3\phi(\delta_n)}$$
$$+ \sum_{m=1}^{\infty}\sum_{n=1}^{\infty} \frac{\sin^2(m\pi a/c)\sin^2(n\pi b/d)}{cd(a/c)^2(b/d)^2(m\pi)^2(n\pi)^2\sqrt{(m\pi/c)^2 + (n\pi/d)^2}\phi(\beta_{mn})} \qquad (4.51)$$

As before, we see that the spreading resistance is simply modified by the general spreading function $\phi$. The total resistance is now the sum of the bulk resistance and the spreading resistance such that

$$R_{total} = R_{1D} + R_s = \frac{1}{4cd}\left(\frac{t}{k} + \frac{1}{h}\right) + R_s \qquad (4.52)$$

**Two-Dimensional Flux Channel**

Going forward, we will refer to the above general results but with appropriate modifications to the bulk resistance and general spreading function which appears in the equation for the spreading resistance. If the problem is specified for a two-dimensional strip, then only the first summation is required in Eq. (4.51). If variable flux is desired on the strip, Eq. (4.20) may be used:

$$kLR_s = \frac{\Gamma(\mu + 3/2)}{\pi^2\epsilon}\sum_{m=1}^{\infty}\left[\frac{2}{m\pi\epsilon}\right]^{\mu+1/2}\frac{\sin(m\pi\epsilon)}{m^2}\frac{J_{\mu+1/2}(m\pi\epsilon)}{\phi(\lambda_m)} \qquad (4.53)$$

where $L = 2d$ and $\lambda_m = m\pi/c$ as before.

**Example 4.1**   Consider a heat sink thermal spreader of dimensions $2c = 30$ cm by $2d = 20$ cm and thickness $t = 1$ cm. The thermal conductivity of the spreader plate is $k = 230$ W/(m K) with an equivalent sink conductance of $h_s = 500$ W/(m² K). A heat source of dimensions $2a = 10$ cm by $2b = 2$ cm and strength $Q = 500$ W is attached at the centroid of the spreader. If the sink temperature is $T_f = 20°$C, what is the mean source temperature?

Assuming a uniform flux heat source, we use Eq. (4.51) to find the spreading resistance. Using the provided dimensions and properties with 100 terms in each summation, we find

$$R_s = 0.06163 \text{ K/W}$$

and

$$R_{1D} = \frac{1}{4cd}\left(\frac{t}{k} + \frac{1}{h}\right) = 0.03406 \text{ K/W}$$

for a total thermal resistance:

$$R_{total} = R_{1D} + R_s = 0.09569 \text{ K/W}$$

Finally, the mean source temperature is found to be

$$T_s = Q \cdot R_{total} + T_f = 67.85°\text{C}$$

**Example 4.2**   A square device is composed of two materials with a thin square heat source embedded between the two layers. The package has dimensions $2c = 2d = 5$ cm. The heat source dimensions are $2a = 2b = 2.5$ cm and has a strength $Q = 40$ W. The heat source is considerably thinner than each layer of the package such that we can ignore its effect. The top layer of the package has $k_1 = 10$ W/(m K) and $t_1 = 2.5$ mm and is cooled with a conductance $h_1 = 250$ W/(m² K). The bottom layer of the package has $k_2 = 5$ W/(m K) and $t_2 = 2.5$ mm and is cooled with a conductance

$h_2 = 50\,\text{W}/(\text{m}^2\,\text{K})$. Find the mean source temperature and the partition of heat between the two layers.

To predict the temperature of the heat source in this system, we will assume that the package can be divided along its interface and treat the system as two spreaders in parallel. In this way, we can calculate the total resistance of each portion of the package using Eqs. (4.51) and (4.52). For each layer, we find

$$R_{s,1} = 1.428\,\text{K/W}$$

$$R_{1D,1} = 1.700\,\text{K/W}$$

$$R_{total,1} = 3.128\,\text{K/W}$$

and

$$R_{s,2} = 3.515\,\text{K/W}$$

$$R_{1D,2} = 8.200\,\text{K/W}$$

$$R_{total,2} = 11.715\,\text{K/W}$$

The energy balance at the heat source requires that

$$Q_1 + Q_2 = 40$$

and, the temperature excess for each layer requires that

$$Q_1 R_{total,1} = Q_2 R_{total,2}$$

or

$$3.128 Q_1 = 11.715 Q_2$$

Solving the system of equations for $[Q_1, Q_2]$ yields: $Q_1 = 31.57\,\text{W}$ and $Q_2 = 8.43\,\text{W}$. Finally, using either result we obtain

$$\theta_{ja} = Q_1 R_{total,1} = Q_2 R_{total,2} = 98.75\,^\circ\text{C}$$

**Example 4.3**  In this example, we will consider a small heat source on a device which is mounted to a thermal spreader. Using the case presented in Lasance (2010), we have a 1 W heat source of dimensions $2a = 2b = 0.8\,\text{mm}$ on a chip (submount) of dimensions $2c = 2d = 5\,\text{mm}$ and thickness $t = 0.1\,\text{mm}$. The chip has a thermal conductivity of $k = 100\,\text{W}/(\text{m K})$ and is mounted to a thermal spreader having outer dimensions of 30 mm by 30 mm and thickness of $t = 2\,\text{mm}$. The thermal conductivity of the spreader is $k = 200\,\text{W}/(\text{m K})$ and is cooled with a sink conductance of $h_s = 250\,\text{W}/(\text{m}^2\,\text{K})$. Determine the junction to ambient temperature excess $\theta_{ja}$ for the heat source.

In this problem, we essentially have two single-layer thermal spreading problems joined together. We may approximate the solution by separating the device from the spreader and treating each as separate problems and use Eqs. (4.51) and (4.52). In the first problem, we have a heat source on a smaller spreader which has an effective conductance acting over its bond area related to the total resistance of the thermal spreader through:

$$\frac{1}{h_{eff} A_{chip}} = R_{spreader}$$

or

$$h_{eff} = \frac{1}{R_{spreader} A_{chip}}$$

In the second problem, the total spreader resistance is determined by considering a heat source with the same dimensions as the chip. In this way, we are assuming that the flux is uniform (which it is not), but is a reasonable approach nonetheless. The total resistance of the spreader assuming a source dimension of $2a = 2b = 5$ mm and a spreader dimension of $2c = 2d = 30$ mm with $t = 2$ mm, $k = 200$ W/(m K) and $h_s = 250$ W/(m$^2$ K) is

$$R_s = 0.6202 \text{ K/W}$$

$$R_{1D} = 4.456 \text{ K/W}$$

$$R_{total} = 5.076 \text{ K/W}$$

The effective conductance for the chip is now found to be

$$h_{eff} = \frac{1}{R_{spreader}A_{chip}} = \frac{1}{5.076 \cdot 2.5 \times 10^{-5}} = 7880.2 \text{ W/(m}^2 \text{ K)}$$

Finally, using the heat source dimensions $2a = 2b = 0.8$ mm, chip (sub-mount) dimensions $2c = 2d = 5$ mm, thickness $t = 0.1$ mm, conductivity $k = 100$ W/(m K), and effective conductance $h_{eff} = 7880.2$ W/(m$^2$ K) in a new calculation, we find

$$R_s = 17.34 \text{ K/W}$$

$$R_{1D} = 5.12 \text{ K/W}$$

$$R_{total} = 22.46 \text{ K/W}$$

For the 1 W input, we find that $\theta_{ja} = Q \cdot R_{total} = 22.46\,°\text{C}$.

## 4.4 Compound Two- and Three-Dimensional Flux Channels

The system of interest is now composed of two layers, which in general, are imperfectly bonded to each other giving rise to an interfacial resistance, see Figure 4.6. Often, a thinner layer with $k_1 > k_2$ is applied to a thermal spreader to promote spreading and reduce the spreading resistance, for example, a heat sink with a copper layer in the source plane. An interfacial contact conductance $h_c$ will be used to model the imperfect contact region. This conductance can also be modeled as an equivalent resistance such that

$$R_i = \frac{t_i}{k_i A_b} = \frac{1}{h_c A_b} \tag{4.54}$$

or

$$h_c = \frac{k_i}{t_i} \tag{4.55}$$

where $k_i$ is the thermal conductivity of the interfacial region or that of an actual bonding agent, and $t_i$ the nominal thickness of this region when these characteristics are known. This approach is adopted since the interfacial layer is usually quite thin relative to the adjacent layers, i.e. $t_i \ll t_1, t_2$.

The system of interest is now modeled using a localized system of coordinates, such that the heat source has an origin at the center of the disk or channel, and the $z$-direction is measured locally within each layer. This approach is chosen as it leads to a simpler solution for the undetermined coefficients. Further, as before we will define the temperature excess, $\theta = T - T_f$, relative to the sink temperature.

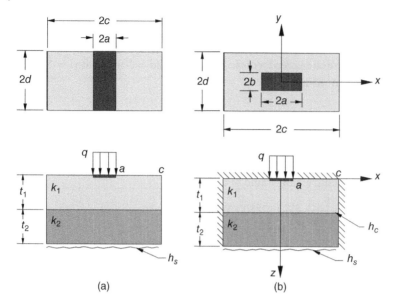

**Figure 4.6** Compound rectangular flux channels: (a) two dimensional finite flux channel and (b) three dimensional finite flux channel, with imperfect interfacial contact.

The problem is now stated as follows: the governing equation for each layer in Cartesian co-ordinates is

$$\frac{\partial^2 \theta_1}{\partial x^2} + \frac{\partial^2 \theta_1}{\partial y^2} + \frac{\partial^2 \theta_1}{\partial z_1^2} = 0, \quad 0 < z_1 < t_1$$

$$\frac{\partial^2 \theta_2}{\partial x^2} + \frac{\partial^2 \theta_2}{\partial y^2} + \frac{\partial^2 \theta_2}{\partial z_2^2} = 0, \quad 0 < z_2 < t_2$$

(4.56)

The following boundary conditions are imposed on the system. In the source plane, we specify a discrete heat flux distribution such that over the region of the heat source a uniform heat flux is specified, while outside of the heat source region the remainder of the surface is taken as adiabatic. Therefore, the following is assumed for a single heat source (and using symmetry):

$$-k_1 \frac{\partial \theta_1}{\partial z_1} = \frac{Q}{4ab}, \quad z = 0 \quad 0 < x < a, \quad 0 < y < b$$

$$\frac{\partial \theta_1}{\partial z_1} = 0, \quad z = 0 \quad a < x < c, \quad b < y < d$$

(4.57)

At the interface, we require the following conditions representing the equality of flux and temperature drop due to the imperfect interface:

$$k_1 \frac{\partial \theta_1}{\partial z_1}\bigg|_{z_1=t_1} = k_2 \frac{\partial \theta_2}{\partial z_2}\bigg|_{z_2=0}$$

$$-k_1 \frac{\partial \theta_1}{\partial z_1}\bigg|_{z_1=t_1} = h_c \left[ \theta_1(x,y,t_1) - \theta_2(x,y,0) \right]$$

(4.58)

Along the heat sink plane, we specify the following condition:

$$-k_2 \frac{\partial \theta_2}{\partial z_2}\bigg|_{z_2=t_2} = h_s \theta_2(x,y,t_2)$$

(4.59)

Along the edges of the plate (using symmetry), the following conditions are also required:

$$\frac{\partial \theta_i}{\partial x}\bigg|_{x=0} = \frac{\partial \theta_i}{\partial x}\bigg|_{x=c} = 0, \ \ i = 1, 2$$

$$\frac{\partial \theta_i}{\partial x}\bigg|_{y=0} = \frac{\partial \theta_i}{\partial y}\bigg|_{y=d} = 0, \ \ i = 1, 2 \tag{4.60}$$

The solution to the above problem was obtained by Muzychka et al. (2013) and Muzychka (2014). We may find a general solution for the thermal spreading resistance using this result and the result for the isotropic disk presented earlier. To extend the solution for the compound channel, we merely need only find the new spreading function $\phi$ through the application of the new boundary conditions in the z-direction for both problems with and without perfect interfacial conductance. With an additional layer and interfacial conductance, the spreading function $\phi$ in the solution for the isotropic channel can now be defined for two cases: (i) finite interfacial conductance and (ii) perfect interfacial contact along with the corresponding case of infinite sink plane conductance, i.e. a constant uniform temperature surface, Muzychka (2014). Muzychka (2014) showed that with the new spreading function, one could easily extend the solution for a single layer to a system with two layers with little additional effort. The thermal spreading resistance is now related to the new spreading function and the thermal conductivity of the layer adjacent to the source plane, i.e. $k = k_1$ in Eq. (4.51):

$$k_1 R_s = \sum_{m=1}^{\infty} \frac{\sin^2(m\pi a/c)}{2d(a/c)^2(m\pi)^3 \phi(\lambda_m)} + \sum_{n=1}^{\infty} \frac{\sin^2(n\pi b/d)}{2c(b/d)^2(n\pi)^3 \phi(\delta_n)}$$

$$+ \sum_{m=1}^{\infty} \sum_{n=1}^{\infty} \frac{\sin^2(m\pi a/c)\sin^2(n\pi b/d)}{cd(a/c)^2(b/d)^2(m\pi)^2(n\pi)^2 \sqrt{(m\pi/c)^2 + (n\pi/d)^2}\phi(\beta_{mn})} \tag{4.61}$$

For all cases given below, $\xi = \lambda_m, \delta_n, \beta_{mn}$ is the eigenvalue in each portion of the solution, i.e. $\lambda_m = m\pi/c$, $\delta_n = n\pi/d$, and $\beta_{mn} = \sqrt{\lambda_m^2 + \delta_n^2}$.

**Finite Interfacial Conductance**
For the general system stated earlier (two isotropic layers with interfacial conductance), the necessary spreading function is

$$\phi(\xi) = \frac{C_1 + C_2 \tanh(\xi t_1)}{C_1 \tanh(\xi t_1) + C_2} \tag{4.62}$$

where

$$C_1 = \left[\xi \tanh(\xi t_2) + h_s/k_2\right] \tag{4.63}$$

and

$$C_2 = k_1/k_2 \left[\xi(1 + h_s/h_c) + (h_s/k_2 + \xi^2 k_2/h_c)\tanh(\xi t_2)\right] \tag{4.64}$$

and

$$R_{1D} = \frac{1}{4cd}\left(\frac{t_1}{k_1} + \frac{1}{h_c} + \frac{t_2}{k_2} + \frac{1}{h_s}\right) \tag{4.65}$$

If the device is in contact with an ideal heat sink, $h_s \to \infty$, the above expression simplifies to yield:

$$\phi(\xi) = \frac{1 + \xi k_1/h_c \tanh(\xi t_1) + k_1/k_2 \tanh(\xi t_2)\tanh(\xi t_1)}{\xi k_1/h_c + k_1/k_2 \tanh(\xi t_2) + \tanh(\xi t_1)} \tag{4.66}$$

and

$$R_{1D} = \frac{1}{4cd}\left(\frac{t_1}{k_1} + \frac{1}{h_c} + \frac{t_2}{k_2}\right)$$ (4.67)

**Perfect Interfacial Contact**

If there is perfect interfacial contact, then $h_c \to \infty$, the solution for the spreading function becomes

$$\phi(\xi) = \frac{\left[\xi\tanh(\xi t_2) + h_s/k_2\right] + k_1/k_2\left[\xi + h_s/k_2\tanh(\xi t_2)\right]\tanh(\xi t_1)}{\left[\xi\tanh(\xi t_2) + h_s/k_2\right]\tanh(\xi t_1) + k_1/k_2\left[\xi + h_s/k_2\tanh(\xi t_2)\right]}$$ (4.68)

and

$$R_{1D} = \frac{1}{4cd}\left(\frac{t_1}{k_1} + \frac{t_2}{k_2} + \frac{1}{h_s}\right)$$ (4.69)

Finally, if the device is in contact with an ideal heat sink, $h_s \to \infty$, the above expression simplifies to yield:

$$\phi(\xi) = \frac{1 + k_1/k_2\tanh(\xi t_2)\tanh(\xi t_1)}{\tanh(\xi t_1) + k_1/k_2\tanh(\xi t_2)}$$ (4.70)

and

$$R_{1D} = \frac{1}{4cd}\left(\frac{t_1}{k_1} + \frac{t_2}{k_2}\right)$$ (4.71)

**Two-Dimensional Flux Channel**

In the special case of a two-dimensional flux channel with strip heat source, Eq. (4.53) may be used with the appropriate spreading defined functions above:

$$k_1 L R_s = \frac{\Gamma(\mu + 3/2)}{\pi^2\epsilon}\sum_{m=1}^{\infty}\left[\frac{2}{m\pi\epsilon}\right]^{\mu+1/2}\frac{\sin(m\pi\epsilon)}{m^2}\frac{J_{\mu+1/2}(m\pi\epsilon)}{\phi(\lambda_m)}$$ (4.72)

where $L = 2d$ and $\lambda_m = m\pi/c$ as before.

### 4.4.1 Special Limiting Cases for Rectangular Flux Channels

Many special cases can be derived from the solutions presented thus far for the three-dimensional compound rectangular flux channel. Limiting our attention to the case of a centrally located rectangular heat source on a compound spreader, the fundamental limits for many problems can be derived from the basic solution for a compound channel. They are summarized in Table 4.3. These include 2D strip limits, half-space limits, and plate limits. Additional special cases may be derived from the general multilayer flux channel in a similar manner.

**Example 4.4** Copper is often added to a heat sink baseplate to help further reduce thermal spreading resistance. Consider a small aluminum heat sink with dimensions $2c = 2d = 5\,\text{cm}$ and thickness $t = 5\,\text{mm}$. The baseplate has a 30 W heat source of dimensions $2a = 2b = 1\,\text{cm}$ at its centroid. The thermal conductivity of the baseplate is $k = 230\,\text{W/(m K)}$, and it has an effective sink conductance of $h_s = 250\,\text{W/(m}^2\,\text{K)}$ with an ambient temperature of $T_f = 25\,°\text{C}$. Determine the spreading resistance, total thermal resistance, and mean source temperature. What is the effect of adding a 2 mm layer of copper assuming a perfect interface $h_c \to \infty$ and a finite conductance between the layers $h_c = 10,000\,\text{W/(m}^2\,\text{K)}$.

**Table 4.3** Summary of solutions for isoflux source on a compound flux channel.

| Configuration | Limiting values |
|---|---|
| *Rectangular heat source* | |
| Finite compound rectangular flux channel | $a, b, c, d, t_1, t_2, k_1, k_2, h$ |
| Semi-infinite compound rectangular flux channel | $t_2 \to \infty$ |
| Finite isotropic rectangular flux channel | $k_1 = k_2$ |
| Semi-infinite isotropic rectangular flux channel | $t_1 \to \infty$ |
| *Strip heat source* | |
| Finite compound rectangular flux channel | $a, c, b = d, t_1, t_2, k_1, k_2, h$ |
| Semi-infinite compound rectangular flux channel | $t_2 \to \infty$ |
| Finite isotropic rectangular flux channel | $k_1 = k_2$ |
| Semi-infinite isotropic rectangular flux channel | $t_1 \to \infty$ |
| *Rectangular source on a semi-infinite plate* | |
| Semi-infinite compound plate | $c, d \to \infty$ |
| Semi-infinite isotropic plate | $c, d \to \infty, t_2 = 0, k_2 = k_1$ |
| *Rectangular source on a half space* | |
| Isotropic half space | $c \to \infty, d \to \infty, t_1 \to \infty$ |
| Compound half space | $c \to \infty, d \to \infty, t_2 \to \infty$ |

For the basic case with no copper layer, we use Eq. (4.51) for the spreading resistance along with Eq. (4.50) for the spreading function and Eq. (4.52) for the total resistance. Using the supplied variables, we find

$$R_s = 0.2088 \text{ K/W}$$

$$R_{1D} = 1.609 \text{ K/W}$$

$$R_{total} = 1.817 \text{ K/W}$$

This total thermal resistance gives a mean source temperature of

$$T_s = T_f + QR_{total} = 79.5\,^\circ\text{C}$$

For the case of the same heat sink variables, but with a copper backing layer, we recalculate the spreading and total resistance using Eqs. (4.61), (4.68), and (4.69) for the perfect interface case, and using Eqs. (4.61)–(4.65) for a finite interface conductance. For the perfect interface, we find

$$R_s = 0.1292 \text{ K/W}$$

$$R_{1D} = 1.611 \text{ K/W}$$

$$R_{total} = 1.740 \text{ K/W}$$

which is approximately a 4% reduction in total thermal resistance. For the imperfect interface, we find

$$R_s = 0.1863 \text{ K/W}$$

$$R_{1D} = 1.651 \text{ K/W}$$

$$R_{total} = 1.837 \text{ K/W}$$

In this case, the imperfect interface has actually led to a poorer system performance. In this case, we would require an interfacial conductance $h_c > 10^5 \text{ W/(m}^2 \text{ K)}$ in order to achieve results better than the baseline case with no copper backing and $h_c > 10^6 \text{ W/(m}^2 \text{ K)}$ to approach the case of a perfect interface. As this example has shown, it is quite important to consider how the addition of a thermal spreading layer will enhance a particular application.

## 4.5 Finite Two- and Three-Dimensional Flux Channels with Eccentric Heat Sources

The two-dimensional eccentric strip heat source for a semi-infinite flux channel was obtained by Veziroglu and Chandra (1969). Muzychka et al. (2003) and Muzychka (2006) obtained a general solution for an eccentric heat source on an isotropic and compound flux channel. Bagnall et al. (2014) obtained a general solution for the three-dimensional multilayered flux channel with an eccentric source and used it to model multiple heat sources. In this section, we will layout the general procedure for finding the thermal spreading resistance of an eccentric heat source for an isotropic rectangular flux channel. The usual modifications for a compound system with and without interface resistance will also apply here.

Due to the lack of symmetry in this application, we use a slightly differing notation than the in Chapter 6. The coordinate axes are placed at the lower left corner of the rectangular flux channel. We also reference the location of the centroid of the heat source as $X_x$, $Y_c$. The governing equation for the system shown in Figure 4.7 is Laplace's equation:

$$\nabla^2 T = \frac{\partial^2 T}{\partial x^2} + \frac{\partial^2 T}{\partial y^2} + \frac{\partial^2 T}{\partial z^2} = 0 \tag{4.73}$$

which is subject to a uniform flux distribution over the source area:

$$\left. \frac{\partial T}{\partial z} \right|_{z=0} = -\frac{(Q/A_s)}{k_1} \tag{4.74}$$

**Figure 4.7**  Eccentric heat source on a rectangular flux channel.

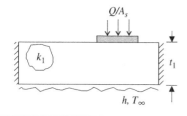

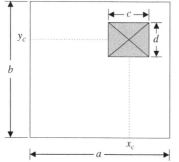

within the heat source area, $A_s = cd$, and

$$\left.\frac{\partial T}{\partial z}\right|_{z=0} = 0 \tag{4.75}$$

outside the heat source area and a convective or mixed boundary condition on the bottom surface

$$\left.\frac{\partial T}{\partial z}\right|_{z=t_1} = -\frac{h_s}{k_1}\left[T(x,y,t_1) - T_f\right] \tag{4.76}$$

In extended surface applications such as heat sinks, the value of $h_s$ is replaced by an effective value which accounts for both the heat transfer coefficient on the fin surface and the increased surface area due to the fins:

$$h_{eff} = \frac{1}{A_b R_{fins}} \tag{4.77}$$

where $A_b = ab$.

Along the edges of the plate, the following conditions are also required

$$\left.\frac{\partial T}{\partial x}\right|_{x=0,a} = 0 \tag{4.78}$$

and

$$\left.\frac{\partial T}{\partial y}\right|_{y=0,b} = 0 \tag{4.79}$$

The general solution for the total thermal resistance and temperature distribution will be obtained for the system shown in Figure 4.7. Later, we may apply the necessary modifications to the solution to account for additional aspects such as additional layers.

The full solution is obtained for the isotropic case following Muzychka et al. (2003). The solution for the isotropic plate may be obtained by means of separation of variables. The solution is assumed to have the form: $\theta(x,y,z) = X(x) \cdot Y(y) \cdot Z(z)$, where $\theta(x,y,z) = T(x,y,z) - T_f$. Applying the method of separation of variables yields the following general solution for the temperature excess in the plate which satisfy the thermal boundary conditions along ($x = 0$, $x = a$) and ($y = 0$, $y = b$):

$$\begin{aligned}
\theta(x,y,z) = A_0 + B_0 z \\
+ \sum_{m=1}^{\infty} \cos(\lambda x)\left[A_m \cosh(\lambda z) + B_m \sinh(\lambda z)\right] \\
+ \sum_{n=1}^{\infty} \cos(\delta y)\left[A_n \cosh(\delta z) + B_n \sinh(\delta z)\right] \\
+ \sum_{m=1}^{\infty}\sum_{n=1}^{\infty} \cos(\lambda x)\cos(\delta y)\left[A_{mn} \cosh(\beta z) + B_{mn} \sinh(\beta z)\right]
\end{aligned} \tag{4.80}$$

where $\lambda = m\pi/a$, $\delta = n\pi/b$, and $\beta = \sqrt{\lambda^2 + \delta^2}$.

The solution contains four components, a uniform flow solution and three spreading (or constriction) solutions which vanish when the heat flux is distributed uniformly over the entire surface $z = 0$. Since the solution is a linear superposition of each component, they may be dealt with separately. Application of the boundary conditions in the $z$ direction will yield solutions for the unknown constants.

Application of the thermal boundary condition at $z = t_1$ for an isotropic rectangular plate yields the following result for the Fourier coefficients:

$$B_i = -\phi(\xi)A_i \quad \text{for } i = m, n, mn \tag{4.81}$$

where

$$\phi(\xi) = \frac{\xi \tanh(\xi t_1) + \dfrac{h_s}{k_1}}{\xi + \dfrac{h_s}{k_1} \tanh(\xi t_1)} \tag{4.82}$$

and $\xi$ is replaced by $\lambda_m$, $\delta_n$, or $\beta_{mn}$, accordingly.

The final Fourier coefficients $A_m$, $A_n$, and $A_{mn}$ are obtained by taking Fourier series expansions of the boundary condition at the surface $z = 0$. This results in

$$A_m = \frac{Q}{b c k_1 \lambda_m \phi(\lambda_m)} \frac{\int_{X_c-c/2}^{X_c+c/2} \cos(\lambda_m x)dx}{\int_0^a \cos^2(\lambda_m x)dx} \tag{4.83}$$

or

$$A_m = \frac{2Q \left[ \sin\left( \frac{(2X_c+c)}{2}\lambda_m \right) - \sin\left( \frac{(2X_c-c)}{2}\lambda_m \right) \right]}{a b c k_1 \lambda_m^2 \phi(\lambda_m)} \tag{4.84}$$

and

$$A_n = \frac{Q}{a d k_1 \delta_n \phi(\delta_n)} \frac{\int_{Y_c-d/2}^{Y_c+d/2} \cos(\delta_n y)dy}{\int_0^b \cos^2(\delta_n y)dy} \tag{4.85}$$

or

$$A_n = \frac{2Q \left[ \sin\left( \frac{(2Y_c+d)}{2}\delta_n \right) - \sin\left( \frac{(2Y_c-d)}{2}\delta_n \right) \right]}{a b d k_1 \delta_n^2 \phi(\delta_n)} \tag{4.86}$$

and

$$A_{mn} = \frac{Q}{c d k_1 \beta_{m,n} \phi(\beta_{m,n})} \cdot \frac{\int_{Y_c-d/2}^{Y_c+d/2} \int_{X_c-c/2}^{X_c+c/2} \cos(\lambda_m x) \cos(\delta_n y)dx\,dy}{\int_0^b \int_0^a \cos^2(\lambda_m x)\cos^2(\delta_n y)dx\,dy} \tag{4.87}$$

or

$$A_{mn} = \frac{16Q \cos(\lambda_m X_c) \sin(\frac{1}{2}\lambda_m c) \cos(\delta_n Y_c) \sin(\frac{1}{2}\delta_n d)}{a b c d k_1 \beta_{mn} \lambda_m \delta_n \phi(\beta_{mn})} \tag{4.88}$$

Finally, values for the coefficients in the uniform flow solution are given by

$$A_0 = \frac{Q}{ab} \left( \frac{t_1}{k_1} + \frac{1}{h} \right) \tag{4.89}$$

and

$$B_0 = -\frac{Q}{k_1 ab} \tag{4.90}$$

The general solution for the mean temperature excess of a single heat source may be obtained by integrating equation (4.80) over the heat source area. Carrying out the necessary integrations leads to the following expression for the mean source temperature:

$$
\begin{aligned}
\bar{\theta} = \bar{\theta}_{1D} \\
&+ 2\sum_{m=1}^{\infty} A_m \frac{\cos(\lambda_m X_c)\sin(\frac{1}{2}\lambda_m c)}{\lambda_m c} \\
&+ 2\sum_{n=1}^{\infty} A_n \frac{\cos(\delta_n Y_c)\sin(\frac{1}{2}\delta_n d)}{\delta_n d} \\
&+ 4\sum_{m=1}^{\infty}\sum_{n=1}^{\infty} A_{mn} \frac{\cos(\delta_n Y_c)\sin(\frac{1}{2}\delta_n d)\cos(\lambda_m X_c)\sin(\frac{1}{2}\lambda_m c)}{\lambda_m c \delta_n d}
\end{aligned}
\tag{4.91}
$$

where

$$
\bar{\theta}_{1D} = \frac{Q}{ab}\left(\frac{t_1}{k_1} + \frac{1}{h}\right)
\tag{4.92}
$$

The total thermal resistance of the heat source may now be computed using the mean temperature excess of the heat source:

$$
R_{total} = \frac{\bar{\theta}}{Q} = R_{1D} + R_s
\tag{4.93}
$$

where $R_{1D}$ is the one-dimensional thermal resistance and $R_s$ is the thermal spreading resistance component. The thermal spreading resistance is defined by the three series solution terms in Eq. (4.91). This leads to the following result for the total thermal resistance for the flux channel:

$$
\begin{aligned}
R_{total} = \frac{1}{ab}\left(\frac{t_1}{k_1} + \frac{1}{h_s}\right) \\
&+ 2\sum_{m=1}^{\infty} B_m \frac{\cos(\lambda_m X_c)\sin(\frac{1}{2}\lambda_m c)}{\lambda_m c} + 2\sum_{n=1}^{\infty} B_n \frac{\cos(\delta_n Y_c)\sin(\frac{1}{2}\delta_n d)}{\delta_n d} \\
&+ 4\sum_{m=1}^{\infty}\sum_{n=1}^{\infty} B_{mn} \frac{\cos(\delta_n Y_c)\sin(\frac{1}{2}\delta_n d)\cos(\lambda_m X_c)\sin(\frac{1}{2}\lambda_m c)}{\lambda_m c \delta_n d}
\end{aligned}
\tag{4.94}
$$

where

$$
B_m = \frac{2\left[\sin\left(\frac{(2X_c+c)}{2}\lambda_m\right) - \sin\left(\frac{(2X_c-c)}{2}\lambda_m\right)\right]}{a\,b\,c\,k_1\,\lambda_m^2\,\phi(\lambda_m)}
\tag{4.95}
$$

$$
B_n = \frac{2\left[\sin\left(\frac{(2Y_c+d)}{2}\delta_n\right) - \sin\left(\frac{(2Y_c-d)}{2}\delta_n\right)\right]}{a\,b\,d\,k_1\,\delta_n^2\,\phi(\delta_n)}
\tag{4.96}
$$

$$
B_{mn} = \frac{16\cos(\lambda_m X_c)\sin(\frac{1}{2}\lambda_m c)\cos(\delta_n Y_c)\sin(\frac{1}{2}\delta_n d)}{a\,b\,c\,d\,k_1\,\beta_{mn}\lambda_m\delta_n\,\phi(\beta_{mn})}
\tag{4.97}
$$

In many applications, an interface material may be added to reduce thermal contact conductance and/or promote thermal spreading. The solution obtained for the isotropic rectangular flux channel may be used for a compound flux channel with only minor modifications, i.e. choosing the appropriate spreading function $\phi$ and bulk resistance discussed earlier in Section 4.4. In Chapter 6, we will adapt this basic solution to the case of modeling multiple heat sources using superposition and develop an efficient method for computations. We also apply the results for systems composed of multiple layers $N > 2$. Application for two-dimensional systems may be easily done, as only the first summation is required when $b = d$, yielding an eccentric strip.

## 4.6 Rectangular Flux Channels with Edge Conductance

The case of a rectangular flux channel with edge conductance $h_e$, as shown in Figure 4.8, was examined by Muzychka et al. (2006). The heat source is placed at the centroid of the flux channel which is symmetrically cooled along the edges as shown with $h_{e,x} \neq h_{e,y}$. Laplace's equation in Cartesian coordinates:

$$\frac{\partial^2 T}{\partial x^2} + \frac{\partial^2 T}{\partial y^2} + \frac{\partial^2 T}{\partial z^2} = 0 \tag{4.98}$$

must be solved with the following boundary conditions prescribed (assuming symmetry) in the $x$- and $y$-directions:

$$x = 0, \quad \frac{\partial \theta}{\partial x} = 0$$

$$x = c, \quad \frac{\partial \theta}{\partial x} + \frac{h_{e,x}}{k}\theta = 0$$

$$y = 0, \quad \frac{\partial \theta}{\partial y} = 0 \tag{4.99}$$

$$y = d, \quad \frac{\partial \theta}{\partial y} + \frac{h_{e,y}}{k}\theta = 0$$

In the source and sink planes we write

$$z = 0, \quad -k\frac{\partial \theta}{\partial z} = \frac{Q}{4ab}, \quad 0 < x < a, \quad 0 < y < b$$

$$\frac{\partial \theta}{\partial z} = 0, \qquad a < x < c, \quad b < y < d \tag{4.100}$$

$$z = t, \quad \frac{\partial \theta}{\partial z} + \frac{h_s}{k}\theta = 0$$

Here, $h_{e,x}$ and $h_{e,y}$ denote the values of the edge heat transfer coefficient, along the $x$-edge and $y$-edge, respectively.

The solution methodology is the same as before and the resulting solution is quite similar, with the exception of the definition of the eigenvalues. The final form of the solution is also somewhat more compact. The solution for an isotropic flux channel with edge cooling may be obtained by means of separation of variables. The solution is assumed to have the form $\theta(x, y, z) = X(x) \cdot Y(y) \cdot Z(z)$, where $\theta(x, y, z) = T(x, y, z) - T_f$. Applying the method of separation of variables yields the following general solution for the temperature excess in the substrate which satisfies the thermal boundary conditions along the two planes of symmetry, $x = 0$ and $y = 0$:

$$\theta(x, y, z) = \sum_{m=1}^{\infty}\sum_{n=1}^{\infty} \cos(\lambda_{xm}x)\cos(\lambda_{yn}y)\left[A_{mn}\cosh(\beta_{mn}z) + B_{mn}\sinh(\beta_{mn}z)\right] \tag{4.101}$$

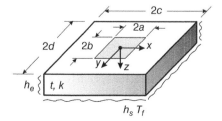

**Figure 4.8** Isotropic rectangular flux channel with edge conductance.

where $\lambda_{xm}$, $\lambda_{yn}$, and $\beta_{mn} = \sqrt{\lambda_{xm}^2 + \lambda_{yn}^2}$ are the eigenvalues. The eigenvalues are obtained from the following equations after application of the edge boundary conditions:

$$\delta_{xm} \sin(\delta_{xm}) = Bi_{e,x} \cos(\delta_{xm}) \tag{4.102}$$

and

$$\delta_{yn} \sin(\delta_{yn}) = Bi_{e,y} \cos(\delta_{yn}) \tag{4.103}$$

where $Bi_{e,x} = h_{e,x}c/k$, $\delta_{xm} = \lambda_{xm}c$, $Bi_{e,y} = h_{e,y}d/k$, and $\delta_{yn} = \lambda_{yn}d$. These equations must be solved numerically for a finite number of eigenvalues for each specified value of the edge cooling Biot numbers. The separation constant $\beta_{mn}$ is now defined as

$$\beta_{mn} = \sqrt{\left(\frac{\delta_{xm}}{c}\right)^2 + \left(\frac{\delta_{yn}}{d}\right)^2} \tag{4.104}$$

In this case, the eigenvalues are no longer simple integer functions as before. Application of the sink plane boundary condition yields the following relation for the spreading function:

$$A_{mn} = -\phi_{mn}B_{mn} \tag{4.105}$$

where

$$\phi_{mn} = \frac{\beta_{mn} + \dfrac{h_s}{k}\tanh(\beta_{mn}t)}{\dfrac{h_s}{k} + \beta_{mn}\tanh(\beta_{mn}t)} \tag{4.106}$$

The final Fourier coefficients are obtained by taking a double Fourier expansion of the upper surface condition. This yields the following expression:

$$B_{mn} = \frac{-Q \displaystyle\int_0^a \cos(\lambda_{xm}x)dx \int_0^b \cos(\lambda_{yn}y)dy}{4kab\beta_{mn} \displaystyle\int_0^c \cos^2(\lambda_{xm}x)dx \int_0^d \cos^2(\lambda_{yn}y)dy} \tag{4.107}$$

Upon evaluation of the integrals, one obtains

$$B_{mn} = \frac{-Q\sin(\delta_{xm}a/c)\sin(\delta_{yn}b/d)}{kab\beta_{mn}[\sin(2\delta_{xm})/2 + \delta_{xm}][\sin(2\delta_{yn})/2 + \delta_{yn}]} \tag{4.108}$$

With both Fourier coefficients now known, the mean source temperature excess is found from Eq. (4.97). Using this result and the definition for thermal resistance, the total resistance is found to be

$$R_{total} = \frac{cd}{ka^2 b^2} \sum_{m=1}^{\infty}\sum_{n=1}^{\infty} \frac{\sin^2(\delta_{xm}a/c)\sin^2(\delta_{yn}b/d)\phi_{mn}}{\delta_{xm}\,\delta_{yn}\,\beta_{mn}[\sin(2\delta_{xm})/2 + \delta_{xm}][\sin(2\delta_{yn})/2 + \delta_{yn}]} \tag{4.109}$$

The total resistance now depends on

$$R_{total} = f(a, b, c, d, t, k, h, h_{e,x}, h_{e,y}) \tag{4.110}$$

The total resistance is nondimensionalized using $R^{\star}_{total} = kR_{total}\sqrt{A_s}$, where $\sqrt{A_s} = 2\sqrt{ab}$ to give

$$
R^{\star}_{total} = \frac{2\sqrt{\epsilon_x \epsilon_y \epsilon_b}}{\epsilon_b \epsilon_x^2 \epsilon_y^2} \sum_{m=1}^{\infty} \sum_{n=1}^{\infty} \frac{1}{\delta_{xm} \delta_{yn} \sqrt{\delta_{xm}^2/\epsilon_b^2 + \delta_{yn}^2}} \cdot
$$

$$
\frac{\sin^2(\delta_{xm}\epsilon_x)\sin^2(\delta_{yn}\epsilon_y)\phi_{mn}}{\left[\dfrac{\sin(2\delta_{xm})}{2} + \delta_{xm}\right]\left[\dfrac{\sin(2\delta_{yn})}{2} + \delta_{yn}\right]} \tag{4.111}
$$

where

$$
\phi_{mn} = \frac{\xi\tau + Bi\tanh(\xi\tau)}{Bi + \xi\tau\tanh(\xi\tau)} \tag{4.112}
$$

and

$$
\xi = 2\sqrt{\epsilon_x \epsilon_y \epsilon_b}\sqrt{\frac{\delta_{xm}^2}{\epsilon_b^2} + \delta_{yn}^2} \tag{4.113}
$$

Thus, the dimensionless total resistance depends upon

$$
R^{\star}_{total} = f(\epsilon_x, \epsilon_y, \epsilon_b, \tau, Bi, Bi_{e,x}, Bi_{e,y}) \tag{4.114}
$$

where $\epsilon_x = a/c$, $\epsilon_y = b/d$, $\epsilon_b = c/d$, $\tau = t/\sqrt{A_s}$, $Bi = h_s t/k$, $Bi_{e,x} = h_{e,x}c/k$, $Bi_{e,y} = h_{e,y}d/k$, and $R^{\star}_t = k\sqrt{A_s}R_t$.

This general result has many geometric special cases. These include semi-infinite flux channels $t \to \infty$, infinite plate $c, d \to \infty$, half-space $t, c, d \to \infty$, three-dimensional strips $b = d$ or $a = c$, and adiabatic edges $h_{ex} \to 0$ and $h_{e,y} \to 0$. A particular interesting property is the case when adiabatic edges are present. It can be shown in this case, that when $Bi_{e,x} \to 0$ and $Bi_{e,y} \to 0$ but not equal to zero, the double summation which represents the total resistance, consists of the one-dimensional resistance when $m = n = 1$ and the spreading resistance when the remaining terms are summed $m = n \geq 2$.

## 4.7 Multilayered Rectangular Flux Channels

A system is composed of $N$ isotropic layers with a constant conductance, $h_s$, at the sink plane and an arbitrarily located heat source in the source plane is now considered. In a multilayered system, we have $N$ layers denoted $i = 1 \ldots N$ and $N - 1$ interfaces denoted $j = 1 \ldots N - 1$ as shown in Figure 4.9. These layers are isotropic and may be perfectly or imperfectly bonded to each other giving rise to an interfacial or contact resistance, $h_{c,j}$.

In Cartesian co-ordinates for $N$ isotropic layers, we have

$$
\frac{\partial^2 \theta_1}{\partial x^2} + \frac{\partial^2 \theta_1}{\partial y^2} + \frac{\partial^2 \theta_1}{\partial z_1^2} = 0, \quad 0 < z_1 < t_1
$$

$$
\frac{\partial^2 \theta_2}{\partial x^2} + \frac{\partial^2 \theta_2}{\partial y^2} + \frac{\partial^2 \theta_2}{\partial z_2^2} = 0, \quad 0 < z_2 < t_2 \tag{4.115}
$$

$$
\vdots \qquad\qquad \vdots
$$

$$
\frac{\partial^2 \theta_N}{\partial x^2} + \frac{\partial^2 \theta_N}{\partial y^2} + \frac{\partial^2 \theta_N}{\partial z_N^2} = 0, \quad 0 < z_N < t_N
$$

**Figure 4.9** Multi-layered rectangular flux channel with arbitrarily located heat source. (a) Top view. (b) Side view.

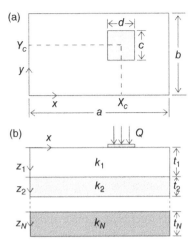

The following boundary conditions are imposed on the system. In the heat source plane, we specify a discrete heat flux distribution such that over the region of the heat source, a uniform heat flux is specified, while outside of the heat source region, the remainder of the surface is taken as adiabatic. Therefore, the following is assumed for a single heat source:

$$-k_1 \frac{\partial \theta_1}{\partial z_1} = \frac{Q}{cd}, \quad z_1 = 0 \text{ Inside Source Region}$$

$$\frac{\partial \theta_1}{\partial z_1} = 0, \qquad z_1 = 0 \text{ Outside Source Region}$$

(4.116)

At the $N-1$ interfaces, we require the following conditions representing the equality of flux and temperature drop due to the imperfect interface:

$$k_i \frac{\partial \theta_i}{\partial z_i}\bigg|_{z_i=t_i} = k_{i+1} \frac{\partial \theta_{i+1}}{\partial z_{i+1}}\bigg|_{z_{i+1}=0}$$

$$-k_i \frac{\partial \theta_i}{\partial z_i}\bigg|_{z_i=t_i} = h_{c,j}\left[\theta_i(x,y,t_i) - \theta_{i+1}(r,0)\right]$$

(4.117)

Finally, along the heat sink plane, we specify the following condition:

$$-k_N \frac{\partial \theta_N}{\partial z_N}\bigg|_{z_N=t_N} = h_s \theta_N(r,t_N)$$

(4.118)

Along the axis and edge of the disk:

$$\frac{\partial \theta_i}{\partial x}\bigg|_{x=0} = \frac{\partial \theta_i}{\partial y}\bigg|_{y=0} \ 0, \ i=1\ldots N$$

$$\frac{\partial \theta_i}{\partial x}\bigg|_{x=c} = \frac{\partial \theta_i}{\partial y}\bigg|_{y=d} \ 0, \ i=1\ldots N$$

(4.119)

The basic solution for a finite isotropic flux channel can be extended to the more general case of a $N$ layer structure strictly through modification of the definition of the spreading function $\phi$. Recently, Bagnall et al. (2014) presented a recursive procedure for defining a new spreading function in terms of spreading functions for the additional layers. The procedure was developed by solving the above equations in the $z$-direction and leads to a recursive formulation for the spreading function $\phi$.

In an $N$ layer system, the mean temperature (or resistance) of the heat source is now calculated from Eqs. (4.80)–(4.97), but with the following modifications:

$$A_0 = \frac{Q}{ab}\left(\sum_{i=1}^{N}\frac{t_i}{k_i} + \sum_{j=1}^{N-1}\frac{1}{h_{c,j}} + \frac{1}{h_s}\right) = \theta_{1D} \tag{4.120}$$

or

$$R_{1D} = \frac{1}{ab}\left(\sum_{i=1}^{N}\frac{t_i}{k_i} + \sum_{j=1}^{N-1}\frac{1}{h_{c,j}} + \frac{1}{h_s}\right) \tag{4.121}$$

to account for the $N$ layers, and $\phi \to \phi_1(\xi)$, where $\xi$ is the appropriate eigenvalue in each term of the solution, i.e. $\xi = \lambda_m, \delta_n, \beta_{mn}$.

In an $N$ layer structure, we can define two different spreading functions, one which assumes perfect interfacial contact and one which assumes imperfect contact. Although the former can be modeled from the latter, both are presented for convenience. The spreading function $\phi$ in Eq. (4.91) is a recursive function when two or more layers are present in a system and is now denoted as $\phi_1$. The recursive relationship is now given by a general expression for $\phi_i$ such that $\phi_1$ is a function of $\phi_2$ which is a function of $\phi_3$..., and so on, until the final spreading function $\phi_N$ is called. Following the work of Bagnall et al. (2014), we may write

**Perfect Interfacial Contact**

$$\phi_i = \frac{\frac{k_i}{k_{i+1}}\tanh(\xi t_i) + \phi_{i+1}}{\frac{k_i}{k_{i+1}} + \phi_{i+1}\tanh(\xi t_i)} \tag{4.122}$$

**Non-perfect Interfacial Contact**

$$\phi_i = \frac{\frac{k_i}{k_{i+1}}\tanh(\xi t_i) + \phi_{i+1}[1 + \frac{k_i}{h_{c,i}}\xi\tanh(\xi t_i)]}{\frac{k_i}{k_{i+1}} + \phi_{i+1}[\tanh(\xi t_i) + \frac{k_i}{h_{c,i}}\xi]} \tag{4.123}$$

In both cases, we denote the $N$th layer spreading function by

$$\phi_N = \left[\frac{\xi\tanh(\xi t_N) + h_s/k_N}{\xi + h_s/k_N\tanh(\xi t_N)}\right] \tag{4.124}$$

Specification of the spreading functions for $N$-layers is now a simple matter. Once defined, it becomes a simple exercise to define sequences of materials in multilayered structures and assessing thermal performance. Bagnall et al. (2014) considered a number of applications of the above solution in addition to also prescribing convection in the source plane.

## 4.8  Rectangular Flux Channel with an Elliptic Heat Source

The solution for a rectangular flux channel having an elliptic shaped heat source was considered by Sadhal (1984). Sadhal (1984) considered both the uniform flux and equivalent isothermal (inverse parabolic flux) sources for both finite and semi-infinite flux channels. Solutions were obtained using the separation of variables method as was done in Section 4.2. We will state the basic problem and its various solutions.

The problem statement is the same as that in Sections 4.3 and 4.4 except for the uniform flow condition at $z \to \infty$ and the specification of the heat source. We also define the flux channel as having dimensions $2a$ by $2b$ and the elliptic source having major and minor axes of $2\alpha$ and $2\beta$, respectively. Beginning with

$$\nabla^2 T = \frac{\partial^2 \theta}{\partial x^2} + \frac{\partial^2 \theta}{\partial y^2} + \frac{\partial^2 \theta}{\partial z^2} = 0 \tag{4.125}$$

subject to the following boundary conditions:

$$\left. \frac{\partial \theta}{\partial x} \right|_{x=0} = \left. \frac{\partial \theta}{\partial y} \right|_{y=0} = 0$$

$$\left. \frac{\partial \theta}{\partial x} \right|_{x=a} = \left. \frac{\partial \theta}{\partial y} \right|_{y=b} = 0 \tag{4.126}$$

at the centroid and edges of the channel, respectively. Over the top surface $z = 0$,

$$\left. -k \frac{\partial \theta}{\partial z} \right|_{z=0} = q(x, y),\ 0 < x < \alpha, \quad 0 < y < \beta$$

$$\left. \frac{\partial \theta}{\partial z} \right|_{z=0} = 0, \qquad \alpha < x < a, \quad \beta < y < b \tag{4.127}$$

where

$$q(x, y) = \begin{cases} \dfrac{Q}{\pi \alpha \beta} & \text{Uniform Flux} \\[2ex] \dfrac{Q}{2 \pi \alpha \beta} \left[ 1 - \left( \dfrac{x}{\alpha} \right)^2 - \left( \dfrac{y}{\beta} \right)^2 \right]^{-1/2} & \text{Equivalent Isothermal} \end{cases} \tag{4.128}$$

The heat source boundary is now defined by the equation for an ellipse:

$$1 - \left( \frac{x}{\alpha} \right)^2 - \left( \frac{y}{\beta} \right)^2 = 0 \tag{4.129}$$

The final Fourier integration over the source area is done with the appropriate limits of integration defined for the elliptic curve.

The sink plane now has the following boundary condition applied to a finite thickness slab:

$$\left. -k \frac{\partial \theta}{\partial z} \right|_{z=t} = h_s \theta(x, y, z = t) \tag{4.130}$$

where $\theta = T - T_f$

We will now summarize the basic solutions from Sadhal (1984) for each source. In each case, the total resistance is defined as

$$R_{total} = \frac{1}{4ab} \left( \frac{t}{k} + \frac{1}{h_s} \right) + R_s \tag{4.131}$$

where $R_s$ is determined from the constriction parameter $\psi$ defined by Sadhal (1984) as

$$\psi = 4k \sqrt{\alpha \beta} R_s \tag{4.132}$$

where $\sqrt{\alpha \beta}$ is referred to as the equivalent radius of the ellipse. In both cases, below the eigenvalues are determined from $\lambda_m = m\pi/a$, $\mu_n = n\pi/b$, and $\gamma_{mn} = \sqrt{\lambda_m^2 + \mu_n^2}$. These are necessary if we are to modify the basic flux channel solutions using the spreading functions for the addition of sink plane conductance or the effect of layers.

The two basic solutions for the semi-infinite flux channel provided by Sadhal (1984) are

**Isoflux Heat Source**

$$\psi^q = \frac{8\sqrt{\alpha\beta}}{\pi^3 ab}\left[\frac{a^3}{\alpha^2}\sum_{m=1}^{\infty}\frac{J_1^2(m\pi\alpha/a)}{m^3} + \frac{b^3}{\beta^2}\sum_{n=1}^{\infty}\frac{J_1^2(n\pi\beta/b)}{n^3}\right.$$
$$\left. + 2\sum_{m=1}^{\infty}\sum_{n=1}^{\infty}\frac{J_1^2(\pi\sqrt{m^2\alpha^2/a^2 + n^2\beta^2/b^2})}{(m^2\alpha^2/a^2 + n^2\beta^2/b^2)\sqrt{m^2/a^2 + n^2/b^2}}\right] \tag{4.133}$$

**Equivalent Isothermal Heat Source**

$$\psi_{eq}^T = \frac{4\sqrt{\alpha\beta}}{\pi^3 ab}\left[\frac{a^3}{\alpha^2}\sum_{m=1}^{\infty}\frac{\sin(m\pi\alpha/a)J_1(m\pi\alpha/a)}{m^3} + \frac{b^3}{\beta^2}\sum_{n=1}^{\infty}\frac{\sin(n\pi\beta/b)J_1(n\pi\beta/b)}{n^3}\right.$$
$$\left. + 2\sum_{m=1}^{\infty}\sum_{n=1}^{\infty}\frac{\sin(\pi\sqrt{m^2\alpha^2/a^2 + n^2\beta^2/b^2})J_1(\pi\sqrt{m^2\alpha^2/a^2 + n^2\beta^2/b^2})}{(m^2\alpha^2/a^2 + n^2\beta^2/b^2)\sqrt{m^2/a^2 + n^2/b^2}}\right]$$
$$\tag{4.134}$$

If the flux channel is finite, compound, multilayered, etc., we may modify the terms in each of the summations by dividing by the appropriate spreading functions $\phi(\xi)$ defined earlier, where $\xi$ are the eigenvalues defined above, i.e. $\phi_m(\lambda_m), \phi_n(\mu_n)$, and $\phi_{mn}(\gamma_{mn})$. This is possible due to the fact that the spreading function is introduced into the solution through application of boundary conditions in the $z$-direction.

## 4.9 Spreading in a Curved Flux Channel (Annular Sector)

The general solution for the spreading resistance of a flux-specified heat source on a compound annular sector with convective or conductive cooling at one boundary will be presented in Figure 4.10. In this system, heat flows through a portion of the outer surface through two layers, having different thermal conductivities, to the interior surface which is convectively cooled by a uniform film conductance. It is assumed that perfect contact is achieved at the interface of the two materials. This problem was solved by Muzychka et al. (2001). Solutions to this problem for both the isotropic and compound configurations will be presented.

A particular application of this solution is in tube/fin heat exchangers where several longitudinal fins are attached at equal spacing to a larger cylindrical tube. The solution for the overall thermal

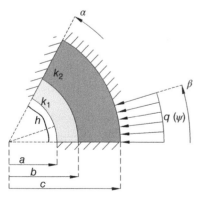

**Figure 4.10** Thermal spreading in a compound annular sector.

resistance in an isotropic or compound annulus would allow the effects of protective coatings or fouling deposits on the larger tube to be analyzed. The solution also considers curvature effects which are not present in other solutions such as the simple 2D compound flux channel.

The general solution will depend upon several dimensionless geometric and thermal parameters. In general, the total resistance is given by

$$R_{total} = R_{1D} + R_s \tag{4.135}$$

where $R_{1D}$ is the one-dimensional composite resistance of the system, and $R_s$ is the spreading resistance component. The total thermal resistance is defined as

$$R_{total} = \frac{\overline{T}_s - T_\infty}{Q} \tag{4.136}$$

where $\overline{T}_s$ is average source temperature.

The governing equation for each layer in the system shown in Figure 4.10 is Laplace's equation

$$\nabla^2 T_i(r, \psi) = \frac{\partial^2 T_i}{\partial r^2} + \frac{1}{r}\frac{\partial T_i}{\partial r} + \frac{1}{r^2}\frac{\partial^2 T_i}{\partial \psi^2} = 0, \quad \text{for } i = 1, 2 \tag{4.137}$$

which is subject to a convective or mixed-boundary condition on the interior surface

$$\left.\frac{\partial T_1}{\partial r}\right|_{r=a} = \frac{h}{k_1}\left(T_1(a, \psi) - T_\infty\right) \qquad 0 \leq \psi \leq \alpha \tag{4.138}$$

equality of the heat flux and temperature at the interface

$$k_1 \left.\frac{\partial T_1}{\partial r}\right|_{r=b} = k_2 \left.\frac{\partial T_2}{\partial r}\right|_{r=b} \qquad 0 \leq \psi \leq \alpha \tag{4.139}$$

$$T_1(b, \psi) = T_2(b, \psi) \qquad 0 \leq \psi \leq \alpha \tag{4.140}$$

and a specified flux distribution on the outer surface

$$\left.\frac{\partial T_2}{\partial r}\right|_{r=c} = \frac{q(\psi)}{k_2} \qquad \psi < \beta \tag{4.141}$$

$$\left.\frac{\partial T_2}{\partial r}\right|_{r=c} = 0 \qquad \beta < \psi \leq \alpha \tag{4.142}$$

The following symmetry conditions are also required

$$\left.\frac{\partial T_i}{\partial \psi}\right|_{\psi=0} = 0 \qquad a \leq r \leq c \tag{4.143}$$

$$\left.\frac{\partial T_i}{\partial \psi}\right|_{\psi=\alpha} = 0 \qquad a \leq r \leq c \tag{4.144}$$

where $\alpha$ varies, depending upon the symmetry of the problem. This allows for any number, $N \geq 1$, of equally spaced heat sources (or sinks) to be considered.

Finally, the heat flux $q(\psi)$ will take the following form:

$$q(\psi) = K\left[1 - \left(\frac{\psi}{\beta}\right)^2\right]^\mu \qquad 0 \leq \psi < \beta \tag{4.145}$$

where $\mu > -1$, and

$$K = \frac{Q}{\beta c}\frac{2}{\sqrt{\pi}}\frac{\Gamma(\mu + 3/2)}{\Gamma(\mu + 1)} \tag{4.146}$$

and $\Gamma(x)$ is the Gamma function. The general solution will be obtained for three heat flux distribution parameter values, $\mu = -1/2$, $0$, $1/2$.

The solution may be obtained by means of separation of variables in the form $\theta(r, \psi) = R(r) \cdot \Psi(\psi)$, where $\theta(r, \psi) = T(r, \psi) - T_\infty$ is the temperature excess. Applying the method of separation of variables yields the following solution which satisfies the thermal boundary conditions in the circumferential direction:

$$\theta_i(r, \psi) = A_i + B_i \ln(r) + \sum_{n=1}^{\infty} \left[ C_i r^{\lambda_n} + D_i r^{-\lambda_n} \right] \cos(\lambda_n \psi) \tag{4.147}$$

where $\lambda_n = n\pi/\alpha$.

The solution contains two parts, a uniform flow solution and a spreading (or constriction) solution which vanishes when the heat flux is distributed over the entire element. Since the solution is a linear superposition of the two parts, they will be dealt with separately.

The Fourier solution method indicates that a uniform flow solution also satisfies the prescribed thermal boundary conditions. This part of the temperature field is always present and leads to a one-dimensional radial thermal resistance.

Application of the boundary conditions yields the following result for the thermal resistance for a full compound annulus on a per unit length $L$ basis:

$$R_{1D} = \frac{\ln(b/a)}{2\pi k_1 L} + \frac{\ln(c/b)}{2\pi k_2 L} + \frac{1}{2\pi h a L} \tag{4.148}$$

or in dimensionless form

$$R_{1D}^* = \kappa \frac{\ln(1/\rho_1)}{2\pi} + \frac{\ln(1/\rho_2)}{2\pi} + \frac{\kappa}{2\pi Bi} \tag{4.149}$$

where $R^* = k_2 RL$, $0 < \rho_1 = a/b < 1$, $0 < \rho_2 = b/c < 1$, $\kappa = k_2/k_1$, and $Bi = ha/k_1$. Later we will adapt the above result accordingly for the basic element under consideration.

The spreading resistance part requires a solution to the two-dimensional eigenvalue problem for each layer. Application of the thermal boundary conditions at the interior surface and at the interface gives the solutions for $C_1$, $C_2$, and $D_1$ in terms of the unknown coefficient $D_2$. The final coefficient $D_2$ is obtained by taking a Fourier expansion of the exterior surface boundary condition.

The thermal spreading resistance is defined as

$$R_s = \frac{\overline{\theta}_s}{Q} \tag{4.150}$$

where

$$\overline{\theta}_s = \frac{1}{\beta} \int_0^\beta \theta_2(c, \psi) d\psi \tag{4.151}$$

It is often convenient to define a dimensionless spreading resistance parameter

$$\psi = k_2 R_s L \tag{4.152}$$

where $L$ is the depth of the region.

Muzychka et al. (2001) obtained the following expression for the dimensionless spreading resistance:

$$\psi = k_2 R_s L = \frac{4}{\sqrt{\pi}} \frac{\Gamma(\mu + 3/2)}{\Gamma(\mu + 1)}$$

$$\times \sum_{n=1}^{\infty} \varphi_n \frac{\sin(n\pi\beta/\alpha)\alpha}{\beta^2 n^2 \pi^2} \int_0^\beta \left[ 1 - \left( \frac{\psi}{\beta} \right)^2 \right]^\mu \cos(n\pi\psi/\alpha) d\psi \tag{4.153}$$

where the parameter $\varphi_n$ determines the effect of shell thicknesses, layer conductivities, and heat transfer coefficient. It is defined as

$$\varphi_n = \left[ \frac{(F_1 Bi + F_2 \lambda_n)\kappa + (F_3 Bi + F_4 \lambda_n)}{(F_4 Bi + F_3 \lambda_n)\kappa + (F_2 Bi + F_1 \lambda_n)} \right] \tag{4.154}$$

where

$$F_1 = 1 - (\rho_1)^{2\lambda_n} + (\rho_2)^{2\lambda_n} - (\rho_1 \rho_2)^{2\lambda_n} \tag{4.155}$$

$$F_2 = 1 + (\rho_1)^{2\lambda_n} + (\rho_2)^{2\lambda_n} + (\rho_1 \rho_2)^{2\lambda_n} \tag{4.156}$$

$$F_3 = 1 + (\rho_1)^{2\lambda_n} - (\rho_2)^{2\lambda_n} - (\rho_1 \rho_2)^{2\lambda_n} \tag{4.157}$$

$$F_4 = 1 - (\rho_1)^{2\lambda_n} - (\rho_2)^{2\lambda_n} + (\rho_1 \rho_2)^{2\lambda_n} \tag{4.158}$$

For an isotropic annulus, $\kappa = 1$, $\varphi_n$ may be simplified to give

$$\varphi_n = \left[ \frac{G_1 Bi + G_2 \lambda_n}{G_2 Bi + G_1 \lambda_n} \right] \tag{4.159}$$

where

$$G_1 = 1 - \rho^{2\lambda_n} \tag{4.160}$$

$$G_2 = 1 + \rho^{2\lambda_n} \tag{4.161}$$

where $\rho = a/c$ and $k_2 = k_1 = k$.

Evaluation of the integral in Eq. (4.153) gives the final general relation for the spreading resistance in the following form:

$$\psi = \frac{2}{\pi^2 \epsilon} \Gamma \left( \mu + \frac{3}{2} \right) \sum_{n=1}^{\infty} \left( \frac{2}{n\pi \epsilon} \right)^{\mu + \frac{1}{2}} \frac{\sin(n\pi\epsilon)}{n^2} J_{\mu + \frac{1}{2}}(n\pi\epsilon)\, \varphi_n \tag{4.162}$$

where $\epsilon = \beta/\alpha$.

The general solution is valid for any heat flux distribution defined by Eqs. (4.145) and (4.146) with $\mu > -1$. However, only three cases of practical interest will be presented. These are the uniform flux ($\mu = 0$), parabolic flux ($\mu = 1/2$), and inverted parabolic flux ($\mu = -1/2$). The inverted parabolic flux distribution is representative of the isothermal boundary condition for values of $\epsilon = \beta/\alpha < 0.5$.

The general solution for the spreading resistance in an annular sector has the same form as that for a finite compound flux channel, with the exception of the parameter $\varphi_n$. The parameter $\varphi_n$, is a function of the radii ratio, conductivity ratio, and Biot number, whereas for the finite compound flux channel, it is a function of the layer thicknesses, conductivity ratio, and Biot number.

The total dimensionless resistance of the compound annulus may now be obtained by combining the uniform flow resistance and spreading resistance. For a compound annulus containing $N$ equally spaced heat sources, the total thermal resistance is obtained from

$$R_{total}^* = \psi/(2N) + R_{1D}^* \tag{4.163}$$

or for the basic element shown in Figure 4.10, the total thermal resistance is

$$R_{total,e}^* = \psi + \frac{2\pi}{\alpha} R_{1D}^* \tag{4.164}$$

Several special cases may be obtained from the general solution, Eq. (4.162). Each of these cases represents a particular heat flux distribution. Three cases of particular interest are inverted parabolic flux distribution $\mu = -1/2$, the uniform flux distribution $\mu = 0$, and the parabolic flux distribution $\mu = 1/2$. In general, the exact flux distribution is not always known. These special cases provide a means to bound the results for the thermal spreading resistance.

The general solution for the compound annulus simplifies for three special cases of the heat flux distribution. The results are given below for each of the special cases:

$$\psi = \frac{2}{\pi^2 \epsilon} \sum_{n=1}^{\infty} \frac{J_0(n\pi\epsilon)\sin(n\pi\epsilon)}{n^2} \varphi_n, \qquad \mu = -1/2 \tag{4.165}$$

$$\psi = \frac{2}{\pi^3 \epsilon^2} \sum_{n=1}^{\infty} \frac{\sin^2(n\pi\epsilon)}{n^3} \varphi_n, \qquad \mu = 0 \tag{4.166}$$

$$\psi = \frac{4}{\pi^3 \epsilon^2} \sum_{n=1}^{\infty} \frac{J_1(n\pi\epsilon)\sin(n\pi\epsilon)}{n^3} \varphi_n, \qquad \mu = 1/2 \tag{4.167}$$

The preceding cases have the following relationship:

$$\psi(\mu = -1/2) < \psi(\mu = 0) < \psi(\mu = 1/2) \tag{4.168}$$

This allows the variation in the thermal resistance to be estimated when the precise heat flux distribution is not known.

When the radius of curvature is large relative to the thickness of the shell, the results for the compound annulus may be modeled using the results for a rectangular flux channel. The general result provided earlier for the simple compound flux channel may be compared with the result for the compound annulus equation (4.164) provided that

$$\psi_{annulus} = 2\psi_{strip} \tag{4.169}$$

since the solution for the strip is for two elements in parallel. Muzychka et al. (2001) also provided comparisons of the cases over a range of parameters and show that curvature is only important for very thick curved sections, provided the equivalent system is defined properly.

## 4.10   Effect of Surface Curvature on Thermal Spreading Resistance in a Two-Dimensional Flux Channel

The effect of a nonflat contact plane for a two-dimensional flux channel was examined by Hodes et al. (2018) who examined thermal spreading effects in super-hydrophobic channels composed of parallel ridges. Hodes et al. (2018) reviewed and compared several results for two-dimensional parallel ridges. Further details of these surfaces are provided in Chapter 9. Here, we only consider the extension of the classical two-dimensional flux channel for an isothermal heat source. Hodes et al. (2018) review several analytical results for the problem illustrated in Figure 4.11.

The interested reader is referred to an original paper for most of the details. We only present the final results for illustration and reference. Figure 4.12 provide the dimensionless thermal spreading resistance defined as follows:

$$\tilde{R}''_{sp} = \frac{k}{d}\left(\frac{T_s - \overline{T}_{cp}}{Q}\right) \tag{4.170}$$

which is plotted versus the contact ratio $\phi = A_s/A_t$. Note, in Sections 4.1–4.9, we use $\epsilon = \phi$ for a two-dimensional channel, and we also nondimensionalized the thermal spreading resistance using the source dimension rather than the flux tube dimension. As can be seen in Figure 4.12, significant deviation from the flat surface result is observed for all contact angles of the adiabatic boundary.

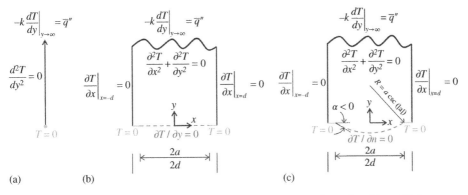

**Figure 4.11** Thermal spreading from a curved boundary: (a) one dimensional flow, (b) flat contact, and (c) non-flat contact with curved surface. Source: Hodes et al. (2018). Used with the permission of the American Society of Mechanical Engineers (ASME).

**Figure 4.12** Dimensionless thermal spreading resistance versus contact ratio for selected angles. Source: Hodes et al. (2018). Used with the permission of the American Society of Mechanical Engineers (ASME).

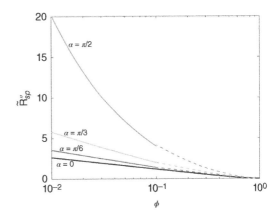

Hodes et al. (2018) reviewed several closed-form formulae for calculating the thermal spreading resistance in two-dimensional channels with a nonflat adiabatic boundary on the source plane. The results presented in Figure 4.12 are from two different formulae valid over a partial range of contact ratio $\phi$.

## References

Abramowitz, M., and Stegun, I.A., *Handbook of Mathematical Functions*, Dover Publishing, New York, 1965.

Bagnall, K., Muzychka, Y.S., and Wang, E., "Analytical Solution for Temperature Rise in Complex, Multi-layer Structures with Discrete Heat Sources", *IEEE Transactions on Components, Packaging and Manufacturing Technologies*, Vol. 4, no. 5, pp. 817–830, 2014.

Ellison, G.N., *Thermal Computations for Electronic Equipment*, Krieger Publishing, Malabar, FL, 1984.

Ellison, G.N., "Maximum Thermal Spreading Resistance for Rectangular Sources and Plates with Nonunity Aspect ratios", *IEEE Transactions on Components and Packaging Technologies*, Vol. 26, no. 2, pp. 439–454, 2003.

Hodes, M., Kirk, T., and Crowdy, D., "Spreading and Contact Resistance Formulae Capturing Boundary Curvature and Contact Distribution Effects", *Journal of Heat Transfer*, Vol. 140, pp. 104503-1–104503-7, 2018.

Krane, M.J.H., "Constriction Resistance in Rectangular Bodies", *Journal of Electronic Packaging*, Vol. 113, pp. 392–396, 1991.

Lasance, C., "How to Estimate Heat Spreading Effects in Practice", *Journal of Electronic Packaging*, Vol. 132, pp. 031004-1–031004-7, 2010.

Mikic, B.B., and Rohsenow, W.M., "Thermal Contact Resistance", M.I.T. Heat Transfer Lab. Report No. 4542-41, Cambridge, MA, September 1966.

Muzychka, Y.S., Influence Coefficient Method for Calculating Discrete Heat Source Temperature on Finite Convectively Cooled Substrates", *IEEE Transactions on Components and Packaging Technologies*, Vol. 29, no. 3, pp. 636–643, 2006.

Muzychka, Y.S., "Spreading Resistance in Compound Orthotropic Flux Tubes and Channels with Interfacial Resistance," *Journal of Thermophysics and Heat Transfer*, doi: http://arc.aiaa .org/doi/abs/10.2514/1.T4203, 2014.

Muzychka, Y.S., Stevanovic, M., and Yovanovich, M.M., "Thermal Spreading Resistance in Compound Annular Sectors", *Journal of Thermophysics and Heat Transfer*, Vol. 15, pp. 354–359, 2001.

Muzychka, Y.S., Culham, J.R., and Yovanovich, M.M., "Thermal Spreading Resistance of Eccentric Heat Sources on Rectangular Flux Channels", *Journal of Electronic Packaging*, Vol. 125, pp. 178–185, 2003.

Muzychka, Y.S., Culham, J.R., and Yovanovich, M.M., "Thermal Spreading Resistances in Compound and Orthotropic Systems", *AIAA Journal of Thermophysics and Heat Transfer*, Vol. 18, pp. 45–51, 2004.

Muzychka, Y.S., Yovanovich, M.M., and Culham, J.R., "Influence of Geometry and Edge Cooling on Thermal Spreading Resistance", *AIAA Journal of Thermophysics and Heat Transfer*, Vol. 20, pp. 247–255, 2006.

Muzychka, Y.S., Bagnall, K., and Wang, E., "Thermal Spreading Resistance and Heat Source Temperature in Compound Orthotropic Systems with Interfacial Resistance", *IEEE Transactions on Components, Packaging and Manufacturing Technologies*, Vol. 3, no. 11, pp. 1826–1841, 2013.

Negus, K.J., Yovanovich, M.M., and Beck, J.V., "On the Nondimensionalization of Constriction Resistance for Semi-infinite Heat Flux Tubes", *Journal of Heat Transfer*, Vol. 111, pp. 804–807, 1989.

Sadhal, S.S., "Exact Solutions for the Steady and Unsteady Diffusion Problems for a Rectangular Prism: Cases of Complex Neumann Conditions", *ASME 84-HT-83*, 1984.

Sexl, R.U., and Burkhard, D.G., "An Exact Solution for Thermal Conduction Through a Two-Dimensional Eccentric Constriction", *Progress in Astronautics and Aeronautics*, Vol. 21, pp. 617–620, 1969.

Smythe, W.R., *Static and Dynamic Electricity*, 3rd Edition, McGraw-Hill Book Company, New York, 1968.

Veziroglu, T.N., and Chandra, S., "Thermal Conductance of Two Dimensional Constrictions", in *Progress in Astronautics and Aeronautics: Thermal Design Principles of Spacecraft and Entry Bodies*, AIAA, Vol. 21, ed. G.T. Bevins, pp. 617–620, 1969.

Yovanovich, M.M., and Marotta, E., "Chapter 4: Thermal Spreading and Contact Resistances", *Heat Transfer Handbook*, eds. A. Bejan and A.D. Kraus, Wiley, 2003.

Yovanovich, M.M., Muzychka, Y.S., and Culham, J.R., "Spreading Resistance of Isoflux Rectangles and Strips on Compound Flux Channels", *AIAA Journal of Thermophysics and Heat Transfer*, Vol. 13, pp. 495–500, 1999.

# 5

# Orthotropic Media

Conduction in orthotropic media is a special case of conduction in an anisotropic solid whereby thermal conductivity is different in each of the principal orthogonal coordinate directions. Since we are only interested in orthotropic media, the reader is referred to Carslaw and Jaeger (1959) or Ozisik (1993) for more details on anisotropic heat conduction. In general, the orthotropic system is the more common application found in microelectronics and hence, most if not all studies cited come from this field.

The first examination of thermal constriction resistance in orthotropic media was by Yovanovich (1970). Yovanovich (1970) obtained the solution for an isothermal circular source on an alternating multilayered half-space. This general analysis illustrated how an orthotropic half-space can be reduced to an isotropic half-space using simple transformations. This allows results for thermal spreading in an orthotropic half-space that can easily be deduced from the solutions in Chapter 2 for many configurations. We consider this problem in detail later and extend the analysis for an isoflux heat source.

Lam and Fischer (1999) and Ying and Toh (1999, 2000) considered orthotropic properties in thermal spreaders for electronics cooling applications. Solutions for finite disks/channels and flux tubes/channels were further developed by Muzychka et al. (2003a,b), while more complex multilayered systems were considered by Muzychka et al. (2013) and Bagnall et al. (2014). In the latter paper, the authors developed a general solution for a multilayered orthotropic rectangular flux channel with and without perfect contact at layer interfaces. Application to devices with multiple discrete heat sources was also developed. In Muzychka (2014, 2015), a general formulation for multilayered disks was obtained considering three different heat source flux distributions.

Gholami and Bahrami (2014) obtained a solution for an orthotropic flux channel with multiple finite sources and finite sinks. Finally, the reader is directed to Al-Khamaiseh et al. (2018) for additional solutions for nonlinear thermal spreading/constriction in isotropic, orthotropic, and multilayered media. Solution for a fully orthotropic and multilayered rectangular spreader is obtained and extended to multiple heat sources.

## 5.1 Heat Conduction in Orthotropic Media

In three-dimensional Cartesian coordinates, Laplace's equation for an orthotropic material is written as

$$k_x \frac{\partial^2 T}{\partial x^2} + k_y \frac{\partial^2 T}{\partial y^2} + k_z \frac{\partial^2 T}{\partial z^2} = 0 \tag{5.1}$$

*Thermal Spreading and Contact Resistance: Fundamentals and Applications*, First Edition.
Yuri S. Muzychka and M. Michael Yovanovich.
© 2023 John Wiley & Sons, Inc. Published 2023 by John Wiley & Sons, Inc.

We may also consider the special case where the in-plane thermal conductivity $k_x = k_y = k_{xy} \neq k_z$ is constant such that

$$k_{xy}\left(\frac{\partial^2 T}{\partial x^2} + \frac{\partial^2 T}{\partial y^2}\right) + k_z\frac{\partial^2 T}{\partial z^2} = 0 \tag{5.2}$$

This is a reasonable assumption for many practical thermal spreading problems in electronics cooling applications where the transverse thermal conductivity is different from the in-plane thermal conductivity, such as the case with printed circuit boards and various semiconductor materials.

Further, under the assumption of two-dimensional axi-symmetric heat conduction, Laplace's equation becomes

$$k_r\left(\frac{\partial^2 T}{\partial r^2} + \frac{1}{r}\frac{\partial T}{\partial r}\right) + k_z\frac{\partial^2 T}{\partial z^2} = 0 \tag{5.3}$$

At first sight, one might assume that thermal spreading/constriction in orthotropic media is more complex. However, by means of a simple coordinate transformation, Laplace's equation for orthotropic media can be transformed into a form with a single value of effective thermal conductivity, thus reducing it to the same form as an isotropic media.

Examination of Eqs. (5.2) and (5.3) indicates that a simple coordinate transformation variable $\xi$ can be introduced such that

$$z = \xi\sqrt{k_z/k_{xy}} \tag{5.4}$$

or

$$z = \xi\sqrt{k_z/k_r} \tag{5.5}$$

We may generalize this for either system of coordinates as

$$z = \xi\sqrt{k_{tp}/k_{ip}} \tag{5.6}$$

where *ip* denotes the "in-plane" conductivity and *tp* denotes the "through-plane" conductivity.

In both instances, the thermal conductivity variables cancel out and the $z$-variable is replaced by the $\xi$-variable, leaving us with Laplace's equation for an isotropic material having constant conductivity:

$$\frac{\partial^2 T}{\partial x^2} + \frac{\partial^2 T}{\partial y^2} + \frac{\partial^2 T}{\partial \xi^2} = 0 \tag{5.7}$$

or

$$\frac{\partial^2 T}{\partial r^2} + \frac{1}{r}\frac{\partial T}{\partial r} + \frac{\partial^2 T}{\partial \xi^2} = 0 \tag{5.8}$$

This constant thermal conductivity will be determined shortly after we consider the transformation of boundary conditions.

With the governing equation transformed, the boundary conditions now need attention. The in-plane conductivity affects the edge boundary conditions, while the through-plane conductivity affects the source and sink plane boundary conditions. We will first address the issue of the source and sink plane conditions. In the source plane, we typically prescribe the constant flux or variable flux condition. This requires

$$-k_z\frac{\partial T}{\partial z}\bigg|_{z=0} = q \tag{5.9}$$

within the source area, while outside the source area we prescribe

$$-k_z \frac{\partial T}{\partial z}\bigg|_{z=0} = 0 \tag{5.10}$$

Introducing the transformation defined by Eq. (5.6) and denoting $k_z = k_{tp}$, we may write for either Cartesian or cylindrical coordinates:

$$-\frac{k_{tp}}{\sqrt{k_{tp}/k_{ip}}} \frac{\partial T}{\partial \xi}\bigg|_{\xi=0} = -\sqrt{k_{tp}k_{ip}} \frac{\partial T}{\partial \xi}\bigg|_{\xi=0} = q \tag{5.11}$$

within the source area, while outside the source area, we prescribe

$$-\frac{k_{tp}}{\sqrt{k_{tp}/k_{ip}}} \frac{\partial T}{\partial \xi}\bigg|_{\xi=0} = -\sqrt{k_{tp}k_{ip}} \frac{\partial T}{\partial \xi}\bigg|_{\xi=0} = 0 \tag{5.12}$$

In the sink plane, we often apply

$$-k_z \frac{\partial T}{\partial z}\bigg|_{z=t} = h_s(T - T_\infty) \tag{5.13}$$

Applying the transformation for the $z$-coordinate (again denoting $k_z = k_{tp}$) gives

$$-\frac{k_{tp}}{\sqrt{k_{tp}/k_{ip}}} \frac{\partial T}{\partial \xi}\bigg|_{\xi=t/\sqrt{k_{tp}/k_{ip}}} = h_s(T - T_\infty) \tag{5.14}$$

or

$$-\sqrt{k_{tp}k_{ip}} \frac{\partial T}{\partial \xi}\bigg|_{\xi=t/\sqrt{k_{tp}/k_{ip}}} = h_s(T - T_\infty) \tag{5.15}$$

This allows us to conclude that we may replace the original values of thermal conductivity and thickness in thermal spreading problems with "effective" values, each defined as follows:

$$t_{eff} = \frac{t}{\sqrt{k_{tp}/k_{ip}}} \tag{5.16}$$

and

$$k_{eff} = \sqrt{k_{tp}k_{ip}} \tag{5.17}$$

leading to

$$-k_{eff} \frac{\partial T}{\partial \xi}\bigg|_{\xi=0} = q \tag{5.18}$$

within the source area, while outside the source area, we prescribe

$$-k_{eff} \frac{\partial T}{\partial \xi}\bigg|_{\xi=0} = 0 \tag{5.19}$$

and in the sink plane

$$-k_{eff} \frac{\partial T}{\partial \xi}\bigg|_{\xi=t_{eff}} = h_s(T - T_\infty) \tag{5.20}$$

All edge boundary conditions with adiabatic conditions remain unchanged since the thermal conductivity does not play a direct factor in determining the eigenvalues in this case. Hence, these

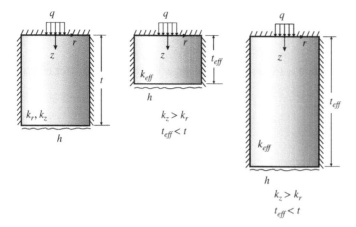

**Figure 5.1** Effect of co-ordinate transformations on effective thermal conductivity and thickness for orthotropic flux tubes and channels.

do not require any special attention. Problems with edge conductance $h_e$ must be dealt with carefully given that the "in-plane" thermal conductivity will now play a factor in determining the eigenvalues. We will address this particular issue in Section 5.3–5.6 for flux tubes and channels as required.

After transformation, the effective thickness of the thermal spreader will have either increased or decreased as shown Figure 5.1, depending upon the ratio of the through-plane conductivity $k_z$ to the in-plane conductivity $k_r, k_{xy}$. Thus, with appropriate care, problem statements for spreading problems in simple orthotropic domains can be reduced to those for an equivalent isotropic system containing a single effective thermal conductivity and effective thickness. The solution for the orthotropic system is then easily obtained from a solution for this equivalent isotropic system. The transformations defined above will be applied to isotropic spreading resistance solutions for finite flux tubes and rectangular channels with adiabatic edges. Further discussion on the transformation of orthotropic systems to isotropic systems may be found in Carslaw and Jaeger (1959) and Ozisik (1993).

In general, while thermal conductivity will vary in all three normal directions for an orthotropic material, many applications can be modeled with only two unique values of thermal conductivity. In orthotropic systems such as printed circuit boards, parallel and series models are often used to define $k_r, k_{xy}$, and $k_z$, respectively, as the effective in-plane and through-plane thermal conductivity. These are defined as

$$k_r, k_{xy} = \frac{\sum_{i=1}^{N} k_i t_i}{t} \tag{5.21}$$

and

$$k_z = \frac{t}{\sum_{i=1}^{N} \dfrac{t_i}{k_i}} \tag{5.22}$$

In general, thermal spreading resistance in multilayered systems is a strong function of the size and distribution of conducting layers. However, if the source size is considerably larger than the thickness of individual layers [Yovanovich (1970)], the above relations for determining effective series and parallel conductivities may be applied.

## 5.2 Circular Source on a Half-Space

We begin our examination of problems with orthotropic properties by reconsidering the classic problem of the isothermal source on a half-space. Yovanovich (1970) first introduced the system of stretched coordinates to solve this fundamental problem for circular isothermal source on a half-space. The problem is formally specified as

$$k_r \left( \frac{\partial^2 T}{\partial r^2} + \frac{1}{r} \frac{\partial T}{\partial r} \right) + k_z \frac{\partial^2 T}{\partial z^2} = 0 \tag{5.23}$$

subject to

$$\begin{aligned}
z = 0, \quad & T = T_0, \ 0 < r < a \\
z = 0, \quad & \frac{\partial T}{\partial z} = 0, \ r > a \\
z \to \infty, \quad & T \to 0
\end{aligned} \tag{5.24}$$

Introducing $\xi = z/\gamma$, where $\gamma = \sqrt{k_z/k_r}$ transforms Eq. (5.23) into

$$\frac{\partial^2 T}{\partial r^2} + \frac{1}{r} \frac{\partial T}{\partial r} + \frac{\partial^2 T}{\partial \xi^2} = 0 \tag{5.25}$$

now subject to

$$\begin{aligned}
\xi = 0, \quad & T = T_0, \ 0 < r < a \\
\xi = 0, \quad & \frac{\partial T}{\partial \xi} = 0, \ r > a \\
\xi \to \infty, \quad & T \to 0
\end{aligned} \tag{5.26}$$

The solution to this problem discussed in Chapter 2 has the form

$$T = \frac{2T_0}{\pi} \int_0^\infty e^{-\lambda \xi} \sin(\lambda a) J_0(\lambda r) \frac{d\lambda}{\lambda} \tag{5.27}$$

The total heat flow from the disk is determined from

$$Q = \int_0^a -k_z \left[ \frac{\partial T}{\partial z} \right]_{z=0} 2\pi r \, dr \tag{5.28}$$

or

$$Q = \int_0^a -\sqrt{k_z k_r} \left[ \frac{\partial T}{\partial \xi} \right]_{\xi=0} 2\pi r \, dr \tag{5.29}$$

Evaluating the above leads to

$$Q = 4\sqrt{k_z k_r} \, a T_0 \tag{5.30}$$

Finally, defining a spreading resistance leads to

$$R_s = \frac{T_0}{Q} = \frac{1}{4\sqrt{k_z k_r} a} = \frac{1}{4k_{eff} a} \tag{5.31}$$

where $k_{eff} = \sqrt{k_z k_r}$ is the effective or equivalent thermal conductivity of the medium. The above result is the same as the result for an isotropic media but with an effective thermal conductivity. Yovanovich (1970) developed his solution for the special case of a composite half-space composed of alternating layers of two materials.

If we now consider the same problem, but with uniform heat flux over the heat source, we must consider Eq. (5.23) with the following new boundary conditions:

$$z = 0, \quad \frac{\partial T}{\partial z} = -\frac{q_0}{k_z}, \quad 0 < r < a$$

$$z = 0, \quad \frac{\partial T}{\partial z} = 0, \quad r > a \tag{5.32}$$

$$z \to \infty, \ T \to 0$$

Introducing $\xi = z/\gamma$ with $\gamma = \sqrt{k_z/k_r}$ once again returns Eq. (5.25), but with the following new boundary conditions:

$$\xi = 0, \quad \frac{\partial T}{\partial \xi} = -\frac{q_0}{\sqrt{k_z k_r}}, \quad 0 < r < a$$

$$\xi = 0, \quad \frac{\partial T}{\partial \xi} = 0, \quad r > a \tag{5.33}$$

$$\xi \to \infty, \ T \to 0$$

The solution for this problem becomes

$$T = \frac{q_0 a}{k_{eff}} \int_0^\infty e^{-\lambda \xi} J_0(\lambda r) J_1(\lambda a) \frac{d\lambda}{\lambda} \tag{5.34}$$

where $k_{eff} = \sqrt{k_r k_z}$. As shown in Chapter 2, the temperature rise in the source area $0 \le r \le a$ for $\xi = 0$ is

$$T(r) = \frac{q_0 a}{k_{eff}} \int_0^\infty J_0(\lambda r) J_1(\lambda a) \frac{d\lambda}{\lambda} \tag{5.35}$$

The area-average source temperature is

$$\overline{T} = \frac{1}{\pi a^2} \frac{q_0 a}{k_{eff}} \int_0^a \left[ \int_0^\infty J_0(\lambda r) J_1(\lambda a) \frac{d\lambda}{\lambda} \right] 2\pi r \, dr \tag{5.36}$$

The integrals can be interchanged giving the result [Carslaw and Jaeger (1959)]:

$$\overline{T} = \frac{2q_0}{k_{eff}} \int_0^\infty J_1^2(\lambda a) \frac{d\lambda}{\lambda^2} = \frac{8}{3\pi} \cdot \frac{q_0 a}{k_{eff}} \tag{5.37}$$

According to the definition of spreading resistance, one obtains for the isoflux circular source, the relation [Carslaw and Jaeger (1959)]:

$$\overline{R}_s = \frac{\overline{T}}{Q} = \frac{8}{3\pi^2} \cdot \frac{1}{k_{eff} a} \tag{5.38}$$

We may also define a spreading resistance based on the maximum centroidal temperature at $r = 0$ in Eq. (5.35):

$$\hat{R}_s = \frac{T(0)}{Q} = \frac{1}{\pi k_{eff} a} \tag{5.39}$$

Thus, in both cases, the thermal spreading resistance for a simple orthotropic half-space can be determined from the solutions for isotropic media by simply using the effective thermal conductivity of the medium.

**Example 5.1** Consider thermal spreading in a silicon dioxide substrate for a micro-contact of $a = 0.1$ mm. Polycrystalline silicon dioxide has a thermal conductivity $k = 1.38$ W/(m K), while crystalline silicon dioxide has $k_z = k_\perp = 6.2$ W/(m K) and $k_r = k_\parallel = 10.4$ W/(m K). Compare values of $R_s$ computed for an isothermal contact.

For the polycrystalline silicon dioxide, we have

$$R_s = \frac{1}{4ka} = \frac{1}{4 \cdot 1.38 \cdot 0.0001} = 1811.6 \text{ K/W}$$

For the crystalline silicon dioxide, we have

$$R_s = \frac{1}{4k_{eff}a} = \frac{1}{4 \cdot \sqrt{6.2 \cdot 10.4} \cdot 0.0001} = 311.3 \text{ K/W}$$

The thermal spreading resistance for crystalline form of silicon dioxide is approximately one-sixth that of the polycrystalline form due to an effective thermal conductivity of $k_{eff} = 8.03 \text{ W/(m K)}$.

## 5.3 Single-Layer Flux Tubes

In a cylindrical orthotropic system, Laplace's equation, Eq. (5.3), may be transformed using $\xi = z/\gamma$, where $\gamma = \sqrt{k_z/k_r}$, and $\theta = T - T_f$ to yield:

$$\frac{\partial^2 \theta}{\partial r^2} + \frac{1}{r}\frac{\partial \theta}{\partial r} + \frac{\partial^2 \theta}{\partial \xi^2} = 0 \tag{5.40}$$

which is subjected to the following transformed boundary conditions:

$$r = 0, b, \quad \frac{\partial \theta}{\partial r} = 0$$

$$\xi = 0, \quad \frac{\partial \theta}{\partial \xi} = -\frac{q}{k_{eff}}, \quad 0 \leq r < a$$

$$\frac{\partial \theta}{\partial \xi} = 0, \quad a < r \leq b \tag{5.41}$$

$$\xi = t_{eff}, \quad \frac{\partial \theta}{\partial \xi} = -\frac{h_s}{k_{eff}}\theta$$

Equations (5.40) and (5.41) are now in the same form as that for an isotropic disk, except that an effective thermal conductivity $k_{eff} = \sqrt{k_r k_z}$ now replaces the isotropic thermal conductivity and an effective thickness, $t_{eff} = t/\gamma$, now replaces the flux tube thickness. The solution for this case is

$$\theta(r, \xi) = \frac{Q}{\pi b^2}\left(\frac{t_{eff}}{k_{eff}} + \frac{1}{h_s}\right) - \frac{Q\xi}{k_{eff}\pi b^2}$$

$$+ \sum_{n=1}^{\infty} \frac{2Q}{\phi \pi a k_{eff}} \frac{J_1(\delta_n a/b)}{\delta_n^2 J_0^2(\delta_n)} J_0(\delta_n r/b)[\cosh(\delta_n \xi/b) - \phi \sinh(\delta_n \xi/b)] \tag{5.42}$$

$$\psi = 4k_{eff}aR_s = \frac{16}{\pi \epsilon}\sum_{n=1}^{\infty} \frac{J_1^2(\delta_n \epsilon)}{\delta_n^3 J_0^2(\delta_n)\phi_n} \tag{5.43}$$

where

$$\phi_n = \frac{\delta_n \tanh(\delta_n \tau_{eff}) + Bi_{eff}}{\delta_n + Bi_{eff}\tanh(\delta_n \tau_{eff})} \tag{5.44}$$

The dimensionless thickness and Biot number now become: $\tau_{eff} = t_{eff}/b$ and $Bi_{eff} = h_s b/k_{eff}$, where the effective conductivity is $k_{eff} = \sqrt{k_r k_z}$, effective thickness is $t_{eff} = t/\gamma$, and $\epsilon = a/b$. The eigenvalues $\delta_n$ are obtained from $J_1(\delta_n) = 0$. The total resistance is now obtained from

$$R_t = R_{1D} + R_s = \frac{1}{\pi b^2}\left(\frac{t_{eff}}{k_{eff}} + \frac{1}{h_s}\right) + R_s \tag{5.45}$$

Note, that $t_{eff}/k_{eff} = t/k_z$ by way of the definitions of the effective thickness and thermal conductivity.

### 5.3.1 Circular Flux Tubes with Edge Cooling

In the case of edge cooling, the only modification required is in the specification of the edge boundary condition:

$$-k_r \frac{\partial \theta}{\partial r}\Big|_{r=b} = h_e \theta(b, z) \tag{5.46}$$

Care must be taken not to use the effective conductivity here. The eigenvalues are now obtained from

$$\delta_n J_1(\delta_n) = \frac{h_e b}{k_r} J_0(\delta_n) \tag{5.47}$$

since the radial or in-plane thermal conductivity appears in the boundary condition. It may be transformed for consistency to give

$$\delta_n J_1(\delta_n) = Bi_e \gamma J_0(\delta_n) \tag{5.48}$$

where $Bi_e = h_e b/k_{eff}$.

The solution of Yovanovich (2003) reported in Chapter 3 becomes

$$\psi = 4ak_{eff}R_{total} = \frac{16}{\pi\epsilon} \sum_{n=1}^{\infty} \left(\frac{2}{\delta_n\epsilon}\right)^{\mu} \frac{\Gamma(2+\mu)J_{1+\mu}(\delta_n\epsilon)J_1(\delta_n\epsilon)}{\delta_n^3[J_0^2(\delta_n) + J_1^2(\delta_n)]\phi_n} \tag{5.49}$$

where $\phi_n$ is defined by Eq. (5.44) and $\tau_{eff} = t_{eff}/b$, $Bi_{eff} = h_s b/k_{eff}$, and $\epsilon = a/b$, and $\delta_n$ are the eigenvalues determined from Eq. (5.48).

## 5.4 Single-Layer Rectangular Flux Channel

In a rectangular orthotropic system with $k_x = k_y = k_{xy} \neq k_z$, Laplace's equation, Eq. (5.2), may also be transformed using $\xi = z/\gamma$, where $\gamma = \sqrt{k_z/k_{xy}}$, and $\theta = T - T_f$, to yield,

$$\frac{\partial^2\theta}{\partial x^2} + \frac{\partial^2\theta}{\partial y^2} + \frac{\partial^2\theta}{\partial \xi^2} = 0 \tag{5.50}$$

which is subjected to the following transformed boundary conditions:

$$x = 0, c, \quad \frac{\partial\theta}{\partial x} = 0$$

$$y = 0, d, \quad \frac{\partial\theta}{\partial y} = 0$$

$$\xi = 0, \quad \frac{\partial\theta}{\partial \xi} = -\frac{q}{k_{eff}}, \quad 0 \leq x < a$$
$$0 \leq y < b$$
$$\frac{\partial\theta}{\partial \xi} = 0, \quad a < x \leq c \tag{5.51}$$
$$b < y \leq d$$
$$\xi = t_{eff}, \quad \frac{\partial\theta}{\partial \xi} = -\frac{h_s}{k_{eff}}\theta$$

Equations (5.50) and (5.51) are now in the same form as those for an isotropic flux tube, except that an effective thermal conductivity $k_{eff} = \sqrt{k_{xy}k_z}$ now replaces the isotropic thermal conductivity and an effective thickness $t_{eff} = t/\gamma$ replaces the flux channel thickness.

This general solution now has the form:

$$\theta(x, y, \xi) = \theta_{1D}(\xi) + \theta_s(x, y, \xi) \tag{5.52}$$

The problem can be solved in two parts with the one-dimensional flow solution easily shown to be

$$\theta_{1D}(\xi) = \frac{Q}{4cd}\left(\frac{t_{eff}}{k_{eff}} + \frac{1}{h_s}\right) - \frac{Q\xi}{4cdk_{eff}} \tag{5.53}$$

The solution for the multidimensional spreading component which satisfies the $x, y$ boundary conditions is

$$\theta_s(x, y, \xi) = \sum_{m=1}^{\infty} \cos(\lambda_m x)[A_m \cosh(\lambda_m \xi) + B_m \sinh(\lambda_m \xi)]$$

$$+ \sum_{n=1}^{\infty} \cos(\delta_n y)[A_n \cosh(\delta_n \xi) + B_n \sinh(\delta_n \xi)]$$

$$+ \sum_{m=1}^{\infty}\sum_{n=1}^{\infty} \cos(\lambda_m x)\cos(\delta_n y)[A_{mn} \cosh(\beta_{mn} \xi) + B_{mn} \sinh(\beta_{mn} \xi)] \tag{5.54}$$

Application of the sink plane boundary condition yields the following general result for each portion of the solution:

$$B_i = -\phi(\zeta)A_i \quad i = m, n, mn \tag{5.55}$$

where $\phi$ is the general spreading function found earlier:

$$\phi(\zeta) = \frac{\zeta \tanh(\zeta t_{eff}) + \dfrac{h_s}{k_{eff}}}{\zeta + \dfrac{h_s}{k_{eff}} \tanh(\zeta t_{eff})} \tag{5.56}$$

where $\zeta = \lambda_m, \delta_n, \beta_{mn}$ is the eigenvalue in each portion of the solution, i.e. $\lambda_m = m\pi/c$, $\delta_n = n\pi/d$, and $\beta_{mn} = \sqrt{\lambda_m^2 + \delta_n^2}$. The final Fourier coefficients are given as before but with $k = k_{eff}$. This gives the final general result for the spreading resistance:

$$k_{eff}R_s = \sum_{m=1}^{\infty} \frac{\sin^2(m\pi a/c)}{2d(a/c)^2(m\pi)^3\phi(\lambda_m)} + \sum_{n=1}^{\infty} \frac{\sin^2(n\pi b/d)}{2c(b/d)^2(n\pi)^3\phi(\delta_n)}$$

$$+ \sum_{m=1}^{\infty}\sum_{n=1}^{\infty} \frac{\sin^2(m\pi a/c)\sin^2(n\pi b/d)}{cd(a/c)^2(b/d)^2(m\pi)^2(n\pi)^2\sqrt{(m\pi/c)^2 + (n\pi/d)^2}\phi(\beta_{mn})} \tag{5.57}$$

As before, we see that the spreading resistance is simply modified by the general spreading function which is a function of the effective thickness and thermal conductivity. The total resistance is now the sum of the bulk resistance and the spreading resistance such that

$$R_{total} = R_{1D} + R_s = \frac{1}{4cd}\left(\frac{t_{eff}}{k_{eff}} + \frac{1}{h_s}\right) + R_s \tag{5.58}$$

The above result is similar to that in Chapter 4, with the exception of the effective variables $t_{eff}$ and $k_{eff}$. Also, note again that in Eq. (5.58), the group $t_{eff}/k_{eff} = t/k_z$, since the effect of orthotropic properties only impacts the multidimensional spreading resistance and not the one-dimensional bulk resistance.

**Example 5.2** Consider a rectangular thermal spreader of dimensions $2c = 2d = 30$ mm consisting of silicon dioxide with a heat source $2a = 2b = 10$ mm. The spreader has thickness $t = 3$ mm and is cooled with a sink conductance of $h_s = 150$ W/(m$^2$ K). Polycrystalline silicon dioxide has a thermal conductivity $k = 1.38$ W/(m K), while crystalline silicon dioxide has $k_z = k_\perp = 6.2$ W/(m K) and $k_{xy} = k_\parallel = 10.4$ W/(m K). Compare values of $R_s$, $R_{1D}$, and $R_{total}$ for each case assuming an isoflux heat source.

For this problem, we use Eqs. (5.56)–(5.58). The resulting effective properties that are required are

$$t_{eff} = \frac{t}{\sqrt{k_z/k_{xy}}} = \frac{0.003}{\sqrt{6.2/10.4}} = 0.00388 \text{ m}$$

and

$$k_{eff} = \sqrt{k_z k_{xy}} = \sqrt{6.2 \cdot 10.4} = 8.03 \text{ W/(m K)}$$

Using the prescribed variables, we find the following thermal resistances for the polycrystalline silicon dioxide:

$$R_s = 22.84 \text{ K/W}$$

$$R_{1D} = 9.82 \text{ K/W}$$

$$R_{total} = 32.66 \text{ K/W}$$

and for the crystalline silicon dioxide, we have

$$R_s = 4.28 \text{ K/W}$$

$$R_{1D} = 7.95 \text{ K/W}$$

$$R_{total} = 12.23 \text{ K/W}$$

Once again, we see that the thermal spreading resistance for crystalline form of silicon dioxide is approximately one-sixth that of the polycrystalline form due to an effective thermal conductivity of $k_{eff} = 8.03$ W/(m K). Significant reduction in the total resistance is seen mainly due to the thermal spreading component of the total resistance.

### 5.4.1 Rectangular Flux Channels with Edge Cooling

In the case of a rectangular flux channel cooled along the edges with different edge conductance $h_{e,x}$ and $h_{e,y}$, we may adapt the solution reported earlier. The solution of Muzychka et al. (2006) for the total system resistance for an edge cooled isotropic flux channel may now be transformed for an orthotropic channel. The required expression for the total thermal resistance is

$$k_{eff} R_T = \frac{cd}{a^2 b^2} \sum_{m=1}^{\infty} \sum_{n=1}^{\infty} \frac{\sin^2(\delta_{xm} a/c)\sin^2(\delta_{yn} b/d)\phi_{mn}}{\delta_{xm} \delta_{yn} \beta_{mn}[\sin(2\delta_{xm})/2 + \delta_{xm}][\sin(2\delta_{yn})/2 + \delta_{yn}]} \tag{5.59}$$

where

$$\phi_{mn} = \frac{t_{eff}\beta_{mn} + \dfrac{h_s t_{eff}}{k_{eff}} \tanh(\beta_{mn} t_{eff})}{\dfrac{h_s t_{eff}}{k_{eff}} + t_{eff}\beta_{mn} \tanh(\beta_{mn} t_{eff})} \tag{5.60}$$

The eigenvalues are obtained from the following equations:

$$\delta_{xm} \sin(\delta_{xm}) = \frac{h_{e,x} c}{k_{xy}} \cos(\delta_{xm})$$

$$\delta_{yn} \sin(\delta_{yn}) = \frac{h_{e,y} d}{k_{xy}} \cos(\delta_{yn}) \tag{5.61}$$

or for consistency, we may write as

$$\delta_{xm} \sin(\delta_{xm}) = Bi_{e,x} \gamma \cos(\delta_{xm})$$

$$\delta_{yn} \sin(\delta_{yn}) = Bi_{e,y} \gamma \cos(\delta_{yn}) \tag{5.62}$$

where $Bi_{e,x} = h_{e,x} c / k_{eff}$, $\delta_{xm} = \lambda_{xm} c$, $Bi_{e,y} = h_{e,y} d / k_{eff}$, and $\delta_{yn} = \lambda_{yn} d$. Finally, the separation constant $\beta_{mn} = \sqrt{\lambda_{xm}^2 + \lambda_{yn}^2}$.

## 5.5 Multilayered Orthotropic Spreaders

We now turn our attention to multilayered orthotropic spreaders. As before, we only consider the special case of simple orthotropic spreaders, i.e. those with an in-plane thermal conductivity which is different from the through-plane thermal conductivity. This is a reasonable assumption for many applications in microelectronics packaging.

A simple system is composed of a single isotropic or orthotropic layer with a constant conductance, $h_s$, in the sink plane. This system forms the basis for all of the solutions presented in this section. In a multilayered system, we have $N$ layers denoted $i = 1 \ldots N$ and $N - 1$ interfaces denoted $j = 1 \ldots N - 1$ as denoted earlier in Figures 3.4 and 4.9. These layers may now be orthotropic or a combination of orthotropic and isotropic and may also be perfectly or imperfectly bonded to each other, giving rise to an interfacial or contact resistance. An interfacial contact conductance $h_c$ is used to model the imperfect contact region by specifying a Robin boundary condition. This conductance can also be modeled as an equivalent resistance such that

$$R_j = \frac{t_j}{k_j A} = \frac{1}{h_{c,j} A} \tag{5.63}$$

or

$$h_{c,j} = \frac{k_j}{t_j} \tag{5.64}$$

where $k_j$ is the thermal conductivity of the $j$th interfacial region and $t_j$ the nominal thickness of this region. This approach is adopted since the interfacial layer is usually thin relative to the adjacent layers, i.e. $t_j \ll t_i$. It can be shown that a two-layer system with a third effective interfacial layer can also be modeled as a two-layer system with contact conductance provided the above condition is reasonably satisfied.

In Chapters 3 and 4, we considered solutions to the basic single-layer problem and also for cases with multiple isotropic layers. These solutions will be extended here for the case when $k_r \neq k_z$ or $k_{xy} \neq k_z$.

### 5.5.1 Circular Flux Tubes

The multilayered flux tube was considered in Chapter 3 for a system of isotropic layers. In cylindrical co-ordinates for a system of $N$ orthotropic layers, we now have [Muzychka (2015)]:

$$k_{1,r}\left(\frac{\partial^2 \theta_1}{\partial r^2} + \frac{1}{r}\frac{\partial \theta_1}{\partial r}\right) + k_{1,z}\frac{\partial^2 \theta_1}{\partial z_1^2} = 0, \qquad 0 < z_1 < t_1$$

$$k_{2,r}\left(\frac{\partial^2 \theta_2}{\partial r^2} + \frac{1}{r}\frac{\partial^2 \theta_2}{\partial r}\right) + k_{2,z}\frac{\partial^2 \theta_2}{\partial z_2^2} = 0, \qquad 0 < z_2 < t_2$$

$$\vdots \qquad\qquad\qquad\qquad \vdots \qquad\qquad (5.65)$$

$$k_{N,r}\left(\frac{\partial^2 \theta_N}{\partial r^2} + \frac{1}{r}\frac{\partial^2 \theta_N}{\partial r}\right) + k_{N,z}\frac{\partial^2 \theta_N}{\partial z_N^2} = 0, \; 0 < z_N < t_N$$

The following boundary conditions are imposed on the system. In the heat source plane, we specify an arbitrary heat flux distribution which may be distributed or uniform, while outside of the heat source region the remainder of the surface is taken as adiabatic. Therefore, the following is assumed for a single heat source:

$$-k_{1,z}\frac{\partial \theta_1}{\partial z_1} = q(r), \; z_1 = 0, \; 0 < r < a$$

$$-k_{1,z}\frac{\partial \theta_1}{\partial z_1} = 0, \qquad z_1 = 0, \; a < r < b \qquad (5.66)$$

At the $N-1$ interfaces, we require the following conditions representing the equality of flux and temperature drop due to the imperfect interface:

$$k_{i,z}\frac{\partial \theta_i}{\partial z_i}\bigg|_{z_i=t_i} = k_{i+1,z}\frac{\partial \theta_{i+1}}{\partial z_{i+1}}\bigg|_{z_{i+1}=0}$$

$$-k_{i,z}\frac{\partial \theta_i}{\partial z_i}\bigg|_{z_i=t_i} = h_{c,j}\left[\theta_i(r,t_i) - \theta_{i+1}(r,0)\right] \qquad (5.67)$$

Finally, along the heat sink plane we specify the following condition:

$$-k_{N,z}\frac{\partial \theta_N}{\partial z_N}\bigg|_{z_N=t_N} = h_s\theta_N(r,t_N) \qquad (5.68)$$

Along the axis and edge of the disk:

$$\frac{\partial \theta_i}{\partial r}\bigg|_{r=0} = 0, \; i = 1\ldots N$$

$$\frac{\partial \theta_i}{\partial r}\bigg|_{r=b} = 0, \; i = 1\ldots N \qquad (5.69)$$

The above statement is now complete. In order to facilitate the solution, we may transform the multilayered orthotropic system into an equivalent multilayered isotropic system using stretched coordinates presented earlier. Muzychka (2015) implemented a system of stretched coordinates for a multilayered orthotropic system where the in-plane thermal conductivity is not equal to the through-plane conductivity within each layer, i.e. $k_{i,r} \neq k_{i,z}$.

Application of the following transformations:

$$\text{Layer 1: } \xi_1 = z_1 / \sqrt{k_{1,z}/k_{1,r}}$$

$$\text{Layer 2: } \xi_2 = z_2 / \sqrt{k_{2,z}/k_{2,r}} \tag{5.70}$$

$$\vdots$$

$$\text{Layer } N: \xi_N = z_N / \sqrt{k_{N,z}/k_{N,r}}$$

for each layer leads to the definition of the following *effective* isotropic layer properties:

$$\text{Layer 1: } k_1 = \sqrt{k_{1,r}k_{1,z}} \quad \bar{t}_1 = t_1/\sqrt{k_{1,z}/k_{1,r}}$$

$$\text{Layer 2: } k_2 = \sqrt{k_{2,r}k_{2,z}} \quad \bar{t}_2 = t_2/\sqrt{k_{2,z}/k_{2,r}} \tag{5.71}$$

$$\vdots$$

$$\text{Layer } N: k_N = \sqrt{k_{N,r}k_{N,z}} \quad \bar{t}_N = t_N/\sqrt{k_{N,z}/k_{N,r}}$$

Equation (5.64) transform to become

$$\left( \frac{\partial^2 \theta_1}{\partial r^2} + \frac{1}{r}\frac{\partial \theta_1}{\partial r} \right) + \frac{\partial^2 \theta_1}{\partial \xi_1^2} = 0, \quad 0 < \xi_1 < \bar{t}_1$$

$$\left( \frac{\partial^2 \theta_2}{\partial r^2} + \frac{1}{r}\frac{\partial^2 \theta_2}{\partial r} \right) + \frac{\partial^2 \theta_2}{\partial \xi_2^2} = 0, \quad 0 < \xi_2 < \bar{t}_2 \tag{5.72}$$

$$\vdots \qquad\qquad \vdots$$

$$\left( \frac{\partial^2 \theta_N}{\partial r^2} + \frac{1}{r}\frac{\partial^2 \theta_N}{\partial r} \right) + \frac{\partial^2 \theta_N}{\partial \xi_N^2} = 0, \quad 0 < \xi_N < \bar{t}_N$$

subject to the same boundary conditions as before, but with effective layer conductivity used in their definitions. For the sake of brevity, we will not rewrite them.

The problem statement is now in an equivalent form as the system solved in Chapter 3 for a multilayer flux tube. In an $N$ layer system, the spreading resistance is now calculated from

$$4k_1 a R_s = \frac{16}{\pi\epsilon} \sum_{n=1}^{\infty} \frac{J_1^2(\delta_n \epsilon)}{\phi_1 \delta_n^3 J_0^2(\delta_n)} \tag{5.73}$$

where $\epsilon = a/b$ is the relative contact size and $\phi_1$ is the spreading function associated with the first layer.

In an $N$ layer structure, we can define two different spreading functions, one which assumes perfect interfacial contact and one which assumes imperfect contact. Although the former can be modeled from the latter, both are presented for convenience. The spreading function $\phi$ in Eq. (5.73) is a recursive function when two or more layers are present in a system and is now denoted $\phi_1$ for the first layer. The recursive relationship is now given by a general expression for $\phi_i$ such that $\phi_1$ is a function of $\phi_2$ which is a function of $\phi_3...$, and so on, until the final spreading function $\phi_N$ is called. Following the work of Bagnall et al. (2014), we may write:

**Perfect Interfacial Contact**

$$\phi_i = \frac{\frac{k_i}{k_{i+1}} \tanh(\delta_n \bar{t}_i/b) + \phi_{i+1}}{\frac{k_i}{k_{i+1}} + \phi_{i+1} \tanh(\delta_n \bar{t}_i/b)} \tag{5.74}$$

**Nonperfect Interfacial Contact**

$$\phi_i = \frac{\frac{k_i}{k_{i+1}} \tanh(\delta_n \bar{t}_i/b) + \phi_{i+1}[1 + \frac{k_i}{h_{c,i}b}\delta_n \tanh(\delta_n \bar{t}_i/b)]}{\frac{k_i}{k_{i+1}} + \phi_{i+1}[\tanh(\delta_n \bar{t}_i/b) + \frac{k_i}{h_{c,i}b}\delta_n]} \tag{5.75}$$

In both cases, we denote the *N*th layer spreading function by

$$\phi_N = \left[\frac{\delta_n \tanh(\delta_n \bar{t}_N/b) + h_s b/k_N}{\delta_n + h_s b/k_N \tanh(\delta_n \bar{t}_N/b)}\right] \tag{5.76}$$

Specification of the spreading functions for *N*-layers is now a simple matter. Once defined, it becomes a simple exercise to define sequences of materials in multilayered structures and assessing thermal performance. The spreading function is a function of the eigenvalues which for the flux tube are

$$\delta_n = \lambda_n b = 3.8317, 7.0156, 10.1735, 13.3237, 16.4706, \ldots \tag{5.77}$$

Successive eigenvalues may be approximated by adding $\pi$ to each new value as the above sequence yields $\delta_i - \delta_{i-1} = \pi$ as $i \to \infty$.

The total resistance of the disk considering the uniform flow portion is

$$R_t = \sum_{i=1}^{N} \frac{\bar{t}_i}{k_i \pi b^2} + \sum_{j=1}^{N-1} \frac{1}{h_{c,j}\pi b^2} + \frac{1}{h_s \pi b^2} + R_s \tag{5.78}$$

The procedure summarized assumes a uniform heat flux source. If a variable axi-symmetric heat flux is desired, Eq. (5.73) may be replaced with

$$4k_1 a R_s = \frac{16(\mu + 1)}{\pi \epsilon} \sum_{n=1}^{\infty} \frac{D_n}{\phi_1} \frac{J_1(\delta_n \epsilon)}{\delta_n^3 J_0^2(\delta_n)} \tag{5.79}$$

where $D_n$ is defined according to

$$\begin{aligned} \mu &= -\frac{1}{2} \quad D_n = \sin(\delta_n \epsilon) \\ \mu &= 0 \quad\; D_n = J_1(\delta_n \epsilon) \\ \mu &= +\frac{1}{2} \quad D_n = \sin(\delta_n \epsilon)\left[\frac{1}{(\delta_n \epsilon)^2} - \frac{1}{(\delta_n \epsilon)\tan(\delta_n \epsilon)}\right] \end{aligned} \tag{5.80}$$

The heat source flux distribution is defined according to

$$q(r) = q_0(1 + \mu)(1 - r^2/a^2)^\mu \tag{5.81}$$

where $q_0 = Q/\pi a^2$ is the mean flux over the source area.

In this manner, when $\mu = -1/2$ an inverse parabolic distribution is obtained which yields an approximately isothermal heat source for $\epsilon < 0.4$, for $\mu = 0$ a uniform heat source is obtained, and when $\mu = 1/2$ a parabolic flux distribution is obtained. This now completes the analysis of a multilayered, orthotropic disk with interfacial resistance and variable heat flux source.

### 5.5.2 Multilayered Orthotropic Flux Channels

We now proceed to rectangular flux channels composed of multiple orthotropic layers which were considered by Bagnall et al. (2014). In Cartesian co-ordinates for $N$ orthotropic layers, we have

$$
k_{1,xy} \left( \frac{\partial^2 \theta_1}{\partial x^2} + \frac{\partial^2 \theta_1}{\partial y^2} \right) + k_{1,z} \frac{\partial^2 \theta_1}{\partial z_1^2} = 0, \quad 0 < z_1 < t_1
$$

$$
k_{2,xy} \left( \frac{\partial^2 \theta_2}{\partial x^2} + \frac{\partial^2 \theta_2}{\partial y^2} \right) + k_{2,z} \frac{\partial^2 \theta_2}{\partial z_2^2} = 0, \quad 0 < z_2 < t_2
$$

$$
\vdots \qquad\qquad\qquad \vdots
$$

$$
k_{N,xy} \left( \frac{\partial^2 \theta_N}{\partial x^2} + \frac{\partial^2 \theta_3}{\partial y^2} \right) + k_{N,z} \frac{\partial^2 \theta_N}{\partial z_N^2} = 0, \; 0 < z_N < t_N
$$

$$(5.82)$$

The following boundary conditions are imposed on the system. In the heat source plane, we specify a discrete heat flux distribution such that over the region of the heat source, a uniform heat flux is specified, while outside of the heat source region the remainder of the surface is taken as adiabatic. Therefore, the following is assumed for a single heat source:

$$
-k_{1,z} \frac{\partial \theta_1}{\partial z_1} = \frac{Q}{4ab}, \; z_1 = 0, \, 0 < x < a, \quad 0 < y < b
$$

$$
-k_{1,z} \frac{\partial \theta_1}{\partial z_1} = 0, \qquad z_1 = 0, \, a < x < c, \quad b < y < d
$$

$$(5.83)$$

At the $N - 1$ interfaces, we require the following conditions representing the equality of flux and temperature drop due to the imperfect interface:

$$
k_{i,z} \frac{\partial \theta_i}{\partial z_i} \bigg|_{z_i = t_i} = k_{i+1,z} \frac{\partial \theta_{i+1}}{\partial z_{i+1}} \bigg|_{z_{i+1} = 0}
$$

$$
-k_{i,z} \frac{\partial \theta_i}{\partial z_i} \bigg|_{z_i = t_i} = h_{c,j} \left[ \theta_i(x, y, t_i) - \theta_{i+1}(x, y, 0) \right]
$$

$$(5.84)$$

Finally, along the heat sink plane, we specify the following condition:

$$
-k_{N,z} \frac{\partial \theta_N}{\partial z_N} \bigg|_{z_N = t_N} = h_s \theta_N(x, y, t_N)
$$

$$(5.85)$$

Along the axis and edges of the disk:

$$
\frac{\partial \theta_i}{\partial x} \bigg|_{x=0} = \frac{\partial \theta_i}{\partial y} \bigg|_{y=0} 0, \; i = 1 \dots N
$$

$$
\frac{\partial \theta_i}{\partial x} \bigg|_{x=c} = \frac{\partial \theta_i}{\partial y} \bigg|_{y=d} 0, \; i = 1 \dots N
$$

$$(5.86)$$

The above statement is now complete. As before, we may transform the multilayered orthotropic system into an equivalent multilayered isotropic system using stretched coordinates for simple orthotropic systems, where the in-plane thermal conductivity is not equal to the through-plane conductivity, i.e. $k_{i,xy} \neq k_{i,z}$.

Application of the following transformations:

Layer 1: $\xi_1 = z_1 / \sqrt{k_{1,z}/k_{1,xy}}$

Layer 2: $\xi_2 = z_2 / \sqrt{k_{2,z}/k_{2,xy}}$

$\vdots$

Layer $N$: $\xi_N = z_N / \sqrt{k_{N,z}/k_{N,xy}}$

$\qquad\qquad\qquad (5.87)$

for each layer leads to the definition of the following *effective* isotropic layer properties as before:

Layer 1: $k_1 = \sqrt{k_{1,xy}k_{1,z}} \quad \bar{t}_1 = t_1 / \sqrt{k_{1,z}/k_{1,xy}}$

Layer 2: $k_2 = \sqrt{k_{2,xy}k_{2,z}} \quad \bar{t}_2 = t_2 / \sqrt{k_{2,z}/k_{2,xy}}$

$\vdots$

Layer $N$: $k_N = \sqrt{k_{N,xy}k_{N,z}} \quad \bar{t}_N = t_N / \sqrt{k_{N,z}/k_{N,xy}}$

$\qquad\qquad\qquad (5.88)$

The thermal spreading resistance obtained in Chapter 4 for a multilayered isotropic system now becomes:

$$k_1 R_s = \sum_{m=1}^{\infty} \frac{\sin^2(m\pi a/c)}{2d(a/c)^2(m\pi)^3 \phi_1(\lambda_m)} + \sum_{n=1}^{\infty} \frac{\sin^2(n\pi b/d)}{2c(b/d)^2(n\pi)^3 \phi_1(\delta_n)}$$
$$+ \sum_{m=1}^{\infty}\sum_{n=1}^{\infty} \frac{\sin^2(m\pi a/c)\sin^2(n\pi b/d)}{cd(a/c)^2(b/d)^2(m\pi)^2(n\pi)^2 \sqrt{(m\pi/c)^2 + (n\pi/d)^2}\phi_1(\beta_{mn})} \qquad (5.89)$$

where we now define the spreading function as before (see Chapter 4):

**Perfect Interfacial Contact**

$$\phi_i = \frac{\frac{k_i}{k_{i+1}}\tanh(\zeta\bar{t}_i) + \phi_{i+1}}{\frac{k_i}{k_{i+1}} + \phi_{i+1}\tanh(\zeta\bar{t}_i)} \qquad\qquad (5.90)$$

**Nonperfect Interfacial Contact**

$$\phi_i = \frac{\frac{k_i}{k_{i+1}}\tanh(\zeta\bar{t}_i) + \phi_{i+1}[1 + \frac{k_i}{h_{c,i}}\zeta\tanh(\zeta\bar{t}_i)]}{\frac{k_i}{k_{i+1}} + \phi_{i+1}[\tanh(\zeta\bar{t}_i) + \frac{k_i}{h_{c,i}}\zeta]} \qquad (5.91)$$

In both cases, we denote the $N$th layer spreading function by

$$\phi_N = \left[ \frac{\zeta\tanh(\zeta\bar{t}_N) + h_s/k_N}{\zeta + h_s/k_N\tanh(\zeta\bar{t}_N)} \right] \qquad\qquad (5.92)$$

where $\zeta = \lambda_m, \delta_n, \beta_{mn}$ is the eigenvalue in each portion of the solution, i.e. $\lambda_m = m\pi/c$, $\delta_n = n\pi/d$, and $\beta_{mn} = \sqrt{\lambda_m^2 + \delta_n^2}$.

The total resistance of the flux channel considering the uniform flow portion is

$$R_t = \sum_{i=1}^{N} \frac{\bar{t}_i}{k_i 4cd} + \sum_{j=1}^{N-1} \frac{1}{h_{c,j} 4cd} + \frac{1}{h_s 4cd} + R_s \qquad (5.93)$$

### 5.5.3 Multilayered Orthotropic Flux Channels with an Eccentric Source

In the case of a single eccentric heat source on a multilayered flux channel, the results in Chapter 4 may be simply extended using the appropriate spreading function, eigenvalues, and effective properties. Care must be taken since the notation for the flux channel with a central heat source is different from that of an eccentric source.

We may also use this basic solution to obtain solutions for more than one source using superposition. In Chapter 6, we will develop the general multisource approach such that the system may be isotropic, orthotropic, or a combination of the both in the multilayered stack.

## 5.6 General Multilayered Rectangular Orthotropic Spreaders

Until now, we have only considered simple orthotropic spreaders, i.e. those with only two distinct values of thermal conductivity. The solution for a true orthotropic rectangular spreader ($k_x \neq k_y \neq k_z$) can be found using the solution approaches in earlier solutions, i.e. separation of variables. Al-Khamaiseh et al. (2018) obtained a general solution for a multilayered rectangular orthotropic spreader with an eccentric source. They also extended the solution using superposition to allow for multiple heat sources. The interested reader is referred to the paper for details and implementation of the solution. Here, we will highlight the essential differences and provide the basic result for a single heat source.

The problem under consideration is a three-dimensional rectangular flux channel consisting of $N$-layers with a single eccentric heat source in the source plane and convective cooling along the sink plane, as shown in Figure 5.2. All the lateral edges are assumed to be adiabatic. Each layer is assumed to be orthotropic with different thermal conductivities in the three spatial directions $(x, y, z)$. An interfacial contact conductance $h_{c_i}$ is considered between the adjacent layers (layer $i$ and $i + 1$) to model the effects of surface roughness, imperfect contact, or the intrinsic phonon mismatch between dissimilar materials joined together. The system is modeled using a local system of coordinates for each layer in which the $xy$-planes have the same coordinates in all the layers with $0 < x < c$ and $0 < y < d$, while the through-plane direction ($z$) is different for each layer. This approach is used as it facilities the stretched coordinate transformations and produces a convenient form of the general solution [Muzychka et al. (2013)].

By defining the temperature excess $\theta = T - T_\infty$ relative to the ambient temperature $T_f$, the governing equation in each layer is Laplace's equation. Hence, the following system of equations represents the governing equations for the $N$-layers:

$$k_{1,x}\frac{\partial^2 \theta_1}{\partial x^2} + k_{1,y}\frac{\partial^2 \theta_1}{\partial y^2} + k_{1,z}\frac{\partial^2 \theta_1}{\partial z_1^2} = 0, \quad 0 < z_1 < t_1$$

$$k_{2,x}\frac{\partial^2 \theta_2}{\partial x^2} + k_{2,y}\frac{\partial^2 \theta_2}{\partial y^2} + k_{2,z}\frac{\partial^2 \theta_2}{\partial z_2^2} = 0, \quad 0 < z_2 < t_2 \tag{5.94}$$

$$\vdots \qquad\qquad \vdots$$

$$k_{N,x}\frac{\partial^2 \theta_N}{\partial x^2} + k_{N,y}\frac{\partial^2 \theta_N}{\partial y^2} + k_{N,z}\frac{\partial^2 \theta_N}{\partial z_N^2} = 0, \quad 0 < z_N < t_N$$

with different thermal conductivities in each direction, i.e. $k_x \neq k_y \neq k_z$ for each layer. The following boundary conditions are considered. In the source plane, a uniform heat flux is specified over the heat source region where the heat source is considered as of rectangular shape with dimensions

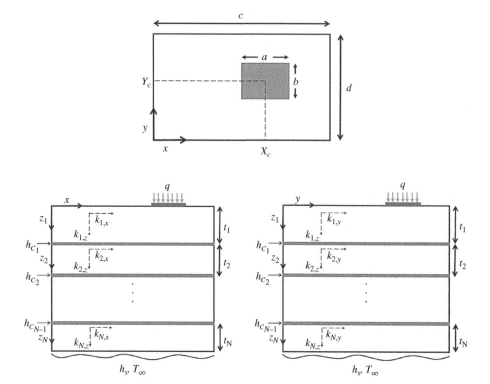

**Figure 5.2** General multi-layered orthotropic rectangular flux channel. Source: Al-Khamaiseh et al. (2018). Used with the permission of the American Society of Mechanical Engineers (ASME).

$a$ and $b$ in $x$ and $y$ directions, respectively, while the remainder of the source plane is considered as adiabatic. Hence, the source plane boundary condition is given by

$$-k_{1,z} \frac{\partial \theta_1}{\partial z_1}\bigg|_{z_1=0} = \begin{cases} q & \text{Inside Source Region} \\ 0 & \text{Outside Source Region} \end{cases} \tag{5.95}$$

At the interface between the adjacent layers, the following conditions are considered (for $i = 1, 2, \dots, N-1$), representing the continuity of heat flux and the temperature drop due to the interfacial conductance, respectively:

$$k_{i,z} \frac{\partial \theta_i}{\partial z_i}\bigg|_{z_i=t_i} = k_{i+1,z} \frac{\partial \theta_{i+1}}{\partial z_{i+1}}\bigg|_{z_{i+1}=0} \tag{5.96}$$

$$-k_{i,z} \frac{\partial \theta_i}{\partial z_i}\bigg|_{z_i=t_i} = h_{c_i} \left[ \theta_i(x,y,t_i) - \theta_{i+1}(x,y,0) \right] \tag{5.97}$$

The temperature drop condition in Eq. (5.97) might be replaced by the following condition in the case of a high value of the interfacial conductance $h_{c_i} \to \infty$:

$$\theta_i(x,y,t_i) = \theta_{i+1}(x,y,0) \tag{5.98}$$

Along the sink plane, a uniform heat transfer coefficient $h_s$ exists and the boundary condition is given by

$$-k_{N,z} \frac{\partial \theta_N}{\partial z_N}\bigg|_{z_N=t_N} = h_s \theta_N(x,y,t_N) \tag{5.99}$$

The lateral edges of the system are assumed to be adiabatic. The lateral edges boundary conditions are

$$\frac{\partial \theta_i}{\partial x}\bigg|_{x=0,c} = 0, \quad \frac{\partial \theta_i}{\partial y}\bigg|_{y=0,d} = 0, \quad i = 1, 2, \ldots, N \tag{5.100}$$

The problem statement along with the governing equations and boundary conditions is now completely defined. We may now proceed to apply the stretched coordinate transformations in order to present the problem in a simpler solvable form. As we shall see, the simple approach we applied earlier will require additional care and attention to obtain a solution.

### 5.6.1  Coordinate Transformations for Fully Orthotropic Media

As we saw earlier in Section 5.1, stretched coordinate transformations can be used as a powerful technique to transform orthotropic systems into equivalent isotropic systems [Ozisik (1993)]. Muzychka et al. (2004, 2013) implemented a system of stretched coordinates for a flux channel consisting of two transversely isotropic layers with equal in-plane thermal conductivities $k_x = k_y$ that are different than the through-plane conductivity, i.e. $k_x = k_y \neq k_z$, in each layer. The application of the following transformations for each layer (for $i = 1, 2, \ldots, N$):

$$\text{Layer } i : \quad y_i = y/\sqrt{k_{i,y}/k_{i,x}}, \quad \zeta_i = z_i/\sqrt{k_{i,z}/k_{i,x}} \tag{5.101}$$

leads to the definition of the following effective isotropic properties:

$$\text{Layer } i : \quad k_i = \sqrt{k_{i,x}k_{i,z}}, \quad \bar{t}_i = t_i/\sqrt{k_{i,z}/k_{i,x}}, \quad \bar{d}_i = d/\sqrt{k_{i,y}/k_{i,x}} \tag{5.102}$$

Under these transformations, the system of governing equations in Eq. (5.94) becomes:

$$\frac{\partial^2 \theta_i}{\partial x^2} + \frac{\partial^2 \theta_i}{\partial y_i^2} + \frac{\partial^2 \theta_i}{\partial \zeta_i^2} = 0, \quad 0 < x < c, \quad 0 < y_i < \bar{d}_i, \quad 0 < \zeta_i < \bar{t}_i. \tag{5.103}$$

for $i = 1 \ldots N$.

Although the direct application of the transformations in Eq. (5.101) is able to transform the governing equations in Eq. (5.94) into an equivalent set of equations given in Eq. (5.103) with isotropic properties, a problem appears when trying to transform the interface boundary conditions given by Eqs. (5.96) and (5.97) using these transformations because we have different stretched coordinates in the $y$-direction for each layer with different dimensions. In other words, the $y_i$ coordinates are different in each of the layers. It is important to note that when the in-plane conductivities are equal, i.e. $k_{i,x} = k_{i,y}$, in each layer, the new stretched coordinates in the $y$-direction are the same for all the layers and equal to the original coordinate, i.e. $y_i = y$; hence, the interface boundary conditions can be transformed directly as in [Muzychka et al. (2004, 2013)]. However, in order to solve the problem with different conductivities in the three directions, a second transformation is necessary. The $y$-direction stretched coordinates $y_i$ in layers $i = 2, 3, \ldots, N$ can be transformed to the stretched coordinate of the first layer $y_1$ by using:

$$y_i = \sqrt{\mu_i}\, y_1, \quad \text{with} \quad \mu_i = \frac{k_{1,x}k_{i,y}}{k_{1,y}k_{i,x}}, \quad i = 2, 3, \ldots, N. \tag{5.104}$$

Hence, the system of equations and boundary conditions given in Eqs. (5.94)–(5.100) can be transformed by using Eqs. (5.101) and (5.104) into the following system:

$$\frac{\partial^2 \theta_1}{\partial x^2} + \frac{\partial^2 \theta_1}{\partial y_1^2} + \frac{\partial^2 \theta_1}{\partial \zeta_1^2} = 0, \qquad 0 < \zeta_1 < \bar{t}_1$$

$$\frac{\partial^2 \theta_i}{\partial x^2} + \frac{1}{\mu_i}\frac{\partial^2 \theta_i}{\partial y_1^2} + \frac{\partial^2 \theta_i}{\partial \zeta_i^2} = 0, \qquad \begin{array}{l} 0 < \zeta_i < \bar{t}_i \\ i = 2, 3, \ldots, N \end{array}$$

(5.105)

with $0 < x < c$ and $0 < y_1 < \bar{d}_1$, and subject to the following boundary conditions:

$$-k_1 \frac{\partial \theta_1}{\partial \zeta_1}\bigg|_{\zeta_1=0} = \begin{cases} q & \text{Inside Transformed Source Region} \\ 0 & \text{Outside Transformed Source Region} \end{cases}$$

(5.106)

in the source plane, while the interfacial boundary conditions are transformed to

$$k_i \frac{\partial \theta_i}{\partial \zeta_i}\bigg|_{\zeta_i=\bar{t}_i} = k_{i+1}\frac{\partial \theta_{i+1}}{\partial \zeta_{i+1}}\bigg|_{\zeta_{i+1}=0}$$

(5.107)

$$-k_i \frac{\partial \theta_i}{\partial \zeta_i}\bigg|_{\zeta_i=\bar{t}_i} = h_{c_i}\left[\theta_i(x, y_1, \bar{t}_i) - \theta_{i+1}(x, y_1, 0)\right]$$

(5.108)

Along the sink plane, we have

$$-k_N \frac{\partial \theta_N}{\partial \zeta_N}\bigg|_{\zeta_N=\bar{t}_N} = h_s \theta_N(x, y_1, \bar{t}_N)$$

(5.109)

and for the lateral edges boundary conditions, we get

$$\frac{\partial \theta_i}{\partial x}\bigg|_{x=0,c} = 0 \quad \frac{\partial \theta_i}{\partial y_1}\bigg|_{y_1=0,\bar{d}_1} = 0 \quad i = 1, 2, \ldots, N$$

(5.110)

The problem is now in a convenient solvable form. To summarize, the system of compound orthotropic layers represented by Eqs. (5.94)–(5.100) has been transformed into an equivalent, simpler system of equations given by Eqs. (5.105)–(5.110) using two transformations. The two transformations associated with Eqs. (5.101) and (5.104), which represent an expansion of the ones introduced by Muzychka et al. (2004, 2013), can be combined by applying only one transformation given by

$$\text{Layer } i: \quad y_1 = y/\sqrt{k_{1,y}/k_{1,x}}, \quad \zeta_i = z_i/\sqrt{k_{i,z}/k_{i,x}},$$

(5.111)

after which, some simple mathematics can be used to obtain the form given in Eqs. (5.105)–(5.110). It is important to note that although the transformed system is not fully isotropic (because of the existence of the parameters $\mu_i$), the general solution can be obtained using the method of separation of variables in the same manner of solving isotropic system with a slightly different form, as we will see in Section 5.6.2.

### 5.6.2 General Solution for $k_x \neq k_y \neq k_z$

The general solution of the first layer temperature excess distribution $\theta_1$ can be found by using the method of separation of variables, where the solution is assumed to have the form $\theta_1(x, y_1, \zeta_1) = X_1(x) \cdot Y_1(y_1) \cdot Z_1(\zeta_1)$. Applying the method of separation of variables to the first governing equation in Eq. (5.105) and using the boundary conditions along ($x = 0$, $x = c$) and ($y_1 = 0$, $y_1 = \bar{d}_1$) yields the following general solution:

$$\theta_1(x, y_1, \zeta_1) = A_{00}^1 + B_{00}^1 \zeta_1$$

$$+ \sum_{m=1}^{\infty} \cos(\lambda_m^1 x) \left[ A_{m0}^1 \cosh(\lambda_m^1 \zeta_1) + B_{m0}^1 \sinh(\lambda_m^1 \zeta_1) \right]$$

$$+ \sum_{n=1}^{\infty} \cos(\delta_n^1 y_1) \left[ A_{0n}^1 \cosh(\delta_n^1 \zeta_1) + B_{0n}^1 \sinh(\delta_n^1 \zeta_1) \right]$$

$$+ \sum_{m=1}^{\infty} \sum_{n=1}^{\infty} \cos(\lambda_m^1 x) \cos(\delta_n^1 y_1) \left[ A_{mn}^1 \cosh(\beta_{mn}^1 \zeta_1) + B_{mn}^1 \sinh(\beta_{mn}^1 \zeta_1) \right] \quad (5.112)$$

where $\lambda_m^1 = m\pi/c$, $\delta_n^1 = n\pi/\overline{d}_1$, and $\beta_{mn}^1 = \sqrt{\lambda_m^{1^2} + \delta_n^{1^2}}$ denote the eigenvalues in the solution for the first layer.

The general solution contains four components: a uniform flow solution, and three spreading solutions represented by the series components that vanish when the heat source area is equal to the sink plane area (the heat flux is distributed over the entire source plane surface $\zeta_1 = 0$). The solution for the temperature excess in the other layers (layer 2, 3, ..., $N$) can be obtained by solving the corresponding governing equations given in Eq. (5.105) also by using the method of separation of variables. It is important to note that in these layers, the governing equations of $\theta_i$ are different in the general form than the first one of $\theta_1$. However, the general solution of $\theta_i$ may be obtained in the same manner with new eigenvalues that can be related to the eigenvalues of the solution of $\theta_1$. This can be done by assuming the general solution to have the form $\theta_i(x, y_1, \zeta_i) = X_i(x) \cdot Y_i(y_1) \cdot Z_i(\zeta_i)$.

Applying the method of separation of variables to the governing equations in Eq. (5.105) and using the boundary conditions along ($x = 0$, $x = c$) and ($y_1 = 0$, $y_1 = \overline{d}_1$) yield the following general solution for the $i$th layer:

$$\theta_i(x, y_1, \zeta_i) = A_{00}^i + B_{00}^i \zeta_i$$

$$+ \sum_{m=1}^{\infty} \cos(\lambda_m^1 x) \left[ A_{m0}^i \cosh(\lambda_m^i \zeta_i) + B_{m0}^i \sinh(\lambda_m^i \zeta_i) \right]$$

$$+ \sum_{n=1}^{\infty} \cos(\delta_n^1 y_1) \left[ A_{0n}^i \cosh(\delta_n^i \zeta_i) + B_{0n}^i \sinh(\delta_n^i \zeta_i) \right]$$

$$+ \sum_{m=1}^{\infty} \sum_{n=1}^{\infty} \cos(\lambda_m^1 x) \cos(\delta_n^1 y_1) \left[ A_{mn}^i \cosh(\beta_{mn}^i \zeta_i) + B_{mn}^i \sinh(\beta_{mn}^i \zeta_i) \right] \quad (5.113)$$

where $\lambda_m^i = \lambda_m^1$, $\delta_n^i = \delta_n^1/\sqrt{\mu_i}$, and $\beta_{mn}^i = \sqrt{\lambda_m^{i^2} + \delta_n^{i^2}} = \sqrt{\lambda_m^{1^2} + \delta_n^{1^2}/\mu_i}$ are the eigenvalues associated with each of the remaining layers of the system.

Equations (5.112) and (5.113) represent the general solution of the temperature excess in the first and $i$th (for $i = 2, 3, ..., N$) layers, respectively, after applying the lateral boundary conditions. The interfacial and sink plane boundary conditions are then used to find a relationship between the Fourier coefficients $A_{mn}^i$ and $B_{mn}^i$ in each layer. Following the work of Muzychka et al. (2013) and Bagnall et al. (2014), the relationship is represented by a spreading function $\phi_i(\gamma^i)$ defined by

$$\phi_i(\gamma^i) = -\frac{B_{mn}^i}{A_{mn}^i} \quad (5.114)$$

where $\gamma^i$ refers to any of the eigenvalues $\lambda_m^i$, $\delta_n^i$, and $\beta_{mn}^i$. First, for $m, n$ not both equal to zero, in order to find the relationship between the $i$th layer's Fourier coefficients $A_{mn}^i$ and $B_{mn}^i$, represented by the spreading function $\phi_i(\gamma^i)$, it is important to note that the Fourier coefficients of $\theta_i$ depend on the Fourier coefficients of $\theta_{i+1}$ (i.e. $A_{mn}^i$ and $B_{mn}^i$ depend on $A_{mn}^{i+1}$ and $B_{mn}^{i+1}$) when applying the interface boundary conditions; hence, the spreading function $\phi_i(\gamma^i)$ depends on the next layer's spreading function $\phi_{i+1}(\gamma^{i+1})$. Thus, we start with finding the spreading function of the $N$th layer solution, and then a backward recursive formula can be obtained for finding $\phi_i(\gamma^i)$. The application of the convection boundary condition at the sink plane ($\zeta_N = \bar{t}_N$) given by Eq. (5.109) leads to

$$\phi_N(\gamma^N) = -\frac{B_{mn}^N}{A_{mn}^N} = \frac{\gamma^N \tanh(\gamma^N \bar{t}_N) + [h_s/k_N]}{\gamma^N + [h_s/k_N]\tanh(\gamma^N \bar{t}_N)} \tag{5.115}$$

Now, the application of the continuity of heat flux and the temperature drop boundary conditions, represented by Eqs. (5.107) and (5.108), leads to the following backward recursive relationship:

$$\phi_i(\gamma^i) = \frac{\left[\left(k_i\gamma^i\right)/\left(k_{i+1}\gamma^{i+1}\right) + \left(k_i\gamma^i/h_{c_i}\right)\phi_{i+1}(\gamma^{i+1})\right]\tanh(\gamma^i\bar{t}_i) + \phi_{i+1}(\gamma^{i+1})}{\left[\left(k_i\gamma^i\right)/\left(k_{i+1}\gamma^{i+1}\right) + \left(k_i\gamma^i/h_{c_i}\right)\phi_{i+1}(\gamma^{i+1})\right] + \phi_{i+1}(\gamma^{i+1})\tanh(\gamma^i\bar{t}_i)} \tag{5.116}$$

which is simplified in the case of continuity of temperature excess boundary condition into (as $h_{c_i} \to \infty$):

$$\phi_i(\gamma^i) = \frac{[\left(k_i\gamma^i\right)/\left(k_{i+1}\gamma^{i+1}\right)]\tanh(\gamma^i\bar{t}_i) + \phi_{i+1}(\gamma^{i+1})}{[\left(k_i\gamma^i\right)/\left(k_{i+1}\gamma^{i+1}\right)] + \phi_{i+1}(\gamma^{i+1})\tanh(\gamma^i\bar{t}_i)} \tag{5.117}$$

Finally, the boundary condition at the source plane is used to find the Fourier coefficients $A_{mn}^i$ after making use of $B_{mn}^i = -\phi_1(\gamma^i)A_{mn}^i$, starting from finding $A_{mn}^1$ and then a forward recursive formula can be used to obtain the $i$th layer's Fourier coefficients $A_{mn}^i$ if desired. The Fourier coefficients in the first layer $A_{mn}^1$ are obtained by taking Fourier series expansions of the boundary condition at the source plane given by Eq. (5.106) and making use of $B_{mn}^1 = -\phi_1(\gamma^1)A_{mn}^1$ to get

$$A_{m0}^1 = \frac{\bar{b}q}{dk_1\lambda_m^1\phi_1(\lambda_m^1)}\frac{\int_{X_c-a/2}^{X_c+a/2}\cos(\lambda_m^1 x)dx}{\int_0^c\cos^2(\lambda_m^1 x)dx} = \frac{4Q\cos(\lambda_m^1 X_c)\sin(\frac{1}{2}\lambda_m^1 a)}{acdk_1\lambda_m^{1^2}\phi_1(\lambda_m^1)} \tag{5.118}$$

and

$$A_{0n}^1 = \frac{aq}{ck_1\delta_n^1\phi_1(\delta_n^1)}\frac{\int_{\bar{Y}_c-\bar{b}/2}^{\bar{Y}_c+\bar{b}/2}\cos(\delta_n^1 y_1)dy_1}{\int_0^{\bar{d}}\cos^2(\delta_n^1 y_1)dy_1} = \frac{4Q\sigma\cos(\delta_n^1\bar{Y}_c)\sin(\frac{1}{2}\delta_n^1\bar{b})}{bcdk_1\delta_n^{1^2}\phi_1(\delta_n^1)} \tag{5.119}$$

and

$$A_{mn}^1 = \frac{q}{k_1\beta_{mn}^1\phi_1(\beta_{mn}^1)}\frac{\int_{\bar{Y}_c-\bar{b}/2}^{\bar{Y}_c+\bar{b}/2}\int_{X_c-a/2}^{X_c+a/2}\cos(\lambda_m^1 x)\cos(\delta_n^1 y_1)dx\,dy_1}{\int_0^{\bar{d}}\int_0^c\cos^2(\lambda_m^1 x)\cos^2(\delta_n^1 y_1)dx\,dy_1}$$

$$= \frac{16Q\sigma\cos(\lambda_m^1 X_c)\sin(\frac{1}{2}\lambda_m^1 a)\cos(\delta_n^1\bar{Y}_c)\sin(\frac{1}{2}\delta_n^1\bar{b})}{abcdk_1\beta_{mn}^1\lambda_m^1\delta_n^1\phi_1(\beta_{mn}^1)} \tag{5.120}$$

where $\sigma = \sqrt{k_{1,y}/k_{1,x}}$, $\bar{b} = b/\sigma$, $\bar{Y}_c = Y_c/\sigma$, and $Q = (ab)q$ is the total heat input of the flux channel. Equations (5.117)–(5.119) represent the Fourier coefficients of the first-layer solution for $m, n$ not both equal to zero. To find the Fourier coefficients of the other layers, the following forward recursive formula can be used:

$$A_{mn}^{i+1} = A_{mn}^i\left(\frac{\cosh(\gamma^i\bar{t}_i) - \phi_i(\gamma^i)\sinh(\gamma^i\bar{t}_i)}{1 + \frac{k_{i+1}\gamma^{i+1}}{h_{c_i}}\phi_{i+1}(\gamma^{i+1})}\right) \tag{5.121}$$

When $m$, $n$ are both zeros, the zeroth-order Fourier coefficients in the first layer $A_{00}^1$ and $B_{00}^1$ can be found by applying the sink plane boundary condition and taking the Fourier expansion in the source plane after relating the coefficients between the adjacent layers to get

$$A_{00}^1 = \frac{Q}{cd} \left[ \sum_{l=1}^{N-1} \left( \frac{\bar{t}_l}{k_l} + \frac{1}{h_{c_l}} \right) + \frac{\bar{t}_N}{k_N} + \frac{1}{h_s} \right]$$

$$B_{00}^1 = -\frac{Q}{cdk_1} \tag{5.122}$$

Moreover, the zeroth-order Fourier coefficients in the other layers $A_{00}^i$ and $B_{00}^i$ can be obtained as

$$A_{00}^i = \frac{Q}{cd} \left[ \sum_{l=i}^{N-1} \left( \frac{\bar{t}_l}{k_l} + \frac{1}{h_{c_l}} \right) + \frac{\bar{t}_N}{k_N} + \frac{1}{h_s} \right]$$

$$B_{00}^i = -\frac{Q}{cdk_i} \tag{5.123}$$

From the previous discussion, the analytical solution for the temperature excess in each layer is illustrated completely along with the proper recursive formulas, which can be used for finding the Fourier coefficients. However, the solution in the first layer $\theta_1(x, y_1, \zeta_1)$ (in particular, the solution in the source plane at $\zeta_1 = 0$) is of most interest for finding the maximum temperature and the total thermal resistance of the flux channel, which is addressed by

$$\theta_1(x, y_1, 0) = A_{00}^1 + \sum_{m=1}^{\infty} A_{m0}^1 \cos(\lambda_m^1 x) + \sum_{n=1}^{\infty} A_{0n}^1 \cos(\delta_n^1 y_1)$$

$$+ \sum_{m=1}^{\infty} \sum_{n=1}^{\infty} A_{mn}^1 \cos(\lambda_m^1 x) \cos(\delta_n^1 y_1) \tag{5.124}$$

and can be transformed back for convenience to the original coordinates, i.e. $x$ and $y$, by making use of Eq. (5.101) to get

$$\theta_1(x, y, 0) = A_{00}^1 + \sum_{m=1}^{\infty} A_{m0}^1 \cos(\lambda_m^1 x) + \sum_{n=1}^{\infty} A_{0n}^1 \cos(\delta_n^1 y/\sigma)$$

$$+ \sum_{m=1}^{\infty} \sum_{n=1}^{\infty} A_{mn}^1 \cos(\lambda_m^1 x) \cos(\delta_n^1 y/\sigma) \tag{5.125}$$

### 5.6.3 Total Thermal Resistance

For a single heat source spreading to a larger extended sink area, the total thermal resistance can be defined by Muzychka et al. (2003b, 2013):

$$R_{total} = \frac{\bar{T}_c - T_\infty}{Q} = \frac{\bar{\theta}_c}{Q} = R_{1D} + R_s \tag{5.126}$$

where $\bar{T}_c$ is the heat source contact mean temperature, $\bar{\theta}_c$ is the mean heat source contact temperature excess, $R_{1D}$ is the 1−D resistance and $R_s$ is the spreading resistance. The mean source temperature excess is given by

$$\bar{\theta}_c = \frac{1}{A_c} \iint_{A_c} \theta_1(x, y, 0) dA_c \tag{5.127}$$

where $A_c = ab$ is the heat source area. The application of Eq. (5.126) to the source plane solution given by Eq. (5.124) yields:

$$\overline{\theta}_c = A_{00}^1 + 2\sum_{m=1}^{\infty} A_{m0}^1 \frac{\cos(\lambda_m^1 X_c)\sin(\frac{1}{2}\lambda_m^1 a)}{a\lambda_m^1} + 2\sum_{n=1}^{\infty} A_{0n}^1 \frac{\sigma\cos(\delta_n^1 \overline{Y}_c)\sin(\frac{1}{2}\delta_n^1 b)}{b\delta_n^1}$$
$$+ 4\sum_{m=1}^{\infty}\sum_{n=1}^{\infty} A_{mn}^1 \frac{\sigma\cos(\lambda_m^1 X_c)\sin(\frac{1}{2}\lambda_m^1 a)\cos(\delta_n^1 \overline{Y}_c)\sin(\frac{1}{2}\delta_n^1 b)}{a\lambda_m^1 b\delta_n^1} \tag{5.128}$$

Thus, the total thermal resistance can be obtained by using Eq. (5.126) as

$$R_{total} = R_{1D} + \sum_{m=1}^{\infty} R_{m0} + \sum_{n=1}^{\infty} R_{0n} + \sum_{m=1}^{\infty}\sum_{n=1}^{\infty} R_{mn} \tag{5.129}$$

where

$$R_{1D} = \frac{1}{cd}\left[\sum_{l=1}^{N-1}\left(\frac{\overline{t}_l}{k_l} + \frac{1}{h_{c_l}}\right) + \frac{\overline{t}_N}{k_N} + \frac{1}{h_s}\right] \tag{5.130}$$

and

$$R_{m0} = \frac{8\cos^2(\lambda_m^1 X_c)\sin^2(\frac{1}{2}\lambda_m^1 a)}{a^2 cd k_1\, \lambda_m^{1^3}\phi_1(\lambda_m^1)} \tag{5.131}$$

and

$$R_{0n} = \frac{8\sigma^2\cos^2(\delta_n^1 \overline{Y}_c)\sin^2(\frac{1}{2}\delta_n^1 b)}{b^2 cd k_1 \delta_n^{1^3}\phi_1(\delta_n^1)} \tag{5.132}$$

and

$$R_{mn} = \frac{64\sigma^2\cos^2(\lambda_m^1 X_c)\sin^2(\frac{1}{2}\lambda_m^1 a)\cos^2(\delta_n^1 \overline{Y}_c)\sin^2(\frac{1}{2}\delta_n^1 b)}{a^2 b^2 cd k_1 \beta_{mn}^1 \lambda_m^{1^2}\delta_n^{1^2}\phi_1(\beta_{mn}^1)} \tag{5.133}$$

This completes the solution for a single heat source on a multilayered general orthotropic flux channel. As we have seen, simply relaxing the earlier assumption of $k_{x,i} = k_{y,i}$ for each layer has led to a significantly more complex solution. Al-Khamaiseh et al. (2018) provide further development for multiple heat sources along with several case studies.

## 5.7 Measurement of Orthotropic Thermal Conductivity

We conclude with a short discussion on the measurement of orthotropic thermal conductivity. Many commercially available measurement systems are designed to measure thermal conductivity using a transient thermal pulse method usually by means of a laser. These methods generally measure the thermal diffusivity $\alpha = k/\rho c_p$ and the resulting thermal conductivity calculated using the known $\rho c_p$ product for the given material. These tools can measure either the in-plane or through-plane thermal conductivity but require samples to be specially prepared. In some cases, this can be difficult for thin-film materials. These systems are also unable to resolve both components simultaneously.

Recent work done at Purdue University [Gaitonde et al. (2023)] has led to an extended Angstrom method. The Angstrom method is used to measure a single value of in-plane thermal conductivity of thin films over the range 0.01 W/(m K) < k < 4000 W/(m K) and 3 μm < t < 10 mm. In this

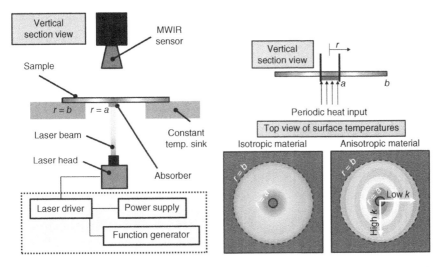

**Figure 5.3** Schematic of test facility developed by Gaitonde et al. (2023). Left image is the side view or equipment while the right shows the top or sample imaging view. Source: Image courtesy of A. Gaitonde

new method, both the in-plane and through-plane thermal conductivity of orthotropic materials can be measured simultaneously with good accuracy in appropriately cut samples, i.e. conductivity varying in both in-plane directions of the thin film. An enhanced infrared microscopy approach for thin films using an Angstrom method is discussed in Hahn et al. (2019). The approach outlined here was developed for two-dimensional fields using a facility shown in Figure 5.3.

The heat diffusion equation in two dimensions can be written as (assuming negligible convection):

$$\frac{\partial}{\partial x}\left(k_x\frac{\partial T}{\partial x}\right) + \frac{\partial}{\partial y}\left(k_y\frac{\partial T}{\partial y}\right) = \rho c_p\frac{\partial T}{\partial t} \tag{5.134}$$

assuming a periodic heat source at the origin:

$$T(0,0,t) = T_0\cos(\omega t) \tag{5.135}$$

The solution takes the form:

$$T(x,y,t) = [P(x,y) + iQ(x,y)]e^{i\omega t} \tag{5.136}$$

The term in square brackets in Eq. (5.136) is the complex amplitude. Substituting Eq. (5.136) into the heat diffusion and taking the real and complex parts yields the following equations:

$$k_x\frac{\partial^2 P}{\partial x^2} + k_y\frac{\partial^2 P}{\partial y^2} = -\rho c_p\omega Q \tag{5.137}$$

and

$$k_x\frac{\partial^2 Q}{\partial x^2} + k_y\frac{\partial^2 Q}{\partial y^2} = \rho c_p\omega P \tag{5.138}$$

For a series of discrete points in the domain, the above equations can be reduced to a linear system of the form:

$$
\begin{bmatrix}
\dfrac{\partial^2 P_1}{\partial x^2} & \dfrac{\partial^2 P_1}{\partial y^2} \\
\vdots & \vdots \\
\dfrac{\partial^2 P_n}{\partial x^2} & \dfrac{\partial^2 P_n}{\partial y^2} \\
\dfrac{\partial^2 Q_1}{\partial x^2} & \dfrac{\partial^2 Q_1}{\partial y^2} \\
\vdots & \vdots \\
\dfrac{\partial^2 Q_n}{\partial x^2} & \dfrac{\partial^2 Q_n}{\partial y^2}
\end{bmatrix}
\begin{bmatrix} k_x \\ k_y \end{bmatrix} = \rho c_p \omega
\begin{bmatrix}
-Q_1 \\ \vdots \\ -Q_n \\ P_1 \\ \vdots \\ P_n
\end{bmatrix}
\tag{5.139}
$$

If convection is a factor, then it may be included using the following modified form of the diffusion equation:

$$
\frac{\partial}{\partial x}\left( k_x \frac{\partial T}{\partial x} \right) + \frac{\partial}{\partial y}\left( k_y \frac{\partial T}{\partial y} \right) - \frac{2h}{\Delta}(T - T_\infty) = \rho c_p \frac{\partial T}{\partial t}
\tag{5.140}
$$

where $\Delta$ is the thickness of the sample. This leads to the following alternate form of Eqs. (5.137) and (5.138):

$$
k_x \frac{\partial^2 P}{\partial x^2} + k_y \frac{\partial^2 P}{\partial y^2} - \left( \frac{2h}{\Delta} \right) P = -\rho c_p \omega Q
\tag{5.141}
$$

and

$$
k_x \frac{\partial^2 Q}{\partial x^2} + k_y \frac{\partial^2 Q}{\partial y^2} - \left( \frac{2h}{\Delta} \right) Q = \rho c_p \omega P
\tag{5.142}
$$

For a series of discrete points in the domain, the above equations can be reduced to a linear system of the form:

$$
\begin{bmatrix}
\dfrac{\partial^2 P_1}{\partial x^2} & \dfrac{\partial^2 P_1}{\partial y^2} & -\dfrac{2P_1}{\Delta} \\
\vdots & \vdots & \vdots \\
\dfrac{\partial^2 P_n}{\partial x^2} & \dfrac{\partial^2 P_n}{\partial y^2} & -\dfrac{2P_n}{\Delta} \\
\dfrac{\partial^2 Q_1}{\partial x^2} & \dfrac{\partial^2 Q_1}{\partial y^2} & -\dfrac{2Q_1}{\Delta} \\
\vdots & \vdots & \vdots \\
\dfrac{\partial^2 Q_n}{\partial x^2} & \dfrac{\partial^2 Q_n}{\partial y^2} & -\dfrac{2Q_n}{\Delta}
\end{bmatrix}
\begin{bmatrix} k_x \\ k_y \\ h \end{bmatrix} = \rho c_p \omega
\begin{bmatrix}
-Q_1 \\ \vdots \\ -Q_n \\ P_1 \\ \vdots \\ P_n
\end{bmatrix}
\tag{5.143}
$$

In both cases, we can approximate the derivatives of the real ($P$) and complex ($Q$) parts of the complex amplitude numerically using the infrared temperature field measurements. This leads to a solution for $k_x$ and $k_y$ (and/or $h$) if we wish to factor this into the analysis. Least squares minimization is applied to obtain optimum values for the entire set of data points used in the analysis.

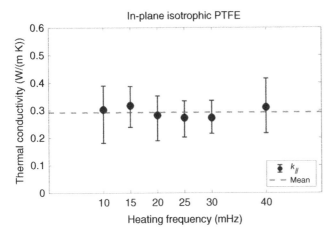

**Figure 5.4** Two dimensional measurement of an isotropic polytetrafluoroethylene (PTFE) sample. Source: Image courtesy of A. Gaitonde.

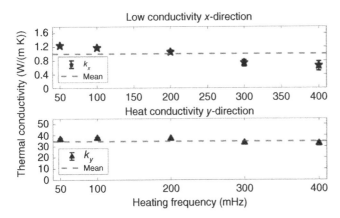

**Figure 5.5** Two dimensional measurement of a Temprion Organic Heat Spreader (OHS). Source: Image courtesy of A. Gaitonde.

Gaitonde et al. (2023) developed the methodology and equipment and applied it to the measurement of selected isotropic and orthotropic materials as shown in Figures 5.4 and 5.5. They also used numerically obtained data to benchmark the method. Further details of the method may be found in the reference.

## References

Al-Khamaiseh, B., Muzychka, Y.S., and Kocabiyik, S., "Spreading Resistance in Compound Orthotropic Flux Channels with Different Conductivities in the Three Spatial Directions", *Journal of Heat Transfer*, Vol. 140, no. 7, p. 071302, 2018.

Bagnall, K., Muzychka, Y.S., and Wang, E., "Analytical Solution for Temperature Rise in Complex Multilayer Structures With Discrete Heat Sources", *IEEE Transactions on Components, Packaging, and Manufacturing Technologies*, Vol. 4, no. 5, pp. 817–830, 2014.

Carslaw, H.S., and Jaeger, J.C., *Conduction of Heat in Solids*, Oxford University Press, Oxford, UK, 1959, pp. 38–49.

Gaitonde, A.U., Candadai, A.A., Weibel, J.A., and Marconnet, A.M., "A Laser-based Angstrom Method for In-Plane Thermal Characterization of Isotropic and Anisotropic Materials using Infrared Imaging", *Review of Scientific Instruments (AIP)*, 2023. DOI: 10.1063/5.0149659.

Gholami, A., and Bahrami, M., "Thermal Spreading resistance Inside Anisotropic Plates with Arbitrarily Located Hotspots", *Journal of Thermophysics and Heat Transfer*, Vol. 28, no. 4, pp. 679–686, 2014.

Hahn, J., Reid, T., and Marconnet, A., "Infrared Microscopy Enhanced Angstrom's Method for Thermal Diffusivity of Polymer Monofilaments and Films", *Journal of Heat Transfer*, Vol. 141, pp. 081601-1–081601-11, 2019.

Lam, T.T., and Fischer, W.D., "Thermal Reistance in Rectangular Orthotropic Heat Spreaders", *ASME Advances in Electronic Packaging*, Vol. 1, pp. 891–898, 1999.

Muzychka, Y.S., "Spreading Resistance in Compound Orthotropic Flux Tubes and Channels with Interfacial Resistance", *AIAA Journal of Thermophysics and Heat Transfer*, Vol. 28, no. 2, pp. 313–319, 2014.

Muzychka, Y.S., "Thermal Spreading Resistance in a Multilayered Orthotropic Circular Disk with Interfacial Resistance and Variable Flux", *Presented at InterPack 2015*, San Francisco, CA, USA, July 6–9, 2015.

Muzychka, Y.S., Culham, J.R., and Yovanovich, M.M., "Thermal Spreading Resistance in Rectangular Flux Channels", AIAA 03-4188, *36th Thermophysics Conference*, Orlando, FL, June 23–26, 2003a.

Muzychka, Y.S., Yovanovich, M.M., and Culham, J.R., "Thermal Spreading Resistance of Eccentric Heat Sources on Rectangular Flux Channels", *Journal of Electronic Packaging*, Vol. 125, pp. 178–185, 2003b.

Muzychka, Y.S., Yovanovich, M.M., and Culham, J.R., "Thermal Spreading Resistances in Compound and Orthotropic Systems", *Journal of Thermophysics and Heat Transfer*, Vol. 18, pp. 45–51, 2004.

Muzychka, Y.S., Yovanovich, M.M., and Culham, J.R., "Influence of Geometry and Edge Cooling on Thermal Spreading Resistance", *Journal of Thermophysics and Heat Transfer*, Vol. 20, pp. 247–255, 2006.

Muzychka, Y.S., Bagnall, K., and Wang, E., "Thermal Spreading Resistance and Heat Source Temperature in Compound Orthotropic Systems with Interfacial Resistance", *IEEE Transactions on Components, Packaging, and Manufacturing Technologies*, Vol. 3, no. 11, pp. 1826–1841, 2013.

Ozisik, N., *Heat Conduction*, Wiley, New York, 1993, pp. 617–628.

Ying, T.M., and Toh, K.C., "A Constriction Resistance Model in Thermal Analysis of Solder Ball Joints in Ball Grid Array Packages", *Proceedings of the 1999 International Mechanical Engineering Congress and Exposition*, Nashville, HTD-Vol. 364-1, pp. 29–36, 1999.

Ying, T.M., and Toh, K.C., "A Heat Spreading Resistance Model for Anisotropic Thermal Conductivity Materials in Electronic Packaging," *Proceedings of the 2000 Inter-Society Conference on Thermal Phenomena*, 2000.

Yovanovich, M.M., "On the Temperature Distribution and Constriction Resistance in Layered Media", *Journal of Composite Materials*, Vol. 4, pp. 567–570, 1970.

Yovanovich, M.M., "Thermal Resistance of Circular Source on Finite Circular Cylinder with Side and End Cooling", *Journal of Electronic Packaging*, Vol. 125, pp. 169–177, 2003.

Yovanovich, M.M., and Marotta, E. "Chapter 4: Thermal Spreading and Contact Resistances", *Heat Transfer Handbook*, eds. A. Bejan and A.D. Kraus, Wiley, 2003.

Yovanovich, M.M., Tien, C.H., and Schneider, G.E., "General Solution of Constriction Resistance within a Compound Disk", *Progress in Astronautics and Aeronautics: Heat Transfer, Thermal Control, and Heat Pipes*, MIT Press, Cambridge, MA, pp. 47–62, 1980.

Yovanovich, M.M., Culham, J.R., and Teertstra, P.M., "Analytical Modeling of Spreading Resistance in Flux Tubes, Half Spaces, and Compound Disks", *IEEE Transactions on Components, Packaging, and Manufacturing Technology - Part A*, Vol. 21, pp. 168–176, 1998.

Yovanovich, M.M., Muzychka, Y.S., and Culham, J.R., "Spreading Resistance of Isoflux Rectangles and Strips on Compound Flux Channels", *Journal of Thermophysics and Heat Transfer*, Vol. 13, pp. 495–500, 1999.

# 6

# Multisource Analysis for Microelectronic Devices

Several approaches for computing the thermal spreading resistance and/or heat source temperature have been developed for a rectangular substrate with single or multiple discrete sources. Several notable methodologies from the literature are discussed below.

In the case of multiple heat sources, several approaches are found in the open literature. Hein and Lenzi (1969) obtained a solution for an IC package using Fourier transforms. In their development, the heat source is specified by means of a Poisson equation using a piecewise function to model discrete heat sources. Both the source plane and sink plane were convectively cooled using uniform heat transfer coefficients. Kokkas (1974) obtained a Fourier/Laplace transform solution for a multilayer substrate containing discrete heat sources. The substrate base was assumed to be attached to a heat sink of fixed temperature similar to that shown in Figure 6.1. Discrete heat sources were dealt with using the source plane boundary condition. Ellison (1984) develops and presents a method referred to as thermal analysis of multilayer structures (TAMS). This method is similar to that of Hein and Lenzi (1969), but considers multiple layers. Culham et al. (2000) developed a three-dimensional Fourier series model for an electronic packaging system. Their model is quite general and allows for the specification of a mixed-boundary condition in the source plane. Heat sources are specified using the source plane boundary condition. Due to the complex nature of the source plane boundary condition, numerical analysis is required to complete the solution. In all of the above methods, significant effort is also required to code the analysis.

In the case of a single discrete heat source, several approaches are readily found in the open literature. Kadambi and Abuaf (1985) obtained steady and transient solutions for a central heat source on an isotropic rectangular substrate which was convectively cooled in the sink plane. Krane (1991) obtained a steady solution for a similar system in which the sink plane is at a constant temperature. Yovanovich et al. (1999) and Muzychka et al. (2003) obtained solutions for a compound convectively cooled rectangular substrate, containing a central and eccentric heat source, respectively. Finally, Muzychka et al. (2004) extended these solutions to orthotropic systems, while Muzychka et al. (2006) obtained a solution for a central heat source on an isotropic convectively cooled rectangular substrate with edge cooling. While none of these solutions considered multiple heat sources, they do offer insights as to how to model issues for multiple heat sources, such as orthotropic properties, compound systems, and edge cooling.

Muzychka (2006) proposed a simple method for predicting heat source temperature using the superposition of the fundamental solution for a single eccentric heat source on a rectangular spreader. This fundamental solution was outlined in Chapter 4 and will be expanded here to facilitate predicting heat source temperatures for applications involving more than one discrete heat source. The approach may also be used to model single and multiple irregular heat source arrangements.

*Thermal Spreading and Contact Resistance: Fundamentals and Applications*, First Edition.
Yuri S. Muzychka and M. Michael Yovanovich.
© 2023 John Wiley & Sons, Inc. Published 2023 by John Wiley & Sons, Inc.

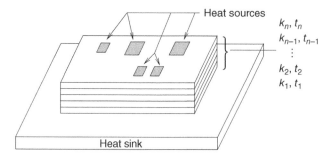

**Figure 6.1** Multiple heat sources on a rectangular flux channel. Source: Culham et al. (2000). Used with the permission of the American Society of Mechanical Engineers (ASME).

Muzychka et al. (2013) and Bagnall et al. (2014) extended the method of Muzychka (2006) to account for multiple heat sources in multilayered orthotropic systems with different thermal conductivity in the transverse and normal directions. This allowed the authors to thermally model gallium nitride high electron mobility transistors (GaN HEMT) microdevices which contain many discrete heat sources. Bagnall et al. (2014) applied this methodology to systems with large numbers of heat discrete heat sources and demonstrated the methodology was significantly quicker than finite element method (FEM) approaches, which require modification and refinement of computational meshes even for small changes in heat source placement.

Finally, Al-Khamaiseh et al. (2018) obtained a solution for a general three-dimensional, multilayered, orthotropic flux channel with all three principal thermal conductivities being different for an eccentric heat source. They also extended the solution to allow multiple heat sources to be considered.

## 6.1 Multiple Heat Sources on Finite Isotropic Spreaders

The system of interest in this chapter is idealized as a rectangular substrate which may be either isotropic, orthotropic, or compound in nature. For the time being, we will only consider an isotropic system, see Figure 6.2. Later, the effects of adding a conductive layer to promote the spreading of

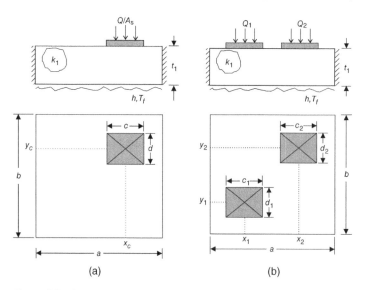

**Figure 6.2** Single eccentric heat source (a) and multiple heat sources (b).

heat and/or system orthotropy will be examined. In the present system, all of the edges are assumed be adiabatic, a reasonable assumption in many electronics applications where edge area is significantly less than the area of the source and sink planes. Finally, there is no heat loss through the source plane, such that all heat is dissipated through the sink plane by means of a uniform film coefficient, i.e. thermal wake effects are neglected. The addition of convection in the source plane is dealt with in a separate Section 6.2. The solution to this problem will become the fundamental building block for all other extensions to the problem.

### 6.1.1 Single Source Surface Temperature Distribution

The temperature distribution for a single arbitrarily located uniform heat source on an isotropic substrate was derived earlier in Chapter 4 and has the following form:

$$\theta(x, y, z) = A_0 + C_0 z$$

$$+ \sum_{m=1}^{\infty} A_m \cos(\lambda_m x) \left[\cosh(\lambda_m z) - \phi(\lambda_m)\sinh(\lambda_m z)\right]$$

$$+ \sum_{n=1}^{\infty} A_n \cos(\delta_n y) \left[\cosh(\delta_n z) - \phi(\delta_n)\sinh(\delta_n z)\right] \tag{6.1}$$

$$+ \sum_{m=1}^{\infty}\sum_{n=1}^{\infty} A_{mn} \cos(\lambda_m x)\cos(\delta_n y) \left[\cosh(\beta_{mn} z) - \phi(\beta_{mn})\sinh(\beta_{mn} z)\right]$$

where $\lambda_m = m\pi/a$, $\delta_n = n\pi/b$, and $\beta_{mn} = \sqrt{\lambda_m^2 + \delta_n^2}$ are the eigenvalues. The origin of the coordinate system is taken to be the lower left corner of the substrate as shown in Figure 6.2.

The general solution contains four components, a uniform flow solution and three spreading (or constriction) solutions which vanish when the heat flux is uniformly distributed over the entire source plane, $z = 0$. The general solution itself is a linear superposition of each component. Application of the boundary conditions in the through plane or $z$-direction yields solutions for one-half of the unknown constants and gives rise to the following expression for the spreading parameter $\phi$:

$$\phi(\xi) = \frac{\xi \tanh(\xi t_1) + h/k_1}{\xi + h/k_1 \tanh(\xi t_1)} \tag{6.2}$$

where $\xi$ is replaced by $\lambda_m$, $\delta_n$, or $\beta_{mn}$, respectively. The spreading parameter accounts for the effects of conductivity, thickness, and convection cooling. If there is more than one layer, we may modify it using appropriate expressions as outlined in Chapter 4.

The three remaining coefficients $A_m$, $A_n$, and $A_{mn}$ were obtained by taking Fourier series expansions of the boundary condition in the source plane, $z = 0$. This yielded the following expressions for the Fourier coefficients:

$$A_m = \frac{2Q\left[\sin\left(\frac{(2X_c+c)}{2}\lambda_m\right) - \sin\left(\frac{(2X_c-c)}{2}\lambda_m\right)\right]}{a\,b\,c\,k_1\,\lambda_m^2\phi(\lambda_m)}$$

$$A_n = \frac{2Q\left[\sin\left(\frac{(2Y_c+d)}{2}\delta_n\right) - \sin\left(\frac{(2Y_c-d)}{2}\delta_n\right)\right]}{a\,b\,d\,k_1\,\delta_n^2\phi(\delta_n)} \tag{6.3}$$

$$A_{mn} = \frac{16Q\cos(\lambda_m X_c)\sin\left(\frac{1}{2}\lambda_m c\right)\cos(\delta_n Y_c)\sin\left(\frac{1}{2}\delta_n d\right)}{a\,b\,c\,d\,k_1\,\beta_{mn}\lambda_m\delta_n\phi(\beta_{mn})}$$

where $X_c$ and $Y_c$ are the coordinates of the centroid of an arbitrarily placed heat source with respect to the lower left corner of the substrate as shown in Figure 6.2.

Finally, values for the coefficients in the uniform flow solution are given by

$$A_0 = \frac{Q}{ab}\left(\frac{t_1}{k_1} + \frac{1}{h}\right) \tag{6.4}$$

and

$$C_0 = -\frac{Q}{k_1 ab} \tag{6.5}$$

### Centroidal Source Temperature

The maximum or centroidal heat source temperature may be determined from Eq. (6.1) when $x = X_c, y = Y_c$, and $z = 0$. This gives rise to

$$\hat{\theta} = A_0 + \sum_{m=1}^{\infty} A_m \cos(\lambda_m X_c) + \sum_{n=1}^{\infty} A_n \cos(\delta_n Y_c)$$

$$+ \sum_{m=1}^{\infty}\sum_{n=1}^{\infty} A_{mn} \cos(\lambda_m X_c) \cos(\delta_n Y_c) \tag{6.6}$$

### Mean Source Temperature

The mean heat source temperature is obtained by integrating the local source temperature over the source area, i.e.

$$\bar{\theta} = \frac{1}{A} \iint \theta(x, y, 0) dA \tag{6.7}$$

This leads to the following result for the mean temperature excess of a single eccentric heat source:

$$\bar{\theta} = A_o + \sum_{m=1}^{\infty} A_m \frac{2\cos(\lambda_m X_c)\sin\left(\frac{1}{2}\lambda_m c\right)}{\lambda_m c}$$

$$+ \sum_{n=1}^{\infty} A_n \frac{2\cos(\delta_n Y_c)\sin\left(\frac{1}{2}\delta_n d\right)}{\delta_n d} \tag{6.8}$$

$$+ \sum_{m=1}^{\infty}\sum_{n=1}^{\infty} A_{mn} \frac{4\cos(\delta_n Y_c)\sin\left(\frac{1}{2}\delta_n d\right)\cos(\lambda_m X_c)\sin\left(\frac{1}{2}\lambda_m c\right)}{\lambda_m c\, \delta_n d}$$

The results given by Eqs. (6.6) and (6.8) may now be used to analyze systems containing multiple heat sources using superposition. These expressions may also be used as a fundamental surface element for analyzing irregularly shaped heat sources, by discretizing the region into several rectangular shaped sources and by using superposition.

### 6.1.2 Multisource Surface Temperature Distribution

The solution for multiple heat sources begins with the basic result developed earlier for $T(x, y, z)$. For $N$ discrete heat sources, the maximum temperatures occur in the source plane. The *surface* temperature distribution is obtained from

$$T_s(x, y, 0) - T_f = \theta_s = \sum_{i=1}^{N} \theta_i(x, y, 0) \tag{6.9}$$

where $\theta_i$ is the temperature excess for each heat source by itself. The surface temperature excess for the $i$th heat source on the surface of the spreader may be computed using Eq. (6.1) evaluated at the surface:

$$\theta_i(x,y,0) = A_0^i + \sum_{m=1}^{\infty} A_m^i \cos(\lambda_m x) + \sum_{n=1}^{\infty} A_n^i \cos(\delta_n y)$$

$$+ \sum_{m=1}^{\infty} \sum_{n=1}^{\infty} A_{mn}^i \cos(\lambda_m x) \cos(\delta_n y) \tag{6.10}$$

The Fourier coefficients are now evaluated at each of the $i$th heat source characteristics, i.e. $c_i, d_i, X_{c,i}, Y_{c,i}$, and $Q_i$.

## Centroidal Source Temperature

If only peak heat source temperature is desired, the maximum or centroidal heat source temperature for an arbitrary heat source may be determined from Eq. (6.10) summed for all sources and evaluated at the centroid of the source of interest. The temperature at the centroid of the $j$th heat source may be written as

$$\hat{\theta}_j = \sum_{i=1}^{N} \hat{\theta}_i(X_{c,j}, Y_{c,j}, 0) = \sum_{i=1}^{N} \hat{\theta}_{ij} \tag{6.11}$$

where

$$\hat{\theta}_{ij} = A_0^i + \sum_{m=1}^{\infty} A_m^i \cos(\lambda_m X_{c,j}) + \sum_{n=1}^{\infty} A_n^i \cos(\delta_n Y_{c,j})$$

$$+ \sum_{m=1}^{\infty} \sum_{n=1}^{\infty} A_{mn}^i \cos(\lambda_m X_{c,j}) \cos(\delta_n Y_{c,j}) \tag{6.12}$$

In the present notation $\theta_{ij}$, denotes the *effect of the $i$th heat source in the region of the centroid of the $j$th heat source.*

## Mean Source Temperature

The mean heat source temperature of an arbitrary rectangular patch of dimensions $c_j$ and $d_j$, i.e. the $j$th heat source, located at $X_{c,j}$ and $Y_{c,j}$, may be computed by integrating Eq. (6.9) over the region $A_j = c_j d_j$, i.e.

$$\overline{\theta}_j = \frac{1}{A_j} \iint_{A_j} \theta_s \, dA_j = \frac{1}{A_j} \iint_{A_j} \sum_{i=1}^{N} \theta_i(x,y,0) dA_j \tag{6.13}$$

which may be written as

$$\overline{\theta}_j = \sum_{i=1}^{N} \frac{1}{A_j} \iint_{A_j} \theta_i(x,y,0) dA_j = \sum_{i=1}^{N} \overline{\theta}_{ij} \tag{6.14}$$

Using Eq. (6.8) for the mean temperature excess of a heat source results in the following expression for the mean temperature excess contribution of the $i$th heat source over the region of the $j$th heat source:

$$\overline{\theta}_{ij} = A_0^i + 2\sum_{m=1}^{\infty} A_m^i \frac{\cos(\lambda_m X_{c,j})\sin(\frac{1}{2}\lambda_m c_j)}{\lambda_m c_j} + 2\sum_{n=1}^{\infty} A_n^i \frac{\cos(\delta_n Y_{c,j})\sin(\frac{1}{2}\delta_n d_j)}{\delta_n d_j}$$

$$+ 4\sum_{m=1}^{\infty} \sum_{n=1}^{\infty} A_{mn}^i \frac{\cos(\delta_n Y_{c,j})\sin(\frac{1}{2}\delta_n d_j)\cos(\lambda_m X_{c,j})\sin(\frac{1}{2}\lambda_m c_j)}{\lambda_m c_j \delta_n d_j} \tag{6.15}$$

Equation (6.14) represents the sum of the effects of all sources over an arbitrary region $c_j d_j$. Equation (6.14) is evaluated over the region of interest $c_j d_j$ located at $X_{c,j}, Y_{c,j}$. The coefficients $A_0^i, A_m^i, A_n^i$, and $A_{mn}^i$ are then evaluated at each of the $i$th source parameters.

## 6.2 Influence Coefficient Method

The methodology presented earlier is convenient for a small number of heat sources. However, if many sources are present, a more convenient approach is desirable. Muzychka (2006) presents a matrix approach using influence coefficients for the above analysis, which is more convenient when there are many heat sources, i.e. $N > 3$. Influence coefficients were first proposed by Negus and Yovanovich (1987a,b) for semi-infinite domains and later for multiple sources on a half space. The concept of an influence coefficient for a finite substrate was also partially addressed by Hein and Lenzi (1969). Influence coefficients offer an insightful assessment of the effect of neighboring heat sources on thermal resistance, and hence, the mean or centroidal temperature is in excess of each discrete heat source.

We begin by examining Eqs. (6.12) and (6.15), for the centroidal and mean temperature excess $\theta$. Beginning first with Eq. (6.11), we may write

$$\hat{\theta}_j = \hat{\theta}_{1j} + \hat{\theta}_{2j} \cdots \hat{\theta}_{Nj} \tag{6.16}$$

which may be written as follows:

$$\hat{\theta}_j = Q_1 \hat{f}_{1j} + Q_2 \hat{f}_{2j} + \cdots + Q_N \hat{f}_{Nj} \tag{6.17}$$

or

$$\hat{\theta}_j = \sum_{i=1}^{N} Q_i \hat{f}_{ij} \tag{6.18}$$

where

$$\hat{f}_{ij} = B_0 + \sum_{m=1}^{\infty} B_m^i \cos(\lambda_m X_{c,j}) + \sum_{n=1}^{\infty} B_n^i \cos(\delta_n Y_{c,j})$$

$$+ \sum_{m=1}^{\infty} \sum_{n=1}^{\infty} B_{mn}^i \cos(\lambda_m X_{c,j}) \cos(\delta_n Y_{c,j}) \tag{6.19}$$

where

$$B_0 = \frac{1}{ab}\left(\frac{t_1}{k_1} + \frac{1}{h}\right)$$

$$B_m^i = \frac{2\left[\sin\left(\frac{(2X_c+c)}{2}\lambda_m\right) - \sin\left(\frac{(2X_c-c)}{2}\lambda_m\right)\right]}{a\,b\,c\,k_1\,\lambda_m^2\phi(\lambda_m)}$$

$$B_n^i = \frac{2\left[\sin\left(\frac{(2Y_c+d)}{2}\delta_n\right) - \sin\left(\frac{(2Y_c-d)}{2}\delta_n\right)\right]}{a\,b\,d\,k_1\,\delta_n^2\phi(\delta_n)} \tag{6.20}$$

$$B_{mn}^i = \frac{16\cos(\lambda_m X_c)\sin(\frac{1}{2}\lambda_m c)\cos(\delta_n Y_c)\sin(\frac{1}{2}\delta_n d)}{a\,b\,c\,d\,k_1\,\beta_{mn}\lambda_m\delta_n\phi(\beta_{mn})}$$

are the modified Fourier coefficients, since $Q_i$ has now been factored out. Once again, it is noted that the coefficients are evaluated at each of the $i$th heat source characteristics, i.e. $c_i, d_i, X_{c,i}$, an $Y_{c,i}$. Thus, the influence coefficients are only functions of the substrate properties and dimensions and of heat source geometry and location.

Similarly, we may obtain an expression for the mean temperature excess of the $j$th heat source using Eq. (6.14):

$$\bar{\theta}_j = \bar{\theta}_{1j} + \bar{\theta}_{2j} \cdots \bar{\theta}_{Nj} \tag{6.21}$$

which may be written as

$$\bar{\theta}_j = Q_1 \bar{f}_{1j} + Q_2 \bar{f}_{2j} + \cdots + Q_N \bar{f}_{Nj} \tag{6.22}$$

or

$$\bar{\theta}_j = \sum_{i=1}^{N} Q_i \bar{f}_{ij} \tag{6.23}$$

where

$$
\bar{f}_{ij} = B_o + \sum_{m=1}^{\infty} B_m^i \frac{2\cos(\lambda_m X_{cj})\sin(\frac{1}{2}\lambda_m c_j)}{\lambda_m c_j}
$$
$$
+ \sum_{n=1}^{\infty} B_n^i \frac{2\cos(\delta_n Y_{cj})\sin(\frac{1}{2}\delta_n d_j)}{\delta_n d_j} + \sum_{m=1}^{\infty}\sum_{n=1}^{\infty} B_{mn}^i \tag{6.24}
$$
$$
\times \frac{4\cos(\delta_n Y_{cj})\sin(\frac{1}{2}\delta_n d_j)\cos(\lambda_m X_{cj})\sin(\frac{1}{2}\lambda_m c_j)}{\lambda_m c_j \delta_n d_j}
$$

When $i = j$, the contribution is a self-effect, i.e. the effect of the source acting alone. When $i \neq j$ the contribution to the temperature excess is an influence effect. The self-effect $f_{ii}$ is merely the single source thermal resistance. The influence effects, $f_{ij}$, are affected by two factors: source strength and the location and size of neighboring sources, i.e. a geometry effect. The influence coefficients $f_{ij}$ are clearly functions only of the location of the neighboring sources.

Finally, we may write the temperature excess in the following matrix form:

$$
\begin{Bmatrix} \theta_1 \\ \theta_2 \\ \theta_3 \\ \vdots \\ \theta_N \end{Bmatrix} = \begin{bmatrix} f_{11} & f_{12} & \cdots & f_{1N} \\ f_{21} & f_{22} & \cdots & f_{2N} \\ f_{31} & f_{32} & \cdots & f_{3N} \\ \vdots & \vdots & \vdots & \vdots \\ f_{N1} & f_{N2} & \cdots & f_{NN} \end{bmatrix} \begin{bmatrix} Q_1 \\ Q_2 \\ Q_3 \\ \vdots \\ Q_N \end{bmatrix} \tag{6.25}
$$

or

$$\{\boldsymbol{\theta}\} = [\mathbf{F}_{ij}][\mathbf{Q}] \tag{6.26}$$

where $\mathbf{F}_{ij}$ is the matrix of influence coefficients.

The influence coefficient method (ICM) offers a number of advantages. First, it becomes obvious what the effect a neighboring heat source has on the thermal resistance of a particular heat source. Examination of Eq. (6.18) or (6.23) reveals that an influence effect arises by virtue of proximity and strength. In other words, a remote and/or weak heat source has little influence on another heat source. Second, it can be shown that the influence coefficients also possess reciprocity for the case when $i \neq j$,

$$f_{ji} = f_{ij} \tag{6.27}$$

This property significantly reduces computation for systems where more than five sources are present. In general, for a system of $N$ sources, a symmetric $N \times N$ matrix results for the influence coefficients. As a result of this symmetry only $N(N+1)/2$ coefficients need be computed. An upper triangular matrix is all that is needed to compute the temperature excesses. Thus, the influence method offers a substantial savings in computation over the use of general superposition approach. The reciprocity is a result of the property of Greens functions [Morse and Feshbach (1953)], i.e. the potential at $X_{c,i}, Y_{c,i}$ due to a unit heat input at $X_{c,j}, Y_{c,j}$ is the same as the potential at $X_{c,j}, Y_{c,j}$ due to a unit heat input at $X_{c,i}, Y_{c,i}$. This property also holds upon integration over a finite region. The reciprocity of the influence coefficients was also observed by Negus and Yovanovich (1987a,b) for semi-infinite regions.

### 6.2.1 Thermal Resistance

Finally, if we consider defining a thermal resistance $R_j = \theta_j / Q_j$, for each heat source, it can be shown that

$$\hat{R}_j = \sum_{i=1}^{N} \frac{Q_i}{Q_j} \hat{f}_{ij} \tag{6.28}$$

or

$$\overline{R}_j = \sum_{i=1}^{N} \frac{Q_i}{Q_j} \overline{f}_{ij} \tag{6.29}$$

The above equations clearly demonstrate that the concept of thermal resistance is not strictly applicable in multiple source applications, since the total resistance of any given source depends on both proximity of the neighboring heat source, i.e. $f_{ij}$, and the relative strength ratio, i.e. $Q_i/Q_j$ of the neighboring sources. Changing location or strength of any one source leads to a new value of thermal resistance for the source of interest.

### 6.2.2 Source Plane Convection

Convection in the source plane may now be dealt with using results of Hein and Lenzi (1969). Comparison of the solution of Muzychka et al. (2003) with that of Hein and Lenzi (1969) shows that coefficient $B_0$ becomes

$$B_0 = \frac{\dfrac{1}{ab}\left(\dfrac{t_1}{k_1} + \dfrac{1}{h_2}\right)}{1 + \dfrac{h_1}{h_2} + \dfrac{h_1 t_1}{k_1}} \tag{6.30}$$

where $h_1$ denotes the film coefficient in the source plane and $h_2$ denotes the film coefficient in the sink plane. Further, the spreading function $\phi$ becomes

$$\phi(\xi) = \frac{\left(\dfrac{h_1}{k_1\xi} + \dfrac{k_1\xi}{h_2}\right)\sinh(\xi t_1) + \left(1 + \dfrac{h_1}{h_2}\right)\cosh(\xi t_1)}{\dfrac{k_1\xi}{h_2}\cosh(\xi t_1) + \sinh(\xi t_1)} \tag{6.31}$$

Both Eqs. (6.30) and (6.31) reduce to the expressions for slab with an adiabatic source plane, i.e. when $h_1 = 0$.

## 6.3 Extension to Compound, Orthotropic, and Multilayer Spreaders

The results developed earlier may be easily adapted to compound and orthotropic systems with little effort. Muzychka et al. (2004) applied the necessary transformations to show the relationship between isotropic and orthotropic systems. Further, using the results of Yovanovich et al. (1999), one may modify the isotropic model to effectively model a resistive or conductive layer placed on a rectangular substrate. Each modification is discussed below.

### 6.3.1 Compound Media

In the case of compound systems, we may apply the results reported in Chapter 4 for the modified spreading functions to be used in the ICM. In these cases all of the expressions for $\phi(\xi)$ are applicable. It should also be noted that the coefficient $B_0 = R_{1D}$. The expressions are repeated here for completeness, as they have the greatest application.

#### Finite Interfacial Conductance

For a compound system (two isotropic layers with interfacial conductance), the necessary spreading function is

$$\phi(\xi) = \frac{C_1 + C_2 \tanh(\xi t_1)}{C_1 \tanh(\xi t_1) + C_2} \tag{6.32}$$

where

$$C_1 = \left[\xi \tanh(\xi t_2) + h_s/k_2\right] \tag{6.33}$$

and

$$C_2 = k_1/k_2 \left[\xi(1 + h_s/h_c) + (h_s/k_2 + \xi^2 k_2/h_c) \tanh(\xi t_2)\right] \tag{6.34}$$

and

$$B_0 = \frac{1}{ab}\left(\frac{t_1}{k_1} + \frac{1}{h_c} + \frac{t_2}{k_2} + \frac{1}{h_s}\right) \tag{6.35}$$

If the device is in contact with an ideal heat sink, $h_s \to \infty$, the above expressions simplify to yield:

$$\phi(\xi) = \frac{1 + \xi k_1/h_c \tanh(\xi t_1) + k_1/k_2 \tanh(\xi t_2) \tanh(\xi t_1)}{\xi k_1/h_c + k_1/k_2 \tanh(\xi t_2) + \tanh(\xi t_1)} \tag{6.36}$$

and

$$B_0 = \frac{1}{ab}\left(\frac{t_1}{k_1} + \frac{1}{h_c} + \frac{t_2}{k_2}\right) \tag{6.37}$$

#### Perfect Interfacial Contact

If there is perfect interfacial contact between layers, then $h_c \to \infty$, the solution for the spreading function becomes

$$\phi(\xi) = \frac{\left[\xi \tanh(\xi t_2) + h_s/k_2\right] + k_1/k_2 \left[\xi + h_s/k_2 \tanh(\xi t_2)\right] \tanh(\xi t_1)}{\left[\xi \tanh(\xi t_2) + h_s/k_2\right] \tanh(\xi t_1) + k_1/k_2 \left[\xi + h_s/k_2 \tanh(\xi t_2)\right]} \tag{6.38}$$

and

$$B_0 = \frac{1}{ab}\left(\frac{t_1}{k_1} + \frac{t_2}{k_2} + \frac{1}{h_s}\right) \tag{6.39}$$

Finally, if the device is in contact with an ideal heat sink, $h_s \to \infty$, the above expression simplifies to yield:

$$\phi(\xi) = \frac{1 + k_1/k_2 \tanh(\xi t_2) \tanh(\xi t_1)}{\tanh(\xi t_1) + k_1/k_2 \tanh(\xi t_2)} \tag{6.40}$$

and

$$B_0 = \frac{1}{ab} \left( \frac{t_1}{k_1} + \frac{t_2}{k_2} \right) \tag{6.41}$$

This modification can only be applied to the case when there is no convection in the source plane.

**Example 6.1** In the first case, a plate or circuit board with the following dimensions is considered: $a = 300$ mm, $b = 300$ mm, $t_1 = 10$ mm, $h = 10$ W/(m² K) and $k = 10$ W/(m K), with $T_f = 25°$C. Two heat sources having dimensions $c = 25$ mm and $d = 25$ mm each. The first source with a power of $Q_1 = 10$ W is located at $X_c = Y_c = 90$ mm and the second having a power of $Q_2 = 15$ W at $X_c = Y_c = 210$ mm. The basic equations may be programmed into any symbolic or numerical mathematics software package. For the present calculations, the symbolic mathematics program Maple V was employed. A total of 100 terms were used in each of the single summations and 50 terms in the double summations. The results for the mean $\overline{T}$ and centroidal $\hat{T}$ temperatures of the first case are presented in the table below. The results are in excellent agreement with the method proposed by Culham et al. (2000).

For this problem, we use Eqs. (6.19) and (6.24) to find the influence coefficients. The appropriate spreading function $\phi$ is also required in each case, i.e. Eq. (6.2) for the simple isotropic plate and Eq. (6.38) for the compound system with perfect contact between layers. The influence coefficient matrices for the $\overline{f}_{ij}$ and $\hat{f}_{ij}$ are provided below for reference, for the single layer case.

$$\overline{f}_{ij} = \begin{bmatrix} 4.781826 & 0.523256 \\ 0.523256 & 4.781826 \end{bmatrix}$$

$$\hat{f}_{ij} = \begin{bmatrix} 5.249928 & 0.521899 \\ 0.521899 & 5.249928 \end{bmatrix}$$

For the same configuration, a thin, $t = 2$ mm, highly conductive layer $k = 350$ W/(m K), is added to the original substrate and the problem reanalyzed. The results are also summarized in the table below. It is clearly seen that adding a conductive thermal spreader has reduced the maximum source temperatures considerably and equalized the temperature.

| One layer | $\hat{T}_1$ | $\overline{T}_1$ | $\hat{T}_2$ | $\overline{T}_2$ |
|---|---|---|---|---|
| Present | 84.33 | 80.67 | 108.97 | 101.96 |
| Culham et al. (2000) | 85.95 | 80.20 | 109.92 | 101.26 |

| Two layers | $\hat{T}_1$ | $\overline{T}_1$ | $\hat{T}_2$ | $\overline{T}_2$ |
|---|---|---|---|---|
| Present | 56.55 | 56.05 | 59.94 | 59.19 |
| Culham et al. (2000) | 56.63 | 56.03 | 60.08 | 59.16 |

## 6.3.2 Orthotropic Spreaders

If the rectangular flux channel is a single orthotropic slab such that the in-plane and through-plane conductivities are different, i.e. $k_{xy} \neq k_z$, then the following transformations may be made to apply the present method for this system:

$$k \rightarrow k_{eff} = \sqrt{k_{xy} k_z} \tag{6.42}$$

where, $k_{xy}$ and $k_z$ represent the *in-plane* and *through-plane* thermal conductivity, and

$$t \rightarrow t_{eff} = \frac{t}{\sqrt{k_z/k_{xy}}} \tag{6.43}$$

The orthotropic transformation is also valid for a substrate which is convectively cooled in the source plane.

**Example 6.2**  For this example, we consider a four-source problem. Three particular cases will be examined: isotropic, compound, and orthotropic. The thermal property and component thicknesses are given the table below.

|  | $t_1$ | $t_2$ | $k_1$ | $k_2$ | $k_{xy}$ | $k_z$ |
|---|---|---|---|---|---|---|
|  | (mm) | | | (W/(m K)) | | |
| Case A | 10 | — | 10 | — | — | — |
| Case B | 2 | 10 | 100 | 10 | — | — |
| Case C | 10 | — | — | — | 100 | 10 |

In all three cases, the heat source layout summarized in the table below is used, along with the following substrate properties: $a = 200$ mm, $b = 100$ mm, and $h = 100$ W/(m$^2$ K).

|  | $Q$ | $c$ | $d$ | $X_c$ | $Y_c$ |
|---|---|---|---|---|---|
|  | (W) | (mm) | (mm) | (mm) | (mm) |
| Source 1 | 10 | 20 | 20 | 40 | 30 |
| Source 2 | 15 | 30 | 40 | 95 | 30 |
| Source 3 | 25 | 30 | 70 | 155 | 45 |
| Source 4 | 10 | 50 | 10 | 55 | 75 |

In the first case, an isotropic substrate which is cooled in the sink plane is considered. Next, the effect of a heat spreader is examined through the addition of a conductive layer. Finally, the effect of orthotropic properties is examined. This gives rise to $t = t_{eff} = 31.62$ mm and $k = k_{eff} = 31.62$ W/(m K) using the given properties. Maple V was used to perform the necessary calculations. The results are provide in the table below for all three cases.

| (°C) | Isotropic | | Compound | | Orthotropic | |
|---|---|---|---|---|---|---|
| | $\hat{\theta}$ | $\overline{\theta}$ | $\hat{\theta}$ | $\overline{\theta}$ | $\hat{\theta}$ | $\overline{\theta}$ |
| Source 1 | 52.59 | 47.37 | 37.55 | 36.16 | 38.06 | 36.67 |
| Source 2 | 51.98 | 47.45 | 40.14 | 38.58 | 37.59 | 36.52 |
| Source 3 | 52.62 | 46.85 | 40.40 | 38.30 | 37.79 | 36.59 |
| Source 4 | 43.57 | 39.88 | 34.66 | 33.56 | 35.65 | 34.79 |

Once again, we use Eqs. (6.19) and (6.24) to find the influence coefficients. The appropriate spreading function $\phi$ is also required in each of the three cases, i.e. Eq. (6.2) for the simple isotropic plate, Eq. (6.38) for the compound system with perfect contact between layers, and Eq. (6.2) using the effective properties of the orthotropic layer.

### 6.3.3 Multilayer Isotropic/Orthotropic Spreaders

Muzychka et al. (2013) and Bagnall et al. (2014) applied the above method to predict temperature in GaN HEMT's in compound and multilayered orthotropic spreaders. The methodology has already been discussed in Chapter 4 for a spreader composed of multiple isotropic layers and in Chapter 5 for multiple orthotropic layers. The equations can also be applied for a system composed of a mix of isotropic and orthotropic layers provided the appropriate material transformations are applied to each orthotropic layer.

In an $N$ layer system, we apply the following modifications:

$$B_0 = \frac{Q}{ab} \left( \sum_{i=1}^{N} \frac{\overline{t}_i}{k_i} + \sum_{j=1}^{N-1} \frac{1}{h_{c,j}} + \frac{1}{h_s} \right) \tag{6.44}$$

for the one-dimensional resistance, and $\phi \rightarrow \phi_1(\xi)$ in Eq. (6.20), where $\xi$ is the appropriate eigenvalue in each term of the solution.

In an $N$ layer structure, we can define two different spreading functions, one which assumes perfect interfacial contact and one which assumes imperfect contact. Although the former can be modeled from the latter, both are presented for convenience. The spreading function $\phi$ in Eq. (6.20) is a recursive function when two or more layers are present in a system and is now denoted $\phi_1$. The recursive relationship is now given by a general expression for $\phi_i$ such that $\phi_1$ is a function of $\phi_2$ which is a function of $\phi_3$ ... and so on, until the final spreading function $\phi_N$ is called. Following the work of Bagnall et al. (2014), we may write:

**Perfect Interfacial Contact**

$$\phi_i = \frac{\frac{k_i}{k_{i+1}} \tanh(\xi \overline{t}_i) + \phi_{i+1}}{\frac{k_i}{k_{i+1}} + \phi_{i+1} \tanh(\xi \overline{t}_i)} \tag{6.45}$$

**Nonperfect Interfacial Contact**

$$\phi_i = \frac{\frac{k_i}{k_{i+1}} \tanh(\xi \overline{t}_i) + \phi_{i+1}[1 + \frac{k_i}{h_{c,i}} \xi \tanh(\xi \overline{t}_i)]}{\frac{k_i}{k_{i+1}} + \phi_{i+1}[\tanh(\xi \overline{t}_i) + \frac{k_i}{h_{c,i}} \xi]} \tag{6.46}$$

In both cases, we denote the $N$th layer spreading function by

$$\phi_N = \left[\frac{\xi \tanh(\xi \bar{t}_N) + h_s/k_N}{\xi + h_s/k_N \tanh(\xi \bar{t}_N)}\right] \tag{6.47}$$

Specification of the spreading functions for $N$-layers is now a simple matter. It becomes a simple exercise to define sequences of materials in multilayered isotropic/orthotropic structures and assessing thermal performance. Using Eqs. (6.42) and (6.43), we may generalize:

$$k_i = \sqrt{k_{i,xy}k_{i,z}} \tag{6.48}$$

**Figure 6.3** Device layout for a comparative thermal analysis of a GaN HEMT device. Source: Bagnall et al. (2014). Used with the permission of the Institute of Electrical and Electronic Engineers (IEEE).

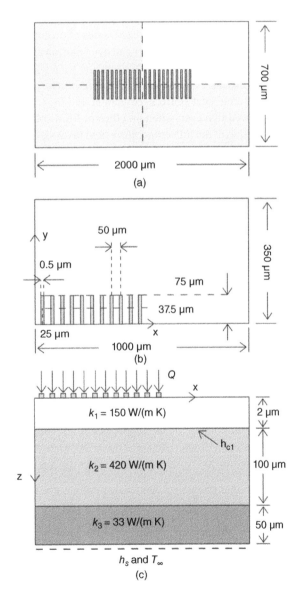

and

$$\bar{t}_i = \frac{t_i}{\sqrt{k_{i,z}/k_{i,xy}}} \tag{6.49}$$

such that:

Layer 1: $k_1 = \sqrt{k_{1,xy}k_{1,z}}$ $\quad \bar{t}_1 = t_1/\sqrt{k_{1,z}/k_{1,xy}}$

Layer 2: $k_2 = \sqrt{k_{2,xy}k_{2,z}}$ $\quad \bar{t}_2 = t_2/\sqrt{k_{2,z}/k_{2,xy}}$ $\qquad\qquad$ (6.50)
$\qquad\quad \vdots$

Layer $N$: $k_N = \sqrt{k_{N,xy}k_{N,z}}$ $\quad \bar{t}_N = t_N/\sqrt{k_{N,z}/k_{N,xy}}$

Since Eqs. (6.48) and (6.49) reduce to isotropic values for materials with $k_{xy} = k_z$, one can simply define the model for a generalized series of orthotropic materials. Bagnall et al. (2014) considered a number of applications of the above solution in addition to also prescribing convection in the source plane for the case of N-Layers.

The ICM can be quite a useful and powerful analysis tool for thermal profiling devices with many discrete heat sources. Bagnall et al. (2014) used the approach to model GaN microdevices and showed that even when an efficient FEM solver is used, the required discretization for accurate solution of the full temperature field takes significantly longer than simply calculating the field and source temperature using the ICM approach. Further, small changes in device location can be considered more easily, since the computational domain using FEM needs to be discretized each time a source is moved. Whereas the ICM approach only requires the location to be changed and temperature recalculated using the influence coefficients.

Figure 6.3 illustrates one case study presented in Bagnall et al. (2014). It is intended to represent a small GaN on SiC HEMT microdevice with 22 heat sources. The device measured 2000 μm by 700 μm and was composed of three layers with an interface resistance between the first two layers

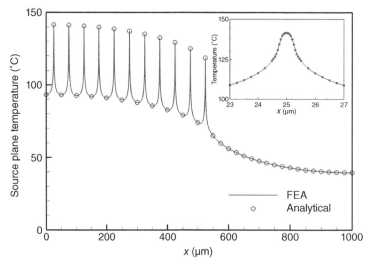

**Figure 6.4** Thermal profile for the device shown in Figure 6.3 along with the profile for the inner most heat source. Source: Bagnall et al. (2014). Used with the permission of the Institute of Electrical and Electronic Engineers (IEEE).

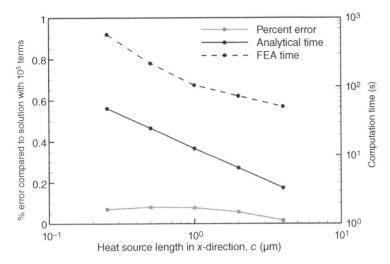

**Figure 6.5** Comparison of computation time for analytical versus finite element based solutions. Source: Bagnall et al. (2014). Used with the permission of the Institute of Electrical and Electronic Engineers (IEEE).

as shown in Figure 6.3. Each heat source was $0.5\,\mu m$ by $150\,\mu m$. The three layers of the device consisted of GaN, SiC, and SnAg die attach. Each heat source had a flux of $10^{10}$ W/m$^2$ representing the typical power dissipation level of 5 W/mm in a GaN power amplifiers (PAs).

Thermal analysis of the device was undertaken using superposition of heat sources as outlined above, with computations for the ICM approach completed in the Matlab mathematical software environment. The device was also modeled in COMSOL Multiphysics (FEM) software. One quarter of the system was used to model the device for computational efficiency. Despite the apparent simplicity of the problem, the full numerical simulation for the device using finite element analysis (FEA) is still somewhat time-consuming. Figure 6.4 provides a comparison of the analytical versus FEM results and shows excellent agreement. Figure 6.5 illustrates the computation time for each model. Bagnall et al. (2014) make a strong case for the use of analytical approaches over full numerical, since small changes in device layout can be assessed more quickly using the analytical approach.

## 6.4 Non-Fourier Conduction Effects in Microscale Devices

In Section 6.3, we showed how analytical models for single heat sources can be combined using superposition to model more complex heat source arrangements in micro-electronics devices. The GaN HEMT micro-device analyzed by Bagnall et al. (2014) is quite small, on the order of a millimeter in its basic dimensions and much less than a millimeter in overall thickness, with heat sources considerably smaller. Many wide bandgap semiconductor devices such as the GaN HEMT, SiC metal semiconductor field effect transistors (MESFETs), and the AlN and $\beta$-Ga$_2$O$_3$ ultrawide bandgap devices are becoming extremely attractive for high-power and high-frequency electronic applications [Shen et al. (2022)]. These devices can experience extremely high junction temperatures and good understanding of their thermal behavior is vital to their reliability.

Shen et al. (2022) outline two prominent features of heat transfer within these devices. One issue is that sources are quite small, and heat generation is confined primarily to the top layer and leading to a large thermal spreading effect. The second issue is that these devices are made up of extremely thin films, often less than 3 μm or less. This dimension is comparable to the mean free path of phonons in many semiconductor materials. Phonons are the primary carriers of heat in these devices and their transport is significantly suppressed by strong boundary scattering, leading to higher junction temperatures. As a result, thermal analysis using classical Fourier law heat conduction models can seriously under predict junction temperature in many of these devices. Without delving into too much detail on micro-scale heat conduction effects, we will show how we may still make reliable predictions of temperature using traditional methods, provided we can define an effective thermal conductivity. The reader is directed to the text of Hahn and Ozisik (2012) for an introduction to microscale conduction or the text of Zhang (2007) on microscale heat transfer.

In microscale heat transfer, we define the limitations of the Fourier conduction law by considering the physical extent of the system, i.e. a characteristic length scale associated with the physical domain, and the mean free path of the energy carrier (phonons or electrons). In simple terms, for the Fourier law to be valid we must satisfy:

$$\ell \gg \Lambda \tag{6.51}$$

where $\ell$ is a characteristic length scale related to the geometry of the microdevice and $\Lambda$ is the mean free path of the energy carrier [Hahn and Ozisik (2012)]. This requirement for Fourier conduction to be the dominant mode of thermal transport is often written as

$$Kn = \frac{\Lambda}{\ell} < 0.1 \tag{6.52}$$

For larger Knudsen numbers, we may still use Fourier's law with an effective thermal conductivity $k_{eff}$ defined such that

$$\frac{k_{eff}}{k_{bulk}} = \frac{1}{1 + \beta \, Kn} \tag{6.53}$$

where the constant $\beta$ typically varies according to

$$\frac{1}{3} < \beta < \frac{5}{4} \tag{6.54}$$

for the range of Knudsen numbers $0.1 < Kn < 1$. For larger Knudsen numbers, i.e. $Kn > 1$, the following behavior is often used [Hahn and Ozisik (2012)]:

$$\frac{k_{eff}}{k_{bulk}} = \frac{4 \ln(2Kn)}{\pi Kn} \tag{6.55}$$

To study the impact of length scale on thermal spreading, Hua et al. (2019) considered the effect of the transverse Knudsen number (i.e. $Kn = \Lambda/t$) on a simple two-dimensional, micro-spreader similar to that shown in Figure 6.6. The authors considered two heat source aspect ratios and a range of Knudsen number $0.2 < Kn_t < 2$. Figure 6.7 shows the effect of Knudsen number and hence, the extent of non-Fourier conduction effects.

In the study by Shen et al. (2022), the authors considered ballistic regime thermal transport in a simple two-dimensional thermal spreader (see Figure 6.6) composed of various semiconductor

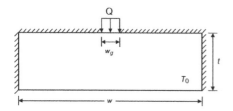

**Figure 6.6** Two dimensional system considered by Hua et al. (2019) and Shen et al. (2022) for studying non-Fourier effects in wide bandgap devices. Source: Adapted from Hua et al. (2019).

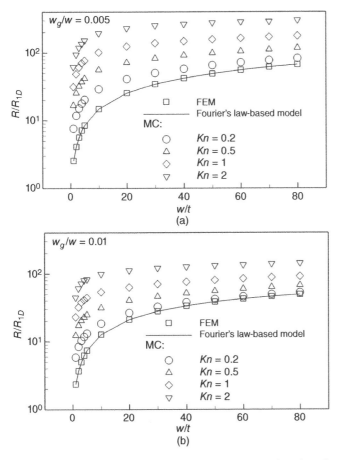

**Figure 6.7** Dimensionless total thermal resistance as a function of spreader size $w/t$ and Knudsen number $Kn = \Lambda/t$ for two source aspect ratios (a) $\epsilon = w_g/w = 0.005$ and (b) 0.01. Source: Modified from Hua et al. (2019).

**Table 6.1** Thermal properties of select semiconductors.

| Material | Mean free path | Bulk thermal conductivity |
|---|---|---|
| | $\Lambda$ (nm) | $k_{bulk}$ (W/(m K)) |
| GaN | 1612.3 | 253 |
| AiN | 3401.4 | 321 |
| SiC | 2506.9 | 490 |
| $\beta\text{-Ga}_2\text{O}_3$ | 450.7 | 21.5–27.0 |

Source: Adapted from Shen et al. (2022).

materials given in Table 6.1. The thermal conductivity of these materials is shown as a function temperature in Figure 6.8. Both experimental data and phonon dispersion model predictions for thermal conductivity are provided. Shen et al. (2022) considered both Dispersion Monte Carlo (MC) predictions and classical Fourier conduction model predictions in their analysis of this simple system. Figure 6.9 provides the total spreader resistance nondimensionalized using

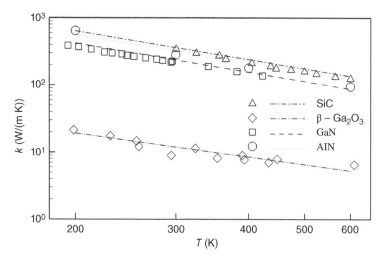

**Figure 6.8** Thermal conductivity of semiconductor materials. Source: Modified from Shen et al. (2022).

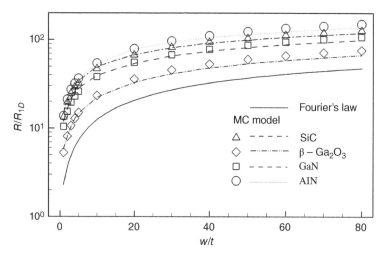

**Figure 6.9** Total thermal resistance of a simple two dimensional thermal spreading problem in wide bandgap semiconductor for materials exhibiting non-Fourier conduction effects. Source: Modified from Shen et al. (2022).

the one-dimensional bulk resistance as a function of the spreader aspect ratio $w/t$ for four semiconductor materials. For the source and spreader dimensions considered, it can be seen that the classical Fourier approach under predicts the Dispersion MC results, though the trend of the Fourier model follows the thermal resistance curves closely. This further illustrates that an "effective" thermal conductivity may be used to predict results using the Fourier method. Since thermal conductivity is reduced due to boundary scattering effects, a lower effective bulk thermal conductivity will increase the overall thermal resistance.

Shen et al. (2022) also provided the temperature field in the source plane for one of their Dispersion MC simulations along with FEM calculated results using an effective thermal conductivity. This is provided in Figure 6.10. It can be seen that excellent results can still be obtained using the classical Fourier method at these dimensional scales provided the appropriate effective thermal

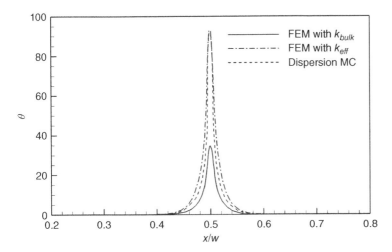

**Figure 6.10** Dimensionless temperature in the source plane for a semiconductor calculated using Dispersion MC methods and FEM methods using the bulk and effective thermal conductivity. Source: After Shen et al. (2022).

conductivity is used in the FEM or analytical solutions. The reader is directed to the papers of Hua et al. (2019) and Shen et al. (2022) for more details.

## 6.5 Application to Irregular-Shaped Heat Sources

We conclude our discussion by considering applications with irregular-shaped heat sources. The basic solution procedure outlined earlier can be used to model irregular-shaped heat sources. For example if a circular- or elliptical-shaped heat source is applied over a rectangular spreader, we can use the basic solution to approximate the source using an arrangement of rectangular strips. With the addition of more strips, the heat source can be approximated to any level of accuracy using $N$ discrete sources.

Consider the case of an elliptic heat source on a finite rectangular spreader. We can use any number of narrow rectangular strip heat sources to approximate the elliptic heat source as shown in Figure 6.11. The elliptic heat source has a constant heat flux $q = Q/A_s$ which can be modeled using $N$ sources, each with a heat input $Q_i = q(c_i d_i)$.

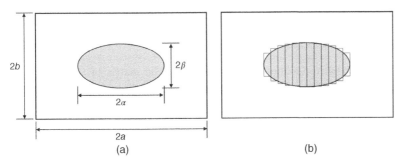

**Figure 6.11** Modeling an elliptic heat source on a rectangular spreader using multiple finite rectangular strip sources. (a) Elliptic heat source. (b) Elliptic heat source approximated with rectangular strips.

The process is easily applied using the ICM to determine the values of mean and centroidal temperature for each source element. If one source is selected to have its centroid coincident with that of the complex source, the maximum source temperature is easily obtained. Mean source temperature can be obtained by taking the area weighted value of all source means, i.e.

$$\bar{\theta}_s = \frac{1}{A_s} \sum_{i=1}^{N} \bar{\theta}_i (c_i d_i) \tag{6.56}$$

and the thermal resistance found by means of

$$R_{total} = \frac{\bar{\theta}_s}{\sum_{i=1}^{N} q(c_i d_i)} = R_{1D} + R_s \tag{6.57}$$

**Example 6.3** We will now illustrate the process by using the ICM outlined earlier to predict the total resistance and the thermal spreading resistance for an elliptic heat source on a finite rectangular spreader. The thermal spreader has dimensions of $2a = 12\,cm$ by $2b = 6\,cm$. The elliptic source is centrally located and has major and minor axes of $2\alpha = 6\,cm$ and $2\beta = 2\,cm$, respectively. The source will be divided into 12 equal width strips of $\Delta x = 5\,mm$ and height defined by $\Delta y = 2\beta \sqrt{1 - x^2/\alpha^2}$, where $x$ is measured from the origin of the ellipse. We will further assume a thermal conductivity of $k = 1\,W/(m\,K)$ and a uniform heat flux $q = 1000\,W/m^2$. We will vary the spreader thickness and consider $t = 0.01, 0.05, 0.1\,m$ and also vary the sink plane conductance using $h_s = 1, 10, 100, 1000\,W/(m^2\,K)$. The 12 strips have the dimensions and locations relative to the lower left corner of the spreader tabulated below:

| Source | $c_i$ | $d_i$ | $X_{c,i}$ | $Y_{c,i}$ |
|--------|-------|-------|-----------|-----------|
| 1 | 0.005 | 0.01993 | 0.0625 | 0.03 |
| 2 | 0.005 | 0.01936 | 0.0675 | 0.03 |
| 3 | 0.005 | 0.01818 | 0.0725 | 0.03 |
| 4 | 0.005 | 0.01624 | 0.0775 | 0.03 |
| 5 | 0.005 | 0.01323 | 0.0825 | 0.03 |
| 6 | 0.005 | 0.00799 | 0.0875 | 0.03 |
| 7 | 0.005 | 0.01993 | 0.0575 | 0.03 |
| 8 | 0.005 | 0.01936 | 0.0525 | 0.03 |
| 9 | 0.005 | 0.01818 | 0.0475 | 0.03 |
| 10 | 0.005 | 0.01624 | 0.0425 | 0.03 |
| 11 | 0.005 | 0.01323 | 0.0375 | 0.03 |
| 12 | 0.005 | 0.00799 | 0.0325 | 0.03 |

Using the equations summarized in Section 6.1, we can define the matrix of $\bar{f}_{ij}$ values and determine the influence coefficients for each of the 12 sources and defining their magnitude $Q_i = q(c_i d_i)$. The method provides the total resistance which must have the bulk resistance subtracted to calculate the spreading resistance and compare with Eq. (4.133). Though, the astute reader should recognize that the $B_0$ coefficient is merely the bulk resistance and if this is set to zero, the spreading resistance is calculated instead of total resistance.

The results provided in the table below show excellent agreement between the approximate approach using the ICM and the exact solution provided by Sadhal (1984), i.e. calculated using

Eq. (4.133). Also take note, we must include the spreading function $\phi(\xi)$ into each term of Eq. (4.133) for thinner spreaders. It is clearly seen that the last column represents a thick spreader or semi-infinite flux channel when $\phi \to 1$. We see from the middle column of values that $t/\sqrt{A_t} = 0.589$ which is the order of magnitude value of the dimensionless spreader thickness for defining a thick spreader, i.e. one which is considered semi-infinite. In the calculations for both approaches, we used 100 terms in all summations.

| | $R_s$ (K/W) | | | | | |
|---|---|---|---|---|---|---|
| **Thickness** | $t = 0.01$ m | | $t = 0.05$ m | | $t = 0.1$ m | |
| $h_s$ (W/(m$^2$ K)) | $R_{s,ellipse}$ | $R_{s,ICM}$ | $R_{s,ellipse}$ | $R_{s,ICM}$ | $R_{s,ellipse}$ | $R_{s,ICM}$ |
| 1 | 11.234 | 11.614 | 7.855 | 7.848 | 7.826 | 7.820 |
| 10 | 10.042 | 10.366 | 7.896 | 7.839 | 7.826 | 7.820 |
| 100 | 7.296 | 7.325 | 7.817 | 7.811 | 7.826 | 7.820 |
| 1000 | 6.110 | 5.864 | 7.799 | 7.793 | 7.826 | 7.820 |

# References

Al-Khamaiseh, B., Muzychka, Y.S., and Kocabiyik, S., "Spreading Resistance in Multilayered Orthotropic Flux Channel with Different Conductivities in the Three Spatial Directions", *Journal of Heat Transfer*, Vol. 140, no. 7, p. 071302, 2018.

Bagnall, K., Muzychka, Y.S., and Wang, E., "Analytical Solution for Temperature Rise in Complex, Multi-layer Structures with Discrete Heat Sources", *IEEE Transactions on Components, Packaging and Manufacturing Technologies*, Vol. 4, no. 5, pp. 817–830, 2014.

Culham, J.R., Yovanovich, M.M., and Lemczyk, T.F., "Thermal Characterization of Electronic Packages Using a Three-Dimensional Fourier Series Solution," *Journal of Electronic Packaging*, Vol. 122, pp. 233–239, 2000.

Ellison, G., *Thermal Computations for Electronic Equipment*, Krieger Publishing, Malabar, FL, 1984.

Hahn, D.W., and Ozisik, N., *Heat Conduction*, 3rd Edition, Wiley, 2012.

Hein, V.L., and Lenzi, V.D., "Thermal Analysis of Substrates and Integrated Circuits", pp. 166–177, Bell Telephone Laboratories, Unpublished Report, 1969.

Hua, Y.C., Li, H.L., and Cao, B.Y., "Thermal Spreading Resistance in Ballistic-Diffusive Regime for GaN HEMTs", *IEEE Transactions on Electron Devices*, Vol. 66, no. 8, pp. 3296–3301, 2019.

Kadambi, V., and Abuaf, N., "An Analysis of the Thermal Response of Power Chip Packages", *IEEE Transactions on Electron Devices*, Vol. ED-32, no. 6, pp. 1024–1033, 1985.

Kokkas, A., "Thermal Analysis of Multiple-Layer Structures", *IEEE Transactions on Electron Devices*, Vol. Ed-21, no. 14, pp. 674–680, 1974.

Krane, M.J.H., "Constriction Resistance in Rectangular Bodies", *Journal of Electronic Packaging*, Vol. 113, pp. 392–396, 1991.

Morse, P.M., and Feshbach, H., *Methods of Theoretical Physics*, McGraw-Hill, New York, 1953.

Muzychka, Y.S., "Influence Coefficient Method for Calculating Discrete Heat Source Temperature on Finite Convectively Cooled Substrates", *IEEE Transactions on Components and Packaging Technologies*, Vol. 29, no. 3, pp. 636–643, 2006.

Muzychka, Y.S., Bagnall, K., and Wang, E., "Thermal Spreading Resistance and Heat Source Temperature in Compound Orthotropic Systems with Interfacial Resistance", *IEEE Transactions on Components, Packaging and Manufacturing Technologies*, Vol. 3, no. 11, pp. 1826–1841, 2013.

Muzychka, Y.S., Culham, J.R., and Yovanovich, M.M., "Thermal Spreading Resistance of Eccentric Heat Sources on Rectangular Flux Channels", *Journal of Electronic Packaging*, Vol. 125, pp. 178–185, 2003.

Muzychka, Y.S., Yovanovich, M.M., and Culham, J.R., "Thermal Spreading Resistances in Compound and Orthotropic Systems", *AIAA Journal of Thermophysics and Heat Transfer*, Vol. 18, pp. 45–51, 2004.

Muzychka, Y.S., Yovanovich, M.M., and Culham, J.R., "Influence of Geometry and Edge Cooling on Thermal Spreading Resistance", *AIAA Journal of Thermophysics and Heat Transfer*, Vol. 20, pp. 247–255, 2006.

Negus, K.J., and Yovanovich, M.M., "Transient Temperature Rise at Surface Due to Arbitrary Contacts on Half Spaces", *Transactions of the CSME*, Vol. 13, pp. 1–9, 1987a.

Negus, K.J., and Yovanovich, M.M., "Thermal Computations in a Semiconductor Die Using Surface Elements and Infinite Images", *International Symposium on Cooling Technology in Electronic Equipment*, Honolulu, HI, USA, pp. 563–574, 1987b.

Sadhal, S.S., "Exact Solutions for the Steady and Unsteady Diffusion Problems for a Rectangular Prism: Cases of Complex Neumann Conditions", *ASME 84-HT-83*, 1984.

Shen, Y., Hua, Y.C., Li, H.L., Sobolev, S.L., and Cao, B.Y., "Spectral Thermal Spreading Resistance of Wide Bandgap Semiconductors in Ballistic-Diffusive Regime", *IEEE Transactions on Electron Devices*, Vol. 69, no. 6, pp. 3047–3054, 2022.

Yovanovich, M.M., Muzychka, Y.S., and Culham, J.R., "Spreading Resistance of Isoflux Rectangles and Strips on Compound Flux Channels", *Journal of Thermophysics and Heat Transfer*, Vol. 13, pp. 495–500, 1999.

Zhang, Z.M., *Nano/Microscale Heat Transfer*, McGraw-Hill, 2007.

# 7

# Transient Thermal Spreading Resistance

Transient spreading resistance occurs during startup, and it is important in many microelectronic devices or applied heating applications. The spreading resistance may be defined with respect to the area-average temperature or with respect to a single point temperature such as the centroid temperature. In many solutions, we may only obtain the result in terms of the centroid value of temperature.

Availability of solutions for transient conditions are significantly more limited, but analytical solutions have been reported for a circular area on an isotropic half-space with isothermal, isoflux, and other heat flux distributions Beck (1979), Blackwell (1972), Dryden et al. (1983), Keltner (1973), Normington and Blackwell (1964, 1972), Schneider et al. (1976), Turyk and Yovanovich (1984), Yovanovich (1997), Lam and Muzychka (2023)]. Various analytical and numerical methods were employed to obtain short and long time solutions. In general, since the solutions are more complicated they are less common. For example, many solutions we found in Chapters 3 and 4 using the Separation of Variables approach will become more complex, requiring double or triple series solutions with the need to find additional eigenvalues. Further complicating matters is that these solutions are also slowly convergent for small time.

In this chapter, we consider several practical solutions for sources on a half-space in addition to some solutions in flux tubes and channels. We also propose simple models for several useful situations using the asymptotic short and long time behavior [Muzychka and Yovanovich (2023)]. These models provide a simple and effective means to calculate the spreading resistance in cases where numerical integration are required or where computational efficiency is compromised in more complex eigen-function solutions.

## 7.1 Transient Spreading Resistance of an Isoflux Source on an Isotropic Half-Space

The transient behavior of thermal spreading in an isotropic half-space can be modeled using the heat diffusion equation:

$$\nabla^2 \theta = \frac{1}{\alpha} \frac{\partial \theta}{\partial t} \tag{7.1}$$

Here, the solution is $\theta(r, t)$, where $r = \sqrt{x^2 + y^2 + z^2}$. We also define the temperature rise $\theta = T - T_\infty$, where $T_\infty$ is the temperature far away from the heat source in the half-space.

*Thermal Spreading and Contact Resistance: Fundamentals and Applications*, First Edition.
Yuri S. Muzychka and M. Michael Yovanovich.
© 2023 John Wiley & Sons, Inc. Published 2023 by John Wiley & Sons, Inc.

Initially, the half-space is assumed to be $\theta(r, 0) = 0$ when $t = 0$. The boundary conditions for $t > 0$ are specified as follows: in the source plane,

$$\frac{\partial \theta}{\partial z} = -\frac{q}{k} \quad \text{for points inside source } A \tag{7.2}$$

and

$$\frac{\partial \theta}{\partial z} = 0 \quad \text{for points outside source } A \tag{7.3}$$

Finally, at positions remote from the source, we require:

$$\theta(r, t) \to 0 \quad \text{as } r \to \infty \tag{7.4}$$

Once $\theta$ is known, we may define the thermal spreading resistance either using the centroid or mean source temperature. In most cases as we shall soon see, the solution is only known at the centroid due to the complexity of integrating over source regions. In either case, we will define the dimensionless thermal spreading resistance as before $\psi = k\mathcal{L}R_s$ and also introduce $Fo_{\mathcal{L}} = \alpha t/\mathcal{L}^2$, where $\mathcal{L}$ is an appropriate length scale. In the case of the circular contact $\mathcal{L} = a$ is used and in the case of a noncircular contact, we often generalize using $\mathcal{L} = \sqrt{A}$.

The transient temperature rise at an arbitrary point:

$$r = \sqrt{(x - x_0)^2 + (y - y_0)^2 + z^2} \tag{7.5}$$

in the source plane due to a point source of strength $qdA$ which is located at the point $(x_0, y_0)$ on the surface of the half-space is given by Carslaw and Jaeger (1959):

$$\theta(r, t) = \frac{qdA}{2\pi kr} \text{erfc}\left(\frac{r}{2\sqrt{\alpha t}}\right) \quad t > 0 \tag{7.6}$$

The solution can be seen to be the product of the steady-state temperature rise (as seen in Chapter 2) and a space–time function which approaches unity as $r/(2\sqrt{\alpha t}) < 0.01$. Solutions using the above approach for a number of half-space problems were considered by Yovanovich (1997).

### 7.1.1 Transient Spreading Resistance of an Isoflux Circular Area

Beck (1979) reported the following integral solution for a circular area of radius $a$ which is subjected to a uniform and constant flux $q$ for $t > 0$. The general solution for the temperature profile was found to be

$$T(r, z, t) = \frac{qa}{2k} \int_0^\infty J_0(\lambda r) J_1(\lambda a) C(z, t) \frac{d\lambda}{\lambda} \tag{7.7}$$

where

$$C(z, t) = \exp(-\lambda z)\left[\text{erfc}\left(\frac{z}{\sqrt{4\alpha t}}\right) - \lambda\sqrt{\alpha t}\right]$$
$$- \exp(\lambda z)\left[\text{erfc}\left(\frac{z}{\sqrt{4\alpha t}}\right) + \lambda\sqrt{\alpha t}\right] \tag{7.8}$$

The mean contact temperature is

$$\overline{T}_c(t) = \frac{2Q}{\pi ka} \int_0^\infty \frac{J_1^2(\lambda a)\text{erf}(\lambda a\sqrt{Fo})d(\lambda a)}{(\lambda a)^2} \tag{7.9}$$

Using the mean contact temperature, we can define the thermal spreading resistance as

$$4kaR_s = \frac{8}{\pi} \int_0^\infty \text{erf}\left(\zeta\sqrt{Fo}\right) J_1^2(\zeta) \frac{d\zeta}{\zeta^2} \tag{7.10}$$

where $\text{erf}(\cdot)$ is the Gaussian error function and $J_1(x)$ is the Bessel function of the first kind of order one [Abramowitz and Stegun (1965)] and $\zeta = \lambda a$ is a dummy variable. The dimensionless time is defined as $Fo = \alpha t/a^2$, where $\alpha$ is the thermal diffusivity of the half-space. The spreading resistance is based on the area average temperature. Steady-state is obtained when $Fo \to \infty$, and the solution goes to $4kaR_s = 32/(3\pi^2)$. Equation (7.10) can be easily integrated in most modern mathematics software tools. However, Beck (1979) gave approximate solutions for short and long times. For short times, where $Fo < 0.6$:

$$4kaR_s = \frac{8}{\pi}\left[\sqrt{\frac{Fo}{\pi}} - \frac{Fo}{\pi} + \frac{Fo^2}{8\pi} + \frac{Fo^3}{32\pi} + \frac{15Fo^4}{512\pi}\right] \tag{7.11}$$

and for long times where $Fo \geq 0.6$:

$$4kaR_s = \frac{32}{3\pi^2} - \frac{2}{\pi^{3/2}\sqrt{Fo}}\left[1 - \frac{1}{3(4Fo)} + \frac{1}{6(4Fo)^2} - \frac{1}{12(4Fo)^3}\right] \tag{7.12}$$

The maximum errors of about 0.18% and 0.07% occur at $Fo = 0.6$ for the short and long time expressions, respectively.

We may also define the dimensionless spreading resistance using the $\sqrt{A}$ as a characteristic length scale. As we have seen this allows us to approximate results for other shapes more easily. Equation (7.10) becomes:

$$k\sqrt{A}R_s = \frac{2}{\sqrt{\pi}} \int_0^\infty \text{erf}\left(\zeta\sqrt{\pi Fo_{\sqrt{A}}}\right) J_1^2(\zeta) \frac{d\zeta}{\zeta^2} \tag{7.13}$$

The above equation can be modeled using the short- and long-time asymptotes, i.e. the leading terms in Eqs. (7.11) and (7.12). Converting the length scale gives [Muzychka and Yovanovich (2023)]:

$$k\sqrt{A}\,R_s = \begin{cases} \sqrt{\dfrac{4Fo_{\sqrt{A}}}{\pi}} & Fo_{\sqrt{A}} \to 0 \\[3mm] \dfrac{8\sqrt{\pi}}{3\pi^2} & Fo_{\sqrt{A}} \to \infty \end{cases} \tag{7.14}$$

We may combine the asymptotes using the Churchill and Usagi (1972) asymptotic modeling approach as follows:

$$k\sqrt{A}R_s = \left[\left(\sqrt{\frac{4Fo_{\sqrt{A}}}{\pi}}\right)^{-3/2} + \left(\frac{8\sqrt{\pi}}{3\pi^2}\right)^{-3/2}\right]^{-2/3} \tag{7.15}$$

The above correlation predicts the transient spreading resistance within 0.49/3.97% error as shown in Figure 7.1 and Table 7.1. It is expected that Eq. (7.15) will predict the thermal spreading resistance for other contact shapes quite well, since only the steady-state limit varies with shape

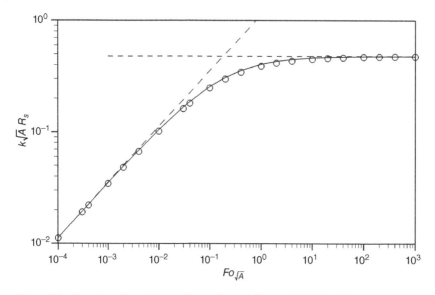

**Figure 7.1** Transient thermal spreading resistance from a circular isoflux heat source on a half-space. Exact solution versus model predictions (points). Short and long time asymptotes are from Eq. (7.14).

**Table 7.1** Comparison of exact and approximate models for $k\sqrt{A}R_s$ for an isoflux circular contact.

| $Fo_{\sqrt{A}}$ | Exact (Eq. (7.13)) $k\sqrt{A}R_s$ | Model (Eq. (7.15)) $k\sqrt{A}R_s$ |
|---|---|---|
| 0.0001 | 0.011171 | 0.011257 |
| 0.001 | 0.034554 | 0.035207 |
| 0.01 | 0.101599 | 0.104979 |
| 0.1 | 0.248939 | 0.256253 |
| 1 | 0.391396 | 0.406972 |
| 10 | 0.450579 | 0.463818 |
| 100 | 0.469922 | 0.476128 |
| 1000 | 0.476059 | 0.478403 |

and aspect ratio of the contact. Later, we will propose a more accurate model for other shapes for both isothermal and isoflux contact areas which include source aspect ratio.

**Example 7.1** A heat 10 W heat source with a 2 mm diameter is attached to a thick silicon spreader having dimensions of 25 mm by 25 mm. The thermal conductivity of the silicon spreader is $k = 148$ W/(m K) with a thermal diffusivity of $\alpha = 89.2 \times 10^{-6}$ m$^2$/s. Determine the steady-state source temperature and the time to reach steady state assuming a $0.99T_{ss}$ criterion for $t_{ss}$.

Since the source to spreader aspect ratio is $\sqrt{A_s/A_t} = 0.071 < 0.1$, we will model the source as a spot on a half-space. Using Eq. (7.14), we find that the steady-state resistance is

$$k\sqrt{A}R_\infty = \frac{8\sqrt{\pi}}{3\pi^2} \approx 0.4789$$

which gives the mean steady-state source temperature rise:

$$\overline{T}_{ss} = QR_{\infty} = 10 \cdot \left( \frac{0.4789}{148 \cdot \sqrt{\pi(0.001)^2}} \right) = 18.25\,°C$$

Finally, equate Eq. (7.15) to $0.99(k\sqrt{A}R_{\infty})$ and solve for the Fourier number:

$$\left[ \left( \sqrt{\frac{4Fo_{\sqrt{A}}}{\pi}} \right)^{-3/2} + \left( \frac{8\sqrt{\pi}}{3\pi^2} \right)^{-3/2} \right]^{-2/3} = 0.99 \left( \frac{8\sqrt{\pi}}{3\pi^2} \right)$$

We find the Fourier number when steady state is reached:

$$Fo_{\sqrt{A}} \approx 47.88$$

Finally, since $Fo_{\sqrt{A}} = \alpha t / A$, we obtain:

$$t_{ss} = \frac{Fo_{\sqrt{A}} A}{\alpha} = \frac{47.88 \cdot \pi(0.001)^2}{89.2 \times 10^{-6}} = 1.69\,s$$

The time to reach steady state clearly depends upon the thermal diffusivity. In electronics devices with much lower thermal diffusivity, the time to reach steady state will increase significantly.

## 7.1.2 Transient Spreading Resistance of an Isoflux Strip on a Half-Space

The solution for an isoflux strip on a half-space is found in Carslaw and Jaeger (1959) and is also developed further in Turyk and Yovanovich (1984). The temperature profile at the surface of the half-space $z = 0$ is found to be

$$T(x, t, z = 0) = \frac{qa}{k} \sqrt{\frac{Fo}{\pi}} f(\xi, Fo) \tag{7.16}$$

where $q = Q'/2a$, $Fo = \alpha t / a^2$, $\xi = x/a$, and

$$f(\xi, Fo) = \text{erf}\left( \frac{1+\xi}{\sqrt{4Fo}} \right) + \text{erf}\left( \frac{1-\xi}{\sqrt{4Fo}} \right)$$

$$+ \frac{1+\xi}{\sqrt{4\pi Fo}} Ei\left( \frac{(1+\xi)^2}{4Fo} \right) + \frac{1-\xi}{\sqrt{4\pi Fo}} Ei\left( \frac{(1-\xi)^2}{4Fo} \right) \tag{7.17}$$

In the above equation, $Ei(\cdot)$ is the exponential integral defined as

$$Ei(x) = \int_{-x}^{\infty} \frac{e^{-t}}{t} dt \tag{7.18}$$

The mean source temperature is found to be

$$\overline{T}_c(t) = \frac{Q'}{2k} \sqrt{\frac{Fo}{\pi}} \int_0^1 f(\xi, Fo) d\xi \tag{7.19}$$

From which we find $\psi^q$ to be

$$\psi^q = R_s k = \frac{1}{2} \sqrt{\frac{Fo}{\pi}} \int_0^1 f(\xi, Fo) d\xi \tag{7.20}$$

The thermal resistance for a two-dimensional strip on a half-space never achieves steady state, but increases linearly with $\log(Fo)$ for $Fo > 1$ [Turyk and Yovanovich (1984)].

### 7.1.3  Transient Spreading Resistance of an Isoflux Hyperellipse

The hyperellipse is defined by $(x/a)^n + (y/b)^n = 1$ with $b \le a$, where $n$ is the shape parameter and $a$ and $b$ are the axes along the $x$- and $y$-axes, respectively. The hyperellipse reduces to many special cases by setting the values of $n$ and the aspect ratio parameter $\gamma = b/a$ which lies in the range: $0 \le \gamma \le 1$. Therefore, the solution developed for the hyperellipse can be used to obtain solutions for many other geometries such as ellipse and circle, rectangle and square, and diamond-like geometries. The transient dimensionless centroid constriction resistance $k\sqrt{A}\hat{R}$, where $\hat{R} = T_0/Q$ is given by the double integral solution [Yovanovich (1997)]:

$$
k\sqrt{A}\hat{R} = \frac{2}{\pi\sqrt{A}} \int_0^{\pi/2} \int_0^{r_0} \mathrm{erfc}\left( \frac{r}{2\sqrt{A}\sqrt{Fo_{\sqrt{A}}}} \right) dr\, d\omega
\tag{7.21}
$$

with $Fo = at/A$, and the area of the hyperellipse is given by $A = (4\gamma/n)B(1+1/n, 1/n)$ and $B(x,y)$ is the Beta function [Abramowitz and Stegun (1965)]. The upper limit of the radius is given by $r_0 = \gamma/\left[(\sin\omega)^n + \gamma^n(\cos\omega)^n\right]^{1/n}$ and the aspect ratio parameter $\gamma = b/a$. The above solution has the following characteristics: (i) for small dimensionless times: $Fo_{\sqrt{A}} \le 4 \times 10^{-2}$, $k\sqrt{A}\hat{R} = \left(2/\sqrt{\pi}\right)\sqrt{Fo_{\sqrt{A}}}$ for all values of $n$ and $\gamma$; (ii) for long dimensionless times: $Fo_{\sqrt{A}} \ge 10^3$, the results are within 1% of the steady-state values which are given by the single integral:

$$
k\sqrt{A}\hat{R} = \frac{2\gamma}{\pi\sqrt{A}} \int_0^{\pi/2} \frac{d\omega}{\left[(\sin\omega)^n + \gamma^n(\cos\omega)^n\right]^{1/n}}
\tag{7.22}
$$

which depends on the aspect ratio $\gamma$ and the shape parameter $n$. The dimensionless spreading resistance depends on the three parameters: $Fo_{\sqrt{A}}, \gamma$, and $n$ in the transition region: $4 \times 10^{-2} \le Fo_{\sqrt{A}} \le 10^3$ in some complicated manner which can be deduced from the solution for the circular area. For this axisymmetric shape, we put $\gamma = 1, n = 2$ into the hyperellipse double integral which yields the following closed-form result valid for all dimensionless time [Yovanovich (1997)]:

$$
k\sqrt{A}\hat{R} = \sqrt{Fo_{\sqrt{A}}} \left[ \frac{1}{\sqrt{\pi}} - \frac{1}{\sqrt{\pi}}\exp\left(-1/\left(4\pi Fo_{\sqrt{A}}\right)\right) \right.
$$
$$
\left. + \frac{1}{2\sqrt{\pi}\sqrt{Fo_{\sqrt{A}}}}\mathrm{erfc}\left( \frac{1}{2\sqrt{\pi}\sqrt{Fo_{\sqrt{A}}}} \right) \right]
\tag{7.23}
$$

where the dimensionless time for the circle of radius $a$ is $Fo_{\sqrt{A}} = at/\left(\pi a^2\right)$.

### 7.1.4  Transient Spreading Resistance of Isoflux Regular Polygons

For regular polygons having sides $N \ge 3$, the area is $A = Nr_i^2 \tan \pi/N$, where $r_i$ is the radius of the inscribed circle. The dimensionless spreading resistance based on the centroid temperature rise $k\sqrt{A}\hat{R}$ is given by the following double integral [Yovanovich (1997)]:

$$
k\sqrt{A}\hat{R} = 2\sqrt{\frac{N}{\tan\dfrac{\pi}{N}}} \int_0^{\pi/N} \int_0^{1/\cos\omega} \mathrm{erfc}\left( \frac{r}{2\sqrt{N\tan\dfrac{\pi}{N}}\sqrt{Fo_{\sqrt{A}}}} \right) dr\, d\omega
\tag{7.24}
$$

where the polygonal area is expressed in terms of the number of sides $N$, and for convenience, the inscribed radius has been set to unity. This double integral solution has identical characteristics as the double integral solution given above for the hyperellipse, i.e. for small dimensionless time: $Fo_{\sqrt{A}} \leq 4 \times 10^{-2}$, $k\sqrt{A}R_0 = \left(2/\sqrt{\pi}\right)\sqrt{Fo_{\sqrt{A}}}$ for all polygons $N \geq 3$; and for long dimensionless times: $Fo_{\sqrt{A}} \geq 10^3$, the results are within 1% of the steady-state values which are given by the following closed-form expression [Yovanovich (1997)]:

$$k\sqrt{A}\hat{R} = \frac{1}{\pi}\sqrt{\frac{N}{\tan(\pi/N)}} \ln\left[\frac{1 + \sin(\pi/N)}{\cos(\pi/N)}\right] \tag{7.25}$$

The dimensionless spreading resistance $k\sqrt{A}\hat{R}$ depends on the parameters: $Fo_{\sqrt{A}}, N$ in the transition region: $4 \times 10^{-2} \leq Fo_{\sqrt{A}} \leq 10^3$ in some complex manner which, as described above, may be deduced from the solution for the circular area. The steady-state solution gives the values $k\sqrt{A}\hat{R} = 0.5617, 0.5611, 0.5642$ for the equilateral triangle, $N = 3$, the square, $N = 4$, and the circle, $N \rightarrow \infty$. The difference between the values for the triangle and the circle is approximately 2.2%, whereas the difference between the values for the square and the circle is less than 0.6%.

### 7.1.5 Universal Time Function

The following procedure [Yovanovich (1997)] is proposed for computation of the centroid-based transient spreading resistance for the range: $4 \times 10^{-2} \leq Fo_{\sqrt{A}} \leq 10^6$. The closed-form solution for the circle is the basis of the proposed method. For any planar, singly connected contact area subjected to a uniform heat flux take

$$\frac{\psi_0}{\psi_0(Fo_{\sqrt{A}} \rightarrow \infty)} = 2\sqrt{Fo_{\sqrt{A}}}\left[1 - \exp\left(-1/\left(4\pi Fo_{\sqrt{A}}\right)\right)\right.$$
$$\left. + \frac{1}{2\sqrt{Fo_{\sqrt{A}}}} \text{erfc}\left(\frac{1}{2\sqrt{\pi}\sqrt{Fo_{\sqrt{A}}}}\right)\right] \tag{7.26}$$

where $\psi_0 = k\sqrt{A}\hat{R}$. The right-hand side of the above equation can be considered to be a *universal* dimensionless time function that accounts for the transition from small times to near steady state. The proposed procedure should provide quite accurate results for any planar, singly connected area. A simpler expression which is based on the Greene (1989) approximation of the complementary error function is proposed [Yovanovich (1997)]:

$$\frac{\psi_0}{\psi_0(Fo \rightarrow \infty)} = \frac{1}{z\sqrt{\pi}}\left[1 - e^{-z^2} + a_1\sqrt{\pi}ze^{-a_2(z+a_3)^2}\right] \tag{7.27}$$

where $z = 1/(2\sqrt{\pi}\sqrt{Fo_{\sqrt{A}}})$, and the three correlation coefficients are $a_1 = 1.5577, a_2 = 0.7182, a_3 = 0.7856$. This approximation will provide values of $\psi_0$ with maximum errors less than 0.5% for $Fo_{\sqrt{A}} \geq 4 \times 10^{-2}$.

## 7.2 Transient Spreading Resistance of an Isothermal Source on a Half-Space

In general, as we have seen in Chapter 2, most thermal spreading results are for the case of a uniform heat flux source region on a half-space. For isothermal contacts of circular or arbitrary shape, there are no simple solutions for transient conditions. Keltner (1973) developed a heat balance integral approach for an isothermal disk on a half-space in oblate spheroidal coordinates while Normington and Blackwell (1972) provided a solution obtained in Oblate Spheroidal coordinates. A solution for the temperature distribution was found using various approximations, but no simple expression was provided for the thermal spreading resistance.

However, we may avail of a more general model proposed by Yovanovich et al. (1995), who considered transient heat conduction from isothermal bodies in a full space. Using symmetry to reduce the problem to a half-space provides the adiabatic region outside of the source area. All that remains is to understand how to adapt the model appropriately for a half-space. The authors had considered ten geometric cases of which three are planar contact regions. For example, the sphere and cuboid shapes degenerate into plates in a full space. As a result of their novel use of the square root of the body surface area as a characteristic length scale $\sqrt{A}$, the scatter in nondimensional results was small, ensuring wider applicability to more arbitrary shapes.

Yovanovich et al. (1995) showed that the nondimensional results for various three-dimensional bodies in full space could be well predicted by the exact analytical solution for a sphere in full space [Ozisik (1968)]. The dimensionless heat transfer rate was composed of two limits which are a linear superposition. The resulting model is given by the exact solution for transient conduction from a sphere into full space:

$$Q^{\star}_{\sqrt{A}} = 2\sqrt{\pi} + \frac{1}{\sqrt{\pi}\sqrt{Fo_{\sqrt{A}}}} \tag{7.28}$$

where

$$Q^{\star}_{\sqrt{A}} = \frac{\overline{q}\sqrt{A}}{k\theta_0} \qquad Fo_{\sqrt{A}} = \frac{\alpha t}{A}$$

where $\mathcal{L} = \sqrt{A}$ is used as a length scale to nondimensionalize both the average surface flux $\overline{q} = Q/A$ and time $t$. In this case, it represents the total surface area of the body in full space. Thus, for plate sources it includes both sides of the plate to be consistent with the total area of three-dimensional bodies. The authors also proposed a more specific and accurate form:

$$Q^{\star}_{\sqrt{A}} = \left[ \left( S^{\star}_{\sqrt{A}} \right)^n + \left( \frac{1}{\sqrt{\pi}\sqrt{Fo_{\sqrt{A}}}} \right)^n \right]^{1/n} \tag{7.29}$$

where $S^{\star}_{\sqrt{A}}$ is the dimensionless shape factor of the body under consideration in full space and $n$ is a blending parameter for correlating the results more accurately. They found $0.87 < n < 1.1$ depends weakly upon the body shape. This provided for more accurate correlation of their numerical results for the ten bodies considered. Overall, they recommended $n \approx 1$ for simplicity. The results of their analysis and simple model using the sphere are shown in Figure 7.2 for a body in full space.

We can now apply this model to develop an expression for planar contacts (or cavities) on a half-space. We begin by rearranging Eq. (7.28) or (7.29) using the above definitions to define the total thermal resistance $R_T$. This leads to

$$k\sqrt{A}R_T = \frac{\theta_0 k\sqrt{A}}{Q} = \frac{1}{2\sqrt{\pi} + \frac{1}{\sqrt{\pi}\sqrt{Fo_{\sqrt{A}}}}} \tag{7.30}$$

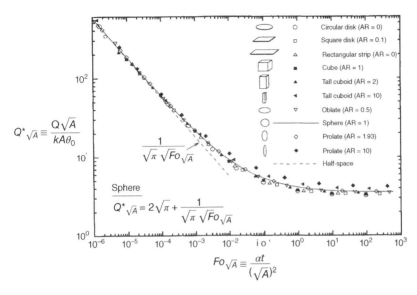

**Figure 7.2** Dimensionless heat flow from convex isothermal bodies in full space. Source: Reproduced from Yovanovich et al. (1995)/with permission of the American Institute of Aeronautics and Astronautics, Inc.

or

$$k\sqrt{A}R_T = \frac{1}{\left[\left(S^{\star}_{\sqrt{A}}\right)^n + \left(\frac{1}{\sqrt{\pi}\sqrt{Fo_{\sqrt{A}}}}\right)^n\right]^{1/n}} \qquad (7.31)$$

Now given that two half-spaces are in parallel when forming a full space, we deduce that the thermal resistance of a half-space is twice that of a full space, i.e. $R_{\frac{1}{2}} = 2R_T$, leaving us with

$$k\sqrt{A}R_{\frac{1}{2}} = \frac{2}{\left[\left(S^{\star}_{\sqrt{A}}\right)^n + \left(\frac{1}{\sqrt{\pi}\sqrt{Fo_{\sqrt{A}}}}\right)^n\right]^{1/n}} \qquad (7.32)$$

Finally, we must consider that the participating area of a body (or plate) in full space is twice that in a half-space, i.e. $A = 2A_{1/2}$, requiring a $\sqrt{2}$ be factored out on the left-hand side dimensionless resistance and in the Fourier number (and dropping the subscript on the $A$):

$$k\sqrt{A}R_s = \frac{2}{\sqrt{2}\left[\left(S^{\star}_{\sqrt{A}}\right)^n + \left(\frac{\sqrt{2}}{\sqrt{\pi}\sqrt{Fo_{\sqrt{A}}}}\right)^n\right]^{1/n}} \qquad (7.33)$$

We do not need to make adjustments to the full space shape factor as it is now simply a geometric constant at this point and varies little with the geometry when nondimensionalized in this manner. For the circular plate in full space $S^{\star}_{\sqrt{A}} = 3.192$ and $n = 1.10$, while for the square plate in full space $S^{\star}_{\sqrt{A}} = 3.343$ and $n = 1.05$ as reported in Yovanovich et al. (1995). This leads to the following simple models for an isothermal circular and square plate:

$$k\sqrt{A}R_s = \frac{1}{\left[(2.257)^{1.10} + \left(\frac{1}{\sqrt{\pi}\sqrt{Fo_{\sqrt{A}}}}\right)^{1.10}\right]^{1/1.10}} \quad -\text{Circle} \qquad (7.34)$$

and

$$k\sqrt{A}R_s = \cfrac{1}{\left[[2.364]^{1.05} + \left[\cfrac{1}{\sqrt{\pi}\sqrt{Fo_{\sqrt{A}}}}\right]^{1.05}\right]^{1/1.05}} - \text{Square} \qquad (7.35)$$

In the study by Yovanovich et al (1995), the universal limit of the sphere $S^{\star}_{\sqrt{A}} = 2\sqrt{\pi}$ was shown to be a good approximation for nearly all bodies and plates in full space. As such we may also propose the simpler form of Eq. (7.33) applied to a half-space when $n = 1$:

$$k\sqrt{A}R_s = \cfrac{1}{\sqrt{2\pi} + \cfrac{1}{\sqrt{\pi}\sqrt{Fo_{\sqrt{A}}}}} \qquad (7.36)$$

where we now denote the surface area of the contact on the half-space problem simply as $A$. Equation (7.36) is similar to the dimensional result presented by Greenwood (1991) for a hemispherical cavity in a half-space. The above equation is an excellent approximate model for transient thermal spreading from isothermal planar contacts or cavities in a half-space. Later, we will develop a more accurate geometry-specific approach to modeling arbitrary-shaped contacts.

The transient limit is universal for all cases as such the maximum error will occur in the steady-state limit. The above equation can be used to model thermal spreading from isothermal planar sources or cavities on a half-space. In other words, Eq. (7.36) is the exact result for a hemispherical cavity but approximates thermal spreading from other source geometries quite well. For example if we consider the limit of steady state, Eq. (7.36) above provides that $k\sqrt{A}R_s \rightarrow 1/\sqrt{2\pi} \approx 0.39894$, while the steady-state limit for the circular disk is $k\sqrt{A}R_s = \sqrt{\pi}/4 \approx 0.44311$ which differs by approximately 10%. Thus, the spreading resistance from a hemispherical cavity is approximately the same as that for a circular disk on a half-space.

We now compare the cases of the circular and square plates considered in Yovanovich et al. (1995) with the universal model above. Using the above correlations, we plot all three cases in Figure 7.3.

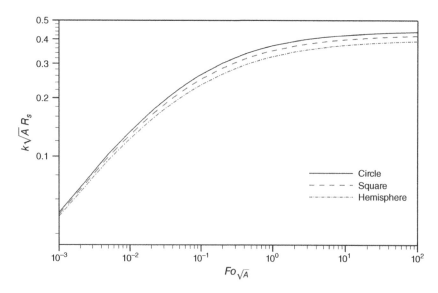

**Figure 7.3** Dimensionless transient spreading resistance of a circular and square isothermal source on a half-space along with the prediction of a hemispherical cavity (exact solution Eq. (7.36)).

The deviation of the general model based on a hemispherical cavity in a half-space is less than 10% for the circular and square planar sources.

## 7.3 Models for Transient Thermal Spreading in a Half-Space

Much can be discerned from the work of Yovanovich et al. (1995). Namely, that for a short time, there is a universal asymptote determined from transient conduction into a half-space. At long time, results vary little with shape of the heat source provided that the characteristic length scale $\mathcal{L} = \sqrt{A}$ is used to nondimensional the thermal resistance. We may now define the simple asymptotic behavior for both isothermal and isoflux heat sources and propose a simpler approach for modeling the full-time behavior for either isothermal or isoflux heat sources based on the work of Muzychka and Yovanovich (2023).

**Isothermal Heat Source**
The asymptotic limits for an isothermal heat source of area $A$ are

$$k\sqrt{A}R_s = \sqrt{\pi Fo_{\sqrt{A}}} \qquad Fo_{\sqrt{A}} \to 0 \tag{7.37}$$

$$k\sqrt{A}R_s = \psi_\infty^T \qquad Fo_{\sqrt{A}} \to \infty \tag{7.38}$$

where $\psi_\infty^T$ represents the nondimensional steady state thermal spreading resistance defined using expressions in Chapter 2 for an isothermal source based on the $\sqrt{A}$ length scale, i.e. Eqs. (2.90) and (2.122) for the rectangle and ellipse, respectively. Equation (7.37) is the result deduced for heat flow into a half-space with an isothermal boundary, e.g.

$$Q = \frac{kA(T_s - T_i)}{\sqrt{\pi \alpha t}} \tag{7.39}$$

A simple model of the form:

$$k\sqrt{A}R_s = \frac{1}{\dfrac{1}{\sqrt{\pi Fo_{\sqrt{A}}}} + \dfrac{1}{\psi_\infty^T}} \tag{7.40}$$

can be used for any heat source provided the steady-state value $\psi_\infty^T = k\mathcal{L}R_s$ (based on $\mathcal{L} = \sqrt{A}$) is accurately known.

**Isoflux Heat Source**
The asymptotic limits for an isoflux heat source of area $A$ are

$$k\sqrt{A}\overline{R}_s = \sqrt{4Fo_{\sqrt{A}}/\pi} \qquad Fo_{\sqrt{A}} \to 0 \tag{7.41}$$

$$k\sqrt{A}\overline{R}_s = \psi_\infty^q \qquad Fo_{\sqrt{A}} \to \infty \tag{7.42}$$

where $\psi_\infty^q$ represents the nondimensional steady state thermal spreading resistance defined using expressions in Chapter 2 for an isoflux source based on the $\sqrt{A}$ length scale, i.e. Eqs. (2.107) and (2.119) for the rectangle and ellipse, respectively. Equation (7.41) is the result deduced for heat flow into a half-space with an isoflux boundary, e.g.

$$T_s - T_i = \frac{2Q\sqrt{\alpha t/\pi}}{kA} \tag{7.43}$$

A simple model of the form

$$k\sqrt{A}\overline{R}_s = \frac{1}{\left[\left(\dfrac{1}{\sqrt{4Fo_{\sqrt{A}}/\pi}}\right)^{3/2} + \left(\dfrac{1}{\psi_\infty^q}\right)^{3/2}\right]^{2/3}}$$ (7.44)

can be used for any heat source provided the steady-state value $\psi_\infty^q = k\mathcal{L}R_s$ (based on $\mathcal{L} = \sqrt{A}$) is accurately known.

We have approximated the blending parameter $n \approx 1$ based on the observations of Yovanovich et al. (1995) and Greenwood (1991) for the isothermal source and $n \approx 3/2$ for the isoflux source based on the solution for the circular contact presented earlier, i.e. Eq. (7.15). Figures 7.4 and 7.5

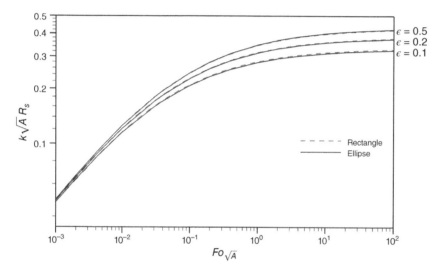

**Figure 7.4** Dimensionless transient thermal spreading resistance for an isothermal elliptic and rectangular source on a half-space.

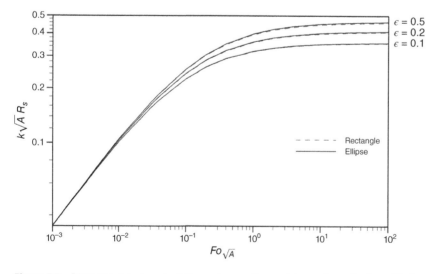

**Figure 7.5** Dimensionless transient thermal spreading resistance for an isoflux elliptic and rectangular source on a half-space.

illustrate the thermal spreading resistance for both cases for an elliptical and rectangular heat source on a half-space for three aspect ratios: $\epsilon = 0.1, 0.2$ and $0.5$. Beyond $\epsilon = 0.5$, the results are very close together since the steady-state solutions are very weak functions of aspect ratio as we have seen in Figure 2.9.

## 7.4 Transient Spreading Resistance Between Two Half-Spaces in Contact Through a Circular Area

When two semi-infinite bodies having different properties and temperatures come into contact with one another, heat is conducted from to the other. If this contact occurs through a small circular contact area, we desire a solution for the thermal spreading resistance between the two bodies in contact. Schneider et al. (1977) solved this problem using a finite volume approach with the solution domain specified in oblate spheroidal coordinates.

The results for the numerical simulations for different material combinations are presented in Figure 7.6. The resistance $R$ is normalized with respect to the steady-state constriction resistance for the two bodies in contact, defined as

$$R_{ss} = \frac{1}{4k_1 a} + \frac{1}{4k_2 a} \tag{7.45}$$

The results were correlated using the following expressions:

$$\frac{R}{R_{ss}} = 0.43 \tanh[0.37 \ln(4X)] + 0.57 \tag{7.46}$$

where

$$X = \frac{1}{2}\left[1 + \sqrt{\alpha_2/\alpha_1}\right]\overline{Fo}, \quad \alpha_2 \leq \alpha_1 \tag{7.47}$$

where $\overline{Fo} = \overline{\alpha}t/a^2$ is defined using the harmonic mean value of thermal diffusivity:

$$\overline{\alpha} = 2\left[\frac{\alpha_1\alpha_2}{\alpha_1 + \alpha_2}\right] \tag{7.48}$$

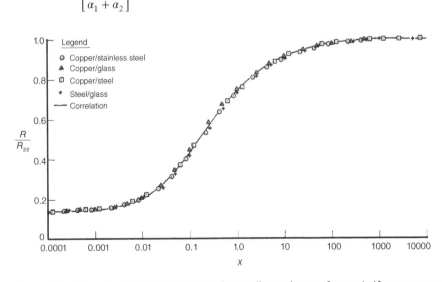

**Figure 7.6** Dimensionless transient thermal spreading resistance for two half-spaces connected through a circular contact area. Source: Reproduced from Schneider et al. (1977)/with permission of Elsevier.

## 7.5 Transient Spreading in a Two-Dimensional Flux Channel

Turyk and Yovanovich (1984) considered the problem for transient thermal spreading resistance in a two-dimensional flux channel. The problem is stated as follows:

$$\frac{1}{\alpha}\frac{\partial T}{\partial t} = \frac{\partial^2 T}{\partial x^2} + \frac{\partial^2 T}{\partial z^2} \tag{7.49}$$

subject to the following boundary conditions:

$$\frac{\partial T}{\partial x}\bigg|_{x=0} = 0$$

$$\frac{\partial T}{\partial x}\bigg|_{x=b} = 0 \tag{7.50}$$

at the centroid and edge of the channel. Over the top surface $z = 0$,

$$\left.\begin{aligned} -k\frac{\partial T}{\partial z}\bigg|_{z=0} &= \frac{Q'}{2a}, \quad 0 < x < a \\[2mm] \frac{\partial T}{\partial z}\bigg|_{z=0} &= 0, \quad a < x < b \end{aligned}\right\} \tag{7.51}$$

where as before for the channel $Q' = Q/L$.

Finally, at $z \to \infty$, we prescribe uniform temperature:

$$T(x, z \to \infty, t) = T_i \tag{7.52}$$

in a region located at a significant distance from the contact plane. The initial condition is also taken as

$$T(x, z, 0) = T_i \tag{7.53}$$

Turyk and Yovanovich (1984) used the Laplace transform method to remove the time variable and then applied separation of variables to obtain the solution in the transform domain. Upon inversion back into the time domain, they obtained the following solution for the temperature distribution:

$$T(x, z, t) = \frac{q_o}{k}\left[B(z, t) + a\sum_{m=1}^{\infty}\frac{\sin(\lambda_m a)}{(\lambda_m a)^2}\cos(\lambda_m x)C_m(z, t)\right] \tag{7.54}$$

where $\lambda_m = m\pi/b$, $q_o = Q'/2b$, and:

$$B(z, t) = \sqrt{4\alpha t}\ \mathrm{ierfc}\left(\frac{z}{\sqrt{4\alpha t}}\right) \tag{7.55}$$

and

$$C_m(z, t) = \exp(-\lambda_m z)\mathrm{erfc}\left(\frac{z}{\sqrt{4\alpha t}} - \sqrt{\alpha\lambda_m^2 t}\right)$$

$$- \exp(\lambda_m z)\mathrm{erfc}\left(\frac{z}{\sqrt{4\alpha t}} + \sqrt{\alpha\lambda_m^2 t}\right) \tag{7.56}$$

The thermal spreading resistance may now be defined as before for a flux channel as

$$R_s = \frac{\overline{T}_s(t) - \overline{T}_{cp}(t)}{Q} \tag{7.57}$$

The mean contact temperature is found by integrating over the source area:

$$\overline{T}_s(t) = \frac{1}{a} \int_0^a T(x, 0, t)dx$$

$$= \frac{q_o}{k}\left[B(0, t) + 2\sum_{m=1}^\infty \frac{b\sin^2(m\pi\epsilon)\text{erf}(m\pi\epsilon\sqrt{Fo})}{\epsilon^2(m\pi)^3}\right] \tag{7.58}$$

where $\epsilon = a/b$ is the source aspect ratio or coverage factor.

The mean contact plane temperature is found by integrating over the channel area:

$$\overline{T}_{cp}(t) = \frac{1}{b}\int_0^b T(x, 0, t)dx = \frac{q_o}{k}B(0, t) = \frac{2q_o}{k}\sqrt{\frac{\alpha t}{\pi}} \tag{7.59}$$

The contact plane temperature can also be written in terms of the source flux such that

$$\overline{T}_{cp}(t) = \frac{q_o}{k}B(0, t) = \frac{2q_a\epsilon}{k}\sqrt{\frac{\alpha t}{\pi}} \tag{7.60}$$

where $q_a = Q'/2a$ which shows the diminished temperature as the applied source flux is essentially smeared over the entire channel surface.

We may now find the thermal spreading resistance using the above results. The dimensionless transient spreading resistance for the isoflux strip as reported by Turyk and Yovanovich (1984):

$$LkR_s = \frac{1}{\pi^3\epsilon^2}\sum_{m=1}^\infty \frac{\sin^2(m\pi\epsilon)\,\text{erf}\left(m\pi\epsilon\,\sqrt{Fo}\right)}{m^3} \tag{7.61}$$

where $\epsilon = a/b < 1$ is the relative size of the contact strip, and the dimensionless time is defined as $Fo = \alpha t/a^2$. There is no steady state half-space solution for the two-dimensional channel. The transient solution is within 1% of the steady-state solution when the dimensionless time satisfies the criterion $Fo \geq 1.46/\epsilon^2$.

# 7.6 Transient Spreading in a Circular Flux Tube from an Isoflux Source

Turyk and Yovanovich (1984) considered the problem for transient thermal spreading resistance in a circular flux tube. The problem is stated as follows:

$$\frac{1}{\alpha}\frac{\partial T}{\partial t} = \frac{\partial^2 T}{\partial r^2} + \frac{1}{r}\frac{\partial T}{\partial r} + \frac{\partial^2 T}{\partial z^2} \tag{7.62}$$

which is subject to

$$\left.\frac{\partial T}{\partial r}\right|_{r=0} = 0$$
$$\left.\frac{\partial T}{\partial r}\right|_{r=b} = 0 \tag{7.63}$$

at the centroid and edge of the disk. Over the top surface $z = 0$,

$$\left.-k\frac{\partial T}{\partial z}\right|_{z=0} = \frac{Q}{\pi a^2}, \quad 0 < r < a$$
$$\left.\frac{\partial T}{\partial z}\right|_{z=0} = 0, \qquad a < r < b \tag{7.64}$$

Finally, at $z \to \infty$, we prescribe uniform temperature:

$$T(r, z \to \infty, t) = T_i \tag{7.65}$$

in a region located at a significant distance from the contact plane. The initial condition is also taken as

$$T(r, z, 0) = T_i \tag{7.66}$$

Turyk and Yovanovich (1984) used the Laplace transform method to remove the time variable and then applied separation of variables to obtain the solution in the transform domain. Upon inversion back into the time domain, they obtained the following solution for the temperature distribution:

$$T(r, z, t) = \frac{q_o}{k} \left[ B(z, t) + \frac{b}{\epsilon} \sum_{m=1}^{\infty} \frac{J_0(\lambda_m b \cdot r/b) J_1(\lambda_m b \epsilon)}{(\lambda_m b)^2 J_0^2(\lambda_m b)} C_m(z, t) \right] \tag{7.67}$$

where $\epsilon = a/b$ is the source aspect ratio or coverage factor and $q_o = Q/\pi b^2$ is the nominal heat flux. The $\lambda_m b$ are the roots of $J_1(\lambda_m b) = 0$ and:

$$B(z, t) = \sqrt{4\alpha t} \ \mathrm{ierfc}\left( \frac{z}{\sqrt{4\alpha t}} \right) \tag{7.68}$$

and

$$C_m(z, t) = \exp(-\lambda_m z)\mathrm{erfc}\left( \frac{z}{\sqrt{4\alpha t}} - \sqrt{\alpha \lambda_m^2 t} \right)$$
$$- \exp(\lambda_m z)\mathrm{erfc}\left( \frac{z}{\sqrt{4\alpha t}} + \sqrt{\alpha \lambda_m^2 t} \right) \tag{7.69}$$

The thermal spreading resistance may now be defined as before for a flux tube as

$$R_s = \frac{\overline{T}_s(t) - \overline{T}_{cp}(t)}{Q} \tag{7.70}$$

The mean contact temperature is found by integrating over the source area:

$$\overline{T}_s(t) = \frac{1}{\pi a^2} \int_0^a T(r, 0, t) 2\pi r \ dr$$
$$= \frac{q_o}{k} \left[ B(0, t) + \frac{4b^2}{\epsilon a} \sum_{m=1}^{\infty} \frac{J_1^2(\lambda_m b \epsilon)\mathrm{erf}(\lambda_m b \epsilon \sqrt{Fo})}{(\lambda_m b)^3 J_0^2(\lambda_m b)} \right] \tag{7.71}$$

The mean contact plane temperature is found by integrating over the channel area:

$$\overline{T}_{cp}(t) = \frac{1}{\pi b^2} \int_0^b T(r, 0, t) 2\pi r \ dr = \frac{q_o}{k} B(0, t) = \frac{2q_o}{k} \sqrt{\frac{\alpha t}{\pi}} \tag{7.72}$$

The contact plane temperature can also be written in terms of the source flux such that

$$\overline{T}_{cp}(t) = \frac{q_o}{k} B(0, t) = \frac{2q_a \epsilon^2}{k} \sqrt{\frac{\alpha t}{\pi}} \tag{7.73}$$

where $q_a = Q/\pi a^2$ which shows the diminished temperature as the applied source flux is essentially smeared over the flux tube surface.

The dimensionless transient spreading resistance for an isoflux circular source of radius $a$ supplying heat to a semi-infinite isotropic flux tube of radius $b$, constant thermal conductivity $k$, and thermal diffusivity $\alpha$ is given by the series solution:

$$4kaR_s = \frac{16}{\pi \, \epsilon} \sum_{n=1}^{\infty} \frac{J_1^2 \left( \delta_n \epsilon \right) \operatorname{erf} \left( \delta_n \epsilon \sqrt{Fo} \right)}{\delta_n^3 \, J_0^2 \left( \delta_n \right)} \tag{7.74}$$

where $\epsilon = a/b < 1$, $Fo = \alpha t/a^2 > 0$, and $\delta_n$ are the positive roots of $J_1(\delta_n) = 0$. The average source temperature rise was used to define the spreading resistance.

The series solution approaches the steady-state solution presented in Section 3.1 when the dimensionless time satisfies the criterion $Fo \geq 1/\epsilon^2$ or when the real time satisfies the criterion $t \geq a^2/(\alpha\epsilon^2)$.

## 7.7 Transient Spreading in a Circular Flux Tube from an Isothermal Source

The solution for transient spreading from an isothermal source on a circular flux tube was considered by Lam et al. (2023). The problem considered the following problem statement:

$$\frac{1}{\alpha} \frac{\partial T}{\partial t} = \frac{\partial^2 T}{\partial r^2} + \frac{1}{r} \frac{\partial T}{\partial r} + \frac{\partial^2 T}{\partial z^2} \tag{7.75}$$

which is subject to

$$\left. \frac{\partial T}{\partial r} \right|_{r=0} = 0 \tag{7.76}$$

$$\left. \frac{\partial T}{\partial r} \right|_{r=b} = 0$$

at the centroid and edge of the disk. Over the top surface $z = 0$,

$$\left. \begin{array}{l} T(r, 0, t) = T_s, \ 0 < r < a \\[2mm] \left. \dfrac{\partial T}{\partial z} \right|_{z=0} = 0, \ a < r < b \end{array} \right\} \tag{7.77}$$

Finally, at $z \to \infty$, we prescribe uniform temperature:

$$T(r, z \to \infty, t) = T_i \tag{7.78}$$

in a region located at a significant distance from the contact plane. The initial condition is also taken as

$$T(r, z, 0) = T_i \tag{7.79}$$

The above problem statement is not straightforward. First, the source plane boundary value problem constitutes a mixed potential boundary value problem. Second, the solution procedure requires a superposition of two solutions, a steady-state solution and a transient solution taking the form:

$$T(r, z, t) = W(r, z, t) + V(r, z) \tag{7.80}$$

The steady state solution was obtained by Gibson (1976) and was used by Lam et al. (2023) to find the transient solution. The solution is somewhat laborious but leads to readily useable results which are summarized below.

The temperature profile was obtained as

$$T(r, z, t) = T_s + \frac{Q}{\pi k b} \left[ -z/b + a_0 + \frac{1}{b} \sum_{n=1}^{\infty} \frac{a_n}{\lambda_n} J_0(\lambda_n r) \left( e^{-\lambda_n z} - \frac{1}{\pi} e^{-\alpha \lambda_n^2 t} D(z, t) \right) \right] \tag{7.81}$$

where

$$D(z, t) = \frac{\pi}{2} e^{\alpha \lambda_n^2 t} \left[ e^{-\lambda_n z} \mathrm{erfc} \left( \frac{2\lambda_n \alpha t - z}{2\sqrt{\alpha t}} \right) - e^{\lambda_n z} \mathrm{erfc} \left( \frac{2\lambda_n \alpha t + z}{2\sqrt{\alpha t}} \right) \right] \tag{7.82}$$

and when $z = 0$, we find

$$D(0, t) = 0 \tag{7.83}$$

The coefficients $a_0$ and $a_n$ are

$$a_0 = \frac{\pi}{2\epsilon} \left[ -\frac{1}{2} + \frac{2\epsilon}{\pi} (1 + I_0) - \frac{4\epsilon^3 I_2}{3\pi} - \frac{32\epsilon^5 I_4}{15\pi} + O\left(\epsilon^7\right) \right] \tag{7.84}$$

where $I_0 = 0.106770, I_2 = 0.398209$, and $I_4 = 0.049997$ are constants and

$$a_n = \frac{1}{\epsilon \delta_n J_0^2(\delta_n)} \left[ \sin \delta_n \epsilon - \frac{8 I_2 \epsilon}{\pi} \left( \frac{\epsilon^2 \sin \delta_n \epsilon}{3} + \frac{\epsilon \cos \delta_n \epsilon}{\delta_n} - \frac{\sin \delta_n \epsilon}{\delta_n^2} \right) \right] + O\left(\epsilon^5\right) \tag{7.85}$$

where $\delta_n$ are the roots of $J_1(\delta_n) = 0$ and $\epsilon = a/b$ is the contact ratio.

The expression for the surface temperature distribution becomes

$$T(r, 0, t) = T_s + \frac{Q}{\pi k b} \left[ a_0 + \frac{1}{b} \sum_{n=1}^{\infty} \frac{a_n}{\lambda_n} J_0(\lambda_n r) \right] \tag{7.86}$$

The spreading resistance can be obtained using

$$R_s = \frac{T_s - \overline{T}_{cp}(t)}{Q} \tag{7.87}$$

After defining the contact plane temperature and integrating the gradient over the source area to find $Q$, the spreading resistance was obtained as

$$R_s = \frac{2}{\pi a^2 k} \frac{\sum_{n=1}^{\infty} \frac{a_n}{\lambda_n^2} J_1(a \lambda_n)}{a + 2 \sum_{n=1}^{\infty} \frac{a_n}{\lambda_n} J_1(a \lambda) \left[ \mathrm{erf}\left( \sqrt{\alpha} \lambda_n \sqrt{t} \right) + \frac{e^{-\lambda_n^2 \alpha t}}{\lambda_n \sqrt{\pi \alpha t}} \right]} \tag{7.88}$$

The spreading resistance is rendered dimensionless by taking $Fo = \alpha t / a^2$, $\epsilon = a/b$ and $\delta_n = \lambda_n b$ resulting in

$$\psi = \frac{8}{\pi \epsilon} \frac{\sum_{n=1}^{\infty} \frac{a_n}{\delta_n^2} J_1(\delta_n \epsilon)}{\epsilon + 2 \sum_{n=1}^{\infty} \frac{a_n}{\delta_n} J_1(\delta_n \epsilon) \left[ \mathrm{erf}\left( \delta_n \epsilon \sqrt{Fo} \right) + \frac{e^{-\delta_n^2 \epsilon^2 Fo}}{\delta_n \epsilon \sqrt{\pi Fo}} \right]} \tag{7.89}$$

where $\psi = 4kaR_s$ is the dimensionless resistance.

Alternatively, we may introduce the square root of source area as a length scale and define the above as

$$
k\sqrt{A}R_s = \frac{2}{\sqrt{\pi}\epsilon} \frac{\displaystyle\sum_{n=1}^{\infty} \frac{a_n}{\delta_n^2} J_1(\delta_n \epsilon)}{\epsilon + 2\displaystyle\sum_{i=1}^{\infty} \frac{a_n}{\delta_n} J_1(\delta_n \epsilon)\left[ \mathrm{erf}\left(\delta_n \epsilon \sqrt{\pi Fo_{\sqrt{A}}}\right) + \dfrac{e^{-\pi\epsilon^2 \delta_n^2 Fo_{\sqrt{A}}}}{\pi\epsilon\delta_n \sqrt{Fo_{\sqrt{A}}}} \right]}
\tag{7.90}
$$

The above solutions for $R_s$ are not computationally efficient and require many terms depending on the aspect ratio and Fourier number. In Section 7.8, we will provide expressions which predict the results more easily.

## 7.8 Models for Transient Thermal Spreading in Circular Flux Tubes

Simple models for the transient spreading resistance in circular flux tubes can be developed using the asymptotic characteristics of the solutions reported earlier [Muzychka and Yovanovich (2023)]. One important point that we must be careful of is the definition of reference temperature and how the short time half-space behavior is captured in the solutions. To better understand this we must carefully consider the basic definitions for transient resistance of a source on a half-space versus on a flux tube.

For the half-space, we used the definition:

$$
R_s = \frac{\overline{T}_s(t) - T_i}{Q}
\tag{7.91}
$$

while for a flux tube we used:

$$
R_s = \frac{\overline{T}_s(t) - \overline{T}_{cp}(t)}{Q}
\tag{7.92}
$$

We will now examine each problem separately, beginning with the isothermal heat source.

**Isothermal Heat Source**
The asymptotic limits for an isothermal heat source of area $A$ are found to be

$$
k\sqrt{A}R_s = \sqrt{\pi Fo_{\sqrt{A}}} \qquad Fo_{\sqrt{A}} \to 0
\tag{7.93}
$$

$$
k\sqrt{A}R_s = C_\infty^T \qquad Fo_{\sqrt{A}} \to \infty
\tag{7.94}
$$

where $C_\infty^T$ represents the nondimensional steady-state thermal spreading resistance defined by:

$$
C_\infty^T = \frac{\sqrt{\pi}}{4}\psi^T
\tag{7.95}
$$

where

$$
\psi^T = 1 - 1.40978\epsilon + 0.34406\epsilon^3 + 0.0435\epsilon^5 + 0.02271\epsilon^7
\tag{7.96}
$$

was provided in Chapter 3.

Equation (7.93) is the result deduced for heat flow into a half-space with an isothermal boundary, e.g.

$$Q = \frac{kA(T_s - T_i)}{\sqrt{\pi \alpha t}} \tag{7.97}$$

The resistance defined from this expression is as follows:

$$R_s = \frac{T_s - T_i}{Q} = \frac{\sqrt{\pi \alpha t}}{kA} = \frac{\sqrt{\pi Fo_{\sqrt{A}}}}{k\sqrt{A}} \tag{7.98}$$

or defining the dimensionless spreading resistance we obtain

$$k\sqrt{A}R_s = \sqrt{\pi Fo_{\sqrt{A}}} \tag{7.99}$$

In this case, we also see that the contact plane temperature for a spot of temperature $T_s$ on a half-space whose short time temperature is essentially $T_i$ everywhere in the source plane outside of the source region, leads us to conclude that $\overline{T}_{cp} \sim T_i$. This means that the Eqs. (7.91) and (7.92) are essentially the same. We may now combine these simple asymptotes using the blended modeling approach of Churchill and Usagi (1972).

A simple model of the form:

$$k\sqrt{A}R_s = \frac{1}{\left[\left(\frac{1}{\sqrt{\pi Fo_{\sqrt{A}}}}\right)^{3/2} + \left(\frac{1}{C_\infty^T}\right)^{3/2}\right]^{2/3}} \tag{7.100}$$

can be used for any isothermal heat source given that the steady value $C_\infty^T$ is not otherwise known for other source shapes/flux tube geometries. The value of $n = 3/2$ was determined by comparing the proposed model with values determined from the exact solution given earlier by Eq. (7.90).

The critical Fourier number can be deduced by equating the asymptotes such that

$$\sqrt{\pi Fo_{\sqrt{A}}} \sim C_\infty^T \tag{7.101}$$

or

$$Fo_{cr} \sim \left[C_\infty^T\right]^2 / \pi \tag{7.102}$$

The simple model is plotted with selected data points determined from the exact solution in Figure 7.7.

**Isoflux Heat Source**

The asymptotic limits for an isoflux heat source of area $A$ are

$$k\sqrt{A}R_s = (1 - \epsilon^2)\sqrt{4Fo_{\sqrt{A}}/\pi} \qquad Fo_{\sqrt{A}} \to 0 \tag{7.103}$$

$$k\sqrt{A}R_s = C_\infty^q \qquad Fo_{\sqrt{A}} \to \infty \tag{7.104}$$

where $C_\infty^q$ represents the nondimensional steady-state thermal spreading resistance defined using expressions for an isoflux source:

$$C_\infty^q = \frac{\sqrt{\pi}}{4}\psi^q \tag{7.105}$$

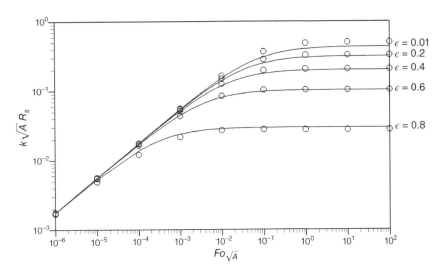

**Figure 7.7** Dimensionless transient thermal spreading resistance for a circular isothermal source on a flux tube. Model (line) versus exact (points) for $\epsilon$ = 0.01, 0.2, 0.4, 0.6, and 0.8.

where

$$\psi^q = 1.08085 - 1.41002\epsilon + 0.259714\epsilon^3 + 0.0188631\epsilon^5 + 0.0420278\epsilon^7 \tag{7.106}$$

was provided in Chapter 3.

Equation (7.103) is the result deduced for heat flow into a half-space with an isoflux boundary, e.g.

$$T_s(t) - T_i = \frac{2Q\sqrt{\alpha t/\pi}}{kA} \tag{7.107}$$

Defining a thermal resistance leads to

$$R_s = \frac{T_s(t) - T_i}{Q} = \frac{\sqrt{4Fo_{\sqrt{A}}/\pi}}{k\sqrt{A}} \tag{7.108}$$

which after nondimensionalizing becomes

$$k\sqrt{A}R_s = \sqrt{4Fo_{\sqrt{A}}/\pi} \tag{7.109}$$

Using the contact plane temperature for the circular flux tube with an isoflux source derived earlier, we may write

$$\overline{T}_{cp}(t) = T_i + \frac{2Q\epsilon^2}{\pi a^2 k}\sqrt{\frac{\alpha t}{\pi}} = T_i + \frac{Q\epsilon^2}{\sqrt{A}k}\sqrt{4Fo_{\sqrt{A}}/\pi} \tag{7.110}$$

or if we define a resistance using $T_i$, we now write

$$\frac{\overline{T}_{cp}(t) - T_i}{Q} = \frac{\epsilon^2}{\sqrt{A}k}\sqrt{4Fo_{\sqrt{A}}/\pi} \tag{7.111}$$

Finally, nondimensionalizing the resistance leads us to

$$k\sqrt{A}R_s = \epsilon^2\sqrt{4Fo_{\sqrt{A}}/\pi} \tag{7.112}$$

Returning to the definition of the spreading resistance in the flux tube which we may write as

$$R_s = \frac{(\overline{T}_s(t) - T_i) - (\overline{T}_{cp}(t) - T_i)}{Q} \tag{7.113}$$

Thus, for short time, we see that

$$k\sqrt{A}R_s = (1 - \epsilon^2)\sqrt{4Fo_{\sqrt{A}}/\pi} \tag{7.114}$$

Now that it is apparent what the correct short time behavior is in relation to the half-space limit for a uniform flux contact, we may now combine the asymptotes as we did before. A simple model of the form:

$$k\sqrt{A}R_s = \frac{1}{\left[\left(\frac{1/(1-\epsilon^2)}{\sqrt{4Fo_{\sqrt{A}}/\pi}}\right)^2 + \left(\frac{1}{C_\infty^q}\right)^2\right]^{1/2}} \tag{7.115}$$

can be used for any heat source shape given that the steady value $C_\infty^q$ is not otherwise known for other source shapes/flux tube geometries. The value of $n = 2$ was determined by comparing the proposed model with values determined from the exact solution given earlier by Eq. (7.74).

The critical Fourier number can be deduced by equation the asymptotes such that

$$(1 - \epsilon^2)\sqrt{4Fo_{\sqrt{A}}/\pi} \sim C_\infty^q \tag{7.116}$$

or

$$Fo_{cr} \sim \frac{\pi}{4}\left[\frac{C_\infty^q}{(1 - \epsilon^2)}\right]^2 \tag{7.117}$$

The simple model is plotted with selected data points determined from the exact solution, Eq. (7.74) which has been redefined using $\sqrt{A}$ as a length scale in Figure 7.8.

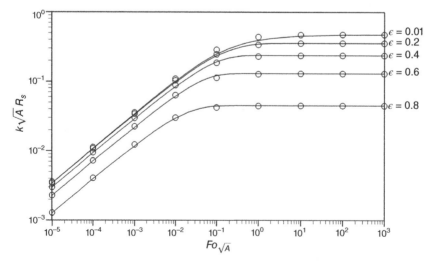

**Figure 7.8** Dimensionless transient thermal spreading resistance for a circular isoflux source on a flux tube. Model (line) versus exact (points) for $\epsilon = 0.01, 0.2, 0.4, 0.6,$ and $0.8$.

# References

Abramowitz, M., and Stegun, I.A., *Handbook of Mathematical Functions*, Dover Publishing, New York, 1965.

Beck, J.V., "Average Transient Temperature within a Body Heated by a Disk Heat Source", in *Heat Transfer, Thermal Control, and Heat Pipes, Progress in Aeronautics and Astronautics*, Vol. 70, AIAA, New York, pp. 3–24, 1979.

Blackwell, J.H., "Transient Heat Flow from a Thin Circular Disk Small-Time Solution", *Journal of the Australian Mathematics Society*, Vol. 14, pp. 433–442, 1972.

Carslaw, H.S., and Jaeger, J.C., *Conduction of Heat in Solids*, Oxford, 1959.

Churchill, S.W., and Usagi, R., "A General Expression for the Correlation of Rates of Transfer and Other Phenomena", *AIChE Journal*, Vol. 18, pp. 1121–1128, 1972.

Dryden, J.R., "The Effect of a Surface Coating on the Constriction Resistance of a Spot on an Infinite Half-Plane", *Journal of Heat Transfer*, Vol. 105, pp. 408–410, 1983.

Gibson, R., "The Contact Resistance for a Semi-infinite Cylinder in a Vacuum", *Applied Energy*, Vol. 2, no. 1, pp. 57–65, 1976.

Greene, P.R., "A Useful Approximation to the Error Function: Applications to Mass, Momentum, and Energy Transport in Shear Layers", *Journal of Fluids Engineering*, Vol. 111, pp. 224–226, 1989.

Greenwood, J.A., "Transient Thermal Contact Resistance", *International Journal of Heat and ass Transfer*, Vol. 34, no. 9, pp. 2287–2290, 1991.

Keltner, N.R., "Transient Heat Flow in Half-Space Due to an Isothermal Disk on the Surface", *Journal of Heat Transfer*, Vol. 95, pp. 412–414, 1973.

Lam, L., Goudarzi, S., and Muzychka, Y.S, "Transient Thermal Spreading Resistance from an Isothermal Source in a Circular Flux Tube", *AIAA Journal of Thermophysics and Heat Transfer*, 2023. (Paper is still under review.)

Muzychka, Y.S, and Yovanovich, M.M., "Compact Models for Transient Thermal Spreading Resistance in Half-space, Flux Tube, and Flux Channel Regions", *IEEE ITHERM 2023*, 2023.

Normington, E.J., and Blackwell, J.H., "Transient Heat Flow from Constant Temperature Spheroids and the Thin Circular Disk", *Quarterly Journal of Mechanics and Applied Mathematics*, Vol. 17, pp. 65–72, 1964.

Normington, E.J., and Blackwell, J.H., "Transient Heat Flow from a Thin Circular Disk: Small Time Solution", *Journal of the Australian Mathematics Society*, Vol. 14, pp. 433–442, 1972.

Ozisik, N., *Boundary Value Problems of Heat Conduction*, Dover Publishing, New York, 1968.

Schneider, G.E., Strong, A.B., and Yovanovich, M.M., "Transient Heat Flow from a Thin Circular Disk", in *AIAA Progress in Astronautics, Radiative Transfer and Thermal Control*, Vol. 49, ed. A.M. Smith, MIT Press, pp. 419–432, 1976.

Schneider, G.E., Strong, A.B., and Yovanovich, M.M., "Transient Thermal Response of Two Bodies Communicating Through a Small Circular Contact Area", *International Journal of Heat and Mass Transfer*, Vol. 20, pp. 301–308, 1977.

Turyk, P.J., and Yovanovich, M.M., "Transient Constriction Resistance for Elemental Flux Channels Heated by Uniform Heat Sources", *ASME-84-HT-52*, ASME, New York, 1984.

Yovanovich, M.M., "Transient Spreading Resistance of Arbitrary Isoflux Contact Areas: Development of a Universal Time Function", AIAA-97-2458, *32nd Thermophysics Conference*, Atlanta, GA, 1997.

Yovanovich, M.M., Teertstra, P.M., and Culham, J., "Modeling Transient Conduction from Isothermal Convex Bodies of Arbitrary Shape", *Journal of Thermophysics and Heat Transfer*, Vol. 9, no. 3, pp. 385–390, 1995.

# 8

# Applications with Nonuniform Conductance in the Sink Plane

Thermal spreading resistance in finite systems such as heat sinks is often modeled using a uniform conductance in the sink plane. In regards to heat sinks containing fins, this requires that the distribution of fins is both uniform in spacing and height and cooled with a constant convection heat transfer coefficient. In many real applications, this is not achieved due to nonuniform fin heights/spacings and the nature of the supplied cooling using fans. In impingement air cooling applications, the local heat transfer coefficient is always distributed over the surface.

For example Figure 8.1a illustrates the nonuniform distribution of pin fins in a CPU cooling device, while Figure 8.1b illustrates the nonuniform distribution of convection cooling provided by an axial flow fan attached to a heat sink. In the first case, since the fins are of nonuniform height, their respective heights lead to a different fin resistance and hence, a distribution of the conductance in the sink plane. In the second case, the impinging fan flow creates a local heat transfer coefficient in addition to a deficit in cooling in the region of the fan hub.

Recent studies by Razavi (2015) and Al-Khamaiseh (2018) have examined the impact of distributed sink plane conductance. The results have shown that the impact is greatest in thin thermal spreaders. As the thermal spreader plate thickness increases, the distribution of conductance begins to have a smaller effect on the spreading resistance and overall resistance. Thus, given that many heat sink spreaders are thin, especially for smaller heat sinks, this issue can be quite important.

In this chapter, we will discuss a number of new solutions for thermal spreading resistance in circular disks and two- or three-dimensional flux channels, namely, the effect of a variable heat transfer coefficient in the heat sink plane and mixed-boundary conditions in the heat source plane. Either of these two elements necessitates the use of an approximate solution technique, since traditional methods such as separation of variables breakdown when applying the usual orthogonal function expansion technique for finding the final Fourier coefficients. A general approach to modeling variable conductance is first presented followed by an overview of the least squares method used to find approximate solutions. Then later, we will consider a number of practical problems which require an approximate solution methodology using the least squares technique for determining the final Fourier coefficients.

## 8.1 Applications with Nonuniform Conductance

Conductance over the heat sink plane can vary due to either the fin distribution on a heat sink or the local distribution of the heat transfer coefficient over the surface or both depending upon the nature of the cooling process since not all processes involve extended surfaces, such as jet impingement cooling. A few examples of variable fin height heat sinks are provided in Figures 8.1a and 8.2.

*Thermal Spreading and Contact Resistance: Fundamentals and Applications*, First Edition.
Yuri S. Muzychka and M. Michael Yovanovich.
© 2023 John Wiley & Sons, Inc. Published 2023 by John Wiley & Sons, Inc.

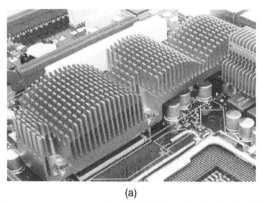

(a)

(b)

**Figure 8.1** Examples of nonuniform conductance in the sink plane of a thermal spreader: (a) nonuniform fin distribution, and (b) impingement air cooling. Source: fir0002/ Wikimedia Commons/GFDL 1.2.

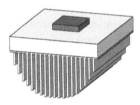

**Figure 8.2** Variable height plate fin heat sink.

Fins may also be nonuniformly spaced in other applications. In either case a local distribution of fin conductance results and can be simply fit to an equation.

If we consider boundary layer flow over a flat plate, we know that the local heat transfer varies inversely with position $x$ from the leading edge of the plate in laminar flow $h_x \propto x^{-1/2}$, while in turbulent flow $h_x \propto x^{-1/5}$. These local distributions can be adapted into the mathematical solution process, but ultimately as we will see, they lead us to develop approximate solutions, as we cannot complete the solutions in the traditional manner. We will examine a small number of useful solutions with a symmetric distribution of conductance, but we are not limited to applying the method to an asymmetric distribution.

### 8.1.1 Distributed Heat Transfer Coefficient Models

In traditional thermal spreading resistance problems, we frequently apply a uniform heat transfer coefficient in the heat sink plane through the application of the following boundary condition:

$$-k\frac{\partial \theta}{\partial z} = h_s \theta \tag{8.1}$$

where $h_s$ is constant. Of particular interest is when the heat transfer coefficient varies symmetrically along the boundary according to

$$h_s(x) = h_0 \left[ 1 - \left( \frac{x}{c} \right)^p \right] \tag{8.2}$$

in Cartesian coordinates, or

$$h_s(r) = h_0 \left[ 1 - \left( \frac{r}{b} \right)^p \right] \tag{8.3}$$

in polar coordinates.

The maximum value of the conductance (or heat transfer coefficient) $h_0$ can be related to an average $\overline{h}$ value through integration over the region. This leads to the following expressions:

$$h_s(x) = \overline{h} \left( \frac{p+1}{p} \right) \left[ 1 - \left( \frac{x}{c} \right)^p \right] \tag{8.4}$$

in Cartesian coordinates, and

$$h_s(r) = \overline{h} \left( \frac{p+2}{p} \right) \left[ 1 - \left( \frac{r}{b} \right)^p \right] \tag{8.5}$$

in polar coordinates.

In the above expressions, the parameter $p$ determines the shape of the heat transfer coefficient distribution as shown in Figure 8.3. As $p \to \infty$, the distribution approaches a uniform value at all points in the heat sink plane. In all cases, the mean coefficient is the same, but how the local value is distributed is what impacts the thermal resistance of the heat spreader.

We may also consider two-dimensional distributions of $h_s(x, y)$ over the sink plane. Al-Khamaiseh (2018) considered the following two-dimensional distributions over a rectangular flux channel:

$$h_1(x, y) = h_0 \sin \left( \frac{\pi x}{c} \right) \sin \left( \frac{\pi y}{d} \right) \tag{8.6}$$

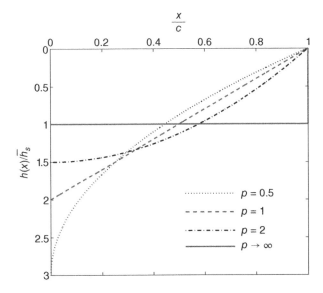

**Figure 8.3** Symmetric distribution of conductance for Eq. (8.4). Source: Al-Khamaiseh (2018)/Memorial University of Newfoundland.

or

$$h_1(x,y) = \frac{\pi^2 \overline{h}_s}{4} \sin\left(\frac{\pi x}{c}\right) \sin\left(\frac{\pi y}{d}\right) \tag{8.7}$$

for the case where conductance is greatest in the central region, and

$$h_2(x,y) = h_0 \left[1 - \sin\left(\frac{\pi x}{c}\right) \sin\left(\frac{\pi y}{d}\right)\right] \tag{8.8}$$

or

$$h_2(x,y) = \frac{\pi^2 \overline{h}_s}{\pi^2 - 4} \left[1 - \sin\left(\frac{\pi x}{c}\right) \sin\left(\frac{\pi y}{d}\right)\right] \tag{8.9}$$

for the case where conductance is minimum near the central region. In both cases, we define the mean value $\overline{h}_s$ using:

$$\overline{h}_s = \frac{1}{cd} \int_0^d \int_0^c h(x,y) dx\, dy \tag{8.10}$$

These distributions are illustrated in Figure 8.4 for both cases. We may consider any distribution of conductance in the sink plane so long as it is easily defined mathematically.

### 8.1.2 Mixed-Boundary Conditions in the Source Plane

We may also wish to solve problems with mixed-boundary conditions in the source plane, such as those with isothermal heat sources or any particular combination of isothermal and isoflux source regions.

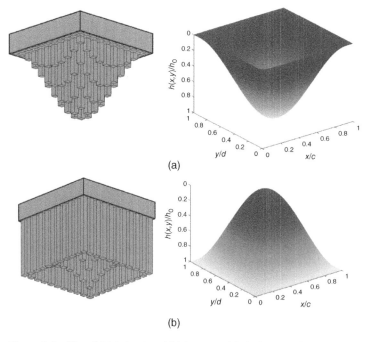

**Figure 8.4** Plot of (a) $h_1(x,y)$ and (b) $h_2(x,y)$ with their approximate heat sink fin distribution. Source: Al-Khamaiseh (2018)/Memorial University of Newfoundland.

When a combination of different types of boundary conditions, i.e.

$$\left.\begin{array}{ll} \text{(i)} & \phi = C_0 \\[2ex] \text{(ii)} & \dfrac{\partial \phi}{\partial n} = C_1 \\[2ex] \text{(iii)} & \dfrac{\partial \phi}{\partial n} + K\phi = C_2 \end{array}\right\} \tag{8.11}$$

are prescribed piecewise on the same boundary, the problem becomes a mixed potential boundary condition, [Sneddon (1966)]. Such boundary conditions cannot be solved using simple expansions of orthogonal functions. In these instances, we must resort to alternate methods. One such method which requires modest effort is the least squares method [Kelman (1979)].

Another type of problem which requires an alternate solution approach is when the applied boundary condition in the sink plane is not constant but depends on one or both spatial coordinates. In these cases, the spreading function $\phi$ becomes a function of $x$ or $y$ or both variables, $\phi \to \phi(x)$ or $\phi \to \phi(x,y)$, or $\phi \to \phi(r)$ in polar coordinates. The separation of variables method breaks down since the independence of the separated solutions is no longer valid, and we require an approximate method to complete the solution.

### 8.1.3  Least Squares Approximation

We briefly review a useful technique for finding approximate analytical solutions to thermal spreading problems which cannot be solved using the traditional orthogonal function expansion. The least squares approach is one such method. The least squares method can be used when there is a mixed-boundary condition or when the spreading function is not a constant or a combination of both. For example, if we have obtained a series type solution and we are applying the final boundary condition, we usually have a solution in the form:

$$\theta(x,y,z) = \sum_{m=1}^{M} \sum_{n=1}^{N} A_{mn} \psi_{mn}(x,y,z) \tag{8.12}$$

where $\psi_{mn}(x,y,z)$ is the resulting combination of eigenfunctions from the separation of variables procedure.

In the case of a two-region mixed-boundary condition in the source plane, we must solve

$$\theta = \sum_{m=1}^{M} \sum_{n=1}^{N} A_{mn} \psi_{mn}(x,y,0) = f(x,y) \quad \text{over } P \tag{8.13}$$

and

$$\frac{\partial \theta}{\partial z} = \sum_{m=1}^{M} \sum_{n=1}^{N} A_{mn} \psi'_{mn}(x,y,0) = g(x,y) \quad \text{over } Q \tag{8.14}$$

We see that we have two equations to satisfy with only one unknown $A_{mn}$. Our traditional Fourier series approach does not lead us to an easy solution. In the least squares approach, we define the following integral:

$$I_{MN} = \iint_P \left[ \sum_{m=1}^{M} \sum_{n=1}^{N} A_{mn} \psi_{mn}(x,y,0) - f(x,y) \right]^2 dx\, dy$$

$$+ \iint_Q \left[ \sum_{m=1}^{M} \sum_{n=1}^{N} A_{mn} \psi'_{mn}(x,y,0) - g(x,y) \right]^2 dx\, dy \tag{8.15}$$

We now find the $A_{mn}$ by minimizing the least squares integral:

$$\frac{\partial I_{MN}}{\partial A_{ij}} = 0 \quad \text{for} \quad i = 1 \ldots M, j = 1 \ldots N \tag{8.16}$$

This leads to a system of $M \times N$ equations which can be solved for the $A_{mn}$ coefficients. While quite simple, the approach can become quite arduous if the integrations are not straightforward. Kelman (1979) outlines the procedure in more detail. In other problems involving nonmixed type source plane boundary conditions, where we only have a nonuniform conductance in the sink plane, we desire the solution to

$$\frac{\partial \theta}{\partial z} = \sum_{m=1}^{M} \sum_{n=1}^{N} A_{mn} \psi'_{mn}(x, y, 0) = f(x, y) \quad \text{over } P \tag{8.17}$$

$$\frac{\partial \theta}{\partial z} = \sum_{m=1}^{M} \sum_{n=1}^{N} A_{mn} \psi'_{mn}(x, y, 0) = g(x, y) \quad \text{over } Q \tag{8.18}$$

where in this case, the spreading function $\phi$ will now be a function of $x$ and/or $y$, and we cannot complete the required orthogonal Fourier series expansion despite the final boundary condition being of the nonmixed type. In this case, we again use the least squares integral to find the unknown coefficients using:

$$I_{MN} = \iint_P \left[ \sum_{m=1}^{M} \sum_{n=1}^{N} A_{mn} \psi'_{mn}(x, y, 0) - f(x, y) \right]^2 dx \, dy$$

$$+ \iint_Q \left[ \sum_{m=1}^{M} \sum_{n=1}^{N} A_{mn} \psi'_{mn}(x, y, 0) - g(x, y) \right]^2 dx \, dy \tag{8.19}$$

In this situation, numerical integration is required to obtain the coefficients, since the presence of the nonconstant spreading function makes the integration more complex. The above procedure is quite amenable in modern mathematics tools such as Matlab, Mathematica, or Maple.

Having covered some of the essentials required to tackle more complex problems, we now proceed to examine several solutions with variable sink plane conductance.

## 8.2 Finite Flux Channels with Variable Conductance

We begin our analysis of flux channels with variable conductance on the simpler two-dimensional problem considered in Chapter 4. Razavi (2014, 2015) and Al-Khamaiseh (2018) each considered several problems involving two- and three-dimensional flux channels to understand the impact of variable conductance in the sink plane. We consider two such problems here: (i) the simple two-dimensional channel with a symmetric conductance distribution and (ii) the more general three-dimensional flux channel with a central heat source and different conductance distributions applied in the sink plane.

### 8.2.1 Two-Dimensional Flux Channel

The first problem under consideration is that of a two-dimensional flux channel with a spatially varying symmetric conductance. Razavi et al. (2015) considered this problem shown in Figure 8.5 and provided a solution approach with validation using finite element analysis.

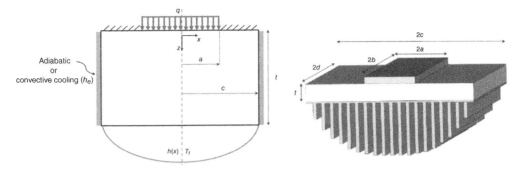

**Figure 8.5** Flux channel with a symmetric distribution of conductance and its equivalent heat sink representation. Source: Razavi et al. (2015)/with permission of The American Society of Mechanical Engineers.

Beginning with the problem statement, we desire to solve:

$$\frac{\partial^2 \theta}{\partial x^2} + \frac{\partial^2 \theta}{\partial z^2} = 0 \tag{8.20}$$

where $\theta = T - T_f$, subject to the following boundary conditions:

$$\frac{\partial \theta}{\partial x}\bigg|_{x=0} = 0$$

$$-\frac{\partial \theta}{\partial x}\bigg|_{x=c} = h_e \theta(c, z) \tag{8.21}$$

at the centroid and edge of the channel, respectively. Over the top surface $z = 0$,

$$\left.\begin{array}{l} -k\dfrac{\partial \theta}{\partial z}\bigg|_{z=0} = q, \ 0 < x < a \\[3mm] \dfrac{\partial \theta}{\partial z}\bigg|_{z=0} = 0, \ a < x < c \end{array}\right\} \tag{8.22}$$

where $q = Q'/2a$ is the applied flux and $Q' = Q/L = Q/2b$ is the heat transfer per unit depth of the flux channel.

Finally, at $z = t$, we prescribe the following condition:

$$-k\frac{\partial T}{\partial z}\bigg|_{z=t} = h_s(x)\theta(x, t) \tag{8.23}$$

where $h_s(x)$ is a prescribed distribution of conductance along the sink plane. This is illustrated in Figure 8.5. Razavi et al. (2015) used Eq. (8.4) to characterize the conductance distribution.

The solution begins by using separation of variables. The general solution which satisfies the two $x$ boundary conditions is found to be

$$\theta(x, z) = \sum_{n=1}^{\infty} \cos(\lambda_n x)[C_n \cosh(\lambda_n z) + D_n \sinh(\lambda_n z)] \tag{8.24}$$

where the eigenvalues are determined from:

$$\lambda_n \sin(\lambda_n c) = \frac{h_e}{k} \cos(\lambda_n c) \tag{8.25}$$

or using $\delta_n = \lambda_n c$ we may write:

$$\delta_n \sin(\delta_n) = Bi_e \cos(\delta_n) \tag{8.26}$$

where $Bi_e = h_e c/k$ is the Biot number for the edge condition.

Application of the sink plane condition yields:

$$D_n = -\phi_n(x)C_n \tag{8.27}$$

where

$$\phi_n(x) = \frac{\delta_n \tanh(\delta_n \tau) + Bi(x)}{\delta_n + Bi(x)\tanh(\delta_n \tau)} \tag{8.28}$$

and $\tau = t/c$ and $Bi(x) = h_s(x)c/k$. Later, we will define the $Bi(x)$ in terms of a reference value for $h_s(x)$, i.e. either the mean or maximum value. The solution is now written as

$$\theta(x, z) = \sum_{n=1}^{\infty} C_n \cos(\delta_n x/c)[\cosh(\delta_n z/c) - \phi_n(x)\sinh(\delta_n z/c)] \tag{8.29}$$

At this point, it should now be clear that the original assumption for the separation of variables method, i.e. that we could devise a solution using independent functions for $X(x)$ and $Z(z)$, is now violated with the term in square brackets now clearly a function of $x$ and $z$, given that the spreading function $\phi$ is no longer a constant. We cannot proceed to a traditional Fourier series expansion to obtain closure to the solution. Though we can proceed assuming an approximate solution can be found using the least squares method. This approach is essentially a modified Kantorovich approach which combines the separation of variables method and the Ritz method [Arpaci (1966), Kantorovich and Krylov (1958)]. In this instance, we avail of an exact solution in one coordinate direction, and an approximate solution method for the remainder of the problem given that Eq. (8.29) satisfies Laplace's equation and the $x$ boundary conditions. We now turn to the least squares method to find the $C_n$ coefficients. The least squares integral for the final boundary condition can now be defined as

$$I_N = \int_0^a \left[ -k\frac{\partial\theta}{\partial z}\bigg|_{z=t} - q \right]^2 dx + \int_a^c \left[ -k\frac{\partial\theta}{\partial z}\bigg|_{z=t} - 0 \right]^2 dx \tag{8.30}$$

or

$$I_N = \int_0^a \left[ k\sum_{n=1}^{N} C_n(\delta_n/c)\phi_n(x)\cos(\delta_n x/c) - q \right]^2 dx$$

$$+ \int_a^c \left[ k\sum_{n=1}^{N} C_n(\delta_n/c)\phi_n(x)\cos(\delta_n x/c) \right]^2 dx \tag{8.31}$$

Finally, the coefficients $C_n$ are determined by solving the system of linear equations defined by

$$\frac{\partial I_N}{\partial C_i} = 0 \quad \text{for} \quad i = 1\ldots N \tag{8.32}$$

The total thermal resistance $R_t$ is determined from the mean source temperature and heat flow per unit depth:

$$R_t = \frac{\overline{\theta}_s}{Q} = \frac{\overline{\theta}_s}{q(2aL)} \tag{8.33}$$

where

$$\overline{\theta}_s = \frac{1}{2a}\int_{-a}^{a} \theta(x, 0)dx = \sum_{n=1}^{N} C_n \frac{\sin(\delta_n \epsilon)}{\delta_n \epsilon} \tag{8.34}$$

Razavi et al. (2015) solved Eq. (8.31) assuming that $h_s(x)$ was defined by Eqs. (8.2) and (8.4) and validated using the finite element method. They considered several values of the parameter $p = 1, 2$, and $\infty$, and examined thin and thick spreaders for various reference $Bi_0$ and $\overline{Bi}$. Depending upon source size, magnitude of $\overline{Bi}$, spreader thickness, and conductance distribution, they found that the total resistance could vary by as much as 20% from the case for a uniform conductance specification.

### 8.2.2 Three-Dimensional Flux Channel

Three-dimensional rectangular flux channels were considered by both Razavi (2015) and Al-Khamaiseh et al. (2018), see Figure 8.6. Razavi (2015) considered the case of a central heat source, while Al-Khamaiseh (2018) considered the more general case of an eccentric heat source.

The heat conduction in the flux channel is governed by Laplace's equation:

$$\frac{\partial^2 \theta}{\partial x^2} + \frac{\partial^2 \theta}{\partial y^2} + \frac{\partial^2 \theta}{\partial z^2} = 0 \tag{8.35}$$

where the temperature excess $\theta = T - T_f$.

Convective cooling boundary conditions are taken along the lateral edges of the system. However, since the symmetry of the system is considered, only a quarter model is required to be solved. The boundary conditions along the planes of symmetry of the system ($x = 0$ and $y = 0$) are as follows:

$$\left.\frac{\partial \theta}{\partial x}\right|_{x=0} = 0$$

$$\left.\frac{\partial \theta}{\partial y}\right|_{y=0} = 0 \tag{8.36}$$

while the boundary conditions along the planes $x = c$ and $y = d$ are given by

$$\left.\frac{\partial \theta}{\partial x}\right|_{x=c} = -\frac{h_e}{k}\theta(c, y, z)$$

$$\left.\frac{\partial \theta}{\partial y}\right|_{y=d} = -\frac{h_e}{k}\theta(x, d, z) \tag{8.37}$$

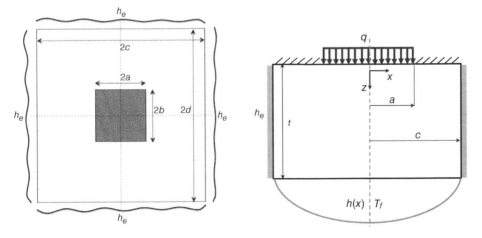

**Figure 8.6** Three-dimensional flux channel with central heat source and symmetric distribution of conductance $h_s(x)$. Source: Al-Khamaiseh et al. (2018)/with permission of IEEE.

These convective cooling boundary conditions can be turned to adiabatic conditions when $h_e \to 0$. In the source plane $z = 0$, we define:

$$
\left.\begin{array}{ll}
-\dfrac{\partial\theta}{\partial z}\bigg|_{z=0} = q/k, & 0 < x < a, \ \ 0 < y < b \\[3mm]
-\dfrac{\partial\theta}{\partial z}\bigg|_{z=0} = \ \ 0, & a < x < c, \ \ b < y < d
\end{array}\right\} = S(x, y) \tag{8.38}
$$

Along the sink plane, a variable heat transfer coefficient varying in $x$-direction exists and the boundary condition is given by

$$
-k\frac{\partial\theta}{\partial z}\bigg|_{z=t} = h_s(x)\theta(x, y, t) \tag{8.39}
$$

Al-Khamaiseh et al. (2020) used a number of conductance distributions defined earlier. We proceed to outline the general approach independent of the type of conductance distribution under the assumption that is symmetric. The general solution of Laplace's equation (8.35) may be found by using the method of separation of variables, where the solution is assumed to have the form $\theta(x, y, z) = X(x) \cdot Y(y) \cdot Z(z)$.

Applying the method of separation of variables and using the boundary conditions along the planes $(x = 0, \ x = c)$ and $(y = 0, \ y = d)$ yields the following general solution:

$$
\theta(x, y, z) = \sum_{m=1}^{\infty}\sum_{n=1}^{\infty}\cos(\lambda_m x)\cos(\delta_n y)\left[C_{mn}\cosh(\beta_{mn}z) + D_{mn}\sinh(\beta_{mn}z)\right] \tag{8.40}
$$

where $\lambda_m$ and $\delta_n$ are the eigenvalues in $x$ and $y$ directions, respectively, which can be obtained by solving the following transcendental equations numerically:

$$
(\lambda_m c)\sin(\lambda_m c) = \frac{h_e c}{k}\cos(\lambda_m c) \qquad m = 1, 2, \ldots
$$
$$
\tag{8.41}
$$
$$
(\delta_n d)\sin(\delta_n d) = \frac{h_e d}{k}\cos(\delta_n d) \qquad n = 1, 2, \ldots
$$

whereas $\beta_{mn}$ is defined by $\beta_{mn} = \sqrt{\lambda_m^2 + \delta_n^2}$. The following result is obtained for the Fourier coefficients when the boundary condition at the sink plane is applied (8.39):

$$
D_{mn} = -\phi(x)C_{mn} \tag{8.42}
$$

where $\phi(x)$ is the spreading function defined by

$$
\phi(x) = \frac{\beta_{mn}\tanh(\beta_{mn}t) + [h(x)/k]}{\beta_{mn} + [h(x)/k]\tanh(\beta_{mn}t)} \tag{8.43}
$$

Thus, the general solution can be rewritten as

$$
\theta(x, y, z) = \sum_{m=1}^{\infty}\sum_{n=1}^{\infty}C_{mn}\cos(\lambda_m x)\cos(\delta_n y)\left[\cosh(\beta_{mn}z) - \phi(x)\sinh(\beta_{mn}z)\right] \tag{8.44}
$$

Finally, the boundary condition at the source plane equation (8.38) is considered in order to find the Fourier coefficients $C_{mn}$. Usually, when solving flux channel problems with a constant heat transfer coefficient, the Fourier coefficients are obtained directly by taking the Fourier series expansions of the boundary condition at the source plane $(z = 0)$ and using the orthogonality of the eigenfunctions. However, since the heat transfer coefficient $h_s(x)$ depends on the variable $x$ and so does the spreading function $\phi(x)$, then the use of the orthogonality of the eigenfunctions in $x$-direction is not possible when following the same procedure for the constant heat transfer coefficient. Instead, the method of least squares is used to obtain the Fourier coefficients $C_{mn}$. The general solution for finite $M$ and $N$ can be written as

$$\theta(x, y, z) = \sum_{m=1}^{M} \sum_{n=1}^{N} C_{mn} \cos(\lambda_m x) \cos(\delta_n y) \left[ \cosh(\beta_{mn} z) - \phi(x) \sinh(\beta_{mn} z) \right] \tag{8.45}$$

The method of least squares can be applied to the general solution given hence the following integral (which represents the residual) is defined:

$$I_{MN} = \int_{0}^{c} \int_{0}^{d} \left[ -\frac{\partial \theta}{\partial z} \Big|_{z=0} - S(x, y) \right]^2 dy\, dx \tag{8.46}$$

where $S(x, y)$ is the function defining the boundary condition at the source plane given by Eq. (8.38).

The first derivative of the general solution equation (8.45) with respect to $z$ at the source plane (at $z = 0$) is

$$\frac{\partial \theta}{\partial z} \Big|_{z=0} = -\sum_{m=1}^{M} \sum_{n=1}^{N} C_{mn} \beta_{mn} \phi(x) \cos(\lambda_m x) \cos(\delta_n y) \tag{8.47}$$

hence, the residual integral in Eq. (8.46) can be rewritten as

$$I_{MN} = \int_{0}^{c} \int_{0}^{d} \left[ \sum_{m=1}^{M} \sum_{n=1}^{N} C_{mn} \beta_{mn} \phi(x) \cos(\lambda_m x) \cos(\delta_n y) - S(x, y) \right]^2 dy\, dx \tag{8.48}$$

The Fourier coefficients are obtained to minimize the residual $I_{MN}$ by using

$$\frac{\partial I_{MN}}{\partial C_{ij}} = 0, \qquad i = 1, 2, \ldots, M, \quad j = 1, 2, \ldots, N \tag{8.49}$$

The application of (8.49) yields

$$\int_{0}^{c} \int_{0}^{d} \left[ \sum_{m=1}^{M} \sum_{n=1}^{N} C_{mn} \beta_{mn} \phi_{mn}(x) \cos(\lambda_m x) \cos(\delta_n y) - S(x, y) \right]$$
$$\cdot \phi_{ij}(x) \cos(\lambda_i x) \cos(\delta_j y) dy\, dx = 0 \tag{8.50}$$

Equation (8.50) can be simplified by using the orthogonality of the eigenfunctions in $y$-direction to get

$$\sum_{m=1}^{M} C_{mj} \beta_{mj} \int_{0}^{c} \phi_{mj}(x) \phi_{ij}(x) \cos(\lambda_m x) \cos(\lambda_i x) dx$$
$$= \frac{q \sin(\delta_j b)}{k \delta_j N(\delta_j)} \int_{0}^{a} \phi_{ij}(x) \cos(\lambda_i x) dx \tag{8.51}$$

where $N(\delta_j)$ is the norm of the $y$-direction eigenfunctions which depend on the specific nature of the $y$-direction eigenvalues:

$$N(\delta_j) = \int_{0}^{d} \cos^2(\delta_j y) dy = \frac{1}{2} \left[ d + \frac{h_e/k}{\delta_j^2 + (h_e/k)^2} \right] \tag{8.52}$$

Thus, in order to find the Fourier coefficients $C_{ij}$, a linear system has to be solved for every $j$ (i.e. for every eigenvalue in $y$-direction). The linear system is as follows:

$$\mathbf{A}^j \mathbf{C}^j = \mathbf{b}^j \tag{8.53}$$

where $\mathbf{A}^j = [a_{im}^j]$ is an $M \times M$ matrix whose entries (represented by row $i$ and column $m$) are given by

$$a_{im}^j = \beta_{mj} \int_0^c \phi_{mj}(x)\phi_{ij}(x)\cos(\lambda_m x)\cos(\lambda_i x)dx \tag{8.54}$$

$\mathbf{C}^j = [C_{1j}\ C_{2j}\ \ldots\ C_{Mj}]^t$ is the unknown Fourier coefficients vector and $\mathbf{b}^j = [b_1^j\ b_2^j\ \ldots\ b_M^j]^t$ represents the right hand side vector whose components are given by

$$b_i^j = \frac{q\sin(\delta_j b)}{k\delta_j N(\delta_j)} \int_0^a \phi_{ij}(x)\cos(\lambda_i x)dx \tag{8.55}$$

It is important to note that the full set of Fourier coefficients $C_{ij}$ can be obtained by solving $N$-linear systems using any mathematical software package (for example, MATLAB) in which numerical integration is used to evaluate the entries of each system.

For a single heat source spreading heat to a larger extended sink area, the total thermal resistance of the system can be defined as

$$R_t = \frac{\overline{T}_s - T_f}{Q} = \frac{\overline{\theta}_s}{Q} \tag{8.56}$$

where $\overline{T}_s$ is the mean temperature over the heat source area, $\overline{\theta}_s$ is the mean heat source temperature excess, and $Q$ is the total heat input of the flux channel. The mean source temperature excess is given by

$$\overline{\theta}_s = \frac{1}{A_s} \int\int_{A_s} \theta(x,y,0)dA_s \tag{8.57}$$

where $A_s$ is the heat source area. The application of Eq. (8.57) yields

$$\overline{\theta}_s = \frac{1}{4ab} \int_{-a}^{a} \int_{-b}^{b} \sum_{m=1}^{M}\sum_{n=1}^{N} C_{mn}\cos(\lambda_m x)\cos(\delta_n y)\,dy\,dx$$

$$= \frac{1}{ab}\sum_{m=1}^{M}\sum_{n=1}^{N} \frac{C_{mn}}{\lambda_m \delta_n}\sin(\lambda_m a)\sin(\delta_n b). \tag{8.58}$$

Hence, the total thermal resistance can be obtained by using Eq. (8.56) to get

$$R_t = \frac{1}{4a^2 b^2 q}\sum_{m=1}^{M}\sum_{n=1}^{N}\frac{C_{mn}}{\lambda_m\ \delta_n}\sin(\lambda_m a)\sin(\delta_n b) \tag{8.59}$$

Al-Khamaiseh (2018) also considered the case of an eccentric and multiple sources for the above problem. Additional details of the solution process and case studies can be found in Al-Khamaiseh et al. (2020).

**Figure 8.7** Flux tube with variable axisymmetric conductance distribution. Source: Al-Khamaiseh et al. (2019)/ with permission of IEEE.

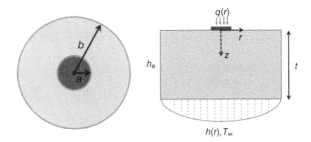

## 8.3 Finite Flux Tube with Variable Conductance

The final problem we consider is that of a finite flux tube with nonuniform conductance, see Figure 8.7. The problem is defined as follows:

$$\frac{\partial^2\theta}{\partial r^2} + \frac{1}{r}\frac{\partial\theta}{\partial r} + \frac{\partial^2\theta}{\partial z^2} = 0 \tag{8.60}$$

which is subject to

$$\left.\frac{\partial\theta}{\partial r}\right|_{r=0} = 0$$
$$\left.\frac{\partial\theta}{\partial r}\right|_{r=b} = -\frac{h_e}{k}\theta(b,z) \tag{8.61}$$

along the axis and edges of the heat spreader, respectively. In the source plane, we specify as before

$$\left.\begin{array}{ll}
\left.\dfrac{\partial\theta}{\partial z}\right|_{z=0} = -\dfrac{q(r)}{k}, & 0 \le r < a \\[3mm]
\left.\dfrac{\partial\theta}{\partial z}\right|_{z=0} = \quad 0, & a < r \le b
\end{array}\right\} \tag{8.62}$$

While in the sink plane, we require the following condition:

$$-k\left.\frac{\partial\theta}{\partial z}\right|_{z=t} = h_s(r)\theta(r,t) \tag{8.63}$$

Al-Khamaiseh et al. (2019) solved this problem assuming that $h(r)$ and $q(r)$ are defined using:

$$h(r) = \overline{h}_s\left(\frac{p+2}{p}\right)\left[1 - \left(\frac{r}{b}\right)^p\right] \tag{8.64}$$

and

$$q(r) = \overline{q}(1 + \mu)\left[1 - \left(\frac{r}{a}\right)^2\right]^{\mu} \tag{8.65}$$

These are both shown in Figure 8.8.

By applying the method of separation of variables and making use of the radial boundary conditions, the following solution can be obtained:

$$\theta(r,z) = \sum_{n=1}^{\infty} J_0(\lambda_n r)\left[C_n\cosh(\lambda_n z) + D_n\sinh(\lambda_n z)\right] \tag{8.66}$$

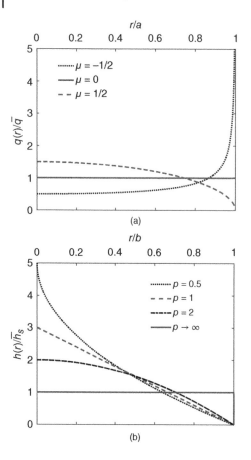

**Figure 8.8** Flux and conductance distributions considered by Al-Khamaiseh et al. (2019). Source: Al-Khamaiseh et al. (2019)/with permission of IEEE.

where $\lambda_n$ are the eigenvalues in the radial direction, which represents the positive roots of the following transcendental equation:

$$\delta_n J_1(\delta_n) = \frac{h_e b}{k} J_0(\delta_n), \qquad n = 1, 2, \ldots \tag{8.67}$$

where $\delta_n = \lambda_n b$.

The functions $J_0(\cdot)$ and $J_1(\cdot)$ correspond to the zeroth-order and first-order Bessel functions of the first kind. Employing the sink-plane boundary condition given, Eq. (8.63), yields the following relationship between the Fourier–Bessel coefficients $C_n$ and $D_n$ (spreading function):

$$\phi(r) = -\frac{D_n}{C_n} = \frac{\delta_n \tanh(\delta_n \tau) + [h_s(r)b/k]}{\delta_n + [h_s(r)b/k]\tanh(\delta_n \tau)} \tag{8.68}$$

As we have seen before, this relationship between the Fourier–Bessel coefficients $\phi(r)$ violates the classical Sturm–Liouville theory since the ratio between the coefficients $C_n$ and $D_n$ is now a function of $r$. Hence, the orthogonality of the eigenfunctions cannot be used in finding these coefficients. In particular, for flux tube problems with a uniform convection along the sink plane, these coefficients are usually found explicitly by imposing the source–plane boundary condition and using the orthogonality of Bessel functions as we did in Chapter 3. However, this is not the case when considering a nonuniform convection as the orthogonality of Bessel functions cannot be employed. Nevertheless, an approximate solution to the problem can be constructed based on the classical Sturm–Liouville theory, in which the source–plane boundary condition is used to find the coefficients by means of some approximate method such as the least squares method. The approximate solution for finite number of terms $N$ can be rewritten as

$$\theta(r, z) = \sum_{n=1}^{N} C_n J_0(\lambda_n r) \left[ \cosh(\lambda_n z) - \phi(r) \sinh(\lambda_n z) \right] \tag{8.69}$$

The constructed approximate solution gives a good approximation to the general solution along the source plane, which is of most interest in finding the maximum temperature and thermal resistance. The final step is to determine the Fourier–Bessel coefficients $C_n$ in the approximate solution by making use of the nonhomogeneous source–plane boundary condition. We follow the same technique as before for a flux channel with a nonuniform convection in which the method of least squares is used to determine the Fourier coefficients. The idea behind using the method of least squares is to determine the coefficients $C_n$ such that they minimize the integral of the squared residual, represented by

$$I_N = \int_0^{2\pi} \int_0^b \left[ -\frac{\partial \theta}{\partial z} \bigg|_{z=0} - S(r) \right]^2 r \, dr \, d\psi \tag{8.70}$$

or

$$I_N = 2\pi \int_0^b \left[ \sum_{n=1}^{N} C_n \lambda_n \phi_n(r) J_0(\lambda_n r) - S(r) \right]^2 r \, dr \tag{8.71}$$

where $S(r)$ is the right-hand side of the source–plane boundary condition. The Fourier–Bessel coefficients $C_n$ are obtained to minimize the integral $I_N$ by setting the gradient of $I_N$ with respect to the coefficients to zero:

$$\frac{\partial I_N}{\partial C_i} = 0, \qquad i = 1, 2, \dots, N \tag{8.72}$$

which leads to the following system of $N$-equation (for $i = 1, 2, \dots, N$):

$$\sum_{n=1}^{N} C_n \lambda_n \int_0^b \phi_n(r) \phi_i(r) J_0(\lambda_n r) J_0(\lambda_i r) r \, dr = \frac{1}{k} \int_0^a q(r) \phi_i(r) J_0(\lambda_i r) r \, dr \tag{8.73}$$

These equations are called the normal equations for the least squares problem which can be represented as a linear system in matrix form by

$$\mathbf{A}\mathbf{C} = \mathbf{b} \tag{8.74}$$

where $\mathbf{A} = [a_{in}]$ is an $N \times N$ matrix whose entries (represented by row $i$ and column $n$) are defined by

$$a_{in} = \lambda_n \int_0^b \phi_n(r) \phi_i(r) J_0(\lambda_n r) J_0(\lambda_i r) r \, dr \tag{8.75}$$

$\mathbf{C} = [C_1 \ C_2 \ \dots \ C_N]^t$ is the unknown Fourier–Bessel coefficients vector, and $\mathbf{b} = [b_1 \ b_2 \ \dots \ b_N]^t$ corresponds to the right-hand-side vector of the linear system whose components are given by

$$b_i = \frac{1}{k} \int_0^a q(r) \phi_i(r) J_0(\lambda_i r) r \, dr \tag{8.76}$$

Al-Khamaiseh et al. (2019) obtained solutions for the $C_n$ and examined the effect of both flux and conductance distribution on the thermal resistance. The thermal resistance is defined as

$$R_t = \frac{\overline{\theta}_s}{Q} \tag{8.77}$$

where $\overline{\theta}_s$ is the averaged temperature excess over the circular heat-source area and $Q = \pi a^2 \overline{q}$ is the total heat input of the flux tube. The averaged temperature excess is defined by

$$\overline{\theta}_s = \frac{1}{A_s} \int\int_{A_s} \theta(r,0) dA_s \tag{8.78}$$

where $A_s$ is the heat-source area, which yields

$$\overline{\theta}_s = \frac{2}{a^2} \int_0^a \theta(r,0) r\, dr = \frac{2}{a} \sum_{n=1}^{N} \frac{C_n}{\lambda_n} J_1(\lambda_n a) \tag{8.79}$$

Hence, the dimensional total thermal resistance is given by

$$R_t = \frac{2}{\pi a^3 \overline{q}} \sum_{n=1}^{N} \frac{C_n}{\lambda_n} J_1(\lambda_n a). \tag{8.80}$$

Al-Khamaiseh et al. (2019) found that for thin spreaders, both the flux and conductance distribution have a significant effect on the total resistance. Increasing the mean sink plane Biot number, $\overline{Bi}_s = \overline{h}_s b/k$ also significantly reduced the effect of the conductance distribution. Similarly, increasing the relative thickness $\tau = t/b$ will also diminish the impact of sink plane conductance distribution. Further details on solution procedure and case studies can be found in Al-Khamaiseh et al. (2019).

# References

Al-Khamaiseh, B., *Analytical Solutions of 3D Heat Conduction in Flux Channels with Nonuniform Properties and Complex Structures*, Ph.D. Thesis, Memorial University of Newfoundland, 2018.

Al-Khamaiseh, B., Razavi, M., Muzychka, Y.S., and Kocabiyik, S., "Thermal Resistance of a Three Dimensional Flux Channel with Non-Uniform Heat Convection in the Sink Plane", *IEEE Transactions on Components, Packaging, and Manufacturing Technology*, Vol. 8, no. 5, pp. 830–839, 2018.

Al-Khamaiseh, B., Muzychka, Y.S., and Kocabiyik, S., "Spreading Resistance in Flux Tubes with Variable Heat Flux and Non-Uniform Convection", *IEEE Transactions on Components, Packaging, and Manufacturing Technology*, Vol. 9, no. 8, pp. 1526–1534, 2019.

Al-Khamaiseh, B., Muzychka, Y.S., and Kocabiyik, S., "Thermal Resistance of a Three Dimensional Flux Channel with Two-Dimensional Variable Convection", *Journal of Thermophyisics and Heat Transfer*, Vol. 34, no. 2, pp. 322–330, 2020.

Arpaci, V.S., *Conduction Heat Transfer*, Addison-Wesley, 1966.

Kantorovich, L.V., and Krylov, V.I., *Approximate Methods of Higher Analysis*, Noordhoff, 1958.

Kelman, R.B., "Least Squares Fourier Series Solutions to Boundary Value Problems", *IIAM Review*, Vol. 21, no. 3, pp. 329–338, 1979.

Razavi, M., *Advanced Thermal Analysis of Microelectronics Using Spreading Resistance Models*, Ph.D. Thesis, Memorial University of Newfoundland, 2015.

Razavi, M., Muzychka, Y.S., and Kocabyik, S., "Thermal Spreading Resistance in a Flux Channel with Arbitrary Heat Convection in the Sink Plane", *2014 International Mechanical Engineering Congress and Exposition (IMECE)*, Montreal, QC, Canada, 2014.

Razavi, M., Muzychka, Y.S., and Kocabyik, S., "Thermal Resistance in a Rectangular Flux Channel with Non-Uniform Heat Convection in the Sink Plane", *ASME Journal of Heat Transfer*, Vol. 137, no. 11, p. 111401, 2015.

Sneddon, I.N., *Mixed Boundary Value Problems in Potential Theory*, North-Holland Publishing, Amsterdam, p. 63, 1966.

# 9

# Further Applications of Spreading Resistance

In this chapter, we consider several special and unique applications where thermal spreading resistance theory plays a major role in transport phenomena; for example in sliding contacts of tribology applications, mass diffusion with and without chemical reaction, transport involving super-hydrophobic surfaces at low Peclet numbers, electronic microdevices where temperature-dependent thermal conductivity plays a major role, applications involving phase change such as freezing droplets on a surface, and finally thermal spreading in solid or hollow spheres for packed-bed applications. These are all applications where performance is controlled by thermal, mass, or viscous spreading principles. In Chapters 10–12, we will consider thermal contact resistance applications in great detail.

## 9.1 Moving Heat Sources

The analysis of heat transfer from sliding and rolling contacts is important in many tribological applications such as ball bearing and gear design. In these applications, heavily loaded contacts are common and knowledge of the nominal contact temperatures which result from frictional heat generation is required for minimizing thermal-related problems such as scoring, lubricant break-down, and adhesive wear due to flash welding. Flash welding occurs when a microasperity reaches its melting point and is instantaneously bonded to another surface. The sliding motion then causes the asperity bond to detach as quickly as it formed, leading to surface damage.

A review of typical tribology books such as the texts by Halling (1975), Williams (1994), and Bhushan (2002), and various handbook sections by Winer and Cheng (1980) and Cowan and Winer (1992) shows that the analysis of heat transfer from sliding or rolling contacts though important, has not been extensively modeled. These reviews generally present equations and results for only one configuration, the circular contact. Although this contact geometry arises quite frequently in tribology applications, others such as the elliptic contact are also quite common in ball bearing and gear applications where nonconforming contacts prevail [Yovanovich (1971, 1978), Bejan (1989), Cameron et al. (1968)]. We will review and extend some of the common solutions and put them in context with the material reported in Chapter 2.

The analysis for real moving (or sliding) heat sources, which are presented in a number of tribology references, is based upon the assumption that one of the contacting surfaces can be modeled with a stationary heat source and the other surface with a moving heat source. Although both surfaces are in mutual contact, each surface is treated differently. The moving surface (the one with the asperity) is treated as a stationary contact from the perspective of the heat conduction into its substrate. While the stationary substrate (assumed smooth relative to the other surface) experiences a

*Thermal Spreading and Contact Resistance: Fundamentals and Applications*, First Edition.
Yuri S. Muzychka and M. Michael Yovanovich.

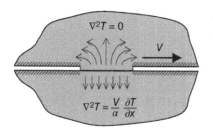

**Figure 9.1** A moving (frictional heat source) in contact with a stationary surface.

sliding region of thermal contact whose temperature field depends on the speed of motion of the source, see Figure 9.1.

In many problems, the assumption of a fast moving heat source may not be strictly valid and the analysis will incorrectly predict the average or maximum contact temperature (often called the flash temperature). With this in mind, Tian and Kennedy (1994) developed accurate correlations for the circular and square heat source which predicts the temperature for any speed. These correlations were then used to formulate models for predicting flash temperatures for sliding asperities in real contact situations where heat is generated through friction and a portion is conducted into each surface.

Neder et al. (1998) developed a hybrid computational method for noncircular heat sources. For this method, a numerical approach based upon the superposition of point heat sources was employed for the stationary source and a transient finite element method was employed for the moving source. This new approach was then used to predict temperatures in a steel/bronze sliding contact problem, with sliding motion normal and parallel to the grinding direction. The primary motivation for the work of Neder et al. (1998) was that the conventional approach adopted in most tribology references was not applicable to noncircular heat sources. Later, we will show the development of a model by Muzychka and Yovanovich (2001) which enable prediction of flash temperatures for arbitrary-shaped contacts.

The present section discusses various aspects of heat transfer in tribological applications involving stationary and sliding contacts. In all cases, heat is either supplied to the contact or is generated through contact friction. The field of tribology has only adopted a simplified approach in the prediction of contact temperatures due to sliding. This approach does not allow for the effect of shape, aspect ratio, and flux distribution to be modeled easily. This was the motivating factor for the development of a hybrid numerical scheme by Neder et al. (1998). The expressions that are summarized here have been validated against a small set of numerical data for real and ideal contacts [Muzychka and Yovanovich (2001)]. The results of Neder et al. (1998) and others were readily computed using simple spreading theory with significantly less effort.

### 9.1.1 Governing Equations

A review of the major literature reveals that extensive analysis of the problem has been undertaken for various contact spot shapes and thermal boundary conditions for both stationary and moving heat sources. The governing equation for a moving heat source may be obtained from the transient heat conduction equation with a transformation of variables [Rosenthal (1946)]. The resulting equation for pseudosteady-state conditions is

$$\frac{\partial^2 T}{\partial x^2} + \frac{\partial^2 T}{\partial y^2} + \frac{\partial^2 T}{\partial z^2} = \frac{V}{\alpha}\frac{\partial T}{\partial x} \tag{9.1}$$

where the coordinate system is fixed to the heat source and the half-space moves beneath it with velocity $V$.

The thermal boundary conditions are constant or zero temperature in regions remote from the source, i.e. $r = \sqrt{x^2 + y^2 + z^2} \to \infty$, $T \to T_b$, or $T(x \to \pm\infty, y \to \pm\infty, z \to \infty) = T_b$ and:

$$
\left.\begin{aligned}
\frac{\partial T}{\partial z}\Bigg|_{z=0} &= -\frac{q(x,y)}{k}, \quad \text{Inside Source Region} \\[2em]
\frac{\partial T}{\partial z}\Bigg|_{z=0} &= 0, \qquad\qquad \text{Outside Source Region}
\end{aligned}\right\}
\tag{9.2}
$$

Solution to Eq. (9.1) is usually obtained by superposition of the moving point heat source [Carslaw and Jaeger (1959)]:

$$
T - T_b = \left(\frac{Q}{2\pi kr}\right) e^{-\frac{V}{2\alpha}(r-x)}
\tag{9.3}
$$

over the region of contact, where $r = \sqrt{x^2 + y^2 + z^2}$. Solution of the moving heat source by means of Eq. (9.3) is rather involved, requiring numerical integration. Solutions for the square and circular contact are tabulated in Tian and Kennedy (1994). Carslaw and Jaeger (1959) provide solutions for a rectangular source and strip source. The general solutions are not easily utilized but are still useful for some engineering applications.

A simpler approach based upon the combination of asymptotic solutions [Muzychka and Yovanovich (2001)] is presented in Section 9.1.4, whereby Eq. (9.1) is solved for the two limiting cases of slow $V \to 0$ and fast $V \to \infty$ sliding velocity. These limiting cases which are easier to obtain are then combined to provide a means to predict the temperature (or resistance) for any velocity $V$.

### 9.1.2 Asymptotic Limits

For the general case of a sliding heat source moving with a speed of $V$, the thermal resistance or temperature for the two special cases will now be considered.

If the velocity of the heat source is small ($V \to 0$), the governing equation reduces to Laplace's equation:

$$
\frac{\partial^2 T}{\partial x^2} + \frac{\partial^2 T}{\partial y^2} + \frac{\partial^2 T}{\partial z^2} = 0
\tag{9.4}
$$

with the same boundary conditions prescribed earlier.

Many solutions for slow moving or stationary heat source problem have been obtained by superposition of the point heat source on a half-space (see Chapter 2):

$$
T - T_b = \frac{Q}{2\pi kr}
\tag{9.5}
$$

where $r = \sqrt{x^2 + y^2 + z^2}$. Solutions for various heat flux distributions and source shapes have been found as shown in Chapter 2. Of particular interest here, are the solutions for the rectangular and elliptical heat sources which contain the limiting cases for the square and circular contacts.

If the velocity of the heat source is large ($V \to \infty$), Eq. (9.1) can be simplified to give:

$$
\frac{\partial^2 T}{\partial z^2} = \frac{V}{\alpha} \frac{\partial T}{\partial x}
\tag{9.6}
$$

which is essentially the one-dimensional diffusion equation for a half-space with $t = x/V$. This equation assumes that heat conduction into the half-space is one-dimensional (thermal penetration is shallow) and the solution may be approximated by the equation for heat flow at the surface of a half space with flux specified boundary conditions [Carslaw and Jaeger (1959)]:

$$T - T_b = \frac{2q}{k} \sqrt{\alpha t / \pi} \tag{9.7}$$

where $t$ must be replaced by the effective traverse time $t = 2x'/V$, and $x'$ is the distance from an arbitrary point within the source to the leading edge of the source. This approach was applied by Jaeger (1942) for the strip and square heat sources by Archard (1958/1959) for the circular source for the uniform heat flux distribution, and by Francis (1970), for the circular heat source having a parabolic heat flux distribution. Muzychka and Yovanovich (2001) applied it to obtain a solution for an elliptical heat source. The solutions for a moving elliptic heat source with uniform and parabolic heat flux distribution will be obtained in Section 9.1.3.

The analysis based on Eq. (9.7) is only valid for large values of the dimensionless group $Pe = Va/\alpha$, or Peclet number. This group may be interpreted as a measure of the relative thermal penetration depth, $\delta/a$, of heat into the half-space if we consider a circular contact of radius $a$. Beginning with the definition, $\delta = \sqrt{\pi \alpha t}$, which is the thermal penetration depth for heat flow into a half-space with isothermal boundary, the relative penetration depth for a circular contact is

$$\frac{\delta}{a} \sim \frac{\sqrt{\pi \alpha t}}{a} \tag{9.8}$$

If the traverse time for a moving circular heat source is taken to be $t = 2a/V$, then Eq. (9.8) may be written as

$$\frac{\delta}{a} \sim \sqrt{\frac{2\pi \alpha}{Va}} \sim \frac{\sqrt{2\pi}}{\sqrt{Pe}} \tag{9.9}$$

Thus, if $Pe \to \infty$, the penetration $\delta$ is small, and the heat diffusion may be taken to be one-dimensional since the spreading of heat into the half-space is negligible, see Figure 9.1. On the other hand, if $Pe \to 0$, the spreading of heat into the substrate will be significant. A solution valid for all values of Peclet number can only be obtained numerically. As we shall see shortly, we may define a slow moving heat source as one with $Pe < 0.1$ and a fast moving heat source as one with $Pe > 10$. In between these limits, an exact analysis or model is required to make more accurate predictions of source resistance.

### 9.1.3 Stationary and Moving Heat Source Limits

A discussion of a number of important solutions for sliding heat sources is now presented. In many cases, gaps existed in the literature, and the present authors have developed new solutions for a number of problems. These are discussed throughout the sections which follow.

#### 9.1.3.1 Stationary Heat Sources ($Pe \to 0$)

As we have seen in Chapter 2, extensive analysis of heat conduction from isolated heat sources on a half-space has been accomplished. The simplest contact geometry is the circular contact. The analysis has been performed for three heat flux distributions: the uniform heat flux, parabolic heat flux, and the inverse parabolic heat flux. The inverse parabolic heat flux represents a uniform temperature distribution over the contact area. The solutions for the dimensionless thermal resistance for these three cases are summarized in Table 2.2.

The thermal resistance may be defined with respect to the average surface temperature such that

$$\overline{R} = \frac{\overline{T}_c - T_b}{Q} \tag{9.10}$$

or with respect to the maximum surface temperature such that

$$\hat{R} = \frac{\hat{T}_c - T_b}{Q} \tag{9.11}$$

A dimensionless thermal resistance may be defined as

$$R^\star = Rk\mathcal{L} \tag{9.12}$$

where $\mathcal{L}$ is an appropriate characteristic length related to the heat source area. This thermal resistance is a spreading resistance due to the transfer of heat through a finite area of contact.

In most tribological applications involving frictional heat generation, the average heat flux $\bar{q} = Q/A$ is known. What may not be known precisely is the distribution of heat flux $q(x,y)$ over the contact region. If the contact loading is Hertzian, the distribution of frictional heat generation may be represented by the parabolic flux distribution. In most analyses, the assumption of a uniform heat flux distribution is often made. The effect of heat flux distribution on the thermal resistance based upon the average contact temperature is small. Thus, the uniform heat flux distribution may be taken as representative of the mean value if the exact flux distribution is not known. In any event, the maximum and minimum values for the average or maximum source temperature are bounded by the solutions for the isothermal and parabolic flux distribution.

If the shape of the heat source is allowed to vary with aspect ratio $\epsilon = b/a$, then the solutions are somewhat more complex than those given in Table 2.2. The solution for the dimensionless thermal resistance of a stationary rectangular uniform heat source (from Chapter 2) is

$$\bar{R}ka = \frac{1}{2\pi} \left\{ \frac{\sinh^{-1}(\epsilon)}{\epsilon} + \sinh^{-1}(1/\epsilon) + \frac{1}{3} \left[ \frac{1}{\epsilon^2} + \epsilon - \frac{(1+\epsilon^2)^{3/2}}{\epsilon^2} \right] \right\} \tag{9.13}$$

for the average contact temperature, and

$$\hat{R}ka = \frac{1}{2\pi} \left\{ \frac{\sinh^{-1}(\epsilon)}{\epsilon} + \sinh^{-1}(1/\epsilon) \right\} \tag{9.14}$$

for the maximum contact temperature.

The solutions for the elliptic heat source (from Chapter 2) are given by

$$\bar{R}ka = \frac{16}{3\pi^3} K(\epsilon') \tag{9.15}$$

and

$$\hat{R}ka = \frac{2}{\pi^2} K(\epsilon') \tag{9.16}$$

for the isoflux contact, where $\epsilon' = \sqrt{1 - \epsilon^2}$, and $K(\epsilon')$ is the complete elliptic integral of the first kind of complementary modulus $\epsilon'$.

The solution for the parabolic heat flux distribution reported in Chapter 2 are

$$\bar{R}ka = \frac{9}{16\pi} K(\epsilon') \tag{9.17}$$

and

$$\hat{R}ka = \frac{3}{4\pi} K(\epsilon') \tag{9.18}$$

The above represent the most practical results that could be applied in sliding heat source applications. As we have seen before, when $\epsilon = 1$ for a circular source, $K(0) = \pi/2$.

### 9.1.3.2 Moving Heat Sources ($Pe \to \infty$)

Solutions for moving heat sources have been obtained for a number of configurations and boundary conditions. All of the moving source solutions are written in terms of the Peclet number. The Peclet number is defined as

$$Pe = \frac{V\mathcal{L}}{\alpha} \tag{9.19}$$

where $\mathcal{L}$ is a characteristic length scale representative of the contact geometry. If the contact geometry is circular or square, then $\mathcal{L} = a$, the radius of the contact or the half side length of the square. However, it has already been shown that if $\mathcal{L} = \sqrt{A}$, the area of the heat source, the effect of shape, and aspect ratio on the dimensionless resistance is quite small.

The effect of heat flux distribution (uniform or parabolic) on the thermal resistance for a moving circular heat source is given in Table 9.1. The solution for the uniform heat flux distribution was obtained by Archard (1958/1959), and the solution for the parabolic heat flux distribution was obtained by Francis (1970). These solutions are only valid for large values of the Peclet number $Pe > 10$. The effect of flux distribution on the thermal resistance based upon the average contact temperature is small being only 1.6%. The relative difference increases to 15.9% for the thermal resistance based upon the maximum source temperature. Finally, Table 9.2 presents a comparison of the asymptotic solutions for the fast moving heat source for the circular and square heat sources. The results are 16.4% and 21.5% higher for the circular heat source for the thermal resistance based upon the average and maximum source temperatures, respectively.

If the contact is rectangular, the thermal resistance will vary with aspect ratio $\epsilon = b/a$, where $0 < \epsilon < \infty$. The solution obtained by Jaeger (1942) for the strip source is applicable to a rectangular heat source since the solution assumes one-dimensional heat flow into the half-space, i.e. the penetration depth is small compared with the characteristic dimension of the contact zone. The solution for the finite rectangular source is

$$\overline{R}\,k\,a = \frac{\sqrt{2}}{3\sqrt{\pi}}\left(\frac{a}{b}\right)\frac{1}{\sqrt{Pe}} \tag{9.20}$$

**Table 9.1** Effect of boundary condition on a moving circular heat source.

| $R^\star$ | Uniform flux | Parabolic flux |
| --- | --- | --- |
| $\overline{R}\,k\,a$ | $\dfrac{0.318}{\sqrt{Pe}}$ | $\dfrac{0.323}{\sqrt{Pe}}$ |
| $\hat{R}\,k\,a$ | $\dfrac{0.508}{\sqrt{Pe}}$ | $\dfrac{0.589}{\sqrt{Pe}}$ |

**Table 9.2** Effect of shape on isoflux moving heat sources.

| $R^\star$ | Circular | Square |
| --- | --- | --- |
| $\overline{R}\,k\,a$ | $\dfrac{0.318}{\sqrt{Pe}}$ | $\dfrac{0.266}{\sqrt{Pe}}$ |
| $\hat{R}\,k\,a$ | $\dfrac{0.508}{\sqrt{Pe}}$ | $\dfrac{0.399}{\sqrt{Pe}}$ |

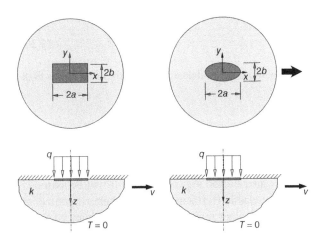

**Figure 9.2** Moving rectangular and elliptic sources on half-space.

for the average contact temperature, and

$$\hat{R}\,k\,a = \frac{\sqrt{2}}{2\sqrt{\pi}}\left(\frac{a}{b}\right)\frac{1}{\sqrt{Pe}} \tag{9.21}$$

for the maximum contact temperature, where $Pe = Va/\alpha$, is based upon the half width of the rectangle in the direction of motion, see Figure 9.2.

No solution exists for the fast moving elliptical contact. In order to obtain a solution for the elliptical contact, the approach developed by Jaeger (1942) for the square source and by Archard (1958/1959) for the circular contact was applied [Muzychka and Yovanovich (2001)]. In this case, the effective contact time is

$$t = \frac{2x'}{V} = \frac{2\sqrt{a^2\left(1 - \frac{y^2}{b^2}\right)}}{V} \tag{9.22}$$

Applying the approach of Jaeger (1942) and Archard (1958/1959) gives:

$$\overline{R}\,k\,a = \frac{1}{\pi}\left(\frac{a}{b}\right)\frac{1}{\sqrt{Pe}} \tag{9.23}$$

for the dimensionless thermal resistance based upon the average contact temperature, and

$$\hat{R}\,k\,a = \frac{2\sqrt{2}}{\pi^{3/2}}\left(\frac{a}{b}\right)\frac{1}{\sqrt{Pe}} \tag{9.24}$$

for the dimensionless thermal resistance based upon the maximum contact temperature. In both cases, $Pe = Va/\alpha$ is based upon the half width of the heat source in the direction of motion, see Figure 9.2.

Comparison of Eqs. (9.20) and (9.21) and Eqs. (9.23) and (9.24) with the solutions for the square and circular heat source provided in Table 9.2, shows that the solutions are identical except for the term $(a/b)$. This factor accounts for the effect of heat source aspect ratio with respect to the direction of motion. These results may be applied to infer the following solutions for a fast-moving elliptic heat source with parabolic heat flux distribution:

$$\overline{R}\,k\,a = 0.323\left(\frac{a}{b}\right)\frac{1}{\sqrt{Pe}} \tag{9.25}$$

and

$$\hat{R} \, k \, a = 0.589 \left( \frac{a}{b} \right) \frac{1}{\sqrt{Pe}} \tag{9.26}$$

In Section 9.1.4, the results will applied to develop new models applicable to a real contacts of noncircular shape.

### 9.1.4 Analysis of Real Contacts

In this section, application of the theory of moving heat sources to real contacts is discussed following the approach of Muzychka and Yovanovich (2001). A simple approach to modeling the effects of shape, aspect ratio, orientation, and heat flux distribution are considered. It will be assumed that the shape of a real contact is elliptic, and that classic Hertzian analysis for elastic contact of nonconforming surfaces may be used to predict the contact zone dimensions (see Chapter 12).

#### 9.1.4.1 Effect of Contact Shape

Hertzian theory may be used to predict the contact size for elastic contact [Yovanovich (1971, 1978, 1986)]. However, this assumes the shape of the contact is elliptic. Depending on the surface topography, this assumption may not be valid. Thus, it is desirable to examine the effect that shape and aspect ratio have on the overall resistance of moving heat sources. In Chapter 2, we considered the effect of the shape and aspect ratio of an isolated stationary contact having a uniform flux distribution. We showed that if the thermal resistance is nondimensionalized using the square root of the contact area $\mathcal{L} = \sqrt{A}$, the solutions are weak functions of shape and aspect ratio.

We now reconsider the solutions for elliptical and rectangular sources using $\mathcal{L} = \sqrt{A}$ as a length scale. In these cases, we will denote the aspect ratio as $0 < \epsilon_s < 1$ to distinguish it from the aspect ratio of a moving heat source $0 < \epsilon_m < \infty$ which depends on the orientation with respect to the direction of motion. The solutions for the isoflux stationary elliptic heat source become

$$\overline{R}_s \, k \, \sqrt{A} = \frac{16}{3\pi^3} \sqrt{\pi \, \epsilon_s} \, K(\epsilon_s') \tag{9.27}$$

for the average contact temperature, and

$$\hat{R}_s \, k \, \sqrt{A} = \frac{2}{\pi^2} \sqrt{\pi \, \epsilon_s} \, K(\epsilon_s') \tag{9.28}$$

for the maximum contact temperature. If the flux distribution is parabolic, then the solutions presented earlier become:

$$\overline{R}_s \, k \, \sqrt{A} = \frac{9}{16\pi} \sqrt{\pi \, \epsilon_s} \, K(\epsilon_s') \tag{9.29}$$

and

$$\hat{R}_s \, k \, \sqrt{A} = \frac{3}{4\pi} \sqrt{\pi \, \epsilon_s} \, K(\epsilon_s') \tag{9.30}$$

If the heat source is rectangular, then the dimensionless thermal resistance, Eqs. (9.13) and (9.14) become

$$\overline{R}_s \, k \, \sqrt{A} = \frac{\sqrt{\epsilon_s}}{\pi} \left\{ \frac{\sinh^{-1}(\epsilon_s)}{\epsilon_s} + \sinh^{-1}(1/\epsilon_s) + \frac{1}{3} \left[ \frac{1}{\epsilon_s^2} + \epsilon_s - \frac{(1 + \epsilon_s^2)^{3/2}}{\epsilon_s^2} \right] \right\} \tag{9.31}$$

for the average contact temperature, and

$$\hat{R}_s \, k \, \sqrt{A} = \frac{\sqrt{\epsilon_s}}{\pi} \left\{ \frac{\sinh^{-1}(\epsilon_s)}{\epsilon_s} + \sinh^{-1}(1/\epsilon_s) \right\} \tag{9.32}$$

for the maximum contact temperature.

In the case of the moving heat source, the Peclet number should also be based upon the square root of the contact area, i.e. $\mathcal{L} = \sqrt{A}$ in Eq. (9.19). Table 9.3 summarizes the results for the rectangular and elliptic heat sources for different heat flux distributions, when the resistance is nondimensionalized using the square root of the contact area. Comparisons of the dimensionless resistance $\hat{R}^{\star}$ are provided in Figure 9.3 using the data for the exact analytical solutions of isoflux square and circular heat sources from Tian and Kennedy (1994). Figure 9.3 shows a comparison of the results for a circular and square heat source are indistinguishable when $\mathcal{L} = \sqrt{A}$. Thus, the effect of the shape of the heat source is not a significant factor, when the results are appropriately nondimensionalized.

If the heat source is rectangular or elliptical, the Peclet number must be replaced with a modified Peclet number $Pe^*$ defined as

$$Pe^*_{\sqrt{A}} = (\epsilon_m)^{1/2} \, Pe_{\sqrt{A}} \tag{9.33}$$

The aspect ratio $\epsilon_m = b/a$ now accounts for the effect of the shape and orientation of the heat source. Since a rectangular or an elliptic heat source may be oriented in the direction of motion

**Table 9.3** Dimensionless resistance of moving heat sources on a half-space.

| Shape (flux distribution) | $\bar{R} k \sqrt{A}$ | $\hat{R} k \sqrt{A}$ |
|---|---|---|
| Rectangular (isoflux) | $\dfrac{0.752}{\sqrt{Pe^*_{\sqrt{A}}}}$ | $\dfrac{1.130}{\sqrt{Pe^*_{\sqrt{A}}}}$ |
| Elliptic (isoflux) | $\dfrac{0.750}{\sqrt{Pe^*_{\sqrt{A}}}}$ | $\dfrac{1.200}{\sqrt{Pe^*_{\sqrt{A}}}}$ |
| Elliptic (parabolic flux) | $\dfrac{0.762}{\sqrt{Pe^*_{\sqrt{A}}}}$ | $\dfrac{1.390}{\sqrt{Pe^*_{\sqrt{A}}}}$ |

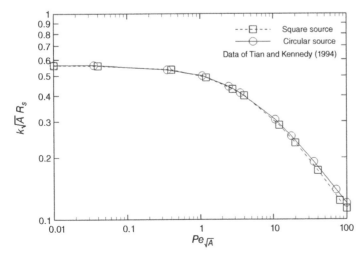

**Figure 9.3** Dimensionless thermal resistance for a square and circular moving heat source. Source: adapted from Muzychka and Yovanovich (2001).

parallel to the short or long axis of the heat source, the resistance must change based on orientation. Given the same source area and velocity, the Peclet number remains unchanged, but the resistance will decrease if the direction of motion is parallel to the short axis of the heat source. Thus, for a moving heat source, $0 < \epsilon_m < \infty$. This is quite important for the moving heat source since the resistance will increase with decreasing $\epsilon_m$, i.e. $a > b$, and decrease with increasing $\epsilon_m$, i.e. $b > a$. If the heat source is stationary, the orientation of the contact is not important and $0 < \epsilon_s < 1$.

### 9.1.4.2 Models for All Peclet Numbers

In Section 9.1.3, the thermal resistance of isolated stationary and fast moving contacts was presented. These solutions represent asymptotic solutions for large and small values of the Peclet number. If the contact is moving at moderate speeds $0.1 < Pe < 10$, a composite solution is required. Tian and Kennedy (1994) combined the asymptotic results for the circular heat source using an equation which is a special case of the more general form:

$$\frac{1}{R_t^n} = \frac{1}{R_s^n} + \frac{1}{R_m^n} \tag{9.34}$$

Equation (9.34) is one form of the asymptotic correlation method proposed by Churchill and Usagi (1972). As we have already seen, this method allows the combination of asymptotic solutions, to generate a model which is valid for all values of the dependent parameter, in this case the Peclet number.

A single value of $n = 2$ was found to give excellent agreement between the approximate model and numerical results of Tian and Kennedy (1994) over the entire range of Peclet numbers for the circular heat source having a uniform or parabolic heat flux distribution and a square heat source having a uniform heat flux distribution. Thus, the parameter $n$ does not appear to depend upon the shape of the source or the flux distribution. The models developed by Tian and Kennedy (1994) are specifically for the circular heat source for uniform and parabolic flux distributions. They are not applicable to elongated contacts such as elliptic or rectangular contacts. In addition, Tian and Kennedy (1994) presented their correlations in terms of contact temperatures, rather than thermal resistance. The use of thermal resistance facilitates the calculation of the partition of heat into the contacting bodies.

A general model for a moving heat source will now be obtained by combining the dimensionless resistances for a stationary and fast moving heat sources in the form of Eq. (9.34). As noted earlier, the definition of aspect ratio is different for the moving and stationary heat sources. The aspect ratio of the stationary heat source is now denoted by $\epsilon_s = b/a$ such that $0 < \epsilon_s < 1$, and the aspect ratio of the moving heat source is now denoted by $\epsilon_m$ such that $0 < \epsilon_m < \infty$. Also, since the effect of shape has been shown to be negligible, only the solution for the elliptic heat source will be considered in the model development.

Combining the stationary and moving heat source solutions for both the average and maximum contact surface temperatures lead to the following models for any source speed (Peclet number):

$$\overline{R}_t\, k\, \sqrt{A} = \frac{0.750}{\sqrt{(\epsilon_m)^{1/2}Pe_{\sqrt{A}} + 6.05/(\epsilon_s K^2(\epsilon_s'))}} \tag{9.35}$$

and

$$\widehat{R}_t\, k\, \sqrt{A} = \frac{1.200}{\sqrt{(\epsilon_m)^{1/2}Pe_{\sqrt{A}} + 11.16/(\epsilon_s K^2(\epsilon_s'))}} \tag{9.36}$$

for the uniform flux distribution, and

$$\overline{R}_t \, k \, \sqrt{A} = \frac{0.762}{\sqrt{(\epsilon_m)^{1/2} Pe_{\sqrt{A}} + 5.77/(\epsilon_s K^2(\epsilon'_s))}} \tag{9.37}$$

and

$$\hat{R}_t \, k \, \sqrt{A} = \frac{1.390}{\sqrt{(\epsilon_m)^{1/2} Pe_{\sqrt{A}} + 10.79/(\epsilon_s K^2(\epsilon'_s))}} \tag{9.38}$$

for the parabolic flux distribution. These expressions can now be applied to arbitrarily shaped heat sources for all values of the Peclet number. Recall, for the special case of a circular source when $\epsilon_s = \epsilon_m = 1$, the elliptic integral simplifies to give $K(0) = \pi/2$ making the above equations much simpler.

### 9.1.5 Prediction of Flash Temperature

The concept of maximum or average flash temperature is discussed in detail in Archard (1958/1959), Blok (1963), Winer and Cheng (1980), and Cowan and Winer (1992). The computation of the flash temperature assumes that one of the contacting surfaces is a stationary heat source and the other a moving heat source. By accounting for the partition of heat into each of the surfaces, an estimate for the average or maximum temperature may be obtained.

In Section 9.1.4, a general model for an isolated moving source for $0 < Pe < \infty$ was developed. This model may now be used to predict the average or maximum flash temperatures. The analysis begins by defining the total heat flow and the partition of each into the two contacting surfaces. The total heat generated by sliding friction is denoted $Q_g = \mu FV$, where $\mu$ is the coefficient of friction, while the heat which flows into the stationary and moving surfaces are denoted $Q_s$ and $Q_m$, respectively. Through conservation of energy, the total heat flow is then

$$Q_g = Q_s + Q_m \tag{9.39}$$

which may be written in terms of the temperature excess and resistance in each surface

$$\mu FV = \frac{(T_s - T_{b,s})}{R_s} + \frac{(T_m - T_{b,m})}{R_m} \tag{9.40}$$

Now, if perfect thermal contact is assumed at the interface, then $T_s = T_m = T_c$ at all points within the contact, and the expression given above may be solved for $T_c$

$$T_c = \frac{\mu FV + \dfrac{T_{b,s}}{R_s} + \dfrac{T_{b,m}}{R_m}}{\dfrac{1}{R_s} + \dfrac{1}{R_m}} \tag{9.41}$$

The general expression given above may be applied to any combination of slow, moderate, or fast-moving sources using the expressions developed earlier. If the bulk temperatures are equal, Eq. (9.41) may be further simplified. In the case of a typical sliding asperity contact, the system is modeled as a stationary heat source and a moving heat source in parallel. Equation (9.41) may be written in terms of the dimensionless thermal resistances $R^* = R \, k \, \sqrt{A}$ to give

$$T_c = \frac{\mu FV + \dfrac{T_{b,s} \sqrt{A} k_s}{R_s^*} + \dfrac{T_{b,m} \sqrt{A} k_m}{R_m^*}}{\dfrac{\sqrt{A} k_s}{R_s^*} + \dfrac{\sqrt{A} k_m}{R_m^*}} \tag{9.42}$$

which may be further simplified to give

$$T_c = \frac{(\mu F V R_s^* R_m^*)/\sqrt{A} + T_{b,s} k_s R_m^* + T_{b,m} k_m R_s^*}{k_s R_m^* + k_m R_s^*} \tag{9.43}$$

where the value for $R_s^*$ is taken to be the appropriate value of the stationary dimensionless resistance Eqs. (9.27)–(9.30) and $R_m^*$ is the appropriate expression for the dimensionless resistance of a moving heat source Eqs. (9.35)–(9.38) which were presented in Section 9.1.4. Thus, $T_c$ may be computed for either the maximum or average value which occurs within the contact for either the isoflux or parabolic flux distribution. The validity of Eq. (9.43) may be questioned on the grounds that the temperature distribution of the real contact will be skewed; however, models for the stationary heat source resistance assume a symmetric temperature profile. For a slow moving contact $Pe < 0.1$, the profile is nearly symmetric and the partition of heat into each of the surfaces is equal assuming that each surface has the same thermal properties. For a fast moving heat source $Pe > 10$, the maximum temperature is located at or near the trailing edge. Most of the heat will be conducted into the surface experiencing the sliding heat source since it has a lower thermal resistance. In the transition region $0.1 < Pe < 10$, the effect of temperature distribution shape should be small since the maximum temperature is located between the centroid and the trailing edge. Thus, Eq. (9.43) may be applied for either the average or maximum contact temperature basis. Also, due to the relatively short contact times and size of asperities, the penetration depth will be small and the assumption of a half-space is then reasonable. We demonstrate the validity of Eq. (9.43) using two case studies from Muzychka and Yovanovich (2001).

**Example 9.1**  Consider the contact for mild steel $k = 60.3$ W/(m K), $\alpha = 17.7 \times 10^{-6}$ m²/s, $F = 400$ g, and $\mu = 0.23$ for a square source with half side length $a = 1 \times 10^{-5}$ m. Table 9.4 presents a comparison of results computed by Jaeger (1942) using the full analytical solution for a square source.

Using the variables and Eq. (9.43) with Eqs. (9.27) and (9.35) for the stationary and moving source resistances, the mean contact temperature is predicted and given in the table below.

The maximum difference between the model and the data of Jaeger (1942) is 2.1% at $V = 15$ m/s. This error is small considering that the values presented by Jaeger (1942) were based upon graphical results which are also subject to round off errors.

| | | Eq. (9.43) | | Jaeger (1942) | |
|---|---|---|---|---|---|
| $V$ (m/s) | $Pe$ | $\overline{T}_c$ (°C) | $\beta = \frac{Q_m}{Q_g}$ | $\overline{T}_c$ (°C) | $\beta = \frac{Q_m}{Q_g}$ |
| 15 | 16.95 | 1399.7 | 0.740 | 1370 | 0.74 |
| 10 | 11.30 | 1057.2 | 0.705 | 1040 | 0.70 |
| 7 | 7.91 | 816.3 | 0.675 | 810 | 0.67 |
| 5 | 5.65 | 632.9 | 0.647 | 630 | 0.64 |
| 2 | 2.26 | 299.5 | 0.582 | 300 | 0.58 |
| 1 | 1.13 | 161.9 | 0.548 | 160 | 0.54 |
| 0.7 | 0.79 | 116.5 | 0.536 | 115 | 0.53 |
| 0.5 | 0.57 | 84.9 | 0.526 | 85 | 0.52 |
| 0.2 | 0.23 | 35.0 | 0.517 | 35 | 0.51 |

**Example 9.2** For this example, the proposed model is compared with numerical results reported by Neder et al. (1998). The system examined by Neder et al. (1998) consisted of a bronze substrate ($\alpha = 13.8 \times 10^{-6}$ m$^2$/s, $k = 50$ W/(m K)) and a steel slider ($\alpha = 20.0 \times 10^{-6}$ m$^2$/s, $k = 62$ W/(m K)). The maximum pressure considered was $P = 450$ MPa and the coefficient of friction $\mu = 0.25$. Neder et al. (1998) considered sliding in directions perpendicular and parallel to the grinding direction of a real surface. The authors reported a range for the equivalent diameter of the largest real contact spot in the direction parallel and perpendicular to the direction of sliding along with the contact width, $2a$, in the sliding direction only. The equivalent diameters which were tabulated by Neder et al. (1998) have different values for each direction.

It is assumed that if the same surface was considered, the equivalent diameter should be the same in both sliding directions, and that the contact spot aspect ratio may be computed assuming an elliptic contact. The area is given by $A = \pi D_e^2/4 = \pi ab$. Given $D_e$ and $2a$, $2b$ is computed using (8 μm $< D_e <$ 10 μm) and (6 μm $< 2a <$ 8 μm).

These dimensions are used for both sliding directions, and the maximum flash temperatures were computed using Eq. (9.43) along with the expressions for the dimensionless resistance based upon the maximum contact temperature.

The results are summarized in the table below. The predicted temperature range is in excellent agreement with the reported values given by Neder et al. (1998). Neder et al. (1998) also reported values of $\hat{T} = 32.8°$C and $\hat{T} = 37.5°$C on two plots for the $V = 10$ m/s case, in the direction perpendicular and parallel to the grinding direction, respectively. These results are also within the range of temperatures computed using the model provided by Eq. (9.43).

| $V$ (m/s) | $D_e$ (μm) | $2a$ (μm) | $2b$ (μm) | Eq. (9.43) (°C) | $\hat{T}$ (°C) |
|---|---|---|---|---|---|
| 1 | 8–10 | 6–8 | 10.7–12.5 | 3.8–4.8 | 4.3 |
| 1 | 8–10 | 10.7–12.5 | 6–8 | 3.9–4.9 | 5.1 |
| 10 | 8–10 | 6–8 | 10.7–12.5 | 32.0–39.2 | 35.4 |
| 10 | 8–10 | 10.7–12.5 | 6–8 | 34.6–42.0 | 40.5 |

## 9.2 Problems Involving Mass Diffusion

Mass diffusion from finite sources can be modeled in a similar manner as thermal spreading resistance. For steady diffusion problems with no chemical reactions, the primary equation is also Laplace's equation:

$$\nabla^2 C = \frac{\partial^2 C}{\partial x^2} + \frac{\partial^2 C}{\partial y^2} + \frac{\partial^2 C}{\partial z^2} = 0 \tag{9.44}$$

in Cartesian coordinates, and

$$\nabla^2 C = \frac{\partial^2 C}{\partial r^2} + \frac{1}{r}\frac{\partial C}{\partial r} + \frac{\partial^2 C}{\partial z^2} = 0 \tag{9.45}$$

in cylindrical coordinates.

In some applications, where chemical reaction occurs, the above equations are modified according to:

$$\frac{\partial^2 C}{\partial x^2} + \frac{\partial^2 C}{\partial y^2} + \frac{\partial^2 C}{\partial z^2} - \frac{k}{D_{AB}}C = 0 \tag{9.46}$$

in Cartesian coordinates, and

$$\frac{\partial^2 C}{\partial r^2} + \frac{1}{r}\frac{\partial C}{\partial r} + \frac{\partial^2 C}{\partial z^2} - \frac{k}{D_{AB}}C = 0 \tag{9.47}$$

in cylindrical coordinates, where $k$ is the reaction rate and $D_{AB}$ is the diffusion coefficient for species $A$ in medium $B$. The flux in the source plane is defined as

$$J_z = -D_{AB}\frac{\partial C}{\partial z}\bigg|_{z=0} \tag{9.48}$$

In most applications, a constant source concentration $C_0$ would be specified as opposed to constant mass flux, and in all applications, we generally specify that the region outside of the source in the source plane is impermeable (similar to the adiabatic condition in thermal transport), $\frac{\partial C}{\partial z} = 0$. This leads to the issue of a mixed-boundary value problem for these applications. We only consider a small number of solutions in this section. Crank (1975) discusses the problem of the constant concentration source on a half-space, while Plawsky (2001) considers the two-dimensional strip problem with chemical reaction. We consider several problems with and without chemical reaction.

### 9.2.1 Mass Transport from a Circular Source on a Half-Space

Given the similarities between heat conduction and steady mass diffusion in a solid, we might wonder about problems involving mass transport from discrete sources. For example, how does one model the diffusion from a chemical spill on the surface of the earth, or given a medicinal patch how does the medicinal ingredient diffuse through the skin.

In the case of a finite circular region of constant concentration where the system can be modeled as a half-space, we consider the following problem [Crank (1975)]:

$$\frac{\partial^2 C}{\partial r^2} + \frac{1}{r}\frac{\partial C}{\partial r} + \frac{\partial^2 C}{\partial z^2} = 0 \tag{9.49}$$

subject to the following boundary conditions

$$\begin{aligned}
C &= C_0 \quad z = 0 \ r \leq a \\
\frac{\partial C}{\partial z} &= 0 \quad z = 0 \ r \geq a \\
C &= 0 \quad r \geq 0 \ z \to \infty \\
C &= 0 \quad z \geq 0 \ r \to \infty
\end{aligned} \tag{9.50}$$

We have dropped the usual subscripts associated with $C$ for convenience. This problem is a mixed-boundary value problem, since both the concentration and gradient are specified along $z = 0$. Mixed-boundary value problems in potential theory pose some difficulties. More information on these types of problems can be found in Sneddon (1966). However, we have seen that analytical solution to the problem for a circular or elliptic contact spot can be found.

The solution is more easily found by using oblate spheroidal coordinates, since in this system of coordinates, the problem is axisymmetric and is actually one-dimensional, Sneddon (1966). The issue of the mixed-boundary condition is naturally resolved in this coordinate system. Fortunately, this is a well-known solution in potential theory. Since we are mainly interested in the relationship between the constant concentration of a circular spot and the molar flux into the half space, we give the solution as reported in Crank (1975) and leave the actual solution as an exercise referring the reader to Chapter 2. The solution similar to that of an isothermal source is

$$C = \frac{2}{\pi}C_0 \int_0^\infty e^{-\lambda z} J_0(\lambda r)\sin(\lambda a)\frac{d\lambda}{\lambda} \tag{9.51}$$

The mass flux over the source area is found to be

$$J_z = -D_{AB} \frac{\partial C}{\partial z}\Big|_{z=0} = \frac{2D_{AB}C_0}{\pi a} \frac{1}{\sqrt{1 - r^2/a^2}} \tag{9.52}$$

To see how the mass spreading problem relates to the thermal spreading problem, we must calculate the total mass flow from the mass flux. Integrating the mass flux over the surface of the concentrated region gives the mass flow rate:

$$\dot{m} = \iint_A J_z \, dA = 4a \, D_{AB} \, C_0 \tag{9.53}$$

where the units of mass in $\dot{m}$ are those used to define the surface concentration, i.e. mol, grams, etc., or in terms of a resistance, we may write

$$R_s = \frac{C_0}{\dot{m}} = \frac{1}{4D_{AB}\, a} \tag{9.54}$$

or nondimensionalzing:

$$4D_{AB}aR_s = 1 \tag{9.55}$$

This is precisely the same as the result reported in Chapter 2 for an isothermal circular heat source except that the thermal conductivity $k$ is replaced by the diffusion coefficient $D_{AB}$. The mean mass flux over the source region is defined as

$$\bar{J} = \frac{\dot{m}}{A} = \frac{4aD_{AB} \, C_0}{\pi a^2} = \frac{4D_{AB}C_0}{\pi a} \tag{9.56}$$

Mass transport problems analogous to the thermal spreading problems are not generally discussed in texts on mass transport. The solution given above is the only such application discussed in the classic text of Crank (1975).

## 9.2.2 Diffusion from Other Source Shapes

We may apply the principles of thermal spreading to mass diffusion problems if we define the problems in an appropriate manner. Many of the problems we have examined for thermal spreading resistance were specified in terms of a constant heat flux over the source region. Therefore, in the mass transfer analogy, the constant heat flux corresponds to a constant mass flux, $J_z$, being prescribed over a portion of the surface. In reality, we are more likely to see a constant concentration, $C_0$, applied to a finite spot, similar to the example above. As such, the solutions for steady-state thermal spreading from an isothermal contact can be converted to mass spreading problems using the following conversions:

$$
\begin{aligned}
\theta &\to C \\
\theta_0 &\to C_0 \\
k &\to D_{AB} \\
q &\to J_z \\
Q &\to \dot{m}
\end{aligned}
\tag{9.57}
$$

If the mass diffusion source is rectangular or elliptical in shape, the solutions from Chapter 2 are easily converted.

Finally, for transient problems, the thermal diffusivity $\alpha \to D_{AB}$ such that the Fourier number becomes $Fo = D_{AB}t/\mathcal{L}^2$. In mass diffusion problems, this is equivalent to setting $\rho C_p = 1$, as shown by Crank (1975). The models developed in Chapter 7 from the solution for an isothermal sphere in full space and the hemisphere in a half-space may be also applied in mass diffusion applications with constant concentration sources.

#### 9.2.2.1 Elliptic Source
The solution for the elliptic source of major and minor axes of $2a$ and $2b$, respectively, becomes

$$D_{AB} \sqrt{A}\, R_s = \frac{\sqrt{\epsilon}}{2\sqrt{\pi}} K(\kappa) \tag{9.58}$$

where $\epsilon = b/a$ and $K(\kappa)$ is the complete elliptic integral of the first kind of modulus $\kappa = \sqrt{1 - \epsilon^2}$.

#### 9.2.2.2 Rectangular Source
The solution for the rectangular source of major and minor axes of $2a$ and $2b$, respectively, becomes

$$D_{AB} \sqrt{A}\, R_s = \sqrt{\frac{1}{\epsilon}} \left[ 0.06588 - \frac{0.00232}{\epsilon} + \frac{0.6786}{1/\epsilon + 0.8145} \right] \tag{9.59}$$

As shown in Chapter 2, the equation for the elliptic source may be used to adequately model other irregular-shaped sources for which no solutions are readily available.

## 9.3 Mass Diffusion with Chemical Reaction

Several problems involving mass transfer with chemical reaction into finite regions will now be considered. In these problems, the reaction rate within the domain is proportional to the internal concentration at a point. Applications of this problem are in the area of pharmaceutical patches where the ingredient reacts within a layer of skin. We model these as finite regions with $C = 0$ in remote regions. Thus, a small patch can be modeled acting over a larger but still finite area. These problems by their very nature are a mixed-boundary value problem in the source plane. For two-dimensional problems in Cartesian or cylindrical coordinates, we will avail of modeling the source using an equivalent flux distribution as we did in Chapters 3 and 4, while in the three-dimensional case, we will consider using a least squares approach for finding the final coefficients. We will examine a circular source, a two-dimensional strip [Plawsky (2001)], and a finite rectangular source [Etminan and Muzychka (2021)]. A typical patch configuration is illustrated in Figure 9.4.

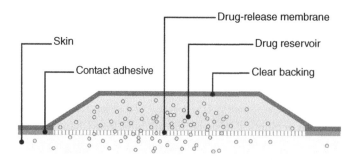

**Figure 9.4** Typical pharmaceutical patch.

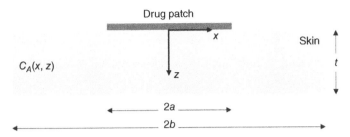

**Figure 9.5** Two-dimensional strip model.

### 9.3.1 Diffusion from a 2D Strip Source with Chemical Reaction

The simplest problem for mass diffusion with chemical reaction is the two-dimensional strip shown in Figure 9.5 which can be posed as follows:

$$\frac{\partial^2 C}{\partial x^2} + \frac{\partial^2 C}{\partial z^2} - \frac{k}{D_{AB}} C = 0 \tag{9.60}$$

subject to the following boundary conditions:

$$
\begin{aligned}
\frac{\partial C}{\partial x} &= 0 & x &= 0 & 0 &\leq z \leq t \\
C &= 0 & x &= \pm b & 0 &\leq z \leq t \\
C &= C_0 & z &= 0 & -a &\leq x \leq a \\
\frac{\partial C}{\partial z} &= 0 & z &= 0 & x &\geq a \ \text{and} \ x \leq -a \\
\frac{\partial C}{\partial z} &= 0 & z &= t & -b &\leq x \leq b
\end{aligned}
\tag{9.61}
$$

The strip subjects the region of interest with a constant concentration $C_0$ which will require us to address the mixed-boundary condition as we have done in Chapter 4. In these types of applications, the domain is assumed to have a finite thickness which is impermeable to the diffusing species along the bottom surface.

The above problem is discussed in Plawsky (2001). However, the author does not apply the proper boundary condition along $x = 0$ by specifying that $C = 0$ rather than $\partial C/\partial x = 0$, which must be applied at a plane of symmetry. Further complicating matters, the author incorrectly applied a Fourier series expansion for the mixed-boundary condition along $z = 0$ by neglecting the impermeable region outside of the source region for this problem. Unlike the two-dimensional thermal spreading problem, the regions far removed from the source (large $x$ values) have zero concentration specified rather than an impermeable boundary condition. This means that the problem will always remain multidimensional as the mass flux will never become uniform at larger contact ratios, i.e. $a/b \to 1$.

To find a useful solution, we first assume that a separation of variables solution of the form $C(x, z) = X(x) \cdot Z(z)$ is possible. Upon separating variables and applying the $x$-boundary conditions and the $z = t$ boundary condition, the correct solution for the concentration field should be

$$C(x, z) = \sum_{n=1}^{\infty} A_n \cos(\lambda_n x)[\cosh(\beta_n z) - \tanh(\beta_n t)\sinh(\beta_n z)] \tag{9.62}$$

where

$$\lambda_n = \frac{(2n-1)\pi}{2b} \tag{9.63}$$

and

$$\beta_n = \sqrt{\lambda_n^2 + k/D_{AB}} \tag{9.64}$$

In order to find the final Fourier coefficient, we must apply the surface condition. This requires that

$$\sum_{n=1}^{\infty} A_n \cos(\lambda_n x) = C_0, \quad 0 < x < a \tag{9.65}$$

and

$$\sum_{n=1}^{\infty} A_n \cos(\lambda_n x) \beta_n \tanh(\beta_n t) = 0, \quad a < x \le b \tag{9.66}$$

Since the above also constitutes a mixed-boundary value problem, we must choose an alternate solution path. Taking a cue from Chapter 4, we may approximate the constant concentration with a distributed mass flux over the contact area in the form of

$$-D_{AB} \frac{\partial C}{\partial z} = \frac{2\bar{J}}{\pi \sqrt{1 - x^2/a^2}} \quad z = 0 \quad 0 < x \le a$$

$$\frac{\partial C}{\partial z} = 0 \qquad z = 0 \quad a < x \le b \tag{9.67}$$

where $\bar{J}$ is the mass flux per unit depth.

The Fourier coefficient is found from

$$A_n = \frac{\dfrac{2\bar{J}}{\pi D_{AB}} \displaystyle\int_0^a (1 - x^2/a^2)^{-1/2} \cos(\lambda_n x) dx}{\beta_n \tanh(\beta_n t) \displaystyle\int_0^b \cos^2(\lambda_n x) dx} \tag{9.68}$$

or

$$A_n = \frac{2\bar{J}a}{D_{AB}} \frac{J_0(\lambda_n a)}{(\beta_n b) \tanh(\beta_n t)} \tag{9.69}$$

The surface distribution is given by

$$C(x, 0) = \frac{2\bar{J}a}{D_{AB}} \sum_{n=1}^{\infty} \frac{\cos(\lambda_n x) J_0(\lambda_n a)}{(\beta_n b) \tanh(\beta_n t)} \tag{9.70}$$

The mean source concentration is found by integrating over the source area ($-a < x < a$), and gives

$$\bar{C} = \sum_{n=1}^{\infty} A_n \frac{\sin(\lambda_n a)}{(\lambda_n a)} = \frac{2\bar{J}a}{D_{AB}} \sum_{n=1}^{\infty} \frac{J_0(\lambda_n a) \sin(\lambda_n a)}{(\beta_n b) \tanh(\beta_n t)(\lambda_n a)} \tag{9.71}$$

The resulting spreading resistance (per unit depth) becomes

$$R_s = \frac{\bar{C}}{\dot{m}} = \frac{\bar{C}}{2a\bar{J}} = \frac{1}{D_{AB}} \sum_{n=1}^{\infty} \frac{J_0(\lambda_n a) \sin(\lambda_n a)}{(\beta_n b) \tanh(\beta_n t)(\lambda_n a)} \tag{9.72}$$

If we nondimensionalize as before using $\epsilon = a/b$, $\tau = t/b$, $Da = kb^2/D_{AB}$, and $\delta_n = \lambda_n b$, we obtain:

$$R_s^\star = R_s D_{AB} = \sum_{n=1}^{\infty} \frac{J_0(\delta_n \epsilon) \sin(\delta_n \epsilon)}{\sqrt{\delta_n^2 + Da} \tanh\left(\sqrt{\delta_n^2 + Da}\,\tau\right)(\delta_n \epsilon)} \tag{9.73}$$

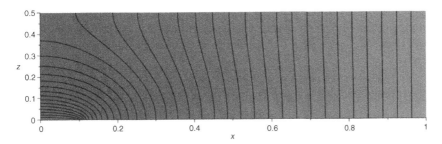

**Figure 9.6** Dimensionless concentration field for $X = x/b$, $Z = z/b$, $\epsilon = 0.1$, $\tau = 0.5$, and $Da = 0$ for a strip source.

where $Da$ is the Damkohler number. The above result contains the following special limit when $Da \to 0$:

$$R_s^{\star} = R_s D_{AB} = \sum_{n=1}^{\infty} \frac{J_0(\delta_n \epsilon) \sin(\delta_n \epsilon)}{\tanh(\delta_n \tau) \delta_n^2 \epsilon} \tag{9.74}$$

This result models the lateral spreading of the diffusing species from an equivalent constant concentration strip. The field is plotted in Figure 9.6 for the special case of $Da = 0$ for a small contact zone in a finite domain. As noted earlier, the spreading field will always be two-dimensional since the diffusing species must turn due to the impermeable boundary at $z = t$ regardless of source aspect ratio, hence, the spreading resistance in this problem will never approach zero.

### 9.3.2 Circular Source on a Disk with Chemical Reaction

A more practical problem may be specified in cylindrical coordinates. This problem as before has practical applications in transdermal diffusion with and without chemical reaction. We will model the system using a finite circular source of constant concentration $C_0$. The general solution will also allow the special case of no chemical reaction to be considered when the reaction constant $k = 0$, i.e. $Da \to 0$.

The diffusion equation including the chemical reaction term and boundary conditions are

$$\frac{\partial^2 C}{\partial r^2} + \frac{1}{r} \frac{\partial C}{\partial r} + \frac{\partial^2 C}{\partial z^2} - \frac{k}{D_{AB}} C = 0 \tag{9.75}$$

subject to the following boundary conditions:

$$\begin{aligned}
\frac{\partial C}{\partial r} &= 0 & r &= 0 & 0 &\leq z \leq t \\
C &= 0 & r &= b & 0 &\leq z \leq t \\
C &= C_0 & z &= 0 & 0 &\leq r \leq a \\
\frac{\partial C}{\partial z} &= 0 & z &= 0 & a &< r \leq b \\
\frac{\partial C}{\partial z} &= 0 & z &= t & 0 &\leq r \leq b
\end{aligned} \tag{9.76}$$

The boundary condition along $r = 0$ models the axis of symmetry, while the condition at the edge of the disk models the lack of the diffusing species near the edge of the region of interest. In this case, we have the freedom to make $b \gg a$ to model a semi-infinite disk if we wish. On the surface, we prescribe a constant concentration over the area of the patch and an impermeable condition along the remainder of the surface. Finally, the condition imposed at $z = t$ denotes the fact that the diffusing species does not penetrate the lower surface. Recognize that once again we have a mixed-boundary problem along $z = 0$. This will pose some problems for the Fourier–Bessel expansion, but we will cross that bridge later. For now, we proceed with the traditional separation of variables approach.

To find a solution, we first assume that a separated solution of the form $C(r, z) = R(r) \cdot Z(z)$ is possible. Applying the procedure results in the following general solution:

$$C(r, z) = [A_1 J_0(\lambda r) + B_1 Y_0(\lambda r)][A_2 \cosh(\beta z) + B_2 \sinh(\beta z)] \tag{9.77}$$

where

$$\beta = \sqrt{\lambda^2 + k/D_{AB}} \tag{9.78}$$

Application of the boundary condition at $r = 0$ requires that $B_1 = 0$, since the solution must remain finite along the disk axis, while application of the condition along $r = b$ requires that

$$J_0(\lambda b) = 0 \tag{9.79}$$

which yields the eigenvalues for the system, i.e.

$$\lambda_n b = 2.404826, 5.520078, 8.653728, 11.791534 \ldots \tag{9.80}$$

Thus, the solution now becomes

$$C(r, z) = \sum_{n=1}^{\infty} J_0(\lambda_n r)[A_n \cosh(\beta_n z) + B_n \sinh(\beta_n z)] \tag{9.81}$$

Application of the boundary condition at $z = t$ requires the following result:

$$B_n = -A_n \tanh(\beta_n t) \tag{9.82}$$

giving:

$$C(r, z) = \sum_{n=1}^{\infty} A_n J_0(\lambda_n r)[\cosh(\beta_n z) - \tanh(\beta_n t) \sinh(\beta_n z)] \tag{9.83}$$

In order to find the final Fourier–Bessel coefficient, we must apply the surface condition. This requires that

$$\sum_{n=1}^{\infty} A_n J_0(\lambda_n r) = C_0, \quad 0 < r < a \tag{9.84}$$

and

$$\sum_{n=1}^{\infty} A_n J_0(\lambda_n r)\beta_n \tanh(\beta_n t) = 0, \quad a < r \le b \tag{9.85}$$

As a result of the mixed-boundary condition, we cannot complete the solution unless we seek an alternative approach such as using a least squares closure. But that will be more challenging given the Bessel function $J_0(\cdot)$. As before, we will define an equivalent flux profile which yields a constant mean surface concentration. This requires that we apply the following alternate form of the surface condition:

$$-D_{AB} \frac{\partial C}{\partial z} = \frac{\bar{J}/2}{\sqrt{1 - r^2/a^2}} \quad z = 0 \quad r \le a$$

$$\frac{\partial C}{\partial z} = 0 \quad z = 0 \quad a \le r \le b \tag{9.86}$$

Using the above surface condition, we can now write the solution to the Fourier–Bessel coefficient as

$$A_n = \frac{\dfrac{\overline{J}}{2D_{AB}} \displaystyle\int_0^a (1 - r^2/a^2)^{-1/2} r J_0(\lambda_n r)\,dr}{\beta_n \tanh(\beta_n t) \displaystyle\int_0^b r J_0^2(\lambda_n r)\,dr} \tag{9.87}$$

Evaluating the integrals gives

$$A_n = \frac{\overline{J}}{D_{AB}(\beta_n b)\tanh(\beta_n t)} \frac{a\sin(\lambda_n a)}{(\lambda_n b)J_1^2(\lambda_n b)} \tag{9.88}$$

The surface concentration is now found to be

$$C(r,0) = \frac{\overline{J}a}{D_{AB}} \sum_{n=1}^{\infty} \frac{\sin(\lambda_n a)J_0(\lambda_n r)}{(\beta_n b)\tanh(\beta_n t)(\lambda_n b)J_1^2(\lambda_n b)} \tag{9.89}$$

Integration over the contact spot now gives the mean source concentration in terms of the mean flux:

$$\overline{C} = \frac{2\overline{J}a}{D_{AB}} \sum_{n=1}^{\infty} \frac{\sin(\lambda_n a)J_1(\lambda_n a)}{(\beta_n b)\tanh(\beta_n t)(\lambda_n b)(\lambda_n a)J_1^2(\lambda_n b)} \tag{9.90}$$

Finally, defining a resistance such that

$$R_s = \frac{\overline{C}}{\dot{m}} = \frac{\overline{C}}{\overline{J}\pi a^2} \tag{9.91}$$

and introducing the dimensionless groups: $\epsilon = a/b$, $\tau = t/b$, $Da = kb^2/D_{AB}$, $\delta_n = \lambda_n b$, and $R_s^\star = R_s D_{AB} a$, we obtain

$$R_s^\star = \frac{2}{\pi\epsilon} \sum_{n=1}^{\infty} \frac{\sin(\delta_n\epsilon)J_1(\delta_n\epsilon)}{\sqrt{\delta_n^2 + Da}\,\delta_n^2 \tanh\left(\sqrt{\delta_n^2 + Da}\,\tau\right)J_1^2(\delta_n)} \tag{9.92}$$

The group $Da$ is called the Damkohler number. For large Damkohler numbers, the process is reaction dominant, while for small values, it is diffusion dominant. The above result contains the following special limit when $Da \to 0$:

$$R_s^\star = \frac{2}{\pi\epsilon} \sum_{n=1}^{\infty} \frac{\sin(\delta_n\epsilon)\,J_1(\delta_n\epsilon)}{\delta_n^3 \tanh(\delta_n\,\tau)\,J_1^2(\delta_n)} \tag{9.93}$$

This result models the lateral spreading of the diffusing species from an equivalent constant concentration spot. The solution is plotted in Figure 9.7 for several values of the source aspect ratio and Damkohler number. Note that due to the nature of the inverse parabolic flux profile which was assumed, the contact is only representative of a constant concentration spot for $\epsilon < 0.4$ as we have seen in Chapter 3.

### 9.3.3 Diffusion from a Rectangular Source with Chemical Reaction

We conclude by considering the problem for a three-dimensional rectangular contact region which can be posed as follows:

$$\frac{\partial^2 C}{\partial x^2} + \frac{\partial^2 C}{\partial y^2} + \frac{\partial^2 C}{\partial z^2} - \frac{k}{D_{AB}}C = 0 \tag{9.94}$$

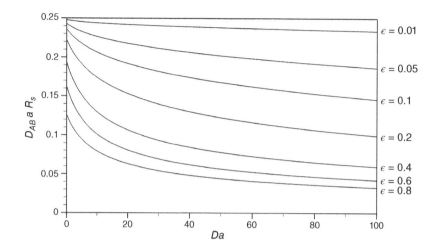

**Figure 9.7** Mass diffusion from a circular source on a semi-infinite flux tube region, $\tau \to \infty$, for $\epsilon = 0.01, 0.05, 0.1, 0.2, 0.4, 0.6$, and $0.8$ as a function of Damkohler number.

subject to the following boundary conditions:

$$\frac{\partial C}{\partial x} = 0, \quad x = 0 \quad 0 \leq z \leq t$$

$$\frac{\partial C}{\partial y} = 0, \quad y = 0 \quad 0 \leq z \leq t$$

$$C = 0, \quad x = c \quad 0 \leq z \leq t$$

$$C = 0, \quad y = d \quad 0 \leq z \leq t \tag{9.95}$$

$$C = C_0, \quad z = 0 \quad 0 \leq x < a, \quad 0 \leq y < b$$

$$\frac{\partial C}{\partial z} = 0, \quad z = 0 \quad a < x \leq c, \quad b < y \leq d$$

$$\frac{\partial C}{\partial z} = 0, \quad z = t \quad 0 \leq x \leq c, \quad 0 \leq y \leq d$$

The solution using separation of variables $C(x, y, z) = X(x) * Y(y) * Z(z)$ can be found to be [Etminan and Muzychka (2021)]:

$$C(x, y, z) = \sum_{m=1}^{\infty} \sum_{n=1}^{\infty} A_{mn} \cos(\lambda_m x) \cos(\delta_n y)[\cosh(\beta_{mn} z) - \tanh(\beta_{mn} t) \sinh(\beta_{mn} z)] \tag{9.96}$$

The above result satisfies the $x$-, $y$-, and $z$-boundary conditions (except the $z = 0$ condition), provided that the eigenvalues are defined as $\lambda_m = (2m - 1)\pi/2c$, $\delta_n = (2n - 1)\pi/2d$, and $\beta_{mn} = \sqrt{\lambda_m^2 + \delta_n^2 + k/D_{AB}}$.

In order to find the final Fourier coefficient, we must apply the surface condition. This requires that

$$\sum_{m=1}^{\infty} \sum_{n=1}^{\infty} A_{mn} \cos(\lambda_m x) \cos(\delta_n y) = C_0, \quad 0 < x < a, \quad 0 < y < b \tag{9.97}$$

and

$$\sum_{m=1}^{\infty}\sum_{n=1}^{\infty}A_{mn}\cos(\lambda_m x)\cos(\delta_n y)\beta_{mn}\tanh(\beta_{mn}t)=0, \quad a<x<c, \quad b<y<d \tag{9.98}$$

As we have seen earlier, this constitutes a mixed-potential boundary value problem. Unlike in the two-dimensional problem where we could use an equivalent flux distribution to approximate the constant concentration $C_0$, we must now resort to a least squares method to solve for the $A_{mn}$. This requires defining the following integral:

$$I = \int_0^a \int_0^b \left[ \sum_{m=1}^{\infty}\sum_{n=1}^{\infty}A_{mn}\cos(\lambda_m x)\cos(\delta_n y) - C_0 \right]^2 dx\, dy$$

$$+ \int_a^c \int_b^d \left[ \sum_{m=1}^{\infty}\sum_{n=1}^{\infty}A_{mn}\cos(\lambda_m x)\cos(\delta_n y)\beta_{mn}\tanh(\beta_{mn}t) \right]^2 dx\, dy \tag{9.99}$$

The coefficients are determined by solving the system of equations defined by

$$\frac{dI}{dA_{mn}}=0, \quad m=1\dots M, \quad n=1\dots N \tag{9.100}$$

This results in an $M \times N$ set of equations which can be solved for the $A_{mn}$. The process is easily implemented in most mathematical software tools. We can now determine the resistance from

$$R_s = \frac{C_0}{\dot{m}} \tag{9.101}$$

where

$$\dot{m} = \iint_A J\, dA = \iint_A \left[ -D_{AB}\frac{\partial C}{\partial z}\Big|_{z=0} \right] dA \tag{9.102}$$

or

$$\dot{m} = 4\int_0^a \int_0^b D_{AB} \sum_{m=1}^{\infty}\sum_{n=1}^{\infty}A_{mn}\cos(\lambda_m x)\cos(\delta_n y)\beta_{mn}\tanh(\beta_{mn}t)dx\, dy \tag{9.103}$$

Completing the integration leads to

$$\dot{m} = 4D_{AB}\sum_{m=1}^{\infty}\sum_{n=1}^{\infty}\frac{A_{mn}\sin(\lambda_m a)\sin(\delta_n b)\beta_{mn}\tanh(\beta_{mn}t)}{\lambda_m \delta_n} \tag{9.104}$$

For simplicity, we will now consider a square source of dimensions $a = b$ and a square domain of $c = d$. This leads to the eigenvalues being defined as: $\lambda_m = (2m-1)\pi/2c$, $\delta_n = (2n-1)\pi/2c$, and $\beta_{mn} = \sqrt{\lambda_m^2 + \delta_n^2 + k/D_{AB}}$. Further if we use $\phi = C/C_0$, $X = x/c$, $Y = y/c$, and $Z = z/c$ in Eq. (9.96), we may write:

$$I = \int_0^\epsilon \int_0^\epsilon \left[ \sum_{m=1}^{\infty}\sum_{n=1}^{\infty}A'_{mn}\cos((\lambda_m c)X)\cos((\delta_n c)Y) - 1 \right]^2 dX\, dY$$

$$+ \int_\epsilon^1 \int_\epsilon^1 \left[ \sum_{m=1}^{\infty}\sum_{n=1}^{\infty}A'_{mn}\cos((\lambda_m c)X)\cos((\delta_n c)Y)(\beta_{mn}c)\tanh((\beta_{mn}c)\tau) \right]^2 dX\, dY \tag{9.105}$$

which can now be solved for the dimensionless coefficients $A'_{mn} = A_{mn}/C_0$ using Eq. (9.100):

$$\frac{dI}{dA'_{mn}}=0, \quad m=1\dots M, \quad n=1\dots N \tag{9.106}$$

Finally, if we introduce the additional dimensionless variables: $\tau = t/c$, $\epsilon = a/c$, and $Da = kc^2/D_{AB}$, we can show that the dimensionless resistance can be written as

$$D_{AB} \, aR_s = \cfrac{C_0}{\cfrac{4}{\epsilon} \displaystyle\sum_{m=1}^{\infty} \sum_{n=1}^{\infty} \cfrac{A_{mn} \sin((\lambda_m c)\epsilon) \, \sin((\delta_n c)\epsilon)(\beta_{mn}c) \tanh((\beta_{mn}c)\tau)}{(\lambda_m c)(\delta_n c)}} \tag{9.107}$$

once the coefficients $A_{mn} = C_0 A'_{mn}$ are determined.

## 9.4 Diffusion Limited Slip Behavior: Super-Hydrophobic Surfaces

In super-hydrophobic channels and surfaces, a fluid flows over a surface composed of micro-engineered pillars or ridges [Ng and Wang (2009a,b), Sbragaglia and Prosperetti (2007a,b), Lauga and Stone (2003), Ybert et al. (2007)]. The liquid is supported by these ridges and pillars leaving a gas (usually air) trapped in the regions between the pillars or ridges. This gives rise to an apparent slip phenomena that leads to a reduction of surface friction relative to a smooth surface. The liquid experiences both viscous and thermal transport through finite points of contact with the pillar surfaces. Depending upon the spacing of the pillars either a half-space contact is formed for large spacings or flux tube/channel contacts for smaller pillar spacings. In the limit of very small Peclet and Reynolds number, the process is diffusion limited by Laplace's equation for thermal transport and Stokes's equation for viscous transport.

Enright et al. (2014) showed that both the thermal and hydrodynamic slip lengths can be obtained from expressions for the thermal spreading resistance for isothermal contacts, i.e. Dirichlet-type boundary conditions. Enright et al. (2014) developed their relationships by using simple scaling arguments applied to a square pillar array. Each pillar of side dimension $w$ forms a square flux channel having a cell dimension of $\ell$ as shown in Figure 9.8.

The heat flux over the source into the liquid region in contact with the pillar surface scales according to

$$q_s \sim k\frac{T_s - T_b}{w} \sim \frac{Q}{w^2} \tag{9.108}$$

while for the nominal channel cross-section the heat flux scales according to

$$q_c \sim \phi_s k\frac{T_s - T_b}{w} \sim \frac{Q}{\ell^2} \tag{9.109}$$

where $\phi_s = A_s/A_t = w^2/\ell^2$ is the area contact ratio. Thus, the pillar aspect ratio is $\epsilon = \sqrt{\phi_s}$.

The thermal slip boundary condition is defined as

$$\overline{T}_s - \overline{T}_c = -b_t \frac{\partial \overline{T}}{\partial n}\bigg|_c \tag{9.110}$$

where the thermal slip length $b_t$ can be written as

$$b_t = \frac{\overline{T}_s - \overline{T}_c}{-\dfrac{\partial \overline{T}}{\partial n}\bigg|_c} \sim \frac{T_s - T_c}{\phi_s(T_s - T_c)/w} \sim \frac{w}{\phi_s} \sim \frac{\ell}{\sqrt{\phi_s}} \tag{9.111}$$

**Figure 9.8** Flux channel formed from a square pillar cell in a super-hydrophobic channel. Source: Adapted from Enright et al. (2014).

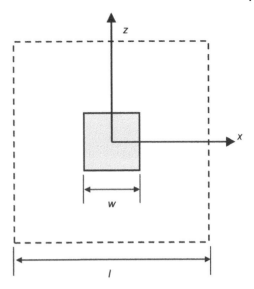

Similarly, the hydrodynamic slip length defined as [Ybert et al. (2007)]:

$$\overline{u}_c = b \left. \frac{\partial \overline{u}}{\partial n} \right|_c \tag{9.112}$$

or

$$b = \frac{\overline{u}_c}{\left. \dfrac{\partial \overline{u}}{\partial n} \right|_c} \sim \frac{U}{\phi_s(U/w)} \sim \frac{w}{\phi_s} \sim \frac{\ell}{\sqrt{\phi_s}} \tag{9.113}$$

Finally, from the above scaling, we deduce that the slip lengths have the same mathematical form, but will differ in their magnitudes requiring their ratio to be a constant [Enright et al. (2014)]:

$$\frac{b}{b_t} = C \tag{9.114}$$

This means that if we know one of the slip lengths, the other is easily deduced. Returning to the thermal slip condition equation (9.110), we can rewrite as

$$\frac{\overline{T}_s - \overline{T}_c}{Q} = -\frac{b_t}{Q} \left. \frac{\partial \overline{T}}{\partial n} \right|_c = \frac{b_t}{Q} \left( \frac{Q}{k\ell^2} \right) = \frac{b_t}{\ell} \frac{\sqrt{\phi_s}}{kw} \tag{9.115}$$

The left-hand side of the above equation is simply the thermal spreading resistance, thus allowing us to write:

$$\frac{b_t}{\ell} = \frac{R_s kw}{\sqrt{\phi_s}} \tag{9.116}$$

for the square pillar.

Enright et al. (2014) proceeded to develop expressions for the thermal slip length using expressions for the thermal spreading resistance for Dirichlet boundary conditions (or their approximations), and further showed that for circular and square pillars, the ratio of the slip lengths for Dirichlet conditions yields $C = 3/4$ in Eq. (9.114). This scaling factor was deduced considering the limiting cases of $\phi_s \to 0$, i.e. the half-space limit.

The thermal spreading resistance for an isothermal source on a half-space was shown in Chapter 2 to be

$$R_s = \frac{T_s - T_c}{Q} = \frac{1}{4ka} \tag{9.117}$$

In the Stokes flow limit of $Re \to 0$, the viscous drag force on a disk of radius $a$ is [Lamb (1932)]:

$$|F| = \frac{32\mu aU}{3} \tag{9.118}$$

Considering only drag on one side of the disc ($|F|/2$), and defining an analogous viscous flow resistance:

$$R_{s,v} = \frac{U}{(|F|/2)} = \frac{3}{16\mu a} \tag{9.119}$$

This leads to

$$\frac{R_{s,v}}{R_s} = \frac{3k}{4\mu} \tag{9.120}$$

Since $b_t/k \propto \Delta T/Q$ and $b/\mu \propto U/F$, we conclude that

$$b \propto \frac{3}{4}b_t \tag{9.121}$$

Enright et al. (2014) validated this approach using published data as well as their own computational data for square pillars, after developing simple expressions using the circular flux tube thermal spreading resistance expressions to model the square pillar over a range of $\phi_s$. In Section 9.4.2, we will further consider the special case of $\phi_s \to 0$ for rectangular and elliptic pillars.

## 9.4.1 Circular and Square Pillars

Using the results for thermal spreading resistance in flux tubes and channels we may now develop expressions for the hydrodynamic and thermal slip lengths for circular and square pillars, as well as ridges oriented parallel and transverse to the direction of flow. Following Enright et al. (2014) we write the slip lengths as follows for square/circular pillars and ridges.

### 9.4.1.1 Circular/Square

The spreading resistance for an equivalent isothermal source on a semi-infinite circular flux tube (see Chapter 3) is

$$k\sqrt{A}R_s = \frac{2}{\sqrt{\pi}\epsilon}\sum_{n=1}^{\infty}\frac{J_1(\delta_n\epsilon)\sin(\delta_n\epsilon)}{\delta_n^3 J_0^2(\delta_n)} \tag{9.122}$$

and, thus since $\Delta T/Q = b_t/(k\ell^2)$ we write [Enright et al. (2014)]:

$$\frac{b_t}{\ell} = \frac{2}{\sqrt{\pi}\phi_s}\sum_{n=1}^{\infty}\frac{J_1(\delta_n\sqrt{\phi_s})\sin(\delta_n\sqrt{\phi_s})}{\delta_n^3 J_0^2(\delta_n)} \tag{9.123}$$

which is valid for contact ratios $\sqrt{\phi_s} \leq 0.4$. We determine the hydrodynamic slip length from Eq. (9.121) to be

$$\frac{b}{\ell} = \frac{3}{2\sqrt{\pi}\phi_s}\sum_{n=1}^{\infty}\frac{J_1(\delta_n\sqrt{\phi_s})\sin(\delta_n\sqrt{\phi_s})}{\delta_n^3 J_0^2(\delta_n)} \tag{9.124}$$

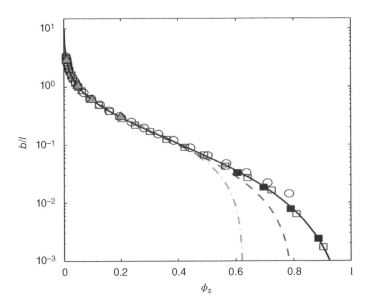

**Figure 9.9** Comparison of hydrodynamic slip length data of Enright et al. (2014) with the result of Eq. (9.126) – Solid line, Eq. (9.124) – Dashed line, and Davis and Lauga (2010) – Dash-Dot line. Source: Enright et al. (2014)/with permission of The American Society of Mechanical Engineers.

In the case of a true isothermal source, we use the following expression from Chapter 3:

$$4kaR_s = 1 - 1.40978\epsilon + 0.34406\epsilon^3 + 0.0435\epsilon^5 + 0.02271\epsilon^7 \tag{9.125}$$

which after rescaling to $k\sqrt{A}R_s$ and converting as before, yields:

$$\frac{b_t}{\ell} = \frac{\sqrt{\pi}}{4\sqrt{\phi_s}}\left(1 - 1.40978\phi_s^{1/2} + 0.34406\phi_s^{3/2} + 0.0435\phi_s^{5/2} + 0.02271\phi_s^{7/2}\right) \tag{9.126}$$

for which we determine the hydrodynamic slip length from Eq. (9.121) to be

$$\frac{b}{\ell} = \frac{3\sqrt{\pi}}{16\sqrt{\phi_s}}\left(1 - 1.40978\phi_s^{1/2} + 0.34406\phi_s^{3/2} + 0.0435\phi_s^{5/2} + 0.02271\phi_s^{7/2}\right) \tag{9.127}$$

Figure 9.9 compares the hydrodynamic slip length, Eqs. (9.127) and (9.124), with numerical data from Enright et al. (2014). It is clear that over the full range of solid ratio $\phi_s$, that Eq. (9.127) accurately models the results for a square (or circular) pillar.

### 9.4.1.2 Ridges
The spreading resistance for an equivalent isothermal source on a semi-infinite flux tube (see Chapter 3) is:

$$kR_s = \frac{1}{\epsilon\pi^2}\sum_{m=1}^{\infty}\frac{\sin(m\pi\epsilon)}{m^2}J_0(m\pi\epsilon) \tag{9.128}$$

which yields the following expression for the thermal slip length:

$$\frac{b_t}{\ell} = \frac{1}{\phi_s\pi^2}\sum_{m=1}^{\infty}\frac{\sin(m\pi\phi_s)}{m^2}J_0(m\pi\phi_s) \tag{9.129}$$

since $\phi_s = \epsilon$ for a two dimensional strip (or ridge). The above result is valid for $\phi_s < 0.4$.

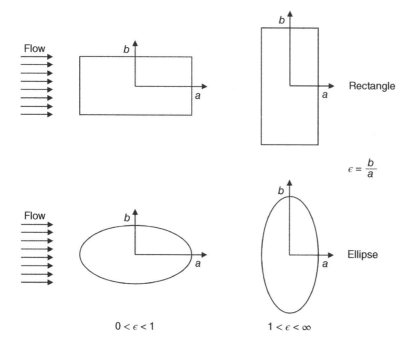

**Figure 9.10** Elliptic and rectangular pillars.

In the case of a true isothermal strip, the following result reported in Chapter 4 is applicable:

$$
\frac{b_t}{l} = \frac{1}{\pi} \ln \left[ \frac{1}{\sin\left(\frac{\phi_s \pi}{2}\right)} \right] \tag{9.130}
$$

In both cases, the hydrodynamic slip length is also equal to the thermal slip length for a ridge.

### 9.4.2 Rectangular and Elliptical Pillars for $\phi_s \to 0$

We now examine the solutions for rectangular and elliptic pillars for the case of small area ratios, i.e. the half-space limit. Enright et al. (2014) found that the ratio of the slip lengths (assuming a circular pillar) yielded $C = 3/4$ in Eq. (9.114). The data for a square pillar supported the validity of this ratio. It is now desirable to determine if this ratio holds for aspect ratios other than unity. The rectangular and elliptic pillar geometries require special care since the elongation of either shape will lead to different values of the thermal and hydrodynamic slip lengths given that viscous drag in Stokes flow is now dependent on orientation of the pillar with respect to the flow where as the equivalent thermal problem it is not (see Figure 9.10). We extend the analysis provided in Enright et al. (2014) by now considering pillars of varying aspect ratio. While the slip lengths $b, b_t \to \infty$ as the solid fraction $\phi_s \to 0$, their ratio which is a constant is what matters in this case.

We begin by considering the creeping flow solution for an elliptic plate can be obtained from the general solution of the ellipsoid. The solution for the ellipsoid can be found in both Lamb (1932) and Happel and Brenner (1965).

In the present case, we consider an ellipsoid defined with axes $[a, b, c]$ in the $[x, y, z]$ directions respectively such that

$$\frac{x^2}{a^2} + \frac{y^2}{b^2} + \frac{z^2}{c^2} = 1 \tag{9.131}$$

The flow is such that it is in the direction of the $x$-axis. The drag force on the ellipsoid is defined in terms of an equivalent spherical radius such that

$$|F| = 6\pi\mu U R \tag{9.132}$$

where

$$R = \frac{8}{3} \frac{abc}{\chi_0 + \alpha_0 a^2} \tag{9.133}$$

The factors $\chi_0$ and $\alpha_0$ are defined as

$$\chi_0 = abc \int_0^\infty \frac{d\lambda}{\sqrt{(a^2 + \lambda)(b^2 + \lambda)(c^2 + \lambda)}} \tag{9.134}$$

and

$$\alpha_0 = abc \int_0^\infty \frac{d\lambda}{(a^2 + \lambda)\sqrt{(a^2 + \lambda)(b^2 + \lambda)(c^2 + \lambda)}} \tag{9.135}$$

For the special case of an elliptic disk, we have $c \to 0$, giving

$$R = \frac{8}{3} \frac{abc}{\chi_0 + \alpha_0 a^2} \tag{9.136}$$

$$\chi_0 = abc \int_0^\infty \frac{d\lambda}{\sqrt{(a^2 + \lambda)(b^2 + \lambda)\lambda}} \tag{9.137}$$

and

$$\alpha_0 = abc \int_0^\infty \frac{d\lambda}{(a^2 + \lambda)\sqrt{(a^2 + \lambda)(b^2 + \lambda)\lambda}} \tag{9.138}$$

Further, we will define the dimensionless shear stress on the surface of the disk as

$$\tau^\star = \frac{\tau_w \sqrt{A}}{\mu U} = \frac{1}{R_{s,v}^\star} \tag{9.139}$$

where $A = \pi ab$ is the area of one side of the elliptic disk. This leads to

$$\tau^\star = \frac{(|F|/2)\sqrt{A}}{A\mu U} = \frac{3\sqrt{\pi}}{\sqrt{\epsilon_v}} \cdot \frac{R}{a} \tag{9.140}$$

where $\epsilon_v = b/a$ is the aspect ratio of the pillar such that $0 < \epsilon < \infty$. The factor $R/a$ is tabulated in Happel and Brenner (1965) for oblate and prolate bodies for both parallel and transverse flows but not for elliptic plates.

In the case of an isothermal elliptic contact on a half-space, we have seen in Chapter 2 that the dimensionless spreading resistance for the elliptic source of major and minor axes of $2a$ and $2b$ respectively, becomes

$$k\sqrt{A}\, R_s = \frac{\sqrt{\epsilon_t}}{2\sqrt{\pi}} K(\kappa) \tag{9.141}$$

where $\epsilon = b/a$ such that $0 < \epsilon_t < 1$ and $K(\kappa)$ is the complete elliptic integral of the first kind of modulus $\kappa = \sqrt{1 - \epsilon_t^2}$.

Using the analogous dimensionless wall heat flux $q^\star$, we find that

$$q^\star = \frac{q\sqrt{A}}{k\Delta T} = \frac{1}{R_s^\star} = \frac{2\sqrt{\pi}}{K(\kappa)\sqrt{\epsilon_t}} \tag{9.142}$$

Returning to the ratio of the hydrodynamic and thermal resistances, Eq. (9.120), we may now write:

$$\frac{R_{s,v}}{R_s} = \frac{b}{b_t} = \frac{q^\star}{\tau^\star} = \Phi \tag{9.143}$$

$$\Phi = \frac{\dfrac{2\sqrt{\pi}}{K(\kappa)\sqrt{\epsilon_t}}}{\dfrac{3\sqrt{\pi}}{\sqrt{\epsilon_v}} \cdot \dfrac{R}{a}} \tag{9.144}$$

Care must be taken when evaluating Eq. (9.144) since the viscous and thermal aspect ratios are defined differently.

Equation (9.144) is tabulated in Table 9.4 for various aspect ratios considering the effect of orientation of the elliptic pillar for viscous drag. In the case of a rectangular pillar, the thermal

**Table 9.4** Dimensionless parameters for an elliptic contact (theory) and rectangular contact (numerical).

| $\epsilon_v$ | $\tau^\star$ Theory (ellip.) | $\tau^\star$ Numerical (rect.) | $\epsilon_t$ | $q^\star$ Theory (ellip.) | $q^\star$ Numerical (rect.) | $\Phi$ Theory (ellip.) | $\Phi$ Numerical (rect.) |
|---|---|---|---|---|---|---|---|
| 0.1 | 3.502 | 3.511 | 0.1 | 3.033 | 2.977 | 0.866 | 0.848 |
| 0.2 | 3.131 | 3.134 | 0.2 | 2.628 | 2.648 | 0.839 | 0.845 |
| 0.3 | 3.003 | 3.008 | 0.3 | 2.463 | 2.503 | 0.820 | 0.832 |
| 0.4 | 2.951 | 2.977 | 0.4 | 2.376 | 2.400 | 0.805 | 0.806 |
| 0.5 | 2.934 | 2.985 | 0.5 | 2.325 | 2.371 | 0.792 | 0.794 |
| 0.6 | 2.935 | 2.986 | 0.6 | 2.294 | 2.353 | 0.782 | 0.788 |
| 0.7 | 2.946 | 3.010 | 0.7 | 2.275 | 2.312 | 0.772 | 0.768 |
| 0.8 | 2.963 | 3.031 | 0.8 | 2.264 | 2.305 | 0.764 | 0.761 |
| 0.9 | 2.985 | 3.058 | 0.9 | 2.258 | 2.309 | 0.757 | 0.755 |
| 1.0 | 3.009 | 3.105 | 1.0 | 2.257 | 2.313 | 0.750 | 0.745 |
| 2 | 3.285 | 3.293 | 1/2 | 2.325 | 2.371 | 0.708 | 0.720 |
| 3 | 3.543 | 3.464 | 1/3 | 2.428 | 2.452 | 0.685 | 0.708 |
| 4 | 3.773 | 3.676 | 1/4 | 2.531 | 2.553 | 0.671 | 0.694 |
| 5 | 3.979 | 3.877 | 1/5 | 2.628 | 2.648 | 0.661 | 0.683 |
| 6 | 4.165 | 4.055 | 1/6 | 2.719 | 2.710 | 0.653 | 0.668 |
| 7 | 4.337 | 4.228 | 1/7 | 2.805 | 2.787 | 0.647 | 0.659 |
| 8 | 4.496 | 4.381 | 1/8 | 2.885 | 2.835 | 0.642 | 0.647 |
| 9 | 4.646 | 4.527 | 1/9 | 2.961 | 2.907 | 0.637 | 0.642 |
| 10 | 4.786 | 4.665 | 1/10 | 3.033 | 2.977 | 0.634 | 0.638 |

spreading resistance for an isothermal source of major and minor axes of $2a$ and $2b$, respectively, becomes:

$$k\sqrt{A}\,R_s = \sqrt{\frac{1}{\epsilon_t}}\left[0.06588 - \frac{0.00232}{\epsilon_t} + \frac{0.6786}{1/\epsilon_t + 0.8145}\right] \qquad (9.145)$$

A correlation for the analogous expression for viscous drag does not appear to exist in the literature, as such we may approximate the solution for the hydrodynamic slip length using the elliptic pillar as a guide, similar to what was done by Enright et al. (2014). In other words,

$$\left(\frac{b}{b_t}\right)_{Rectangle} \approx \Phi_{Elliptic} \qquad (9.146)$$

The factor $\Phi$ for both elliptic and rectangular pillars was determined by Kane (2016) using numerical methods for the rectangular pillar and theory for the elliptic pillar and is tabulated below in Table 9.4 as a function of $\epsilon_v$ and $\epsilon_t$ along with $\tau^\star$ and $q^\star$ for each case. It is easily seen that the results are approximately equal for both shapes at all aspect ratios. For the special case of $\epsilon_t = \epsilon_v = 1$ for the square and circular pillars, we see that the factor is almost the same as that derived for the circular plate. Note that the plate aspect ratio can vary in both directions for the viscous flow, i.e. $0 < \epsilon_v < \infty$, while for the thermal spreading problem, it is such that $0 < \epsilon_t < 1$. While for $\Phi \approx 3/4$ for square and circular pillars, we find that $\Phi$ varies $\pm 15\%$ with aspect ratio over the range considered which represents pillars varying in aspect ratios of $1\,:\,1$ to $10\,:\,1$ in both flow orientations.

### 9.4.3 Effect of Meniscus Curvature

In Sections 9.4.1 and 9.4.2, the effect of meniscus curvature was not considered [Enright et al. (2014), Kane (2016)]. Rather, the fluid lay flat across the ridges and pillars. The effect of meniscus curvature can have a significant effect on the thermal spreading resistance (and hence, the analogous slip lengths) as shown by Hodes et al. (2018) and Mayer et al. (2019). Hodes et al. (2018) considered the effect of a curved meniscus for parallel ridges. Mayer et al. (2019) considered the solution to the axi-symmetric meniscus boundary for a cylindrical pillar. The key results of these studies for thermal spreading resistance are discussed in Chapters 3 and 4.

## 9.5 Problems with Phase Change in the Source Region (Solidification)

When a droplet or an array of droplets undergoes phase change on a surface such as during solidification, evaporation, or condensation, the thermal spreading resistance associated with the droplet/surface interface is one of an isothermal contact, see Figure 9.11. In evaporation processes, the size of the droplet is decreasing with time while for condensation, it increases with time. In solidification processes, it either remains nearly constant for a sessile drop or undergoes a complex process if the droplet impacts the surface and spreads and recoils. Predicting the size of the droplet or the time to progress through phase change is often of interest. Understanding the instantaneous or time-averaged thermal spreading resistance is necessary for modeling the transport phenomena throughout the process, whether it be solidification, evaporation, or condensation. Additionally, understanding all of the paths for heat, mass, and/or momentum transport is also quite important.

One interesting application we consider here is the freezing of impinging droplets. In these applications, droplets are propelled toward a surface, impact, and spread, and recoil, during which they

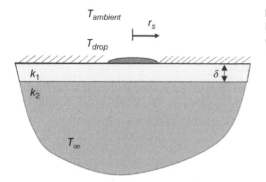

**Figure 9.11** Solidification of a drop on a layered half-space. Source: Reproduced from Lam et al. (2020)/ with permission of Elsevier.

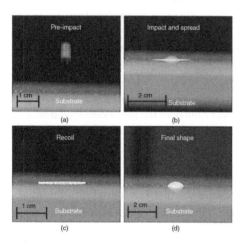

**Figure 9.12** Impact dynamics of a drop on a layered half-space. (a) Pre-impact, (b) impact and spread, (c) recoil, (d) final shape. Source: Lam et al. (2020)/Elsevier.

begin to undergo solidification as shown in Figure 9.12. During this process, the heat transfer is generally dominated by the thermal spreading resistance in the substrate. Lam et al. (2020) considered the freezing of impacting droplets on painted and unpainted surfaces in order to model the effects of various marine coatings on the complete process of drop impact, drop spread, drop retraction, and drop solidification. Two parameters were of interest: final drop radius and time to freeze.

Lam et al. (2020) developed a resistive model consisting of multiple resistances including the transient thermal spreading resistance in a half-space. Using predicted impact velocity and predicted spread radius of the droplet, Lam et al. (2020) integrated the model over time to provide a time to freeze.

The experimental data are predicted well using the simple model for the final dimensionless time to freeze $Fo_{\sqrt{A}f}$:

$$Fo_{\sqrt{A}f} = \frac{4}{3} \frac{\overline{R}^{\star}_{\sqrt{A},total}}{\sqrt{\pi \xi^3 Ste}} \tag{9.147}$$

where $Ste = c_p \Delta T / L_f$ is the Stefan number and $L_f$ is the latent heat of fusion, and $\xi = \overline{r}_s / r_d$ is the spread factor accounting for the size of the droplet as a function of Reynolds and Weber numbers, and the preimpact radius. $\overline{R}^{\star}_{\sqrt{A},total}$ is the total dimensionless resistance which is a function of internal droplet resistance, thermal spreading resistance, and coating resistances depending upon how the system is made up [Lam et al. (2020)].

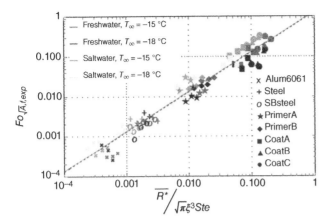

**Figure 9.13** Experimental and theoretical time to freeze as a function of spread factor, Stefan number, and average thermal resistance for a layered half-space. Source: Reproduced from Lam et al. (2020)/ with permission of Elsevier.

The data in Figure 9.13 clearly illustrate the effect of coatings and hence, thermal spreading resistance, which shows that the unpainted surfaces having the lowest overall time to freeze and increasing with the effect of the paint/primer coating thermo-physical properties, as well as the nature of the substrate which was either aluminum or steel. With all other variables being equal except for surface properties, the thermal spreading resistance increases with addition of coatings and hence increases the time to freeze.

## 9.6 Thermal Spreading with Temperature-Dependent Thermal Conductivity

We now examine the effect temperature-dependent thermal conductivity has on the overall temperature field and thermal spreading resistance. The issue of temperature-dependent thermal conductivity in thermal spreading problems has been examined sparingly in the literature, namely [Bagnall et al. (2014), Rahmani and Shokouhmand (2012), Al-Khamaiseh et al. (2018, 2019)] who have studied various thermal conductivity relationships in flux channels and half-space problems.

We will only dwell on a few studies and attempt to lay the foundations for undertaking a Kirchhoff analysis for thermal spreading problems. Essentially, when one understands the Kirchhoff transform and how to apply it to a basic thermal spreading problem, the solutions developed earlier can be utilized to predict the effect of temperature-dependent thermal conductivity.

Most heat conduction texts introduce the Kirchhoff transform [Carslaw and Jaeger (1959), Arpaci (1966), Ozisik (1968)], but few illustrate the application of theory. We will present as succinctly as possible the required approach for a handful of relationships modeling temperature effects in thermal spreaders.

### 9.6.1 Kirchoff Transform

Beginning with the Laplace equation in vector form, we write

$$\nabla \cdot (k \nabla T) = 0 \tag{9.148}$$

where $k = k(T)$ is a temperature-dependent thermal conductivity.

To facilitate solving steady-state conduction problems with temperature-dependent thermal conductivity, Kirchhoff (1894) introduced an integral transform defined as

$$U = k\{T\} = \int^{T} k(\tau)d\tau \tag{9.149}$$

where the lower bound of the integral can be any reference temperature. Kirchhoff demonstrated that

$$\frac{dU}{dT} = k \tag{9.150}$$

and that

$$\nabla U = k\nabla T \tag{9.151}$$

which leads to

$$\nabla^2 U = 0 \tag{9.152}$$

such that Eq. (9.148) is transformed in terms of the new variable $U$. Note that from Eq. (9.150), the new variable does not have units of temperature.

A number of variations of Eq. (9.149) have been proposed [e.g. Carslaw and Jaeger (1959), Arpaci (1966)], but the more widely used form of the Kirchhoff transform may be defined [Bagnall et al. (2014), Al-Khamaiseh et al. (2019)] as

$$\theta = K\{T\} = T_0 + \frac{1}{k_0}\int_{T_0}^{T} k(\tau)d\tau. \tag{9.153}$$

where $k_0$ is the thermal conductivity evaluated at $T_0$. In most thermal spreading applications, $T_0$ is taken as the temperature in the sink plane.

As a result, when applying the Kirchhoff transform given by Eq. (9.153) to the nonlinear heat conduction equation (9.148), the nonlinear equation is transformed into the linear Laplace's equation:

$$\nabla^2 \theta = 0 \tag{9.154}$$

where we now refer to $\theta$ as the apparent temperature. The process appears complex, but can be rather straightforward depending upon how the thermal conductivity model is defined.

For example, if we consider a simple thermal conductivity relationship such as

$$k(T) = a + bT \tag{9.155}$$

the apparent temperature is given by

$$\theta = T_0 + \frac{1}{k_0}\int_{T_0}^{T}(a + b\tau)d\tau \tag{9.156}$$

or

$$\theta = T_0 + \frac{1}{k_0}\left(aT + \frac{1}{2}bT^2 - aT_0 - \frac{1}{2}bT_0^2\right) \tag{9.157}$$

which may be solved in this case for $T$ as a function of the apparent temperature distribution $\theta$. The apparent temperature distribution is determined from the linear Laplace equation with appropriately transformed boundary conditions applied. Thus, once the apparent temperature distribution is known, the actual temperature field is determined from an inversion process similar to what is shown above. The complication here is that it is not often possible to find an explicit relationship for $T$. In many cases, numerical integration may be required to construct the actual temperature field point by point. Additionally, the Kirchhoff transform is generally limited to the Dirichlet- and Neumann-type boundary conditions, but with some care can be applied to a Robin type condition using an equivalent boundary temperature.

In the case of a Dirichlet boundary condition defined as

$$T = T_s \tag{9.158}$$

we write:

$$\theta = K\{T\} = T_0 + \frac{1}{k_0} \int_{T_0}^{T_s} k(\tau)d\tau. \tag{9.159}$$

whereby, if we are using the linear thermal conductivity model defined by Eq. (9.155), we obtain

$$\theta_b = T_0 + \frac{1}{k_0}\left(aT_s + \frac{1}{2}bT_s^2 - aT_0 - \frac{1}{2}bT_0^2\right) \tag{9.160}$$

as the transformed boundary condition.

In the case of a Neumann type boundary condition:

$$-k(T)\frac{\partial T}{\partial \hat{n}} = q \tag{9.161}$$

the Kirchhoff transform returns the following:

$$-k_0\frac{\partial \theta}{\partial \hat{n}} = q \tag{9.162}$$

Finally, if we consider the Robin type condition defined as

$$-k(T)\frac{\partial T}{\partial \hat{n}} = h_s(T_{z=t} - T_\infty) \tag{9.163}$$

Taking the Kirchhoff transform of the above boundary condition gives:

$$-k_0\frac{\partial \theta}{\partial \hat{n}} = h_s(K^{-1}\{\theta_{z=t}\} - T_\infty) \tag{9.164}$$

which is still a nonlinear condition, since $K^{-1}\{\theta\}$ is a nonlinear function of $\theta$, i.e. $K^{-1}\{\theta\} \neq \theta$.

We can alleviate this issue and solve approximately assuming a constant sink plane temperature defined as

$$T_{z=t} = T_0 = \frac{q}{h_s} + T_\infty \tag{9.165}$$

When the thermal spreader is sufficiently thick to allow for complete spreading of the heat, Eq. (9.165) is an excellent approximation in this case, as the sink plane temperature is nearly uniform. In cases of thinner spreaders, where the surface temperature is nonuniform, it is an acceptable approximation for most situations. Both Bagnall et al. (2014) and Al-Khamaiseh et al. (2019) applied this approach with excellent agreement with finite element simulations.

## 9.6.2 Thermal Conductivity Models

The thermal conductivity of most materials are temperature-dependent that vary with temperature according to specific functional relationships between the thermal conductivity and the temperature $k(T)$. In some materials, the thermal conductivity increases with increasing the temperature, while in other materials, the thermal conductivity decreases with increasing the temperature. Different expressions for the thermal conductivity varying with temperature can be found in the literature. In this chapter, we will focus on three general forms of the thermal conductivity which is considered by Al-Khamaiseh et al. (2019):

$$k_1(T) = k_0(1 + \omega_1(T - T_0))^p$$
$$k_2(T) = k_0 \exp[\omega_2(T - T_0)] \qquad\qquad (9.166)$$
$$k_3(T) = k_0\left(\frac{T_0}{T}\right)^s$$

where $k_0$ is a reference constant thermal conductivity; $\omega_1, \omega_2$ are constants called the temperature coefficients of the thermal conductivity, and $p, s$ are real numbers representing the exponents in the corresponding functions. It is important to note that the reference temperature $T_0$ is included in the definition of the temperature-dependent thermal conductivities in order to get the same reference thermal conductivity at $T_0$, i.e. $k_i(T_0) = k_0$, for comparison reasons.

Considering the Kirchhoff transform in Eq. (9.153), the functional relationship between the apparent temperature $\theta$ and the actual temperature $T$ that corresponds to each of the three general forms of thermal conductivity functions given in Eq. (9.166) can be obtained explicitly, and then by solving these relationships for $T$, the actual temperature $T$ can be obtained in terms of the apparent temperature $\theta$ as

$$T = K_1^{-1}\{\theta\} = \begin{cases} T_0 + \frac{1}{\omega_1}\left\{\exp[\omega_1(\theta - T_0)] - 1\right\}, & p = -1 \\ T_0 + \frac{1}{\omega_1}\left[(\omega_1(p+1)(\theta - T_0) + 1)^{1/(p+1)} - 1\right], & p \neq -1 \end{cases}$$

$$T = K_2^{-1}\{\theta\} = T_0 + \frac{1}{\omega_2}\ln(1 + \omega_2(\theta - T_0)) \qquad\qquad (9.167)$$

$$T = K_3^{-1}\{\theta\} = \begin{cases} T_0 \exp(\theta/T_0 - 1), & s = 1 \\ T_0\left[\frac{(1-s)\theta}{T_0} + s\right]^{1/(1-s)}, & s \neq 1 \end{cases}$$

The above relationships can now be applied to appropriately obtained thermal spreading problems.

### 9.6.3 Application for Thermal Spreading Resistance in a Rectangular Flux Channel

Al-Khamaiseh et al. (2018) examined all three models for the thermal conductivity reported earlier. They considered the rectangular flux channel solution presented in Chapter 4 for an isotropic flux channel with convection in the sink plane. The system under consideration is shown in Figure 9.14. Al-Khamaiseh et al. (2019) considered two cases with the system parameters shown in Figure 9.14, one had $h_s \to \infty$ and the other $h_s = 500\,\text{W}/(\text{m}^2\,\text{K})$. Results of the analysis are provided in Figures 9.15 and 9.16.

Additionally, Al-Khamaiseh (2018) also tabulated the exact results for the source centroid and mean temperatures and also developed a more efficient approach to predicting the mean source temperature using a first order Taylor series approximation of the Kirchhoff transform relationships for the apparent temperature. The approximate results for the mean source temperature were in excellent agreement with the exact analytical and finite element method approaches.

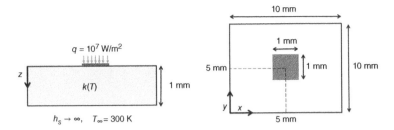

**Figure 9.14** System considered by Al-Khamaiseh et al. (2019). Source: Reproduced from Al-Khamaiseh et al. (2019)/ with permission of American Institute of Aeronautics and Astronautics.

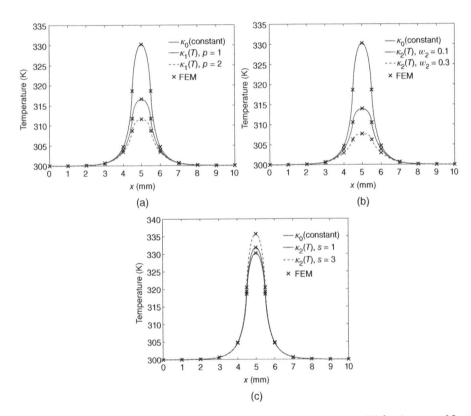

**Figure 9.15** Temperature profile along $x$-axis in the source plane (at $y = Y_c$) for the case of fixed-sink temperature, i.e. $h_s \rightarrow \infty$, by considering the three $k(T)$ models: (a) $k_1(T)$ with $\omega_1 = 0.1$, (b) $k_2(T)$, and (c) $k_3(T)$. All three figures compare the analytical solution for a rectangular flux channel and finite element based solutions, Al-Khamaiseh et al. (2019). Source: Reproduced from Al-Khamaiseh et al. (2019) / with permission of American Institute of Aeronautics and Astronautics.

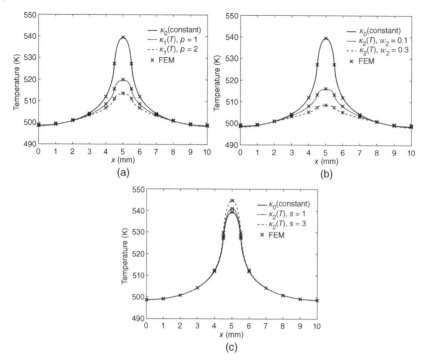

**Figure 9.16** Temperature profile along $x$-axis in the source plane (at $y = Y_c$) for the case of fixed-sink convection, $h_s = 500\,\text{W}/(\text{m}^2\,\text{K})$, by considering the three $k(T)$ models: (a) $k_1(T)$ with $\omega_1 = 0.1$, (b) $k_2(T)$, and (c) $k_3(T)$. All three figures compare the analytical solution for a rectangular flux channel and finite element based solutions, Al-Khamaiseh et al. (2019). Source: Reproduced from Al-Khamaiseh et al. (2019)/ with permission of American Institute of Aeronautics and Astronautics.

## 9.7 Thermal Spreading in Spherical Domains

We conclude the current chapter by considering thermal spreading through spherical shells. Yovanovich et al. (1978) considered thermal spreading in hollow spherical shells with arbitrary flux specified over the poles. The application in Yovanovich et al. (1978) was driven by heat conduction in a packed bed of solid or hollow spheres in a vacuum. The general solution which we will present here is also applicable to solid spheres with appropriate care. In electronics cooling, thermal spreading can occur through solder ball joints connecting devices to circuit boards. More recently, Elsafi and Bahrami (2021) considered thermal spreading in a hollow hemisphere with convection on the interior boundary for applications in spherical pressure vessels.

Both problems will be considered as the solution procedure for both have much in common.

### 9.7.1 Thermal Spreading in Hollow Spherical Shells

The solution for thermal spreading in a spherical region begins with Laplace's equation in spherical coordinates:

$$\frac{\partial^2 T}{\partial r^2} + \frac{2}{r}\frac{\partial T}{\partial r} + \frac{1}{r^2}\frac{\partial^2 T}{\partial \theta^2} + \frac{\cot \theta}{r^2}\frac{\partial T}{\partial \theta} = 0 \tag{9.168}$$

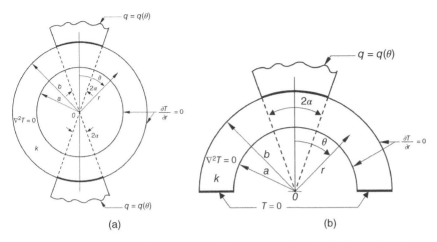

**Figure 9.17** Thermal spreading between polar contact regions in a hollow sphere: (a) full region and (b) equivalent model. Source: Yovanovich et al. (1978)/ American Institute of Aeronautics and Astronautics.

The problem is depicted in Figure 9.17a along with its equivalent model in Figure 9.17b. The following boundary conditions are applied:

$$r = a, \quad \frac{\partial T}{\partial r} = 0, \qquad 0 \leq \theta \leq \frac{\pi}{2}$$

$$r = b, \quad \frac{\partial T}{\partial r} = \frac{q(\theta)}{k}, \quad 0 \leq \theta \leq \alpha$$

$$r = b, \quad \frac{\partial T}{\partial r} = 0, \qquad \alpha < \theta \leq \frac{\pi}{2} \tag{9.169}$$

$$\theta = 0, \quad \frac{\partial T}{\partial \theta}, \qquad a \leq r \leq b$$

$$\theta = \frac{\pi}{2}, \quad T = 0, \qquad a \leq r \leq b$$

As the heat enters one pole and leaves another through the similar-sized contact areas with the same subtended angle, the boundary condition taken in the diametric plane $\theta = \pi/2$ is taken to be $T = 0$ for convenience.

The solution can be found using separation of variables and has the form [Yovanovich et al. (1978)]:

$$T(r, \theta) = \sum_{n=0}^{\infty} \left[ A_n r^n + B_n r^{-(n+1)} \right] P_n(\cos \theta) \tag{9.170}$$

The function $P_n(\cos \theta)$ is a Legendre polynomial of the first kind of degree $n$. The Legendre polynomial of the second kind $Q_n(\cos \theta)$ which also appears in the general solution are inadmissible because they become singular at $\theta = 0$ and will not satisfy this boundary condition. The reader is referred to the Appendix on Special Functions for further details on the Legendre functions.

The Legendre polynomials satisfy the $\theta = 0$ boundary condition for all positive values of $n$, but in the case of the boundary condition at $\theta = \pi/2$, we must limit the solution to only the odd values of $n$ because the functions have the property:

$$P_n(0) = 0, \quad n = 1, 3, 5, 7, \ldots, \text{ odd integers} \tag{9.171}$$

$$P_n(0) \neq 0, \quad n = 2, 4, 6, 8, \ldots, \text{odd integers} \tag{9.172}$$

Thus, after excluding the even degree Legendre polynomials, we now write:

$$T(r, \theta) = \sum_{n,odd}^{\infty} \left[ A_n r^n + B_n r^{-(n+1)} \right] P_n(\cos \theta) \tag{9.173}$$

Turning now, the first boundary condition, we see that

$$\frac{\partial T}{\partial r} = \sum_{n,odd}^{\infty} \left[ n A_n r^{n-1} - (n+1) B_n r^{-(n+2)} \right] P_n(\cos \theta) \tag{9.174}$$

which requires that

$$B_n = \frac{n}{n+1} A_n a^{2n+1} \tag{9.175}$$

The final constant $A_n$ is obtained using the piecewise condition defining the applied flux at the pole. Taking advantage of the orthogonality property of Legendre polynomials [Yovanovich et al. (1978)] obtained the following result:

$$A_n = \frac{1}{k} \frac{(2n+1)b^{1-n}}{n(1 - \epsilon^{2n+1})} \int_{\cos \alpha}^{1} q(\theta) P_n(\mu) d\mu \tag{9.176}$$

where $\mu = \cos \theta$ and $\epsilon = a/b$ is the radii ratio. This completes the general solution. Yovanovich et al. (1978) considered two special heat flux distributions obtained from the general distribution below:

$$q = q_0 (\cos \theta - \cos \alpha)^\nu \tag{9.177}$$

where $q_0$ is a convenient heat flux level determined by the parameter $\nu$. When $\nu = 0$, we have a uniform heat flux, while for $\nu = -1/2$, we approximate the flux distribution that results for an isothermal circular source on a half-space.

The heat flux level over the source is defined as

$$q_0 = \frac{Q}{2\pi b^2 \sqrt{1 - \cos \alpha}}, \quad \nu = 0$$

$$\tag{9.178}$$

$$q_0 = \frac{Q}{4\pi b^2 \sqrt{1 - \cos \alpha}}, \quad \nu = -\frac{1}{2}$$

for each case.

Yovanovich et al. (1978) define the thermal spreading resistance as

$$R_s = \frac{\overline{T}_{source} - \overline{T}_{sink}}{Q} = \frac{\overline{T}_{source}}{Q} \tag{9.179}$$

for the equivalent problem shown in Figure 9.16b since $\overline{T}_{sink} = 0$. For the contact region the mean source temperature is defined as:

$$\overline{T}_{source} = \frac{1}{A_s} \iint_{A_s} T(b, \theta) dA_s \tag{9.180}$$

and

$$Q = \iint_{A_s} q(\theta) dA_s = 2\pi b^2 \int_{\cos \alpha}^{1} q(\theta) d\mu \tag{9.181}$$

The elemental area for integration over the contact spot is $dA_s = 2\pi b^2 \sin\theta \, d\theta$ and $A_s = 2\pi b^2(1 - \cos\alpha)$. For the full sphere, the resistance is twice the value determined by Eq. (9.179). The results for both the temperature field and thermal resistance are given below. Using the prescribed heat flux distribution, Yovanovich et al. (1978) report the following results the two special flux distributions.

When $v = 0$ the temperature distribution is found to be

$$T(r, \theta) = \frac{q_0 b}{k} \sum_{n,odd}^{\infty} \frac{E(r, \epsilon, n)}{n} \left[P_{n-1}(\cos\alpha) - P_{n+1}(\cos\alpha)\right] P_n(\cos\theta) \qquad (9.182)$$

where

$$E(r, \epsilon, n) = \frac{\left(\frac{r}{n}\right)^n + \left(\frac{n}{n+1}\right)\epsilon^{2n+1}\left(\frac{r}{b}\right)^{-(n+1)}}{1 - \epsilon^{2n+1}} \qquad (9.183)$$

The total sphere resistance for $v = 0$ is

$$kcR_s = \frac{\sin\alpha}{\pi(1 - \cos\alpha)^2} \sum_{n,odd}^{\infty} \frac{E(b, \epsilon, n)\left[P_{n-1}(\cos\alpha) - P_{n+1}(\cos\alpha)\right]^2}{n(2n+1)} \qquad (9.184)$$

When $v = -1/2$ the temperature distribution is found to be

$$T(r, \theta) = \frac{q_0 b}{k} \sum_{n,odd}^{\infty} \frac{2E(r, \epsilon, n)}{n} \frac{\left[T_n(\cos\alpha) - T_{n+1}(\cos\alpha)\right]}{\sqrt{1 - \cos\alpha}} P_n(\cos\theta) \qquad (9.185)$$

The total sphere resistance for $v = -1/2$ is

$$kcR_s = \frac{\sin\alpha}{\pi(1 - \cos\alpha)^2} \sum_{n,odd}^{\infty} E(b, \epsilon, n) \frac{\left[\cos\alpha - \cos(n+1)\alpha\right]\left[P_{n-1}(\cos\alpha) - P_{n+1}(\cos\alpha)\right]}{n(2n+1)} \qquad (9.186)$$

In both cases, $q_0$ is defined in Eq. (9.178) and $c$ is the half chord length of the contact area segment subtended by the angle $\alpha$ defined as $c = b \sin\alpha$.

### 9.7.2 Thermal Spreading in a Hollow Hemispherical Shell with Convection on the Interior Boundary

Elsafi and Bahrami (2021) considered the previous problem, but modified the interior surface boundary condition at $r = a$ to be

$$k\frac{\partial T}{\partial r} = hT \qquad (9.187)$$

where $h$ is now an internal convection coefficient. All other conditions specified in Eq. (9.169) remained the same.

The solution proceeds the same as in Section 9.7.1 except that new boundary condition, Eq. (9.185) leads to the following result:

$$B_n = \frac{na^{(n-1)} - \frac{h}{k}a^n}{(n+1)a^{-(n+2)} + \frac{h}{k}a^{-(n+1)}}A_n \qquad (9.188)$$

The general solution may now be written as

$$T(r, \theta) = \sum_{n,odd}^{\infty} A_n \left[r^n + \phi_n r^{-(n+1)}\right] P_n(\cos\theta) \qquad (9.189)$$

where

$$\phi_n = b^{2n+1} \epsilon^{2n+1} \Psi \tag{9.190}$$

and

$$\Psi = \left[ \frac{n - Bi}{(n+1) + Bi} \right] \tag{9.191}$$

where $Bi = ha/k$ and $\epsilon = a/b$. The final coefficient is obtained in the same manner as before using the orthogonality of Legendre polynomials and becomes

$$A_n = \frac{1}{k} \frac{(2n+1)b^{1-n}}{[n - (n+1)\epsilon^{2n+1}\Psi]} \int_{\cos \alpha}^{1} q(\theta) P_n(\mu) d\mu \tag{9.192}$$

The interested reader is referred to the paper by Elsafi and Bahrami (2021) for the remainder of the details and case studies for this particular problem.

## References

Archard, J.F., "The Temperature of Rubbing Surfaces", *Wear*, Vol. 2, pp. 438–455, 1958/1959.

Arpaci, V., *Conduction Heat Transfer*, Addison-Wesley, New York, 1966.

Al-Khamaiseh, B., *Analytical Solutions of 3D Heat Conduction in Flux Channels with Nonuniform Properties and Complex Structures*, PhD Thesis, Memorial University of Newfoundland, 2018.

Al-Khamaiseh, B., Muzychka, Y.S., and Kocabiyik, S., "Spreading Resistance in Multilayered Orthotropic Flux Channel with Temperature Dependent Thermal Conductivities", *AIAA Journal of Thermophysics and Heat Transfer*, Vol. 32, no. 2, pp. 392–400, 2018.

Al-Khamaiseh, B., Muzychka, Y.S., and Kocabiyik, S., "Effect of Temperature Dependent Thermal Conductivity on Spreading Resistance in Flux Channels", *AIAA Journal of Thermophysics and Heat Transfer*, Vol. 33, no. 1, pp. 23–32, 2019.

Bagnall, K.R., Muzychka, Y.S., and Wang, E.N., "Application of the Kirchhoff Transform to Thermal Spreading Problems with Convection Boundary Conditions", *IEEE Transactions on Components, Packaging and Manufacturing Technology*, Vol. 4, no. 3, pp. 408–420, doi: 10.1109/TCPMT.2013.2292584, 2014.

Bejan, A., "Theory of Rolling Contact Heat Transfer", *Journal of Heat Transfer*, Vol. 111, pp. 257–263, 1989.

Bhushan, B., *Introduction to Tribology*, Wiley, New York, 2002.

Blok, H., "The Flash Temperature Concept", *Wear*, Vol. 6, pp. 483–494, 1963.

Cameron, A., Gordon, A.N., and Symm, G.T., "Contact Temperatures in Rolling/Sliding Surfaces", *Proceedings of the Royal Society*, Vol. A268, pp. 45–61, 1968.

Carslaw, H.S., and Jaeger, J.C., *Conduction of Heat in Solids*, Oxford University Press, 1959.

Churchill, S.W., and Usgai, R., "A General Expression for the Correlation of Rates of Transfer and Other Phenomena", *American Institute of Chemical Engineers*, Vol. 18, pp. 1121–1128, 1972.

Cowan, R.S., and Winer, W.O., "Frictional Heating Calculations", in *ASM Handbook, Volume 18 Friction, Lubrication, and Wear Technology*, ASM International, Materials Park, OH, pp. 39–44, 1992.

Crank, J., *Mathematics of Diffusion*, Oxford University Press, 1975.

Davis, A.M.J., and Lauga, E., "Hydrodynamic Friction of Fakir-Like Superhydrophobic Surfaces", *Journal of Fluid Mechanics*, Vol. 661, pp. 402–411, 2010.

Elsafi, A.M., and Bahrami, M., "Thermal Spreading Resistance of Hollow Hemisphere with Internal Convective Coolling", *International Journal of Heat and Mass Transfer*, Vol. 170, p. 120959, 2021.

Enright, R., Hodes, M., Salamon, T., and Muzychka, Y., "Isoflux Nusselt Number and Slip Length Formulae for Super-hydrophobic Micro-channels", *Journal of Heat Transfer*, Vol. 136, p. 012402, 2014.

Etminan, A., and Muzychka, Y.S., "Three Dimensional Mathematical Analysis of the Diffusion of Reactive Hormone from a Transdermal Drug Patch", *CSME Congress*, June 27–30, PEI, Charlottetown, 2021.

Francis, H.A., "Interfacial Temperature Distribution Within a Sliding Hertzian Contact", *ASLE Transactions*, Vol. 14, pp. 41–54, 1970.

Halling, J., *Principles of Tribology*, MacMillan Education Ltd., 1975.

Happel, J., and Brenner, H., *Low Reynolds Number Hydrodynamics*, Prentice-Hall, 1965.

Hodes, M., Kirk, T., and Crowdy, D., "Spreading and Contact Resistance Formulae Capturing Boundary Curvature and Contact Distribution Effects", *Journal of Heat Transfer*, Vol. 140, pp. 104503-1–104503-7, 2018.

Jaeger, J.C., "Moving Sources of Heat and Temperature at Sliding Contacts", *Proceedings of the Royal Society, New South Wales*, Vol. 76, pp. 203–224, 1942.

Kane, D., *On Spreading Resistances and Apparent Slip Lengths for Rectangular and Elliptical Pillars*, M.Sc. Thesis, Tufts University, 2016.

Kirchhoff, G., *Vorlesungen Uber die Theorie der Varme*, Tuebner, Leipzig, Germany, pp. 1–13, 1894.

Lam, L., Sultana, K., Pope, K., and Muzychka, Y.S., "Effect of Thermal Transport on Solidification of Salt and Freshwater Water Droplets on Marine Surfaces", *International Journal of Heat and Mass Transfer*, vol. 153, p. 119452, 2020.

Lamb, H., *Hydrodynamics*, Cambridge University Press, Cambridge, 1932.

Lauga, E., and Stone, H.A., "Effective Slip in Pressure Driven Stokes Flow", *Journal of Fluid Mechanics*, Vol. 489, pp. 55–77, 2003.

Mayer, M., Hodes, M., Kirk, T., and Crowdy, D., "Effect of Surface Curvature on Contact Resistance Between Cylinders", *Journal of Heat Transfer*, Vol. 141, pp. 032002-1–032002-12, 2019.

Muzychka, Y.S., and Yovanovich, M.M., "Thermal Resistance Models for Non-Circular Moving Heat Sources on a Half Space", *Journal of Heat Transfer*, Vol. 123, pp. 624–632, 2001.

Neder, Z., Varadi, K., Man, L., and Friedrich, K., "Numerical and Finite Element Contact Temperature Analysis of Steel-Bronze Real Surfaces in Dry Sliding Contact", *ASME/STLE Tribology Conference*, Toronto, Canada, 1998.

Ng, C.O., and Wang, C.Y., "Stokes Flow a Grating: Implications for Superhydrophobic Slip", *Physics of Fluids*, Vol. 21, pp. 013602-1–013602-12, 2009a.

Ng, C.O., and Wang, C.Y., "Apparent Slip Arising from Stokes Shear Flow Over a Bidimensional Patterned Surface", *Microfluidics and Nanofluidics*, doi: https://doi.org/10.1007/s10404-009-0466-x, 2009b.

Ozisik, M.N., *Boundary Value Problems of Heat Conduction*, International Textbook, Scranton, PA, 1968.

Plawsky, J., *Transport Phenomena Fundamentals*, Dekker, 2001.

Rahmani, Y., and Shokouhmand, H., "Assessment of Temperature-Dependent Conductivity Effects on the Thermal Spreading/Constriction Resistance of Semiconductors", *AIAA Journal of Thermophysics and Heat Transfer*, Vol. 26, no. 4, pp. 638–643, 2012.

Rosenthal, D., "The Theory of Moving Sources of Heat and Its Application to Metal Treatments", *Transactions of the ASME*, Vol. 68, pp. 849–866, 1946.

Sbragaglia, M., and Prosperetti, A., "A Note on the Effective Slip Properties for Mico-channel Flows with Ultrahydrophobic Surfaces", *Physics of Fluids*, Vol. 19, pp. 043603-1–043603-8, 2007a.

Sbragaglia, M., and Prosperetti, A., "Effective Velocity Boundary Condition at a Mixed Slip Surface", *Journal of Fluid Mechanics*, Vol. 578, pp. 435–451, 2007b.

Sneddon, I., *Mixed Boundary Value Problems in Potential Theory*, North Holland Publishing, Amsterdam, 1966.

Tian, X., and Kennedy, F.E., "Maximum and Average Flash Temperatures in Sliding Contacts", *Journal of Tribology*, Vol. 116, pp. 167–174, 1994.

Williams, J.A., *Engineering Tribology*, Oxford University Press, 1994.

Winer, W.O., and Cheng, H.S., "Film Thickness, Contact Stress and Surface Temperatures", in *Wear Control Handbook*, pp. 81–141, ASME Press, 1980.

Ybert, C., Barentin, C., Cottin-Bizonne, C., Joseph, P., and Bocquet, L., "Achieving Large Slip with Superhydrophobic Surfaces: Scaling Laws for Generic Geometries", *Physics of Fluids*, Vol. 19, pp. 123601-1–123601-10, 2007.

Yovanovich, M.M., "Thermal Constriction Resistance Between Contacting Metallic Paraboloids: Application to Instrument Bearings", *AIAA Progress in Astronautics and Aeronautics: Heat Transfer and Spacecraft Control, 24*, ed. J.W. Lucas, MIT Press, pp. 337–358, 1971.

Yovanovich, M.M., "Simplified Explicit Elastoconstriction Resistance Expression for Ball/Race Contacts", *Paper 78-84 AIAA 16th Aerospace Sciences Meeting*, 1978.

Yovanovich, M.M., "Recent Developments in Thermal Contact, Gap, and Joint Conductance Theories and Experiment", *Heat Transfer 1986, Proceedings of the Eight International Heat Transfer Conference*, Vol. 1, pp. 35–45, 1986.

Yovanovich, M.M., Tien, C.L., and Schneider, G.E., "Thermal Resistance of Hollow Spheres Subjected to Arbitrary Flux Over Their Poles", AIAA Paper 78-872, *2nd AIAA/ASME Thermophysics and Heat Transfer Conference*, Palo Alto, CA, USA, May 24–26, pp. 120–134, 1978.

# 10

# Introduction to Thermal Contact Resistance

Contact resistance has been an area of interest in engineering applications dating as far back as 1930. Yovanovich (2005) provides a comprehensive historical overview of the field with emphasis on the research-intensive period spanning the early 1960s to early 2000s. Figure 10.1 presents the timeline of major activities which drove early and later research. Today, nearly all of the research in thermal contact resistance are driven by microelectronic, micro-device, and nanoscale technologies. The early era was dominated primarily by empirical studies, while from the 1960s onwards was led by aerospace and nuclear application that provided a balance of empirical and theoretical studies with much rigorous analysis.

In this chapter, we will introduce the reader to the nuances of this highly interdisciplinary application area. Thermal contact resistance combines the mechanics of contacting surfaces and thermal transport physics, along with the material and surface properties of the mechanical joint.

One historical application of thermal contact resistance is IBM's Thermal Conduction Module (TCM) shown in Figure 10.2. The TCM used aluminum pistons with smooth hemispherical ends which were spring-loaded to make light mechanical contact with an array of micro-chips. This system removed the heat generated by the microchips through the mechanical contact and later dissipated in a water-cooled cold plate. The hemispherical contact and its helium-filled macro-gap region are easily modeled using appropriate physics. This application is dealt with later in Chapter 12 dealing with nonconforming contacts.

## 10.1 Thermal Contact Resistance

When two solids are joined, imperfect joints (interfaces) are formed. The imperfect joints occur because "real" surfaces are not "perfectly" smooth and flat. A mechanical joint consists of numerous, discrete micro-contacts that may be distributed in a random pattern over the apparent contact area if the contacting solids are nominally flat (conforming) and rough, or they may be distributed over a certain portion of the apparent contact area, called the contour area, if the contacting solids are nonconforming and rough, see Figure 10.3. The contact spot size and density depends on surface roughness parameters, physical properties of the contacting asperities, and the apparent contact pressure. The distribution of the contact spots over the apparent contact area depends on the local out-of-flatness of the two solids, their elastic, or plastic or elastic–plastic properties and the mechanical load. Micro-gaps and macro-gaps appear whenever there is absence of solid-to-solid contact. The micro-gaps and macro-gaps are frequently occupied by a third substance such as gas (e.g. air), liquid (e.g. oil, water), or grease whose thermal conductivities are frequently much smaller than those of the contacting solids.

*Thermal Spreading and Contact Resistance: Fundamentals and Applications*, First Edition.
Yuri S. Muzychka and M. Michael Yovanovich.
© 2023 John Wiley & Sons, Inc. Published 2023 by John Wiley & Sons, Inc.

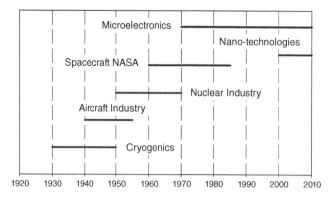

**Figure 10.1** Timeline of thermal contact resistance research. Source: Yovanovich (2005)/ with permission of IEEE.

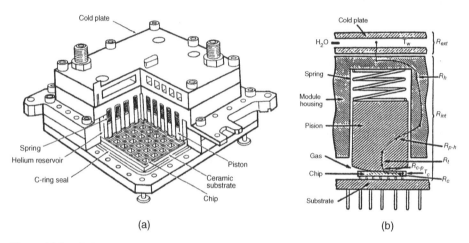

(a)  (b)

**Figure 10.2** IBM's Thermal Conduction Module (TCM). (a) TCM module, (b) TCM piston/contact element. Source: Yovanovich (2005)/ with permission of IEEE.

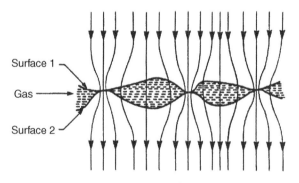

**Figure 10.3** Thermal contact resistance between a region of two contacting surfaces. Source: Hegazy (1985)/Adel Abdel-Halim Hegaz.

The joint formed by explosive bonding may appear to be perfect because there is metal-to-metal contact at all points in the interface which is not perfectly flat and perpendicular to the local heat flux vector. When two metals are brazed, soldered, or welded, a joint is formed that has a small, but finite, thickness and it consists of a complex alloy whose thermal conductivity is lower than that of the joined metals. A complex joint is formed when the solids are bonded or epoxied.

**Figure 10.4** Thermal contact resistance TRIAD. Source: Modified from Yovanovich (2005).

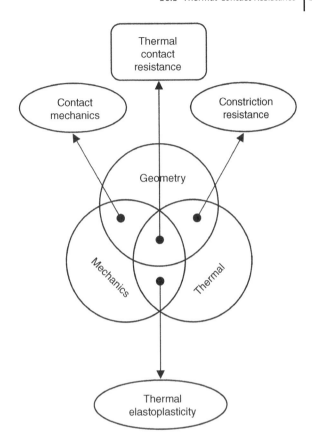

As a result of the "imperfect" joint, whenever heat is transferred across the joint there is a measurable temperature drop across the joint that is directly related to the joint resistance and the heat transfer rate.

Before accurate thermal joint conductance (resistance) models can be developed, it is important to examine the TRIAD for thermal contact resistance which is shown in Figure 10.4. Figure 10.4 shows three basic problems: (i) geometry, (ii) mechanics, and (iii) thermal. The intersection of geometry and mechanics constitutes the contact mechanics problem, the intersection of geometry and thermal constitutes the constriction (spreading) resistance problem, and the intersection of mechanics and thermal constitutes the thermal elastoplasticity problem which will not be discussed in the subsequent sections. The intersection of geometry, mechanics and thermal constitutes the thermal contact resistance problem. Before a thermal contact resistance problem can be developed, the contact mechanics problem must be solved. There are three types of contact mechanics problems and solutions: (a) pure elastic contact, (b) pure plastic contact, and (c) the more difficult elastoplastic contact. The TRIAD will be used as a guide in the development of different types of contact resistance and conductance models.

There is abundance of literature on thermal contact resistance. Many of the significant contributions to the field are found in the heat transfer Handbooks [Yovanovich (1998), Yovanovich and Marotta (2003)]. There are also several comprehensive review articles by Fletcher (1972, 1988, 1990), Madhusudana and Fletcher (1986), Madhusudana (1996), Yovanovich and Antonetti (1988), and Yovanovich (1986, 1991, 2005) that should also be consulted for details of thermal joint resistance and conductance of different types of joints.

## 10.2 Types of Joints or Interfaces

Several definitions are required to define heat transfer across joints (interfaces) formed by two solids which are brought together under a static mechanical load. The heat transfer across the joint is frequently related to *contact* resistances or *contact* conductances and the effective temperature drop across the joint (interface). The definitions are based on the type of joint (interface) which depends on the macro- and micro-geometry of the contacting solids, the physical properties of the substrate and the contacting asperities, and the applied load or apparent contact pressure.

Figure 10.5 illustrates six types of joints that are characterized by whether the contacting surfaces are (i) smooth and nonconforming (Figure 10.5a), (ii) rough and conforming (nominally flat) (Figure 10.5c), and (iii) rough and nonconforming (Figure 10.5b). One or more layers may also be present in the joint as shown in Figure 10.5d–f.

If the contacting solids are nonconforming (e.g. convex solids) and their surfaces are smooth (Figure 10.5a,d), the joint will consist of a single macro-contact and a macro-gap. The macro-contact may be formed by elastic, plastic, or elastic–plastic deformation of the substrate (bulk). The presence of a single "layer" will alter the nature of the joint according to its physical and thermal properties relative to those of the contacting solids. Thermomechanical models are available for finding the joint resistance of these types of joints.

The surfaces of the solids may be conforming (nominally flat) and rough (Figure 10.5c,f). Under a static load, elastic, plastic, or elastic–plastic deformation of the contacting surface asperities occurs. The joint (interface) is characterized by many discrete micro-contacts with associated micro-gaps that are more or less *uniformly* distributed in the apparent (nominal) contact area. The sum of the micro-contact areas is called the *real* area of contact which is a small fraction of the apparent contact area. Thermomechanical models are available for obtaining the contact, gap, and joint conductances (or resistances) of these types of joints.

A third type of joint is formed when nonconforming solids with surface roughness on one or both solids (Figure 10.5b,e) are brought together under load. In this more complex case, the micro-contacts with associated micro-gaps are formed in a region called the *contour* area which is some fraction of the apparent contact area. The substrate may undergo elastic, plastic, or elastic-plastic deformation while the micro-contacts may experience elastic, plastic, or elastic-plastic deformation. A few thermo-mechanical models have been developed for this type of joint.

The substance in the micro-gaps and macro-gaps may be a gas (air, helium, etc.), a liquid (water, oil, etc.), grease, or some compound that consists of grease filled with many micron-size solid particles (zinc oxide, etc.) that increase its effective thermal conductivity and alters its rheology. The interstitial substance is assumed to wet completely the surfaces of the bounding solids, and its effective thermal conductivity is assumed to be isotropic.

If one (or more) layers are present in the joint, the contact problem is much more complex, and the associated mechanical and thermal problems are more difficult to model because the layer thickness, and its physical and thermal properties, and surface characteristics must be taken into account.

The total (joint) heat transfer rate across the interface may take place by conduction through the micro-contacts and conduction through the interstitial substance and by radiation across the micro-gaps and macro-gaps if the interstitial substance is transparent to radiation.

Definitions of thermal *contact, gap, and joint* resistances and *contact, gap, and joint* conductances for several types of joints will be given below.

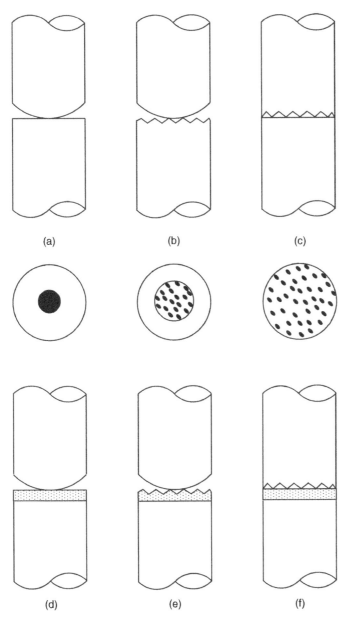

**Figure 10.5** Schematics of six types of joints. (a) Smooth-smooth non-conforming contact, (b) smooth-rough non-conforming contact, (c) rough-smooth conforming contact, (d) smooth-smooth layered non-conforming contact, (e) smooth-rough layered non-conforming contact, (f) rough-smooth layered conforming contact

### 10.2.1 Conforming Rough Solids

If the solids are conforming and their surfaces are rough (Figure 10.5c,f), heat transfer across the joint (interface) occurs by conduction through the contacting micro-contacts and through the micro-gap substance and by radiation across the micro-gap if the substance is transparent (e.g. dry air). The total or joint heat transfer rate $Q_j$, in general, is the sum of three separate heat transfer rates:

$$Q_j = Q_c + Q_g + Q_r \tag{10.1}$$

where $Q_j$, $Q_c$, $Q_g$, and $Q_r$ represent the joint, contact, gap, and radiative heat transfer rates, respectively. The heat transfer rates are generally coupled in some complex manner; however, in many important problems, the coupling is relatively weak. The joint heat transfer rate is related to the effective temperature drop across the joint $\Delta T_j$, the nominal contact area $A_a$, and joint resistance $R_j$, and joint conductance $h_j$ by the definitions:

$$Q_j = h_j A_a \Delta T_j \quad \text{and} \quad Q_j = \frac{\Delta T_j}{R_j} \tag{10.2}$$

These definitions result in the following relationships between joint conductance and joint resistance:

$$h_j = \frac{1}{A_a R_j} \quad \text{and} \quad R_j = \frac{1}{A_a h_j} \tag{10.3}$$

The component heat transfer rates are defined by the relationships:

$$Q_c = h_c A_a \Delta T_j, \qquad Q_g = h_g A_g \Delta T_j, \qquad Q_r = h_r A_g \Delta T_j \tag{10.4}$$

which are all based on the effective joint temperature drop $\Delta T_j$ and their respective heat transfer areas: $A_a$ and $A_g$, the apparent and gap areas, respectively. It is the convention to use the apparent contact area in the definition of the contact conductance. Since $A_g = A_a - A_c$ and $A_c/A_a \ll 1$, then $A_g \approx A_a$. Finally, using the relationships given above, one can write the following relationships between the resistances and the conductances:

$$\frac{1}{R_j} = \frac{1}{R_c} + \frac{1}{R_g} + \frac{1}{R_r} \tag{10.5}$$

and

$$h_j = h_c + h_g + h_r \tag{10.6}$$

If the gap substance is opaque, then $R_r \to \infty$ and $h_r \to 0$, and the relationships reduce to

$$\frac{1}{R_j} = \frac{1}{R_c} + \frac{1}{R_g} \tag{10.7}$$

and

$$h_j = h_c + h_g \tag{10.8}$$

For joints (interfaces) placed in a vacuum and there is no substance in the micro-gaps, then $R_g \to \infty$ and $h_g \to 0$ and the relationships become

$$\frac{1}{R_j} = \frac{1}{R_c} + \frac{1}{R_r} \tag{10.9}$$

and

$$h_j = h_c + h_r \tag{10.10}$$

In all cases, there is heat transfer through the contacting asperities and $h_c$ and $R_c$ are present in the relationships. This heat transfer path is, therefore, very important. For most applications where the joint (interface) temperature level is below 600°C, radiation heat transfer becomes negligible, and, therefore, it is frequently ignored.

### 10.2.2 Nonconforming Smooth Solids

If two smooth, nonconforming solids are in contact (Figure 10.5a,d), heat transfer across the joint can be described by the relationships given in Section 10.2.1. The radiative path becomes more complex because the enclosure and its radiative properties must be considered. If the apparent contact area is difficult to define, then the use of conductances should be avoided and resistances should be used. The joint resistance, neglecting radiation, is

$$\frac{1}{R_j} = \frac{1}{R_c} + \frac{1}{R_g} \tag{10.11}$$

### 10.2.3 Nonconforming Rough Solids

If two rough, nonconforming solids make contact (Figure 10.5b,e), heat transfer across the joint is much more complex when a substance "fills" the micro-gaps associated with the micro-contacts and the macro-gap associated with the contour area. The joint resistance, neglecting radiative heat transfer, is defined by the following relationship:

$$\frac{1}{R_j} = \frac{1}{R_{ma,c} + \left[\dfrac{1}{R_{mi,c}} + \dfrac{1}{R_{mi,g}}\right]^{-1}} + \frac{1}{R_{ma,g}} \tag{10.12}$$

where the component resistances are $R_{mi,c}, R_{mi,g}$, the micro-contact and micro-gap resistances, respectively, and $R_{ma,c}, R_{ma,g}$, the macro-contact and macro-gap resistances, respectively. If there is no interstitial substance in the micro-gaps and macro-gap, and the contact is in a vacuum, then the joint resistance (neglecting radiation) consists of the macro-and micro-resistances in series:

$$R_j = R_{ma,c} + R_{mi,c} \tag{10.13}$$

### 10.2.4 Single Layer Between Two Conforming Rough Solids

If a single, thin metallic or nonmetallic layer of uniform thickness is placed between the contacting rough solids, the mechanical and thermal problems become more complex. The layer thickness, its thermal conductivity and physical properties must also be included in the development of the joint resistance (conductance) models. There are now two interfaces formed which are generally different.

The presence of the layer can increase or decrease the joint resistance depending on several geometric, physical and thermal parameters. A thin, isotropic silver layer bonded to one of the solids can decrease the joint resistance because the layer is relatively soft and it has a high thermal conductivity. On the other hand, a relatively thick oxide coating, which is hard and has a low thermal conductivity, can increase the joint resistance. The joint resistance, neglecting radiation, is given by the general relationship:

$$R_j = \left[\frac{1}{R_{mi,c1}} + \frac{1}{R_{mi,g1}}\right]^{-1} + R_{layer} + \left[\frac{1}{R_{mi,c2}} + \frac{1}{R_{mi,g2}}\right]^{-1} \tag{10.14}$$

where $R_{mi,c1}, R_{mi,g1}$, and $R_{mi,c2}, R_{mi,g2}$ are the micro-contact and micro-gap resistances at the two interfaces formed by the two solids which are separated by the layer. The thermal resistance of the layer is modeled as

$$R_{layer} = \frac{t}{k_{layer}A_a} \tag{10.15}$$

where $t$ is the layer thickness under loading conditions. Except for very soft metals (e.g. indium, lead, and tin) at or above room temperature, the layer thickness under load conditions is close to the thickness before loading. If the layers are nonmetallic, such as elastomers, the thickness under load may be smaller than the preload thickness and elastic compression should be included in the mechanical model.

To develop thermal models for the component resistances, it is necessary to consider single contacts on a half-space and on semi-infinite flux tubes to find solutions for the spreading/constriction resistances. Many of the fundamental solutions for the half-space and flux tube are discussed in Chapters 2 and 3.

## 10.3 Parameters Influencing Contact Resistance or Conductance

Real surfaces are not perfectly smooth (specially prepared surfaces such as those found in ball and roller bearings can be considered to be almost ideal surfaces) but consist of microscopic peaks and valleys. Whenever two real surfaces are placed in contact, intimate solid-to-solid contact occurs only at discrete parts of the joint (interface) and the real contact area will represent a very small fraction (<2%) of the nominal contact area. The real joint (interface) is characterized by several important factors:

(1) Intimate contact occurs at numerous discrete parts of the nominal contact area.
(2) The ratio of the real contact area to the nominal contact area is usually much less than 2%.
(3) The pressure at the real contact area is much greater than the apparent contact pressure. The real contact pressure is related to the flow pressure of the contacting asperities.
(4) A very thin gap exists in the regions in which there is no solid–solid contact, and it is usually occupied by a third substance.
(5) The third substance can be air, other gases, liquid, grease, grease filled with very small solid particles, and another metallic or nonmetallic substance.
(6) The joint (interface) is idealized as a line; however, the actual "thickness" of the joint (interface) ranges from 0.5 μm for very smooth surfaces to about 60–80 μm for very rough surfaces.
(7) Heat transfer across the interface can take place by conduction through the real contact area, by conduction through the substance in the gap, or by radiation across the gap if the substance in the gap is transparent to radiation or if the gap is under a vacuum. All three modes of heat transfer may occur simultaneously, but usually they occur in pairs, with solid–solid conduction always present.

The process of heat transfer across a joint (interface) is complex because the joint resistance may depend upon many geometrical, thermal, and mechanical parameters of which the following are very important:

(a) Geometry of the contacting solids: surface roughness, asperity slope, and out-of-flatness or waviness.
(b) Thickness of the gap (noncontact region).
(c) Type of interstitial fluid: gas, liquid, grease, or vacuum.
(d) Interstitial gas pressure.
(e) Thermal conductivities of the contacting solids and the interstitial substance.
(f) Micro-hardness or flow pressure of the contacting asperities: plastic deformation of the highest peaks of the softer solid.

(g) Modulus of elasticity and Poisson's ratio of the contacting solids: elastic deformation of the wavy parts of the joint.

(h) Average temperature of the joint influences radiation heat transfer as well as the thermophysical properties.

(i) Load or apparent contact pressure.

## 10.4 Assumptions for Resistance and Conductance Model Development

Since thermal contact resistance is such a complex problem, it is necessary to develop simple thermophysical models which can be analyzed and experimentally verified. To achieve these goals, the following assumptions have been made in the development of the several contact resistance models which will be discussed later:

(a) Contacting solids are isotropic: thermal conductivity and physical parameters are constant.

(b) Contacting solids are thick relative to the roughness or the waviness.

(c) Surfaces are clean: no oxide effect.

(d) Contact is static: no vibration effects.

(e) First loading cycle only: no hysteresis effect.

(f) Relative apparent contact pressure ($P/H_p$ for plastic deformation and $P/H_e$ for elastic deformation) is not too small ($>10^{-6}$) nor too large ($<10^{-1}$).

(g) Radiation is small or negligible.

(h) Heat flux at micro-contacts is steady and not too large ($<10^7$ W/m$^2$).

(i) Contact is in a vacuum or the interstitial fluid can be considered to be a continuum if it is not a gas.

(j) Interstitial fluid perfectly wets both contacting solids.

In Chapters 11 and 12, we will consider details of nonconforming and conforming surfaces and the additional inputs required to determine thermal contact resistance (conductance).

## 10.5 Measurement of Joint Conductance and Thermal Interface Material Resistance

Since the remaining two chapters focus on thermal contact resistance models for conforming and nonconforming surface contact, we provide a brief overview of a typical experimental apparatus used to measure contact conductance and thermal interface material (TIM) resistance. In both cases, the apparatus is similar depending upon whether tests are done with or without interface materials, and also whether tests are conducted in a vacuum or with a particular interstitial gas such as helium or nitrogen.

A schematic of a typical apparatus is shown in Figure 10.6. A close-up of the test section showing the heat flux meters is given in Figure 10.7. Details related to the design and assembly of highly precise test facilities can be found in the papers by Culham et al. (2002) and Kempers et al. (2009). Designs typically follow the ASTM standard D 5470-95 (2007).

A typical tester contains two instrumented flux meters in which an interface material is placed in between or no material at all, if the purpose is to measure the contact resistance between two

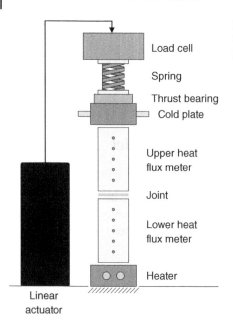

**Figure 10.6** Schematic of a contact resistance measurement facility. Source: Culham et al. (2002)/ with permission of IEEE.

**Figure 10.7** Thermal interface material test facility. Source: Culham et al. (2002)/ Used with permission from IEEE.

contacting surfaces. In the latter case, the flux meters themselves are the contacting samples whose surfaces are treated in some desired manner, i.e. machined, sandblasted, polished, etc. The purpose of each flux meter is to measure the one-dimensional heat flow in each section and to allow the thermal gradient to be extrapolated back to the contacting surfaces to measure the temperature drop $\Delta T_j$ across the joint as shown in Figure 10.8 giving a joint resistance:

$$R_j = \frac{\Delta T_j}{Q} \tag{10.16}$$

The heat flow rate is determined from the measured heat flux:

$$Q = -kA\frac{dT}{dx} \tag{10.17}$$

where the temperature gradient is determined from the instrumented heat flux meter as shown in Figure 10.8.

**Figure 10.8** Extrapolation of flux meter temperature gradient to obtain joint surface temperature. Source: Culham et al. (2002)/ with permission of IEEE.

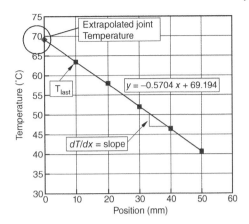

The apparatus is also equipped with a load cell to measure applied loads and a means to measure the displacement of the flux meters to ensure accurate measurement of the in situ thickness of the TIM or the displacement of the contacting surfaces with applied loads. These measurements are done using either direct measurements, a linear variable differential transformer (LVDT), or a laser displacement measurement system. In the case of contact resistance measurements, additional components are needed to allow testing in a vacuum by placing the facility in a bell jar allowing the space to be evacuated or for tests to be conducted using various gases rather than air.

A properly designed facility such as the one shown in Figure 10.7 will provide highly accurate results to be obtained with low uncertainty. Further details related to thermal contact resistance measurements using these types of facilities can be found in many of the past studies such as those of Hegazy (1985), Song (1988), or Bahrami (2004).

In Chapters 11 and 12, various data and models will be discussed for conforming and nonconforming surface contact problems. In virtually all cases, the data are obtained using a facility similar to that described above.

## References

ASTM, "Standard Test Method for Thermal Transmission Properties of Thermally Conductive Electrical Insulation Materials", *ASTM D 5470-95*, Conshohocken, PA, 2007.

Bahrami, M., *Modeling of Thermal Joint Resistance for Sphere-Flat Contacts in a Vacuum*, Ph.D. Thesis, University of Waterloo, Waterloo, Ontario, Canada, 2004.

Culham, J.R., Teertstra, P.M., Savija, I., and Yovanovich, M.M., "Design, Assembly, and Commissioning of a Test Apparatus for Characterizing Thermal Interface Materials", *Proceedings of IEEE Inter-Society Conference on Thermal Phenomena ITHERM*, San Diego, CA, USA, 2002.

Fletcher, L.S., "A Review of Thermal Control Materials for Metallic Junctions", *Journal of Spacecraft and Rockets*, Vol. 9, no. 12, pp. 849–850, 1972.

Fletcher, L.S., "Recent Developments in Contact Conductance Heat Transfer", *Journal of Heat Transfer*, Vol. 110, pp. 1059–1070, 1988.

Fletcher, L.S., "A Review of Thermal Enhancement Techniques for Electronic Systems", *IEEE Transactions on Components and Hybrids Manufacturing Technology*, Vol. 13, no. 4, pp. 1012–1021, 1990.

Hegazy, A., *Thermal Joint Resistance of Conforming Rough Surfaces: Effect of Surface Microhardness Variation*, Ph.D. Thesis, University of Waterloo, Waterloo, Ontario, Canada, 1985.

Kempers, R., Kolodner, P., Lyons, A., and Robinson, A.J., "A High Precision Apparatus for the Characterization of Thermal Interface Materials", *Review of Scientific Instruments*, Vol. 80, 095111-1–095111-11, 2009.

Madhusudana, C.V., *Thermal Contact Conductance*, Springer-Verlag, New York, 1996.

Madhusudana, C.V., and Fletcher, L.S., "Contact Heat Transfer: The Last Decade", *AIAA Journal*, Vol. 24, no. 3, pp. 510–523, 1986.

Song, S., *Analytical and Experimental Study of Heat Transfer through Gas Layers of Contact Interfaces*, Ph.D. Thesis, University of Waterloo, Waterloo, Ontario, Canada, 1988.

Yovanovich, M.M., "Recent Developments in Thermal Contact, Gap and Joint Conductance Theories and Experiments", *Proceedings of the 8th International Heat Transfer Conference*, San Francisco, CA, USA, Vol. 1, pp. 35–45, 1986.

Yovanovich, M.M., "Theory and Applications of Constriction and Spreading Resistance Concepts for Microelectronic Thermal Management", in *Cooling Techniques for Computers*, ed. W. Aung, Hemisphere Publishing, New York, pp. 277–332, 1991.

Yovanovich, M.M., "Chapter 3. Conduction and Thermal Contact Resistances (Conductances)", in *Handbook of Heat Transfer*, eds. W.M. Rohsenow, J.P. Hartnett, and Y.I. Cho, McGraw-Hill, New York, 1998.

Yovanovich, M.M., "Four Decades of Research on Thermal Contact, Gap, and Joint Resistance in Microelectronics", *IEEE Transactions on Components and Packaging Technologies*, Vol. 28, no. 2, pp. 182–206, 2005.

Yovanovich, M.M., and Antonetti, V.W., "Application of Thermal Contact Resistance Theory to Electronic Packages", in *Advances in Thermal Modeling of Electronic Components and Systems*, Vol. 1, eds. A. Bar-Cohen and A.D. Kraus, Hemisphere Publishing, New York, pp. 79–128, 1988.

Yovanovich, M.M., and Marotta, E., "Chapter 4. Thermal Spreading and Contact Resistances", in *Heat Transfer Handbook*, eds. A. Bejan and A.D. Kraus, Wiley, New York, 2003.

# 11

# Conforming Rough Surface Models

In this chapter, we consider thermal contact resistance between conforming rough surfaces. Both elastic and plastic contacts are considered. Gap conductance models for filled and unfilled joints are discussed. Joint conductance models considering all resistances (contact and gap) that are presented. Finally, the methods for enhancing contact conductance using coatings, foils, polymers, greases, pastes, and phase change materials are examined. We only present the most useful approaches but provide discussion for other problems and models where applicable.

Comprehensive reviews of the microgeometric, micromechanical, and thermal models are given for conforming rough surfaces. The contacting surfaces are either smooth and flat or they are rough and flat. The rough surfaces consist of numerous asperities which have Gaussian height distribution over the apparent contact area and a Gaussian distribution with respect to the mean plane which passes through the asperities. The surface roughness is characterized by either the center-line-average (CLA) asperity height or the root-mean-square (RMS) $(\sigma)$ of the asperity heights. The two roughness parameters are related as $\sigma = \sqrt{\pi/2}$ CLA. Another important surface parameter is the absolute mean asperity slope $(m)$ which determines the density $(n = N/A_a)$ of contacting asperities $(N)$ on the apparent contact area $(A_a)$. The relative contact area is defined as $(\epsilon = \sqrt{A_c/A_a})$.

For two rough contacting surfaces, there are two planes which pass through the respective asperities. The relative distance between the mean planes $(\lambda = Y/\sigma)$ is an important microgeometric parameter which determines the contact spot density $(n)$ and mean contact spot radius $(a)$. The real contact area is given by $(A_c = n\pi a^2 A_a)$. The relative mean plane separation depends on the deformation of the contacting asperities. As the mean contact pressure increases, the relative mean plane separation decreases slowly and the contact spot density and contact spot radius increase.

Two types of joints are considered in this chapter: (i) a smooth surface in contact with a rough surface or (ii) two rough surfaces in contact. The contacting asperities deform elastically or they undergo plastic deformation. The more general and complex elastic–plastic deformation problem is also considered. For both elastic and plastic contact, the resulting real contact area consists of numerous microcontact areas which are modeled as circular microareas. The real contact area is equal to the sum of the microcontact areas. Outside the contact spots, the surfaces are separated by a gap which is in a vacuum or it may be filled with a gas, liquid, grease, or some soft substance which conducts heat. The contact spot density and the mean contact spot radius depend on the mode of deformation (elastic or plastic).

The contact conductance $(h_c)$ can be found from the relation $h_c = 2k_s na/\psi(\epsilon)$, where the effective thermal conductivity associated with each contact spot is $k_s = 2k_1 k_2/(k_1 + k_2)$ whenever solids with different thermal conductivities make contact, and $\psi(\epsilon) = (1 - \epsilon)^{1.5}$ is the local thermal constriction parameter for an isothermal circular contact spot. The constriction parameter is applicable

*Thermal Spreading and Contact Resistance: Fundamentals and Applications*, First Edition.
Yuri S. Muzychka and M. Michael Yovanovich.
© 2023 John Wiley & Sons, Inc. Published 2023 by John Wiley & Sons, Inc.

for contacts with $\epsilon = \sqrt{A_c/A_a} < 0.3$. Contact conductance is applicable whenever contacts occur in a vacuum and radiation heat transfer across the gap is negligible.

In this chapter, a plastic contact model based on Vickers micro-hardness tests and correlations is used to model the contact spot density and mean contact spot radius which are essential micro-geometric parameters for the contact conductance model. The Vickers micro-hardness is also correlated with respect to Brinell macro-hardness which is available in many materials handbooks. Additional reference material is presented in Appendix B on hardness.

An elastic contact model is also presented to account for observed discrepancies between certain data and the plastic model predictions. The elastic contact model has similarities with the plastic contact model. The major differences are that the elastic contact model is based on an equivalent elastic micro-hardness ($H_e$) which is related to a geometric constant ($C = 1/\sqrt{2}$), the mean asperity slope ($m$), and the effective Young's modulus defined as $1/E' = (1 - v_1^2)/E_1 + (1 - v_2^2)/E_2$, where ($v_1, v_2$) are the Poisson's ratios and ($E_1, E_2$) are the Young's modulus of the contacting asperities.

Dimensionless contact conductance models for plastic contact, elastic contact, and elastic–plastic contact are also given in this chapter. A general model is presented for conduction across a large narrow gap ($Y/\sigma = 1.6 - 4.0 \times 10^{-4}$) filled with a gas which completely wets the two contacting solid surfaces. The model accounts for rarefaction effects at the gas–solid interface. Very good agreements are observed for different gases (argon, helium, and nitrogen) and contacting solids (Ni 200, SS 304) for gas pressures from 1 to 760 Torr.

A general radiation model for heat transfer across the gap is presented. A study is given to ascertain the relative importance of radiation heat transfer compared with contact conductance.

Methods of enhancing joint conductances are presented in some detail. Mechanical and thermal models are reviewed for metallic coatings and foils. Thin coatings such as lead, tin, silver, and aluminum were used to increase the contact conductance. Elastomeric (Delrin B, Polycarbonate A, PVC A, and Teflon A) inserts were employed to enhance contact conductance. Mechanical and thermal models were developed for the effect of elastomeric inserts on contact conductance. Mechanical and thermal models for thermal greases and pastes are presented and compared with experimental data. Phase change materials (PCM) are also found to be effective interstitial substances for enhancement of contact conductances.

Finally, the contact resistance of bolted joints is discussed. These systems are often found in aerospace applications.

## 11.1 Conforming Rough Surface Models

There are models for predicting contact, gap, and joint conductances between conforming (nominally flat) rough surfaces developed by Greenwood and Williamson (1966), Greenwood (1967), Greenwood and Tripp (1967), Cooper et al. (1969), Mikic (1974), and Yovanovich (1982).

The three mechanical models, elastic, plastic, or elastic–plastic deformation of the contacting asperities, are based on the assumptions that (i) the surface asperities have Gaussian height distributions about some mean plane passing through each surface and (ii) the surface asperities are distributed "randomly" over the apparent contact area $A_a$. Figure 11.1 shows a very small portion of a typical joint formed between two nominally flat rough surfaces under a mechanical load.

Each surface has a mean plane, and the distance between them, denoted as $Y$, is related to the effective surface roughness, the apparent contact pressure, and the plastic or elastic physical properties of the contacting asperities.

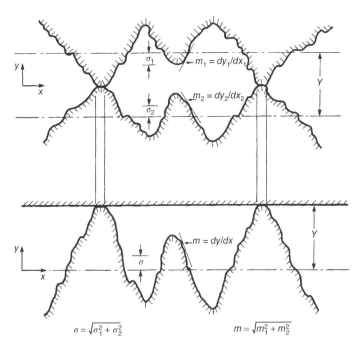

**Figure 11.1** Schematic of typical joint between conforming rough surfaces. Source: Hegazy (1985)/ Adel Abdel-Halim Hegazy.

A very important surface roughness parameter is the surface roughness: either the RMS roughness or the CLA roughness which are defined as [Whitehouse and Archard (1970), Thomas (1982)]:

$$CLA \ roughness = \frac{1}{L} \int_0^L |y(x)| dx \tag{11.1}$$

and

$$RMS \ roughness = \sqrt{\frac{1}{L} \int_0^L y^2(x) dx} \tag{11.2}$$

where $y(x)$ is the distance of points in the surface from the mean plane (Figure 11.1) and $L$ is the length of a trace that contains a sufficient number of asperities. For Gaussian asperity heights with respect to the mean plane, these two measures of surface roughness are related [Mikic and Rohsenow (1966)]

$$\sigma = \sqrt{\frac{\pi}{2}} \ CLA \tag{11.3}$$

A second very important surface roughness parameter is the absolute mean asperity slope which is defined as [Cooper et al. (1969), Mikic and Rohsenow (1966)]

$$m = \frac{1}{L} \int_0^L \left| \frac{dy(x)}{dx} \right| dx \tag{11.4}$$

The "effective" RMS surface roughness and the "effective" absolute mean asperity slope for a typical joint formed by two conforming rough surfaces are defined as [Cooper et al. (1969), Mikic (1974), Yovanovich (1982)]

$$\sigma = \sqrt{\sigma_1^2 + \sigma_2^2} \quad \text{and} \quad m = \sqrt{m_1^2 + m_2^2} \tag{11.5}$$

Antonetti et al. (1991) reported approximate relationships for the $m$ as a function of $\sigma$ for several metal surfaces that were bead-blasted.

The three deformation models (elastic, plastic, or elastic–plastic) give relationships for three important geometric parameters of the joint: (i) the relative real contact area $A_r/A_a$, (ii) the contact spot density $n$, and (iii) the mean contact spot radius $a$ in terms of the relative mean plane separation defined as $\lambda = Y/\sigma$. The mean plane separation $Y$ and the effective surface roughness are illustrated in Figure 11.1 for the joint formed by the mechanical contact of two nominally flat rough surfaces.

The models differ in the mode of deformation of the contacting asperities. The three modes of deformation are (i) plastic deformation of the softer contacting asperities, (ii) elastic deformation of all contacting asperities, and (iii) elastic–plastic deformation of the softer contacting asperities.

For all the three deformation models, there is one thermal contact conductance model that is given as [Cooper et al. (1969), Yovanovich (1982)]

$$h_c = \frac{2nak_s}{\psi(\epsilon)} \tag{11.6}$$

where $n$ is the contact spot density, $a$ is the mean contact spot radius, and the "effective" thermal conductivity of the joint is

$$k_s = \frac{2k_1 k_2}{k_1 + k_2} \tag{11.7}$$

and the spreading/constriction parameter $\psi$, based on isothermal contact spots, is approximated by

$$\psi(\epsilon) = (1 - \epsilon)^{1.5} \quad \text{for} \quad 0 < \epsilon < 0.3 \tag{11.8}$$

where the relative contact spot size is $\epsilon = \sqrt{A_r/A_a}$. The geometric parameters $n, a$ and $A_r/A_a$ are related to the relative mean plane separation $\lambda = Y/\sigma$.

## 11.2 Plastic Contact Model for Asperities

The original plastic deformation model of Cooper et al. (1969) has undergone significant modifications during the past 50 years. First, a new, more accurate correlation equation was developed by Yovanovich (1982). Then Yovanovich et al. (1982b) and Hegazy (1985) introduced the micro-hardness layer which appears in most *worked* metals. Figures 11.2 and 11.3 show plots of measured micro-hardness and macro-hardness versus the penetration depth $t$ or the Vickers diagonal $d_V$. These two measures of indenter penetration are related through $d_v/t = 7$. Figure 11.2 shows the measured Vickers micro-hardness versus indentation diagonal for four metal types (Ni 200, stainless steel 304, Zr-4, and Zr-2.5 wt% Nb). The four sets of data show the same trends, i.e. that as the load on the indenter increases, the indentation diagonal increases, and the Vickers micro-hardness decreases with increasing diagonal (load). The indentation diagonal was between 8 and 70 μm.

Figure 11.3 shows the Vickers micro-hardness measurements and the Brinell and Rockwell macro-hardness measurements versus indentation depth. The Brinell and Rockwell macro-hardness values are very close because they correspond to large indentations and, therefore, they are a measure of the bulk hardness which does not change with load. According to Figure 11.3, the penetration depths for the Vickers micro-hardness measurements are between 1

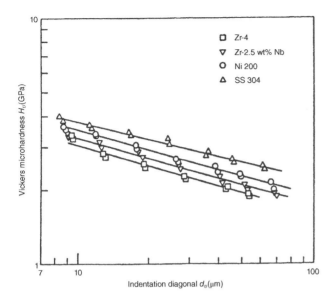

**Figure 11.2** Vickers micro-hardness versus indentation diagonal for four metal types. Source: Hegazy (1985)/ Adel Abdel-Halim Hegazy.

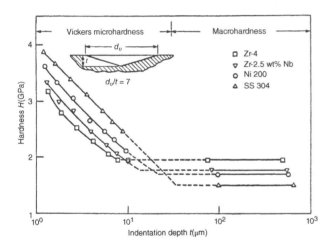

**Figure 11.3** Vickers, Brinell, and Rockwell hardness versus indentation depth for four metal types. Source: Hegazy (1985)/ Adel Abdel-Halim Hegazy.

and 10 μm, whereas the larger penetration depths for the Brinell and Rockwell macro-hardness measurements lie between approximately 100 and 1000 μm.

The micro-hardness layer may be defined by means of the Vickers micro-hardness measurements which relate the Vickers micro-hardness $H_V$ to the Vickers average indentation diagonal $d_V$ [Yovanovich et al. (1982b), Hegazy (1985)]:

$$H_V = c_1 \left( \frac{d_V}{d_0} \right)^{c_2} \tag{11.9}$$

where $d_0$ represents some convenient reference value for the average diagonal, and $c_1$ and $c_2$ are correlation coefficients. It is conventional to set $d_0 = 1$ µm. Hegazy (1985) found that $c_1$ is closely related to the metal bulk hardness such as the Brinell hardness denoted as $H_B$.

The original mechanical contact model [Yovanovich et al. (1982b), Hegazy (1985)] required an iterative procedure to calculate the appropriate micro-hardness for a given surface roughness $\sigma$ and $m$, given the apparent contact pressure $P$, and the coefficients $c_1$ and $c_2$.

Song and Yovanovich (1988) developed an explicit relationship for the micro-hardness $H_p$ which is presented below. Recently, Sridhar and Yovanovich (1996b) developed correlation equations between the Vicker's correlation coefficients $c_1$, $c_2$ and Brinell hardness $H_B$ over a wide range of metal types. These relationships are also presented below.

**Plastic Contact Geometric Parameters**

For plastic deformation of the contacting asperities, the contact geometric parameters are obtained from the following relationships [Cooper et al. (1969), Yovanovich (1982)]:

$$\frac{A_r}{A_a} = \frac{1}{2} \text{ erfc}(\lambda/\sqrt{2}) \tag{11.10}$$

$$n = \frac{1}{16} \left(\frac{m}{\sigma}\right)^2 \frac{\exp(-\lambda^2)}{\text{erfc}(\lambda/\sqrt{2})} \tag{11.11}$$

$$a = \sqrt{\frac{8}{\pi}} \left(\frac{\sigma}{m}\right) \exp(\lambda^2/2) \text{ erfc}(\lambda/\sqrt{2}) \tag{11.12}$$

$$na = \frac{1}{4\sqrt{2\pi}} \left(\frac{m}{\sigma}\right) \exp(-\lambda^2/2) \tag{11.13}$$

where $n$ is the contact spot density and $a$ is the mean contact spot radius. The relative mean plane separation for plastic deformation is given by

$$\lambda = \sqrt{2} \text{ erfc}^{-1} \left(\frac{2P}{H_p}\right) \tag{11.14}$$

where $H_p$ is the micro-hardness of the softer contacting asperities. Here, $\text{erfc}^{-1}(\cdot)$ represents the inverse of the $\text{erfc}(\cdot)$ function. In other words, the mean plane separation is an implicit function of the dimensionless contact pressure, i.e. $2P/H_p = \text{erfc}(\lambda/\sqrt{2})$.

**Correlation of Geometric Parameters**

To make calculations easier, correlations for the plastic contact parameters have been developed for the above equations:

$$\frac{A_r}{A_a} = \exp\left(-0.8141 - 0.61778\,\lambda - 0.42476\,\lambda^2 - 0.004353\,\lambda^3\right) \tag{11.15}$$

$$n = \left(\frac{m}{\sigma}\right)^2 \exp\left(-2.6516 + 0.6178\,\lambda - 0.5752\,\lambda^2 + 0.004353\,\lambda^3\right) \tag{11.16}$$

$$a = \left(\frac{\sigma}{m}\right) \left(1.156 - 0.4526\,\lambda + 0.08269\,\lambda^2 - 0.005736\,\lambda^3\right) \tag{11.17}$$

and for the relative mean plane separation:

$$\lambda = 0.2591 - 0.5446 \left[\ln\left(\frac{P}{H_p}\right)\right] - 0.02320 \left[\ln\left(\frac{P}{H_p}\right)\right]^2 - 0.0005308 \left[\ln\left(\frac{P}{H_p}\right)\right]^3 \tag{11.18}$$

**Relative Contact Pressure**

The appropriate micro-hardness may be obtained from the relative contact pressure $P/H_p$. For plastic deformation of the contacting asperities, the explicit relationship is [Song and Yovanovich (1988)]

$$\frac{P}{H_p} = \left[ \frac{P}{c_1 (1.62 \, \sigma/m)^{c_2}} \right]^{\frac{1}{1 + 0.071 c_2}} \tag{11.19}$$

where $\sigma$ is given in microns (μm) and the coefficients $c_1, c_2$ are obtained from Vickers micro-hardness tests. The Vickers micro-hardness coefficients are related to the Brinell hardness for a wide range of metal types.

## 11.2.1   Vickers Micro-hardness Correlation Coefficients

The correlation coefficients: $c_1$ and $c_2$ are obtained from Vickers micro-hardness measurements. Sridhar and Yovanovich (1996b) developed correlation equations for the Vickers coefficients:

$$\frac{c_1}{3178} = \left[ 4.0 - 5.77 H_B^* + 4.0 \left( H_B^* \right)^2 - 0.61 \left( H_B^* \right)^3 \right] \tag{11.20}$$

and

$$c_2 = -0.370 + 0.442 \left( \frac{H_B}{c_1} \right) \tag{11.21}$$

where $H_B$ is the Brinell hardness in (MPa) [Johnson (1985), Tabor (1951)] and $H_B^* = H_B/3178$. The correlation equations are valid for the Brinell hardness range: 1300–7600 MPa. The above correlation equations were developed for a range of metal types (e.g. Ni 200, SS 304, Zr alloys, Ti alloys, and tool steel). Sridhar and Yovanovich (1996b) also reported a correlation equation that relates the Brinell hardness number to the Rockwell C hardness number:

$$BHN = 43.7 + 10.92 \, HRC - \frac{HRC^2}{5.18} + \frac{HRC^3}{340.26} \tag{11.22}$$

for the range: $20 \le HRC \le 65$.

## 11.2.2   Dimensionless Contact Conductance: Plastic Deformation

The dimensionless contact conductance $C_c$ is given by

$$C_c \equiv \frac{h_c \, \sigma}{k_s m} = \frac{1}{2\sqrt{2\pi}} \frac{\exp(-\lambda^2/2)}{\left[ 1 - \sqrt{\frac{1}{2} \, \mathrm{erfc}(\lambda/\sqrt{2})} \right]^{1.5}} \tag{11.23}$$

The correlation equation of the dimensionless contact conductance obtained from theoretical values for a wide range of $\lambda$ and $P/H_p$ is [Yovanovich (1982)]:

$$C_c \equiv \frac{h_c \, \sigma}{k_s \, m} = 1.25 \left( \frac{P}{H_p} \right)^{0.95} \tag{11.24}$$

which agrees with the theoretical values to within $\pm 1.5\%$ in the range: $2 \le \lambda \le 4.75$.

It has been demonstrated that the above plastic contact conductance model predicts accurate values of $h_c$ for a range of surface roughness $\sigma/m$, a range of metal types (e.g. Ni 200, SS 304, and Zr alloys), and a range of the relative contact pressure $P/H_p$ [Antonetti (1983), Hegazy (1985), Sridhar

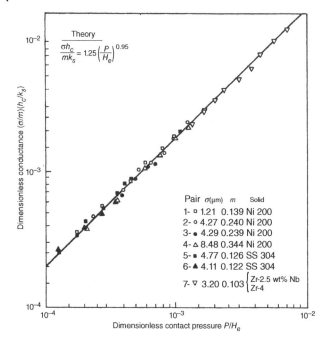

**Figure 11.4** Comparison of plastic contact conductance model and vacuum data. Source: Yovanovich and Hegazy (1983)/ American Institute of Aeronautics and Astronautics.

(1994), Sridhar and Yovanovich (1994b, 1996b,c,d)]. The very good agreement between the contact conductance models and experiments is seen in Figure 11.4.

In Figure 11.4, the dimensionless contact conductance model and the vacuum data for different metal types and a range of surface roughness are compared over two decades of the relative contact pressure defined as $P/H_e$, where $H_e$ was called the effective micro-hardness of the joint. The agreement between the theoretical model developed for conforming rough surfaces that undergo plastic deformation of the contacting asperities is very good over the entire range of dimensionless contact pressure. Because of the relatively high contact pressures and high thermal conductivity of the metals, the effect of radiation heat transfer across the gaps was found to be negligible for all tests.

## 11.3 Elastic Contact Model for Asperities

The conforming rough surface model proposed by Mikic (1974) for elastic deformation of the contacting asperities is summarized below [Sridhar (1994), Sridhar and Yovanovich (1996c)].

**Elastic Contact Geometric Parameters**

The elastic contact geometric parameters are [Mikic (1974)]

$$\frac{A_r}{A_a} = \frac{1}{4}\,\mathrm{erfc}(\lambda/\sqrt{2}) \tag{11.25}$$

$$n = \frac{1}{16}\left(\frac{m}{\sigma}\right)^2 \frac{\exp(-\lambda^2)}{\mathrm{erfc}(\lambda/\sqrt{2})} \tag{11.26}$$

$$a = \frac{2}{\sqrt{\pi}} \left(\frac{\sigma}{m}\right) \exp(\lambda^2/2)\, \mathrm{erfc}(\lambda/\sqrt{2}) \tag{11.27}$$

$$na = \frac{1}{8\sqrt{\pi}} \left(\frac{m}{\sigma}\right) \exp(-\lambda^2/2) \tag{11.28}$$

The relative mean plane separation is given by

$$\lambda = \sqrt{2}\, \mathrm{erfc}^{-1}\left(\frac{4P}{H_e}\right) \tag{11.29}$$

As before, $\mathrm{erfc}^{-1}(\cdot)$ represents the inverse of the $\mathrm{erfc}(\cdot)$ function. In other words, the mean plane separation is an implicit function of the dimensionless contact pressure, i.e. $4P/H_p = \mathrm{erfc}(\lambda/\sqrt{2})$.

The equivalent elastic micro-hardness according to Mikic (1974) is defined as

$$H_e = C\, mE' \quad \text{where} \quad C = \frac{1}{\sqrt{2}} = 0.7071 \tag{11.30}$$

where the "effective" Young's modulus of the contacting asperities is

$$\frac{1}{E'} = \frac{1 - v_1^2}{E_1} + \frac{1 - v_2^2}{E_2} \tag{11.31}$$

Greenwood and Williamson (1966), Greenwood (1967), and Greenwood and Tripp (1970) developed more complex elastic contact model that gives a dimensionless elastic micro-hardness $H_e/(m\,E')$ that depends on the surface roughness bandwidth $\alpha$ and the separation between the mean planes of the asperity "summits" denoted as $\lambda_s$. For a typical range of values of $\alpha$ and $\lambda_s$ [McWaid and Marschall (1992a)], the value of Mikic (1974), i.e. $H_e/(m\,E') = 0.7071$ lies in the range obtained with the Greenwood and Williamson (1966) model. There is, at present, no simple correlation for the model of Greenwood and Williamson (1966).

**Correlation Equations for Surface Parameters**
The correlation equations for $A_r/A_a$, $n$ and $a$ for the relative contact pressure range $10^{-6} \leq P/H_e \leq 0.2$ are

$$\frac{A_r}{A_a} = \frac{1}{2} \exp\left(-0.8141 - 0.61778\,\lambda - 0.42476\,\lambda^2 - 0.004353\,\lambda^3\right) \tag{11.32}$$

$$n = \left(\frac{m}{\sigma}\right)^2 \exp\left(-2.6516 + 0.6178\,\lambda - 0.5752\,\lambda^2 + 0.004353\,\lambda^3\right) \tag{11.33}$$

$$a = \frac{1}{\sqrt{2}} \left(\frac{\sigma}{m}\right) \left(1.156 - 0.4526\,\lambda + 0.08269\,\lambda^2 - 0.005736\,\lambda^3\right) \tag{11.34}$$

and the relative mean planes separation:

$$\lambda = -0.5444 - 0.6636\left[\ln\left(\frac{P}{H_e}\right)\right] - 0.03204\left[\ln\left(\frac{P}{H_e}\right)\right]^2 - 0.000771\left[\ln\left(\frac{P}{H_e}\right)\right]^3 \tag{11.35}$$

### 11.3.1 Dimensionless Contact Conductance: Elastic Deformation

The dimensionless contact conductance for conforming rough surfaces whose contacting asperities undergo elastic deformation is [Mikic (1974), Sridhar and Yovanovich (1996a)]

$$\frac{h_c\,\sigma}{k_s m} = \frac{1}{4\sqrt{\pi}} \frac{\exp(-\lambda^2/2)}{\left[1 - \sqrt{\frac{1}{4}\,\mathrm{erfc}(\lambda/\sqrt{2})}\right]^{1.5}} \tag{11.36}$$

The power law correlation equation based on calculated values obtained from the theoretical relationship is [Sridhar and Yovanovich (1996a)]

$$\frac{h_c \, \sigma}{k_s m} = 1.54 \left( \frac{P}{H_e} \right)^{0.94} \tag{11.37}$$

has an uncertainty of about $\pm 2\%$ for the relative contact pressure range: $10^{-5} \le P/H_e \le 0.2$.

Sridhar and Yovanovich (1996a) reviewed the plastic and elastic deformation contact conductance correlation equations and compared them against vacuum data [Mikic and Rohsenow (1966), Antonetti (1983), Hegazy (1985), Nho (1989), McWaid and Marschall (1992a,b)] for several metal types having a range of surface roughness, over a wide range of apparent contact pressure.

Sridhar and Yovanovich (1996a) showed that the elastic deformation model was in better agreement with the vacuum data obtained for joints formed by conforming rough surfaces of tool steel which is very hard.

The elastic asperity contact and thermal conductance models of Greenwood and Williamson (1966), Greenwood (1967), Greenwood and Tripp (1967, 1970), Bush et al. (1975), and Bush and Gibson (1979) are different from the Mikic (1974) elastic contact model presented in this chapter. However, they predict similar trends of contact conductance as a function apparent contact pressure.

## 11.4 Conforming Rough Surface Model: Elastic–Plastic Asperity Deformation

Sridhar and Yovanovich (1996a) developed an elastic–plastic contact conductance model which is based on the plastic contact model of Cooper et al. (1969) and the elastic contact model of Mikic (1974). The results are summarized below in terms of the geometric parameters: (i) $A_r/A_a$, the real to apparent area ratio, (ii) $n$, the contact spot density, (iii) $a$, the mean contact spot radius, and (iv) $\lambda$, the dimensionless mean plane separation:

$$\frac{A_r}{A_a} = \frac{f_{ep}}{2} \operatorname{erfc}\left( \lambda / \sqrt{2} \right) \tag{11.38}$$

$$n = \frac{1}{16} \left( \frac{m}{\sigma} \right)^2 \frac{\exp\left( -\lambda^2 \right)}{\operatorname{erfc}\left( \lambda / \sqrt{2} \right)} \tag{11.39}$$

$$a = \sqrt{\frac{8}{\pi}} \, \sqrt{f_{ep}} \, \frac{\sigma}{m} \, \exp\left( \lambda^2 / 2 \right) \operatorname{erfc}\left( \lambda / \sqrt{2} \right) \tag{11.40}$$

$$na = \frac{1}{8} \sqrt{\frac{2}{\pi}} \, \sqrt{f_{ep}} \, \frac{m}{\sigma} \, \exp(-\lambda^2 / 2) \tag{11.41}$$

$$\frac{h_c \sigma}{k_s m} = \frac{1}{2\sqrt{2\pi}} \frac{\sqrt{f_{ep}} \, \exp\left( -\lambda^2 / 2 \right)}{\left[ 1 - \sqrt{\dfrac{f_{ep}}{2}} \operatorname{erfc}\left( \lambda / \sqrt{2} \right) \right]^{1.5}} \tag{11.42}$$

$$\lambda = \sqrt{2} \operatorname{erfc}^{-1} \left( \frac{1}{f_{ep}} \frac{2P}{H_{ep}} \right) \tag{11.43}$$

The important elastic–plastic parameter $f_{ep}$ is a function of the dimensionless contact strain $\epsilon_c^*$ which depends on the amount of work hardening. This physical parameter lies in the range: $0.5 \le$

$f_{ep} \leq 1.0$. The smallest and largest values correspond to zero and infinitely large contact strain, respectively. The elastic–plastic parameter is related to the contact strain:

$$f_{ep} = \frac{\left[1 + (6.5/\epsilon_c^*)^2\right]^{1/2}}{\left[1 + (13.0/\epsilon_c^*)^{1.2}\right]^{1/1.2}}, \qquad 0 < \epsilon_c^* < \infty \tag{11.44}$$

The dimensionless contact strain is defined as

$$\epsilon_c^* = 1.67 \left(\frac{mE'}{S_f}\right) \tag{11.45}$$

where $S_f$ is the material yield or flow stress [Johnson (1985)] which is a complex physical parameter that must be determined by experiment for each metal.

The elastic–plastic micro-hardness $H_{ep}$ can be determined by means of an iterative procedure which requires the following relationship:

$$H_{ep} = \frac{2.76 \, S_f}{\left[1 + (6.5/\epsilon_c^*)^2\right]^{1/2}} \tag{11.46}$$

The elasto–plastic contact conductance model "moves" smoothly between the elastic contact model of Mikic (1974) and the plastic contact conductance model of Cooper et al. (1969) which was modified by Yovanovich (1982), Yovanovich et al. (1982), and Song and Yovanovich (1988) to include the effect of work hardened layers on the deformation of the contacting asperities. The dimensionless contact pressure for elastic–plastic deformation of the contacting asperities is obtained from the following approximate explicit relationship:

$$\frac{P}{H_{ep}} = \left[\frac{0.9272 \, P}{c_1 (1.43 \, \sigma/m)^{c_2}}\right]^{\frac{1}{1 + 0.071 c_2}} \tag{11.47}$$

where the coefficients $c_1, c_2$ are obtained from Vickers micro-hardness tests. Once again, $\sigma$ is specified in microns (μm) in Eq. (11.47). The Vickers micro-hardness coefficients are related to Brinell and Rockwell hardness for a wide range of metals.

### 11.4.1 Correlation Equations for Dimensionless Contact Conductance: Elastic–Plastic Model

The complex elastic–plastic contact model proposed by Sridhar and Yovanovich (1996a) may be approximated by the following correlation equations for the dimensionless contact conductance:

$$C_c = 1.54 \left(\frac{P}{H_{ep}}\right)^{0.94}, \qquad 0 < \epsilon_c^* < 5 \tag{11.48}$$

$$C_c = 1.245 \, b_1 \left(\frac{P}{H_{ep}}\right)^{b_2}, \qquad 5 < \epsilon_c^* < 400 \tag{11.49}$$

$$C_c = 1.25 \left(\frac{P}{H_{ep}}\right)^{0.95}, \qquad 400 < \epsilon_c^* < \infty \tag{11.50}$$

where the elastic–plastic correlation coefficients $b_1, b_2$ depend on the dimensionless contact strain:

$$b_1 = \left(1 + \frac{46690.2}{(\epsilon_c^*)^{2.49}}\right)^{1/30} \tag{11.51}$$

and

$$b_2 = \left[\frac{1}{1 + 2086.9/(\epsilon_c^*)^{1.842}}\right]^{1/600} \tag{11.52}$$

**Example 11.1** A joint is formed by two nominally flat rough Ni 200 surfaces. One surface is lapped and its surface parameters are $\sigma_1 = 0.19\,\mu m$, $m_1 = 0.0240$. The second surface was glass bead-blasted after the lapping process, and its surface parameters are $\sigma_2 = 1.19\,\mu m$, $m_2 = 0.137$. The Vickers micro-hardness indentations are correlated by

$$H_V = c_1 \left(\frac{d_V}{d_0}\right)^{c_2}$$

where $H_V$ is the Vickers micro-hardness in GPa, $d_V$ is the average indentation diagonal whose units are $\mu m$, and $d_0 = 1\,\mu m$ is a reference value. The Vickers micro-hardness correlation coefficients for Ni 200 are $c_1 = 6.30\,GPa$ and $c_2 = -0.264$. The bulk hardness or Brinell hardness is $H_B = 170.4\,kg/mm^2$.

(a) Calculate the values of the contact micro-hardness $H_p$ for the Ni 200 joint for the apparent contact pressures: $P = 0.1, 1, 10\,MPa$. Use the Song–Yovanovich correlation equation:

$$\frac{P}{H_p} = \left[\frac{P}{c_1(1.62\,\sigma/m)^{c_2}}\right]^{1/(1+0.071c_2)}$$

The effective roughness $\sigma$ and slope $m$ are found using

$$\sigma = \sqrt{\sigma_1^2 + \sigma_2^2} = 1.21\,\mu m$$

$$m = \sqrt{m_1^2 + m_2^2} = 0.139$$

Using the given values for $c_1$ and $c_2$ and applied loading values $P$, we find

| P (MPa) | P/H$_p$ | H$_p$ (GPa) |
|---------|---------|-------------|
| 0.1 | $2.62 \times 10^{-5}$ | 3.82 |
| 1 | $2.74 \times 10^{-4}$ | 3.65 |
| 10 | $2.86 \times 10^{-3}$ | 3.50 |

(b) Calculate the value of the contact micro-hardness using the Hegazy approximation which was developed for the relative contact pressure $P/H_p = 10^{-3}$ for the Ni 200, SS 304, and zirconium alloys (Zr-4, Zr-Nb):

$$H_p = (12.2 - 3.54 H_B)\left(\frac{\sigma}{m}\right)^{-0.26}$$

where the units of $H_B$ must be GPa and $\sigma$ in $\mu m$.

Using the calculated values of $\sigma$ and $m$ along with the Brinell hardness converted to $H_B = 1.671$ GPa:

$$H_p = (12.2 - 3.54 \cdot 1.671)(1.21/0.139)^{-0.26} = 3.58 \text{ GPa}$$

This value is very close to the values calculated in the table above.

(c) Use the correlation equations of Sridhar–Yovanovich to calculate values of the Vickers micro-hardness correlation coefficients: $c_1$ and $c_2$ for Ni 200 given the Brinell hardness $H_B$. Using Eqs. (11.17) and (11.18) with $H_B^\star = H_B/3170 = 1671/3170 = 0.525$, we find:

$$c_1 = 6305 \text{ MPa}$$
$$c_2 = -0.253$$

which are very close to the original values used in part (a).

**Example 11.2** A joint is formed by two nominally flat rough SS 304 surfaces. The effective surface roughness parameters of the joint are $\sigma = 0.478 \,\mu\text{m}$, $m = 0.072$. The apparent contact area of the joint is $A_a = 640 \times 10^{-6} \text{ m}^2$. The Brinell hardness is $H_B = 1470 \text{ MPa}$. The elastic properties of SS 304 are $E = 207 \text{ GPa}$ and $v = 0.3$.

The thermal conductivity of SS 304 was correlated with temperature:

$$k_s = 17.02 + 0.0152 \, T, \quad 60 \leq T \leq 250°\text{C}$$

(a) For the apparent contact pressures $P_1 = 80 \text{ kPa}$ and $P_2 = 240 \text{ kPa}$, calculate the contact conductance values $h_c$. Use the extended Cooper–Mikic–Yovanovich (CMY) contact conductance model. Assume that the mean temperature of the joint is $T_m = 330 \text{ K}$.

The thermal conductivity of the SS 304 is found to be $k = 17.89 \text{ W/(m K)}$. Further using Eqs. (11.17) and (11.18) with $H_B^\star = H_B/3170 = 1470/3170 = 0.464$, we find

$$c_1 = 6758 \text{ MPa}$$
$$c_2 = -0.2739$$

Next, we use

$$\frac{P}{H_p} = \left[\frac{P}{c_1(1.62 \, \sigma/m)^{c_2}}\right]^{1/(1+0.071c_2)}$$

where $\sigma$ is in $\mu$m, and

$$C_c = \frac{h_c \, \sigma}{k_s \, m} = 1.25 \left(\frac{P}{H_p}\right)^{0.95}$$

where $\sigma$ is in $m$. The results are summarized in the table below:

| P (kPa) | P/H_p | C_c | h_c (W/(m² K)) |
|---------|-------|-----|----------------|
| 80 | $1.84 \times 10^{-5}$ | $3.957 \times 10^{-5}$ | 106.6 |
| 240 | $5.62 \times 10^{-5}$ | $1.146 \times 10^{-4}$ | 309.2 |

(b) If the micro-gaps are filled with an oil whose thermal conductivity $k_g = 0.145\,\text{W/(m K)}$, calculate the gap conductance $h_g$ values for the two contact pressures given in (a). Calculate the corresponding values of the joint conductance $h_j$.

We assume that the oil is fully wetting and does not interfere with the mechanical contact. To calculate the gap conductance for the oil, we need the relative mean plane separation $Y/\sigma$. We may use an approximation developed by Yovanovich (1982) for Eq. (11.18) or solve Eq. (11.14):

$$\frac{Y}{\sigma} = 1.184[-\ln(3.132 \cdot P/H_p)]^{0.547}$$

Once we have the mean plane separation, the gap conductance is determined from

$$h_g = \frac{k_g}{Y}$$

and the joint conductance from

$$h_j = h_c + h_g$$

Using the above expression for $P/H_p$, we find the following results:

| P (kPa) | P/H_p | λ = Y/σ | Y (μm) | $h_g$ (W/(m² K)) | $h_j$ (W/(m² K)) |
|---------|-------|---------|--------|------------------|------------------|
| 80 | $1.84 \times 10^{-5}$ | 4.117 | 1.968 | 73,679 | 73,785 |
| 240 | $5.62 \times 10^{-5}$ | 3.852 | 1.841 | 78,761 | 79,070 |

(c) For the two contact pressures, calculate the values of the joint resistance $R_j = 1/(A_a h_j)$ when oil occupies the micro-gaps.

Using the joint conductance from (b) and apparent contact area, we find the joint resistance $R_j = 0.0211\,\text{K/W}$ and $R_j = 0.0198\,\text{K/W}$ for each of the two contact pressures, respectively.

## 11.5 Radiation Resistance and Conductance for Conforming Rough Surfaces

The radiation heat transfer across gaps formed by conforming rough solids and filled with a transparent substance (or it is in a vacuum) is complex because the geometry of the micro-gaps is very difficult to characterize and the temperatures of the bounding solids vary in some complex manner because they are coupled to heat transfer by conduction through the micro-contacts.

The radiative resistance and the conductance can be estimated by modeling the heat transfer across the micro-gaps as equivalent to radiative heat transfer between two gray infinite isothermal smooth plates. The radiative heat transfer is given by

$$Q_r = \sigma A_a \mathcal{F}_{12}\left(T_{j1}^4 - T_{j2}^4\right) \tag{11.53}$$

where $\sigma = 5.67 \times 10^{-8}\,\text{W/(m}^2\,\text{K}^4)$ is the Stefan–Boltzmann constant and $T_{j1}, T_{j2}$ are the absolute joint temperatures of the bounding solid surfaces. These temperatures are obtained by extrapolation of the temperature distributions within the bounding solids. The radiative parameter is given by

$$\frac{1}{\mathcal{F}_{12}} = \frac{1}{\epsilon_1} + \frac{1}{\epsilon_2} - 1 \tag{11.54}$$

where $\epsilon_1, \epsilon_2$ are the emissivities of the bounding surfaces. The radiative resistance is given by

$$R_r = \frac{T_{j1} - T_{j2}}{Q_r} = \frac{T_{j1} - T_{j2}}{\sigma A_a \mathcal{F}_{12} \left( T_{j1}^4 - T_{j2}^4 \right)} \tag{11.55}$$

and the radiative conductance by

$$h_r = \frac{Q_r}{A_a \left( T_{j1} - T_{j2} \right)} = \frac{\sigma \mathcal{F}_{12} \left( T_{j1}^4 - T_{j2}^4 \right)}{\left( T_{j1} - T_{j2} \right)} \tag{11.56}$$

The radiative conductance is seen to be a complex parameter which depends on the emissivities $\epsilon_1, \epsilon_2$ and the joint temperatures $T_{j1}, T_{j2}$. For many interface problems, the following approximation can be used to calculate the radiative conductance:

$$\frac{\left( T_{j1}^4 - T_{j2}^4 \right)}{\left( T_{j1} - T_{j2} \right)} \approx 4 \left( \overline{T}_j \right)^3 \tag{11.57}$$

where the mean joint temperature is defined as

$$\overline{T}_j = \frac{1}{2} \left( T_{j1} + T_{j2} \right) \tag{11.58}$$

If we assume black-body radiation across the gap, then $\epsilon_1 = 1, \epsilon_2 = 1$ gives $\mathcal{F}_{12} = 1$. This assumption gives the upper bound on the radiation conductance across gaps formed by conforming rough surfaces. If one further assumes that $T_{j2} = 300$ K and $T_{j1} = T_{j2} + \Delta T_j$, then one can calculate the radiation conductance for a range of values of $\Delta T_j$ and $\overline{T}_j$. The values of $h_r$ for black surfaces represent the maximum radiative heat transfer across the micro-gaps. For micro-gaps formed by real surfaces, the radiative heat transfer rates may by smaller. Table 11.1 shows that when the joint temperature is $\overline{T}_j = 800$ K and $\Delta T_j = 1000$ K, the maximum radiation conductance is approximately $161.5$ W/(m² K). This value is much smaller than the contact and gap conductances for most applications, where $\overline{T}_j < 600$ K and $\Delta T_j < 200$ K. The radiation conductance becomes relatively important when the interface is formed by two very rough, very hard, low conductivity solids under very light contact pressures. Therefore, for many practical applications, the radiative conductance can be neglected, but not forgotten.

**Table 11.1** Radiative conductances for black surfaces.

| $\Delta T_j$ | $\overline{T}_j$ | $h_r$ |
| --- | --- | --- |
| 100 | 350 | 9.92 |
| 200 | 400 | 15.42 |
| 300 | 450 | 22.96 |
| 400 | 500 | 32.89 |
| 500 | 550 | 45.53 |
| 600 | 600 | 61.24 |
| 700 | 650 | 80.34 |
| 800 | 700 | 103.2 |
| 900 | 750 | 130.1 |
| 1000 | 800 | 161.5 |

## 11.6 Gap Conductance for Large Parallel Isothermal Plates

Two infinite isothermal surfaces form a gap of uniform thickness $d$ which is much greater than the roughness of both surfaces $d \gg \sigma_1$ and $\sigma_2$. The gap is filled with a stationary monatomic or diatomic gas. The boundary temperatures are $T_1$ and $T_2$, where $T_1 > T_2$. The Knudsen number for the gap is defined as $Kn = \Lambda/d$, where $\Lambda$ is the molecular mean free path of the gas which depends on the gas temperature and its pressure.

The gap can be separated into three zones: two boundary zones which are associated with the two solid boundaries and a central zone. The boundary zones have thicknesses which are related to the molecular mean free paths $\Lambda_1$ and $\Lambda_2$, where

$$\Lambda_1 = \Lambda_0 \left( \frac{T_1}{T_0} \right) \left( \frac{P_{g,0}}{P_g} \right) \quad \text{and} \quad \Lambda_2 = \Lambda_0 \left( \frac{T_2}{T_0} \right) \left( \frac{P_{g,0}}{P_g} \right) \tag{11.59}$$

and $\Lambda_0, T_0$ and $P_{g,0}$ represent the molecular mean free path and the reference temperature and gas pressure. In the boundary zones, the heat transfer is due to gas molecules that move back and forth between the solid surface and other gas molecules located at distances $\Lambda_1$ and $\Lambda_2$ from both solid boundaries. The energy exchange between the gas and solid molecules is *imperfect*. At the hot solid surface at temperature $T_1$, the gas molecules which leave the surface after contact are at some temperature $T_{g,1} < T_1$, and at the cold solid surface at temperature $T_2$, the gas molecules which leave the surface after contact are at some temperature $T_{g,2} > T_2$. The two boundary zones are called *slip* regions.

In the central zone, whose thickness is modeled as $d - \Lambda_1 - \Lambda_2$, and whose temperature range is $T_{g,1} \geq T \geq T_{g,2}$, heat transfer occurs primarily by molecular diffusion. Fourier's law of conduction can be used to determine heat transfer across the central zone.

There are two heat flux asymptotes corresponding to very small and very large Knudsen numbers. They are

$$Kn \to 0, \quad q \to q_0 = k_g \frac{T_1 - T_2}{d}, \qquad \text{continuum} \tag{11.60}$$

and

$$Kn \to \infty, \quad q \to q_\infty = k_g \frac{T_1 - T_2}{M}, \qquad \text{free molecules} \tag{11.61}$$

where

$$M = \alpha\beta\Lambda = \left( \frac{2 - \alpha_1}{\alpha_1} + \frac{2 - \alpha_2}{\alpha_2} \right) \left( \frac{2\gamma}{(\gamma + 1)Pr} \right) \Lambda \tag{11.62}$$

and

$k_g$ = thermal conductivity
$\alpha_1, \alpha_2$ = accommodation coefficients
$\gamma$ = ratio of specific heats
$Pr$ = Prandtl number

The gap conductance defined as $h_g = q/(T_1 - T_2)$ has two asymptotes:

$$\text{for} \quad Kn \to 0, \quad h_g \to \frac{k_g}{d}, \qquad \text{for} \quad Kn \to \infty, \quad h_g \to \frac{k_g}{M}$$

For the entire range of the Knudsen number, the gap conductance is given by the relationship:

$$h_g = \frac{k_g}{d + M} \quad \text{for} \quad 0 < Kn < \infty \tag{11.63}$$

**Figure 11.5** Gap conductance model and data for two large parallel isothermal plates. Source: Song et al. (1992b)/ with permission of The American Society of Mechanical Engineers.

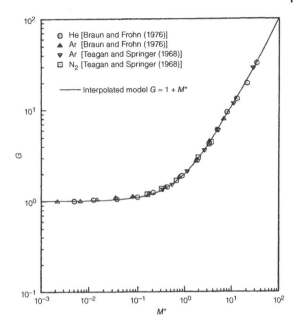

This relatively simple relationship covers the continuum: $0 < Kn < 0.1$, slip: $0.1 < Kn < 10$, and free molecule: $10 < Kn < \infty$, regimes. Song (1988) introduced the dimensionless parameters:

$$G = \frac{k_g}{h_g d} \quad \text{and} \quad M^\star = \frac{M}{d} \tag{11.64}$$

and recast the above relationship as

$$G = 1 + M^\star \quad \text{for} \quad 0 < M^\star < \infty \tag{11.65}$$

The accuracy of the simple parallel plate gap model was compared against the data (argon and nitrogen) of Teagan and Springer (1968) and the data (argon and helium) of Braun and Frohn (1976). The excellent agreement between the simple gap model and all data are shown in Figure 11.5.

The simple gap model forms the basis of the gap model for the joint formed by two conforming rough surfaces.

## 11.7 Gap Conductance for Joint Between Conforming Rough Surfaces

If the gap between two conforming rough surfaces as shown in Figure 11.1 is occupied by a gas, then conduction heat transfer will occur across the gap. This heat transfer is characterized by the gap conductance defined as

$$h_g = \frac{\Delta T_j}{Q_g} \tag{11.66}$$

with $\Delta T_j$ as the "effective" temperature drop across the gas gap, and $Q_g$ is the heat transfer rate across the gap. Because the local gap thickness and the local temperature drop vary in very complicated ways throughout the gap, it is difficult to develop a simple gap conductance model.

**Table 11.2** Models and correlation equations for gap conductance for conforming rough surfaces.

| Authors | Models and correlations |
|---|---|
| Cetinkale and Fishenden (1951) | $h_g = \dfrac{k_g}{0.305 b_t + M}$ |
| Rapier et al. (1963) | $h_g = k_g \left[ \dfrac{1.2}{2b_t + M} + \dfrac{0.8}{2b_t} \ln\left( 1 + \dfrac{2b_t}{M} \right) \right]$ |
| Shlykov (1965) | $h_g = \dfrac{k_g}{b_t} \left[ \dfrac{10}{3} + \dfrac{10}{X} + \dfrac{4}{X^2} - 4\left\{ \dfrac{1}{X^3} + \dfrac{3}{X^2} + \dfrac{2}{X} \ln(1 + X) \right\} \right]$ |
| Veziroglu (1967) | $h_g = \dfrac{k_g}{0.264\, b_t + M} \quad$ for $\quad b_t > 15\ \mu\text{m}$ <br><br> $h_g = \dfrac{k_g}{1.78\, b_t + M} \quad$ for $\quad b_t < 15\ \mu\text{m}$ |
| Lloyd et al. (1973) | $h_g = \dfrac{k_g}{\delta + \beta \Lambda/(\alpha_1 + \alpha_2)} \quad \delta$ not given |
| Garnier and Begej (1979) | $h_g = k_g \left[ \dfrac{\exp(-1/Kn)}{M} + \dfrac{1 - \exp(-1/Kn)}{\delta + M} \right] \quad \delta$ not given |
| Loyalka (1982) | $h_g = \dfrac{k_g}{\delta + M + 0.162\left( 4 - \alpha_1 - \alpha_2 \right) \beta \Lambda} \quad \delta$ not given |
| Yovanovich et al. (1982a,b) | $h_g = \dfrac{k_g/\sigma}{\sqrt{2\pi}} \displaystyle\int_0^\infty \dfrac{\exp\left[ -(Y/\sigma - t/\sigma)^2/2 \right]}{t/\sigma + M/\sigma}\, d(t/\sigma)$ <br><br> $\dfrac{Y}{\sigma} = \sqrt{2}\,\mathrm{erfc}^{-1}\left( \dfrac{2P}{H_p} \right)$ <br><br> $\dfrac{P}{H_p} = \left[ \dfrac{P}{c_1(1.62\sigma/m)^{c_2}} \right]^{\frac{1}{1+0.071 c_2}}$ |

Source: Adapted from Song (1988).

Several gap conductance models and correlation equations have been presented by a number of researchers [Cetinkale and Fishenden (1951), Rapier et al. (1963), Shlykov (1965), Veziroglu (1967), Lloyd et al. (1973), Garnier and Begej (1979), Loyalka (1982), Yovanovich et al. (1982a,b)]; they are given in Table 11.2.

The parameters that appear in Table 11.2 are $b_t = 2\left( CLA_1 + CLA_2 \right)$, where $CLA_i$ is the center-line-average surface roughness of the two contacting surfaces, $M = \alpha\beta\Lambda$, $X = b_t/M$, $\sigma = \sqrt{\sigma_1^2 + \sigma_2^2}$, where the units of $\sigma$ are $\mu$m. The Knudsen number $Kn$ that appears in the Garnier and Begej (1979) correlation equation is not defined.

Song and Yovanovich (1987), Song (1988), and Song et al. (1993b) reviewed the models and correlation equations given in Table 11.2. They found that for some of the correlation equations, the required gap thickness $\delta$ was not defined, and for other correlation equations, an empirically based average gap thickness was specified that is constant, independent of variations of the apparent contact pressure. The gap conductance model developed by Yovanovich et al. (1982a,b) is the only one that accounts for the effect of mechanical load and physical properties of the contacting asperities on the gap conductance. This model is presented below.

The gap conductance model for conforming rough surfaces was developed, modified, and verified by Yovanovich et al. (1982b), Hegazy (1985), Song and Yovanovich (1987), Negus and Yovanovich (1988), and Song et al. (1992b, 1993b).

The gap contact model is based on surfaces having Gaussian height distributions and it also accounts for the mechanical deformation of the contacting surface asperities. The development of the gap conductance model is presented in Yovanovich (1982, 1986), Yovanovich et al. (1982b), and Yovanovich and Antonetti (1988).

The gap conductance model is expressed in terms of an integral:

$$h_g = \frac{k_g}{\sigma} \frac{1}{\sqrt{2\pi}} \int_0^\infty \frac{\exp\left[-(Y/\sigma - u)^2/2\right]}{u + M/\sigma} \, du = \frac{k_g}{\sigma} I_g \tag{11.67}$$

where $k_g$ is the thermal conductivity of the gas "trapped" in the gap and $\sigma$ is the "effective" surface roughness of the joint, and $u = t/\sigma$ is the dimensionless local gap thickness. The integral depends on two independent dimensionless parameters: (i) $Y/\sigma$, the mean plane separation, and (ii) $M/\sigma$, the relative gas "rarefaction" parameter.

The relative mean planes separation for plastic and elastic contact are given by the relationships:

$$\left(\frac{Y}{\sigma}\right)_{\text{plastic}} = \sqrt{2} \, \text{erfc}^{-1}\left(\frac{2P}{H_p}\right)$$
$$\left(\frac{Y}{\sigma}\right)_{\text{elastic}} = \sqrt{2} \, \text{erfc}^{-1}\left(\frac{4P}{H_e}\right) \tag{11.68}$$

The relative contact pressures: $P/H_p$ for plastic deformation and $P/H_e$ for elastic deformation can be determined by means of appropriate relationships.

The gas "rarefaction" parameter is $M = \alpha\beta\Lambda$, where the gas parameters are defined as

$$\alpha = (2 - \alpha_1)/\alpha_1 + (2 - \alpha_2)/\alpha_2 \tag{11.69}$$
$$\beta = 2\gamma/[(\gamma + 1)Pr] \tag{11.70}$$
$$\Lambda = \Lambda_0 \left(T_g/T_{g,0}\right) \left(P_g/P_{g,0}\right) \tag{11.71}$$

where $\alpha$ is the accommodation coefficient which accounts for the efficiency of gas–surface energy exchange. There is a large body of research dealing with experimental and theoretical aspects of $\alpha$ for various gases in contact with metallic surfaces under various surface conditions and temperatures [Wiedmann and Trumpler (1946), Hartnett (1961), Wachman (1962), Thomas (1967), Semyonov et al. (1984), Thomas and Loyalka (1982)]. Song and Yovanovich (1987) and Song et al. (1992a, 1993b) examined the several gap conductance models available in the literature and the experimental data and models for the accommodation coefficients.

Song and Yovanovich (1987) developed a correlation for the accommodation for "engineering" surfaces (i.e. surfaces with adsorbed layers of gases and oxides). They proposed a correlation which is based on experimental results of numerous investigators for monatomic gases. The relationship was extended by the introduction of a "monatomic equivalent molecular weight" to diatomic and polyatomic gases. The final correlation is

$$\begin{aligned} \alpha = \exp\left(C_0 T\right) & \left[M_g/\left(C_1 + M_g\right)\right. \\ & \left. + \left\{1 - \exp\left(C_0 T\right)\right\} \left\{2.4\mu/(1 + \mu)^2\right\}\right] \end{aligned} \tag{11.72}$$

with

$$C_0 = -0.57$$
$$T = (T_s - T_0)/T_0$$
$$M_g = M_g \text{ for monatomic gases}$$
$$\qquad = 1.4 \, M_g \text{ for diatomic and polyatomic gases}$$

$$C_1 = 6.8, \text{units of } M_g \ (\text{g/mol})$$

$$\mu = M_g/M_s$$

where $T_s$ and $T_0 = 273\text{K}$ are the absolute temperatures of the surface and the gas, and $M_g$ and $M_s$ are the molecular weights of the gas and the solid, respectively.

The agreement between the predictions according to the above correlation and the published data for diatomic and polyatomic gases was within $\pm 25\%$.

The gas parameter $\beta$ depends on the specific heat ratio: $\gamma = C_p/C_v$ and the Prandtl number $Pr$. The molecular mean free path of the gas molecules $\Lambda$ depends on the type of gas, the gas temperature $T_g$ and the gas pressure $P_g$, and the reference values of the mean free path $\Lambda_0$, the gas temperature $T_{g,0}$, and the gas pressure $P_{g,0}$, respectively.

Wesley and Yovanovich (1986) compared the predictions of the gap conductance model and experimental measurements of gaseous gap conductance between the fuel and clad of a nuclear fuel rod. The agreement was very good, and the model was recommended for fuel pin analysis codes.

The gap integral can be computed accurately and easily by means of Computer Algebra Systems. Negus and Yovanovich (1988) developed the following correlation equations for the gap integral:

$$I_g = \frac{f_g}{\frac{Y}{\sigma} + \frac{M}{\sigma}} \tag{11.73}$$

In the range: $2 \leq Y/\sigma \leq 4$

$$f_g = 1.063 + 0.0471 \, (4 - Y/\sigma)^{1.68} \left[\ln (\sigma/M)\right]^{0.84}, \qquad 0.01 \leq M/\sigma \leq 1$$

$$f_g = 1 + 0.06(\sigma/M)^{0.8}, \qquad 1 \leq M/\sigma < \infty \tag{11.74}$$

The correlation equations have a maximum error of approximately 2%.

## 11.8 Joint Conductance for Conforming Rough Surfaces

The joint conductance for a joint between two conforming rough surfaces is

$$h_j = h_c + h_g \tag{11.75}$$

when radiation heat transfer across the gap is neglected. The relationship is applicable to joints that are formed by elastic, plastic, or elastic–plastic deformation of the contacting asperities. The mode of deformation will influence $h_c$ and $h_g$ through the relative mean plane separation parameter $Y/\sigma$.

The gap and joint conductances are compared against data [Song (1988)] obtained for three types of gases: argon, helium, and nitrogen over a gas pressure range between 1 and 700 Torr. The gases occupied gaps which were formed by conforming rough Ni 200 and stainless steel type 304 metals. In all tests, the metals forming the joint were identical, and one surface was flat and lapped, while the other surface was flat and glass bead-blasted.

The gap and joint conductance models were compared against data obtained for relatively light contact pressures where the gap and contact conductances were comparable. Figure 11.6 shows plots of the joint conductance data and the model predicts for very rough stainless steel type 304 surfaces at $Y/\sigma = 1.6 \times 10^{-4}$. The agreement between the data for argon, helium, and nitrogen is very good for gas pressures between approximately 1 and 700 Torr. At the low gas pressure of 1 Torr, the measured and predicted joint conductance values for the three gases differ by a few percent because $h_g \ll h_c$ and $h_j \approx h_c$. As the gas pressure increases, there is a large increase in the

**Figure 11.6** Joint conductance model and data for conforming rough stainless steel 304 surfaces. Source: Adapted from Song (1988).

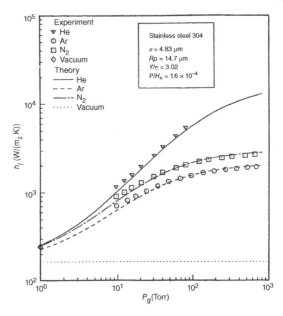

**Figure 11.7** Gap conductance model and data for conforming rough Ni 200 surfaces. Source: Adapted from Song (1988).

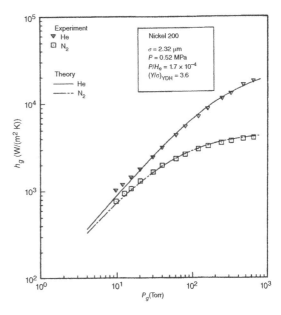

joint conductances because the gap conductances are increasing rapidly. The joint conductances for argon and nitrogen approach asymptotes for gas pressures approaching 1 atmosphere. The joint conductances for helium are greater than for argon and nitrogen, and the values do not approach an asymptote in the same pressure range. The asymptote for helium is approached at gas pressures greater than 1 atm.

Figure 11.7 shows the experimental and theoretical gap conductances as points and curves for nitrogen and helium for gas pressures between approximately 10 and 700 Torr. The relative contact pressure is $1.7 \times 10^{-4}$ based on the plastic deformation model. The joint was formed by Ni 200 surfaces (one flat and lapped and the second flat and glass bead-blasted). The data were obtained

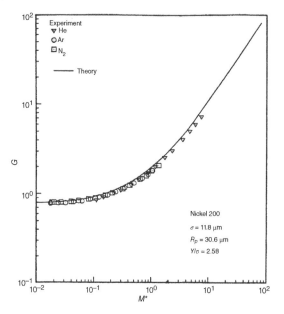

**Figure 11.8** Dimensionless gap conductance model and data for conforming rough Ni 200 surfaces. Source: Song et al. (1993b)/ with permission of The American Society of Mechanical Engineers.

by subtracting the theoretical value of $h_c$ from the measured values of $h_j$ to get the values of $h_g$ that appear on the plots. The agreement between the data and the predicted curves is very good.

Figure 11.8 shows the experimental data for argon, nitrogen, and helium, and the dimensionless theoretical curve for the gap model recast as [Song et al. (1993b)]:

$$G = 1 + M^\star \tag{11.76}$$

where

$$G = \frac{k_g}{h_g Y} \quad \text{and} \quad M^\star = \frac{M}{Y} = \frac{\alpha \beta \Lambda}{Y} \tag{11.77}$$

There is excellent agreement between the model and the data over the entire range of the "gas-gap" parameter $M^\star$. The joint was formed by very rough conforming Ni 200 surfaces. The plastic deformation model was used to calculate $Y$.

The points for $M^\star < 0.01$ correspond to the high gas pressure tests (near 1 atmosphere), and the points for $M^\star > 2$ correspond to the low-gas pressure tests.

**Example 11.3**   The micro-gaps of the joint described in Example 11.2 are occupied with dry air at one atmosphere gas pressure where the mean joint temperature is $T_g = 350$ K.

(a) Calculate the values of the gap conductance for the two contact pressures. The values of the thermal conductivity of the air can be calculated by means of the following Sutherland relation:

$$k_g = k_0 \left( \frac{T_g}{T_0} \right)^{3/2} \left( \frac{T_0 + S}{T_g + S} \right)$$

with $k_0 = 0.0241$ W/(m K), $T_0 = 273$ K, and the Sutherland factor is $S = 194$ K. Use the following air properties:

$$\alpha_1 = 0.87, \quad \alpha_2 = 0.90, \quad \gamma = 1.61, \quad Pr = 0.70, \quad \Lambda_0 = 64 \text{ nm}$$

The reference gas temperature and pressure for the molecular mean free path are $T_0 = 288$ K and $P_{g,0} = 760$ Torr.

The thermal conductivity of the gas (air) is determined from the above equation (also provided in Appendix C):

$$k_g = 0.0241 \left( \frac{350}{288} \right)^{3/2} \left( \frac{288 + 194}{350 + 194} \right) = 0.0286 \text{ W/(m K)}$$

Next, we need the intermediate parameters defined below from Eqs. (11.67) and (11.69)–(11.71):

$$\alpha = \frac{2 - \alpha_1}{\alpha_1} + \frac{2 - \alpha_2}{\alpha_2} = 2.521$$

$$\beta = \frac{2\gamma}{(\gamma + 1)Pr} = 1.762$$

$$\Lambda = \Lambda_0 \frac{T_g}{T_{g,0}} \frac{P_g}{P_{g,0}} = 77.78 \text{ nm}$$

$$M = \alpha\beta\Lambda = 345.6 \text{ nm}$$

Finally, using Eqs. (11.73) with (11.74) we find:

$$I_g = \frac{f_g}{Y/\sigma + M/\sigma}$$

and

$$h_g = \frac{k_g}{\sigma} I_g$$

The results of the calculations are provided in the table below:

| P (kPa) | Y/σ | $f_g$ | $I_g$ | $h_g$ (W/(m² K)) |
|---------|-----|-------|-------|------------------|
| 80 | 4.117 | 1.0635 | 0.2197 | 13,789 |
| 240 | 3.852 | 1.0634 | 0.232 | 14,560 |

(b) Estimate the gap radiation conductance if the emissivities of the two surfaces are $\epsilon_1 = 0.80$ and $\epsilon_2 = 0.88$, and the temperature drop across the joint is $\Delta T_j = 80$ K.

To find the radiation conductance of the joint, we use Eq. (11.56):

$$h_r = \frac{\sigma \mathcal{F}_{12} \left( T_{j1}^4 - T_{j2}^4 \right)}{\left( T_{j1} - T_{j2} \right)} \approx \sigma \mathcal{F}_{12} 4(\overline{T}_j)^3$$

where

$$\mathcal{F}_{12} = \left[ \frac{1}{\epsilon_1} + \frac{1}{\epsilon_2} - 1 \right]^{-1} = 0.721$$

such that

$$h_r \approx \sigma \mathcal{F}_{12} 4(\overline{T}_j)^3 \approx 5.67 \times 10^{-8}(0.721)(4)(350)^3 \approx 7.01 \text{ W/(m}^2 \text{ K)}$$

## 11.9 Joint Conductance for Conforming Rough Surfaces: Scale Analysis Approach

A novel scaling analysis approach was employed by Bahrami (2004) to develop simpler models for joint conductance of conforming rough surfaces. The approach taken is given in Bahrami et al. (2004a,b). The general assumptions used in Section 11.8 are also employed again to obtain relations for the contact, gap, and joint conductances $(h_s, h_g, h_j)$ or the contact, gap, and joint resistances: $(R_s, R_g, R_j)$. The conductances and resistances are related through the associated areas: $(A_s, A_g, A_j)$ and the areas are related as

$$A_a = A_s + A_g \qquad (11.78)$$

and the resistances and conductances are related by the general expression:

$$R = \frac{1}{hA} \qquad (11.79)$$

It is also assumed as before that radiation heat transfer across the gaps is negligible. The joint is formed by the mechanical contact of a smooth flat surface and a flat rough surface which consists of numerous asperities (produced by micro-glass bead-blasting) which have Gaussian height distribution and random distribution over the apparent contact area. The micro-geometry is characterized by the asperity roughness $\sigma$ and asperity slope $m$. The smooth flat surface is highly polished, and its assumed that the asperities have negligible heights so that $\sigma \approx 0$ and $m = 0$. Whenever contact occurs between the rough surface and the smooth flat, the contacting asperities deform plastically and the contact micro-hardness is taken to be related to the Vickers micro-hardness of the softer surface.

The gap which is formed as a result of the plastic deformation of the contacting asperities is characterized by the length $Y$ which is the distance from the flat surface to the plane that passes through the contacting asperities. Whenever the gap is filled with a substance, such as gas, the overall thermal resistance is given by the simple relation:

$$R_g = \frac{Y}{k_g A_g} \qquad (11.80)$$

where $k_g$ is the thermal conductivity of the gas and $A_g < A_a$ is the gas conduction area.

In the present approach, the effective thickness of the gas gap $d$ is approximated as $d = Y + M$, where the gas parameter that accounts for temperature jump at the gas–solid interfaces is $M = \alpha\beta\Lambda$ which depends on several gas properties such as the thermal accommodation coefficients $(\alpha_1, \alpha_2)$, Prandtl number $(Pr)$, and the molecular mean free path of the gas $(\Lambda)$ which depends on the gas pressure and temperature. Material in Section 11.9 should be consulted for details of heat conduction across gas micro-gaps.

The development and validation of the scale model of Bahrami (2004) are based on selected experimental data of Hegazy (1985) and Song (1988). Hegazy (1985) conducted several tests primarily under vacuum conditions and three tests with helium or nitrogen in the microscopic gaps at a gas pressure of $P_g = 40$ Torr. The surface roughness ranged from ($\sigma = 4.02$ μm and $m = 0.168$) to ($\sigma = 6.29$ μm and $m = 0.195$).

In the tests conducted by Hegazy (1985) and Song (1988), the joint temperature levels and the temperature drops across the gaps were such that radiation heat transfer across the gaps was negligible and therefore radiation was ignored in the development of the new joint resistance model.

Details of the development of the general scaling model are given in Chapter 12 and in Bahrami (2004). The salient parts of the new model are given below.

Application of the scaling model to the plastic deformation of the contacting asperities yields the following relation [Bahrami (2004)]:

$$R^\star = \frac{c}{P^\star} \tag{11.81}$$

where the dimensionless contact pressure is defined in terms of the micro-hardness as $P^\star = P/H^\star$, where the unknown scaling parameter $c$ must be found from extensive experimental data obtained for joints formed by contacting conforming rough surfaces. The micro-resistance can be recast as [Bahrami (2004)]:

$$R_s = \frac{\pi c (\sigma/m) H^\star}{2 k_s F} \tag{11.82}$$

where the micro-hardness of the contacting asperities is defined as [Bahrami (2004)]:

$$H^\star = c_1 \left( \sigma'/m \right)^{c_2} \tag{11.83}$$

where $\sigma' = \sigma/\sigma_0$ and $\sigma_0 = 1$ µm. The Vickers micro-hardness correlation coefficients are $c_1$ and $c_2$. The Vickers micro-hardness correlation coefficients for the metals of interest to us are given in Appendix B. The total mechanical load is $F$.

In order to find the value of the empirical constant, Bahrami (2004) chose data obtained from tests conducted under vacuum conditions from Antonetti (1983), Hegazy (1985), Milanez et al. (2003), McWaid and Marshall (1992a,b), and Nho (1989) to calculate the corresponding dimensionless contact pressure $P^\star = P/H^\star$ and dimensionless micro-resistance $R_s^\star = 2 k_s L R_{exp}$, with the dimensionless length scale $L = b_L^2/(\sigma/m)$ with $b_L$ is the radius of the circular apparent contact area. Approximately, 610 data points are shown plotted in Figure 11.9 with $c = 0.36$. The light load data

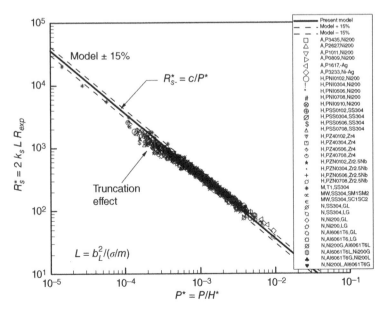

**Figure 11.9** Comparison between scale analysis model and data for conforming rough limit. Source: Bahrami et al. (2004b)/ with permission of The American Society of Mechanical Engineers.

are located near $P^\star = 10^{-4}$ which is near the truncation region where the experimental data frequently lie above the theoretical predictions. This is attributed to the fact that the asperity heights do not reach the theoretical maximum values.

As the dimensionless contact pressure increases, the truncation effect becomes negligible as more and more asperities come into contact and the data approach the prediction of the proposed model which is given by the following:

$$R_s = \frac{0.565 H^\star (\sigma/m)}{k_s F} \tag{11.84}$$

The overall gap resistance is given by the following simplified relation:

$$R_g = \frac{d + M}{k_g A_g} \tag{11.85}$$

where $d = Y/2\sqrt{\pi}$.

The overall joint resistance is given by

$$\frac{1}{R_j} = \frac{1}{R_s} + \frac{1}{R_g} \tag{11.86}$$

Since the scaling model depends on plastic deformation of the contacting asperities, it is assumed that the contribution of the micro-contacts is primary and that it is present for all gas pressures. The joint and gas resistances are normalized with respect to $R_s$.

We multiply by $R_s$ to get

$$\frac{R_s}{R_j} = 1 + \frac{R_s}{R_g} \tag{11.87}$$

Now, we invert the foregoing relation and define the normalized joint resistance as $R_j^\star = R_j/R_s$ to get

$$R_j^\star = \frac{1}{1 + R_s/R_g} \tag{11.88}$$

The normalized joint resistance is plotted in Figure 11.10 as $R_j/R_s$ versus $R_s/R_g$ over four decades from $10^{-2}$ to $10^2$. As $R_s/R_g \to 0$ corresponding to vacuum conditions, $R_j/R_s \to 1$ or $R_j = R_s$. As $R_s/R_g \to \infty$ corresponding to very light loads (contact pressures), then $R_j/R_s \to R_g/R_s$ and $R_j = R_g$. The ratio of $R_s/R_g$ decreases either by a decrease in the gas pressure or an increase of the external load. The trends of the $R_j/R_s$ curve are bounded by the curves corresponding to the $R_g/R_s$ and $R_s/R_s$ curves. The transition region lies in the range: $0.10 < R_s/R_g < 10$.

The effect of mechanical load on the joint resistance is shown in Figure 11.11, and the effect of gas pressure on the joint resistance is shown in Figure 11.12 for four decades of contact pressure $P(\text{MPa})$ and six decades of gas pressure $P_g(\text{Torr})$, respectively. For both sets of experiments, the test parameters are shown in the legend of each plot.

The foregoing simplified joint resistance model for the conforming rough surface contact was compared with experimental data of Hegazy (1985) and Song (1988). The test parameters of the Hegazy (1985) experiments are given in Table 11.3, and the test parameters of the Song (1988) experiments are given in Table 11.4. The test conditions for the stainless steel–nitrogen joint are given in Table 11.5, and the ranges of the test parameters are found in Table 11.6. For the tests with gases (argon, helium, and nitrogen) in the gaps, the gas properties are given in Table 11.7.

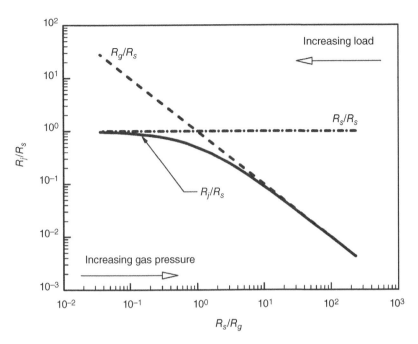

**Figure 11.10** Nondimensional thermal joint resistance. Source: Bahrami et al. (2004a)/American Institute of Aeronautics and Astronautics.

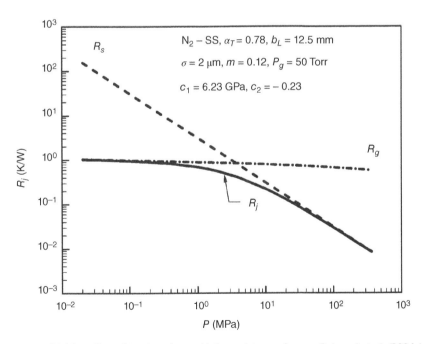

**Figure 11.11** Effect of load on thermal joint resistance. Source: Bahrami et al. (2004a)/ American Institute of Aeronautics and Astronautics.

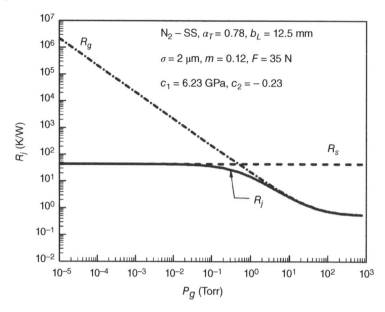

**Figure 11.12** Effect of nitrogen pressure on thermal joint resistance. Source: Bahrami et al. (2004a)/ American Institute of Aeronautics and Astronautics.

**Table 11.3** Summary of Hegazy (1985) experiments.

| Test | Gas | $P_g$ (Torr) | $\sigma$ (μm) | $m$ |
|------|------|------|------|------|
| T1 | $N_2$ | 562–574 | 5.65 | 0.153 |
| T2 | $N_2$ He | Vacuum, 40 | 5.61 | 0.151 |
| T3 | $N_2$ He | Vacuum, 40 | 6.29 | 0.195 |
| T4 | $N_2$ He | Vacuum, 40 | 4.02 | 0.168 |

Source: Adapted from Hegazy (1985).

**Table 11.4** Summary of Song (1988) experiments.

| Test | Solid–Gas | $P$ (MPa) | $\sigma$ (μm) | $m$ |
|------|------|------|------|------|
| T1 | SS–$N_2$, Ar, He | 0.595–0.615 | 1.53 | 0.090 |
| T2 | SS–$N_2$, Ar, He | 0.467–0.491 | 4.83 | 0.128 |
| T3 | Ni–$N_2$, Ar, He | 0.511–0.530 | 2.32 | 0.126 |
| T4 | Ni–$N_2$, Ar, He | 0.371–0.389 | 11.8 | 0.206 |
| T5 | SS–$N_2$, He | 0.403–7.739 | 6.45 | 0.132 |
| T6 | SS–$N_2$, He | 0.526–8.713 | 2.09 | 0.904 |
| T7 | Ni–$N_2$, He | 0.367–6.550 | 11.8 | 0.206 |

Source: Adapted from Song (1988).

**Table 11.5** Input parameters for a typical stainless steel nitrogen joint.

| Parameter | Value |
|-----------|-------|
| $\alpha\,(\text{SS–N}_2)$ | 0.78 |
| $b_L$ | 12.5 mm |
| $\sigma$ | 2 μm |
| $m$ | 0.12 |
| $F$ | 35 N |
| $\Lambda_0$ | 62.8 nm |
| $k_g, k_s$ | 0.31, 20 W/(m K) |
| $c_1, c_2$ | 6.23 GPa, −0.23 |

**Table 11.6** Range of parameters for experimental data.

| Parameter | Range |
|-----------|-------|
| $F$ | 69.7–4357 N |
| $P$ | 0.14–8.8 MPa |
| $k_s$ | 19.2–72.5 W/(m K) |
| $m$ | 0.08–0.205 |
| $P_g$ | $10^{-5}$–760 Torr |
| $\alpha_T$ | 0.55–0.9 |
| $\sigma$ | 1.52–11.8 μm |

Source: Adapted from Song (1988).

**Table 11.7** Properties of interstitial gases.

| Gas | $k_g$ (W/(m K)) | Pr | $\alpha_T$ | $\gamma$ | $\Lambda_0$ (nm) |
|-----|-----------------|-----|-----------|----------|------------------|
| Ar | $0.018 + 4.05 \times 10^{-5}T$ | 0.67 | 0.90 | 1.67 | 66.6 |
| He | $0.147 + 3.24 \times 10^{-4}T$ | 0.67 | 0.55 | 1.67 | 186.0 |
| $N_2$ | $0.028 + 5.84 \times 10^{-5}T$ | 0.69 | 0.78 | 1.41 | 62.8 |

The 516 experimental vacuum data points from Hegazy (1985) shown in Figure 11.13 and Song (1988) shown in Figure 11.14 are plotted as $R_j$ versus $R_{exp}$ over three decades of $R_{exp}$ as seen in Figure 11.15. The RMS difference between the present conforming rough surface model and the vacuum data is approximately 7.3%.

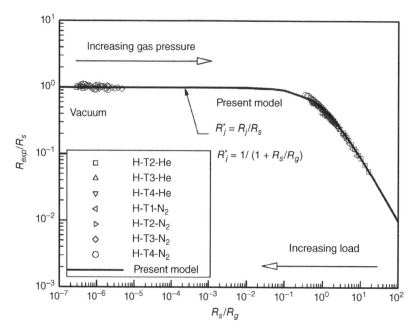

**Figure 11.13** Comparison of model with Hegazy (1985) data. Source: Bahrami et al. (2004a)/American Institute of Aeronautics and Astronautics.

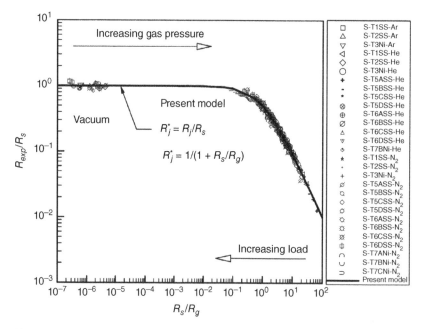

**Figure 11.14** Comparison of model with Song (1988) data. Source: Bahrami et al. (2004a)/American Institute of Aeronautics and Astronautics.

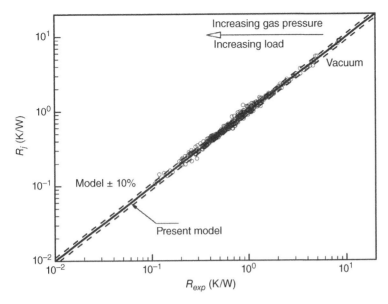

**Figure 11.15** Comparison of model with Song (1988) and Hegazy (1985) data. Source: Bahrami et al. (2004a)/American Institute of Aeronautics and Astronautics.

## 11.10 Joint Conductance Enhancement Methods

In many electronics packages, the thermal joint conductance across a particular joint must be improved for the thermal design to meet its performance objectives. If the joint cannot be made permanent because of servicing or other considerations, the joint conductance must be "enhanced," i.e. it must be improved above the "bare" joint situation utilizing one of several known techniques such as application of thermal interface materials (TIM) such as thermal grease, grease filled with particles (also called paste), oils, and phase change materials (PCM), for example. Enhancement of the joint conductance has also been achieved by the insertion of soft metallic foils into the joint, or by the use of a relatively soft metallic coating on one or both surfaces. More recently, soft nonmetallic materials such as polymers and rubber have been used.

One may consult review articles by Fletcher (1988, 1990), Madhusudana and Fletcher (1986), Madhusudana (1996), Marotta and Fletcher (1996), Prasher (2001), and Savija et al. (2002a,b). Other pertinent references may be found above reviews.

### 11.10.1 Metallic Coatings and Foils

An effective method for enhancement of joint conductance consists of vapor deposition of a very thin soft metallic layers on the surface of the substrate. The layer thickness is often less than 100 μm; it is in "perfect" thermal and mechanical contact with the substrate; and its bulk resistance is negligibly small relative to the contact resistance. The thermal resistance at the layer-substrate interface is also negligible.

A comprehensive treatment of the theoretical development and experimental verification of the thermo-mechanical model can be found in Antonetti (1983) and Antonetti and Yovanovich (1983, 1985). In the following discussion, therefore, only those portions of the theory needed to apply the model to a thermal design problem will be presented. The general expression for the contact conductance of the coated joint operating in a vacuum is

$$h'_c = h_c \left(\frac{H_S}{H'}\right)^{0.93} \left[\frac{k_1 + k_2}{Ck_1 + k_2}\right] \tag{11.89}$$

where $h_c$ is the uncoated contact conductance, $H_S$ is the micro-hardness of the softer substrate, $H'$ is the "effective" micro-hardness of the layer-substrate combination, $C$ is a spreading/constriction parameter correction factor which accounts for the heat spreading in the coated substrate, and $k_1, k_2$ are the thermal conductivities of the two substrates, respectively.

The coated contact conductance relationship consists of the product of three quantities: (i) the uncoated contact conductance $h_c$, (ii) the mechanical modification factor $(H_S/H')^{0.93}$, and (iii) the thermal modification factor in brackets. The uncoated (bare) contact conductance may be determined by means of the conforming, rough surface correlation equation based on plastic deformation:

$$h_c = 1.25 \left(\frac{m}{\sigma}\right) \left(\frac{2k_1 k_2}{k_1 + k_2}\right) \left(\frac{P}{H_S}\right)^{0.95} \tag{11.90}$$

where $H_S$ is the flow pressure (micro-hardness) of the softer substrate, $m$ is the combined average absolute asperity slope, and $\sigma$ is the combined RMS surface roughness of the joint.

For a given joint, the only unknowns are the effective micro-hardness $H'$ and the spreading/constriction parameter correction factor $C$. Thus, the key to solving coated contact problems is the determination of these two quantities.

**Mechanical Model**

The substrate micro-hardness can be obtained from the following approximate relationship [Hegazy (1985)]:

$$H_S = \left(12.2 - 3.54 H_B\right) \left(\frac{\sigma}{m}\right)^{-0.26} \tag{11.91}$$

which requires the combined surface roughness parameters $\sigma$ and $m$, and the bulk hardness of the substrate $H_B$. In the correlation equation, the units of the joint roughness parameter $\sigma/m$ are μm. For Ni 200 substrates, $H_B = 1.67$ GPa.

The "effective" micro-hardness must be obtained empirically for the particular layer (coating)-substrate combination under consideration. This requires a series of Vickers micro-hardness measurements which will result in an "effective" micro-hardness plot similar to that shown in Figure 11.16 (e.g. a silver layer on a Ni 200 substrate).

The "effective" Vickers micro-hardness measurements, denoted $H'$, are plotted against the relative indentation depth $t/d$, where $t$ is the layer thickness and $d$ is the indentation depth. The three micro-hardness regions were correlated as

$$H' = H_S \left(1 - \frac{t}{d}\right) + 1.81 H_L \left(\frac{t}{d}\right) \quad \text{for} \quad 0 \leq \frac{t}{d} < 1.0 \tag{11.92}$$

$$H' = 1.81 H_L - 0.21 H_L \left(\frac{t}{d} - 1\right) \quad \text{for} \quad 1.0 \leq \frac{t}{d} \leq 4.90 \tag{11.93}$$

$$H' = H_L \quad \text{for} \quad \frac{t}{d} > 4.90 \tag{11.94}$$

where $H_S$ and $H_L$ are the substrate and layer micro-hardness, respectively. The Ni 200 substrate micro-hardness is found to be $H_S = 2.97$ GPa for the joint roughness parameter values: $\sigma = 4.27$ μm and $m = 0.236$ radians. The Vickers micro-hardness of the silver layer is approximately $H_L = 40$ kg/mm$^2$ = 0.394 GPa.

**Figure 11.16** Vickers micro-hardness of silver layer on nickel substrate. Source: Antonetti and Yovanovich (1985)/ with permission of The American Society of Mechanical Engineers.

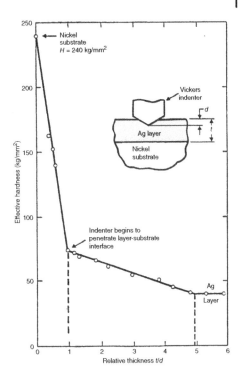

The relative indentation depth is obtained from the following approximate correlation equation [Antonetti and Yovanovich (1983, 1985)]:

$$\frac{t}{d} = 1.04 \left(\frac{t}{\sigma}\right) \left(\frac{P}{H'}\right)^{-0.097} \tag{11.95}$$

To implement the procedure [Antonetti and Yovanovich (1983, 1985)] for finding $H'$ from the three correlation equations, an iterative method is required.

To initiate the iterative method, the first guess is based on the arithmetic average of the substrate and layer micro-hardness values:

$$H'_1 = \frac{H_S + H_L}{2}$$

For a given value of $t$ and $P$, the first value of $t/d$ can be computed. From the three correlation equations, one can find a new value for $H'$, say $H'_2$. The new micro-hardness value, $H'_2$ is used to find another value for $t/d$ which leads to another value $H'_3$. The procedure is continued until convergence occurs. This usually occurs within three to four iterations [Antonetti and Yovanovich (1983, 1985)]. One may also solve the equations simultaneously for $H'$ and $t/d$.

### Thermal Model
The spreading/constriction resistance parameter correction factor $C$ is defined as the ratio of the spreading/constriction resistance parameter for a substrate with a layer to a bare substrate, for the same value of the relative contact spot radius $\epsilon'$:

$$C = \frac{\Psi(\epsilon', \phi_n)}{\Psi(\epsilon')} \tag{11.96}$$

The dimensionless spreading/constriction resistance parameter is defined as

$$\Psi(\epsilon', \phi_n) = 4k_2 a' R'_c \tag{11.97}$$

where $k_2$ is the thermal conductivity of the substrate that is coated, $a'$ is the contact spot radius for the layer on the substrate, and $R'_c$ is the spreading/constriction resistance of the contact spot.

The spreading/constriction resistance parameter with a layer on the substrate is [Antonetti and Yovanovich (1983, 1985)]

$$\Psi(\epsilon', \phi_n) = \frac{16}{\pi \epsilon'} \sum_{n=1}^{\infty} \frac{J_1^2(\delta'_n \epsilon')}{(\delta'_n)^3 J_0^2(\delta'_n)} \phi_n \gamma_n \rho_n \tag{11.98}$$

The first of these, $\phi_n$, accounts for the effect of the layer though its thickness and thermal conductivity; the second $\gamma_n$, accounts for the contact temperature basis used to determine the spreading/constriction resistance; and the third $\rho_n$, accounts for the contact spot heat flux distribution. For contacting surfaces, it is usual to assume that the contact spots are "isothermal." The modification factors in this case are $\gamma_n = 1.0$, and

$$\phi_n = K \left[ \frac{(1+K) + (1-K)e^{-2\delta'_n \epsilon' \tau'}}{(1+K) - (1-K)e^{-2\delta'_n \epsilon' \tau'}} \right] \tag{11.99}$$

where $K$ is the ratio of the substrate to layer thermal conductivity, and $\tau' = t/a'$ is the layer thickness to contact spot radius ratio, and

$$\rho_n = \frac{\sin(\delta'_n \epsilon')}{2 J_1(\delta'_n \epsilon')} \tag{11.100}$$

The parameter $\delta'_n$ are the eigenvalues which are roots of $J_1(\delta'_n) = 0$.

Tabulated values of $C$ were reported by Antonetti (1983) for a wide range of the parameters $K$ and $\tau'$. Details of the thermo-mechanical model development are found in Antonetti (1983) and Antonetti and Yovanovich (1983, 1985).

The thermo-mechanical model of Antonetti and Yovanovich (1983, 1985) has been verified by extensive tests. First, the bare joint was tested to validate that part of the model. Figure 11.17 shows the dimensionless joint conductance data and theory plotted versus the relative contact pressure

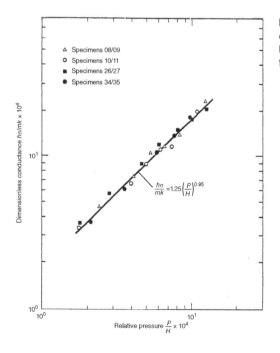

**Figure 11.17** Dimensionless contact conductance versus relative contact pressure for bare Ni 200 surfaces in vacuum. Source: Adapted from Antonetti (1983).

**Figure 11.18** Effect of layer thickness and contact pressure on joint conductance: vacuum data and theory. Source: Antonetti and Yovanovich (1985)/ with permission of The American Society of Mechanical Engineers.

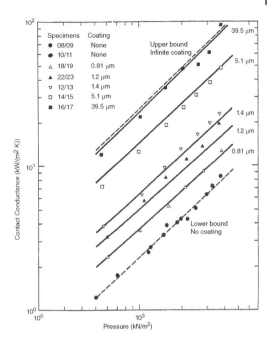

for three joints having three levels of surface roughness. The two surfaces were flat; one was lapped, and the other was glass bead-blasted. All tests were conducted in a vacuum.

The agreement between the model given by the correlation equation, and all data are very good over the entire range of relative contact pressure.

The bare surface tests were followed by three sets of tests for joints having three levels of surface roughness. Figure 11.18 shows the effect of the vapor-deposited silver layer thickness on the measured joint conductance plotted against the contact pressure. For these tests, the average values of the combined surface roughness parameters were $\sigma = 4.27$ μm and $m = 0.236$ radians. For the contact pressure range, the substrate micro-hardness was estimated to be $H_S = 2.97$ GPa. The layer thickness was between 0.81 and 39.5 μm. The lowest set of data and the theoretical curve correspond to the bare surface tests. Agreement between data and model is very good. The highest set of data for layer thickness of $t = 39.4$ μm corresponds to the so-called "infinitely" thick layer, where the thermal spreading occurs in the layer only, and the layer micro-hardness controls the formation of the micro-contacts. Again, the agreement between experiment and theory is good.

The difference between the highest and lowest joint conductance values is approximately a factor of 10. The enhancement is clearly significant.

The agreement between the measured values of joint conductance and the theoretical curves for the layer thicknesses: $t = 0.81, 1.2, 1.4, 5.1$ μm is also very good as seen in Figure 11.18.

All of the test points for bare and coated surfaces are plotted in Figure 11.19 as dimensionless joint conductance versus relative contact pressure. The agreement between experiment and theory is very good for all points.

A parametric study was conducted to calculate the enhancement that can be achieved when different metal types are used. The theory outlined in Section 11.10.1 will now be applied to a common problem in electronics packaging : heat transfer across an aluminum joint. What is required is a parametric study showing the variation in joint conductance as a function of metallic coating type and thickness, for fixed surface roughness and contact pressure.

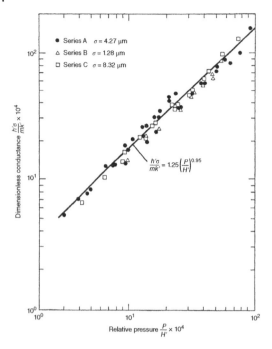

Dimensionless conductance $\frac{h'\sigma}{mk'} \times 10^4$

Relative pressure $\frac{P}{H'} \times 10^4$

Series A $\sigma = 4.27\ \mu m$
Series B $\sigma = 1.28\ \mu m$
Series C $\sigma = 8.32\ \mu m$

$$\frac{h'\sigma}{mk'} = 1.25\left(\frac{P}{H'}\right)^{0.95}$$

**Figure 11.19** Dimensionless joint conductance for bare and silver layer on Ni 200 substrates versus relative contact pressure. Source: Antonetti and Yovanovich (1985)/ with permission of The American Society of Mechanical Engineers.

**Table 11.8** Assumed nominal property values of four coatings.

| | $k$ (W/(m K)) | $H$ (kg/mm$^2$) |
| --- | --- | --- |
| Lead | 32.4 | 3.0 |
| Tin | 58.4 | 8.5 |
| Silver | 406.0 | 40.0 |
| Aluminum | 190.0 | 85.0 |

The thermophysical properties of the coatings and the aluminum substrate material are presented in Table 11.8.

Figure 11.20 shows the effect of the metallic layers on the joint conductance.

As seen from this figure, except for a very thin layer (about one micrometer), the performance curves are arranged according to layer micro-hardness. Lead with the lowest micro-hardness has the highest contact conductance, and silver with the highest micro-hardness has the lowest contact conductance. The thermal conductivity of the coating appears to play a secondary role.

The unusual shape of the curves is attributable to the fact that the assumed effective hardness curve shown in Figure 11.16 has three distinct zones. Moreover, because the micro-hardness of silver is much closer to aluminum than are the micro-hardness of lead and tin, respectively, the transition from one region to the next is not abrupt in the silver on aluminum effective micro-hardness curve, and this is reflected in the smoother contact conductance plot for the silver layer shown in Figure 11.16.

It should be noted that in the model which has been used, the load is assumed to be uniformly applied over the apparent contact area.

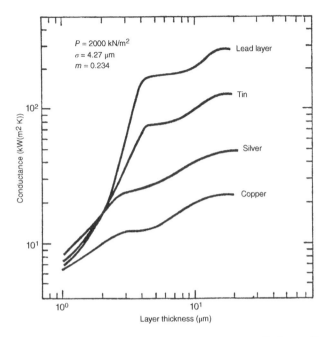

**Figure 11.20** Effect of layer thickness for four metallic layers. Source: Adapted from Antonetti (1983).

**Example 11.4** Two Ni 200 nominally, flat rough surfaces form a joint. One surface is lapped, and the second surface is lapped and glass bead-blasted. The effective surface roughness is $\sigma = 4.26\,\mu m$ and $m = 0.234$ radian. The apparent contact area is $A_a = 641 \times 10^{-6}\,m^2$. The thermal conductivity of the Ni 200 surfaces are $k_1 = k_2 = 64.4\,W/(m\,K)$. The contact micro-hardness at the contact pressure $P = 2\,MPa$ is determined to be $H_c = 2.94\,GPa$. The joint is in a vacuum. Radiation heat transfer is assumed to be negligible.

(a) Calculate the contact conductance $h_c$ for the joint using the CMY model.
We begin with the CMY model equation which for $P/H_p = P/H_c$ gives:

$$C_c = \frac{h_c}{k_s}\frac{\sigma}{m} = 1.25\left(\frac{P}{H_p}\right)^{0.95} = 1.25\left(\frac{2.0}{2940}\right)^{0.95} = 0.001224$$

This gives a contact conductance (with $k_s = k_1 = k_2$) equal to

$$h_c = C_c\frac{k_s m}{\sigma} = 0.001224\left(\frac{64.4 \cdot 0.234}{4.26 \times 10^{-6}}\right) = 4331.6\,W/(m^2\,K)$$

(b) Calculate the mean contact spot radius $a$, the number of micro-contact spots $N$ for the given apparent contact area, the joint resistance $R_j$, and the joint conductance $h_j$.
The joint resistance assuming a vacuum and no radiation is determined from

$$R_j = \frac{1}{h_c A_a} = \frac{1}{4331.1 \cdot 641 \times 10^{-6}} = 0.360\,K/W$$

The mean contact spot radius, spot density, and number of spots can be found once the mean separation $Y/\sigma$ is found. Equation (11.18) yields:

$$\lambda = \frac{Y}{\sigma} = 3.203$$

The spot radius is found from Eq. (11.12):

$$a = \sqrt{\frac{8}{\pi}} \left(\frac{\sigma}{m}\right) \exp(\lambda^2/2)\, \mathrm{erfc}(\lambda/\sqrt{2}) = 4.72\ \mu\text{m}$$

and the contact spot density using Eq. (11.11):

$$n = \frac{1}{16}\left(\frac{m}{\sigma}\right)^2 \frac{\exp(-\lambda^2)}{\mathrm{erfc}(\lambda/\sqrt{2})} = 4.86 \times 10^6\ \text{m}^{-2}$$

which for the apparent area of contact gives:

$$N = nA_a = 4.86 \times 10^6 \cdot 641 \times 10^{-6} = 3115$$

(c) A thick layer of silver is vapor deposited on the lapped surface. A new joint is now formed between the glass bead-blasted surface and the layer of silver whose thermal conductivity is $k_3 = 405\ \text{W/(m K)}$. Assume the surface roughness parameters are unchanged. The contact hardness of the silver layer is $H_c = 0.392\ \text{GPa}$. Repeat the calculations of (b).

We now assume a thick silver layer coats the surface such that the substrate has little role in the problem. With $k_3 = 405\ \text{W/(m K)}$, we have

$$k_s = \frac{2k_1 k_3}{k_1 + k_3} = \frac{2 \cdot 64.4 \cdot 405}{64.4 \cdot 405} = 111.13\ \text{W/(m K)}$$

With a new contact hardness of $H_c = 0.392$, we find $P/H = 0.005102$, and we find

$$C_c = 1.25(0.005102)^{0.95} = 0.008304$$

and

$$h_c = C_c \frac{k_s m}{\sigma} = 0.008304 \left(\frac{111.13 \cdot 0.234}{4.26 \times 10^{-6}}\right) = 50,687.9\ \text{W/(m}^2\ \text{K)}$$

which gives a joint resistance of

$$R_j = \frac{1}{h_c A_a} = \frac{1}{50,687.9 \cdot 641 \times 10^{-6}} = 0.0308\ \text{K/W}$$

Finally, recalculating the mean relative separation gives $Y/\sigma = 2.565$, and the spot radius and number of spots are

$$a = 5.687\ \mu\text{m}$$
$$n = 2.537 \times 10^7\ \text{m}^{-2}$$
$$N = nA_a = 2.537 \times 10^7 \cdot 641 \times 10^{-6} = 16,261$$

(d) If the thickness of the silver layer is $t = 2\ \mu\text{m}$, calculate the effective contact micro-hardness $H'$, the effective thermal conductivity of the joint $k'$, the mean contact spot radius $a'$, and the joint resistance $R_j$. Use the results of (b) and (c) as a guide to find the new conductance $h_c'$.

We must iterate or solve directly from the two equations below to find $H'$:

$$\frac{t}{d} = 1.04 \left(\frac{t}{\sigma}\right)\left(\frac{P}{H'}\right)^{-0.097}$$

$$H' = H_S \left(1 - \frac{t}{d}\right) + 1.81\, H_L \left(\frac{t}{d}\right)$$

Solving yields a value $H' = 0.960\ \text{GPa}$ and $t/d = 0.889$.

Antonetti (1983) proposed the following equation for determining the contact spot radius as a function of the effective Hardness $H'$:

$$a' = 0.77(\sigma/m)(P/H')^{0.097}$$

which gives $a' = 7.144$ μm.

In order to determine $h'_c$ from Eqs. (11.89) and (11.90), we now need to find $C$ for Eq. (11.89). The constant $C$ is determined from Eqs. (11.96)–(11.100):

$$C = 0.457$$

Combination of the Eqs. (11.89) and (11.90) yields an effective thermal conductivity for the contacting substrates:

$$k' = \frac{2k_1 k_2}{Ck_1 + k_2} = \frac{2 \cdot 64.4 \cdot 64.4}{0.457 \cdot 64.4 + 64.4} = 88.4 \text{ W/(m K)}$$

The new conductance is found from the combined Eqs. (11.89) and (11.90):

$$h'_c = 1.25(m/\sigma)(P/H_s)^{0.95}(H_s/H')^{0.93}k' = 16,868.2 \text{ W/(m K)}$$

Finally, the new joint resistance (assuming no radiation and a vacuum, $h_j \approx h'_c$) is

$$R_j = \frac{1}{h'_c A_a} = \frac{1}{16,868.2 \cdot 641 \times 10^{-6}} = 0.0925 \text{ K/W}$$

## 11.10.2 Ranking Metallic Coating Performance

Yovanovich (1972), in his research on the effects of soft metallic foils on joint conductance, proposed that the performance of different foil materials may be ranked according to the parameter $k/H$, using the properties of the foil material. He showed empirically that the higher the value of this parameter, the greater was the improvement in the joint conductance over a bare joint. Following this thought, in Antonetti and Yovanovich (1983) it was proposed, although not proven experimentally, that the performance of coated joints can be ranked by the parameter $k'/(H')^{0.93}$. Table 11.9 shows the variation in this parameter as the layer thickness is increased. Table 11.9

**Table 11.9** Ranking the effectiveness of coatings [$k'/(H')^{0.93}$].

| Coating thickness (μm) | Lead | Tin | Silver |
|---|---|---|---|
| 0 | 3.05 | 3.05 | 3.05 |
| 1 | 3.72 | 3.96 | 3.53 |
| 2 | 7.05 | 6.81 | 3.98 |
| 4 | 19.6 | 10.5 | 4.68 |
| 8 | 18.0 | 10.8 | 6.24 |
| 16 | 21.0 | 12.9 | 8.16 |
| ∞ | 19.9 | 12.2 | 8.38 |

$P = 2000$ kN/m$^2$, $\sigma = 4.0$ μm, $m = 0.20$ radians

suggests as well, that even if the "effective" micro-hardness of the layer-substrate combinations being considered are not known, the relative performance of coating materials can be estimated by assuming an infinitely thick coating (where the "effective" micro-hardness is equal to the micro-hardness of the layer).

This section has shown how a thermo-mechanical model for coated substrates can be used to predict the enhancement in thermal joint conductance. For the particular case considered, an aluminum to aluminum joint, it was demonstrated that up to an order of magnitude improvement in the contact conductance is possible, depending on the choice of coating material and the thickness employed. It should also be noted that aluminum substrates are relatively soft and have a relatively high thermal conductivity, and if the joint in question had been, for example, steel against steel, the improvement in the joint conductance would have been even more impressive.

### 11.10.3 Elastomeric Inserts

Thin polymers and organic materials are being used to a greater extent in power generating systems. Frequently, these thin layers of relatively low thermal conductivity are inserted between two metallic rough surfaces which are assumed to be nominally flat. The joint which is formed consists of a single layer whose initial, "unloaded," thickness is denoted as $t_0$, and has thermal conductivity called $k_p$. There are two mechanical interfaces that consist of numerous micro-contacts with associated gaps that are generally occupied with air. The overall joint conductance is, if radiation heat transfer across the two gaps is negligible,

$$\frac{1}{h_j} = \frac{1}{h_{c,1} + h_{g,1}} + \frac{t}{k_p} + \frac{1}{h_{c,2} + h_{g,2}} \tag{11.101}$$

where $t$ is the polymer thickness under mechanical loading. The contact and gap conductances at the mechanical interfaces are denoted as $h_{c,i}$ and $h_{g,i}$, respectively, and $i = 1, 2$. Marotta and Fletcher (1996) conducted experiments to determine the effect of selected polymeric inserts on the contact resistance of joints in a vacuum. Several inserts having a range of thermal conductivities were tested. They essentially gave similar resistances. The results were graphed, and no model was proposed.

Since this joint is too complex to study, Fuller (2000) and Fuller and Marotta (2000) choose to investigate the simpler joint which consisted of thermal grease at interface 2, and the joint was placed in a vacuum. Under these conditions, they assumed that $h_{g,2} \to \infty$, and $h_{g,1} \to 0$. They further assumed that the compression of the polymer layer under load may be approximated by the relationship:

$$t = t_0 \left[1 - \frac{P}{E_p}\right] \tag{11.102}$$

where $E_p$ is Young's modulus of the polymer. Under these assumptions, the joint conductance reduces to the simpler relationship:

$$h_j = \left[\frac{1}{h_{c,1}} + \frac{t_0}{k_p}\left(1 - \frac{P}{E_p}\right)\right]^{-1} \tag{11.103}$$

On further examination of the physical properties of polymers, Fuller (2000) concluded that the polymers will undergo elastic deformation of the contacting asperities. Fuller examined the use of the elastic contact model of Mikic (1974) and found that the disagreement between the data and the predictions was large. To bring the model into agreement with the data, it was found that the elastic hardness of the polymers should be defined as

$$H_{ep} = \frac{E_p m}{2.3} \tag{11.104}$$

where $m$ is the combined mean absolute asperity slope. The dimensionless contact conductance correlation equation was expressed as

$$\frac{h_c \, \sigma}{k_s \, m} = a_1 \left( \frac{2.3 \, P}{m \, E_p} \right)^{a_2} \tag{11.105}$$

where $a_1$ and $a_2$ are correlation coefficients. Fuller and Marotta (2000) chose the coefficient values: $a_1 = 1.49$ and $a_2 = 0.935$, compared with the values that Mikic (1974) reported: $a_1 = 1.54$ and $a_2 = 0.94$. In the Mikic (1974) elastic contact model, the elastic hardness was defined as

$$H_e = \frac{mE'}{\sqrt{2}} \tag{11.106}$$

where $E'$ is the "effective" Young's modulus of the joint. For most polymer-metal joints, $E' \approx E_p$ because $E_p \ll E_{metal}$.

Fuller (2000) conducted a series of vacuum tests for validation of the joint conductance model. The thickness, surface roughness, and the thermophysical properties of the polymers and the aluminum alloy are given in Table 11.10. The polymer thickness in all cases is two to three orders of magnitude larger than the surface roughness, i.e. $t/\sigma > 100$.

The dimensionless joint conductance data for three polymers (Delrin, polycarbonate, and PVC) are plotted against the dimensionless contact pressure in Figure 11.21 over approximately three decades.

Two sets of data are reported for Delrin. The joint conductance model and the data show similar trends with respect to load. At the higher loads, the data and the model approach asymptotes corresponding to the bulk resistance of the polymers. The dimensionless joint conductance goes to different asymptotes because the bulk resistance is defined by the thickness of the polymer layers.

In general, there is acceptable agreement between the vacuum data and the model predictions.

**Table 11.10** Thickness, surface roughness, and thermophysical properties of test specimens.

| Material | $t_0$ $(10^3 \text{ m})$ | $\sigma$ $(10^6 \text{ m})$ | $m$ (radians) | $k$ (W/(m K)) | $E$ (GPa) | $v$ |
|---|---|---|---|---|---|---|
| Delrin B | 1.88 | 1.02 | 0.492 | 0.38 | 3.59 | 0.38 |
| Polycarbonate A | 1.99 | 0.773 | 0.470 | 0.22 | 2.38 | 0.38 |
| PVC A | 1.83 | 0.650 | 0.436 | 0.17 | 4.14 | 0.38 |
| Teflon A | 1.89 | 0.622 | 0.305 | 0.25 | 0.135 | 0.38 |
| Aluminum | — | 0.511 | 0.267 | 183 | 72.0 | 0.32 |

Source: Adapted from Fuller (2000).

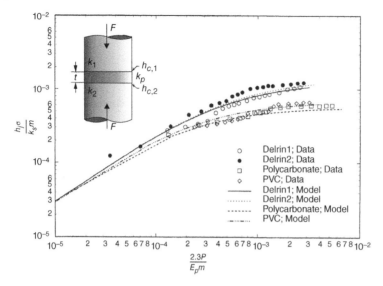

**Figure 11.21** Dimensionless joint conductance versus dimensionless contact pressure for polymer layers. Source: Adapted from Fuller and Marotta (2000).

### 11.10.4 Thermal Greases and Pastes

There is much interest today in the use of TIM such as thermal greases and pastes to enhance joint conductance. The recent publications of Prasher (2001) and Savija et al. (2002a,b) review the use of TIMs, and the models that are available to predict joint conductance.

The thermal joint resistance or conductance of a joint formed by two nominally, flat rough surfaces which is filled with a grease as shown in Figure 11.1 depend on several geometric, physical, and thermal parameters. The resistance and conductance relations are obtained from a model which is based on the following simplifying assumptions:

- Nominally flat, rough surfaces with Gaussian height distributions
- Load is supported by the contacting asperities only
- Load is light; nominal contact pressure is small; $P/H_c \approx 10^{-3} - 10^{-5}$
- Plastic deformation of the contacting asperities of the softer solid
- Grease is homogeneous, completely fills the interstitial gaps, and perfectly wets the bounding surfaces

In general the joint conductance $h_j$ and joint resistance $R_j$ depend on the contact and gap components. The joint conductance is modeled as

$$h_j = h_c + h_g \tag{11.107}$$

where $h_c$ represents the contact conductance and $h_g$ represents the gap conductance. The joint resistance is modeled as

$$\frac{1}{R_j} = \frac{1}{R_c} + \frac{1}{R_g} \tag{11.108}$$

where $R_c$ is the contact resistance, and $R_g$ is the gap resistance. For very light contact pressures, it is assumed that $h_c \ll h_g$ and $R_j \gg R_g$. The joint conductance and resistance depend on the gap only;

therefore,

$$h_j = h_g \quad \text{and} \quad \frac{1}{R_j} = \frac{1}{R_g} \quad \text{where} \quad h_j = \frac{1}{A_a R_j} \tag{11.109}$$

Based on the assumptions given above the gap conductance is modeled as an equivalent layer of thickness $t = Y$ filled with grease having thermal conductivity $k_g$. The joint conductance is given by

$$h_j = \frac{k_g}{Y} \tag{11.110}$$

The gap parameter $Y$ is the distance between the mean planes passing through the two rough surfaces. This geometric parameter is related to the combined surface roughness $\sigma = \sqrt{\sigma_1^2 + \sigma_2^2}$ where $\sigma_1, \sigma_2$ are the RMS surface roughness of the two surfaces, and the contact pressure $P$ and the effective micro-hardness of the softer solid, $H_c$. The mean plane separation $Y$, shown in Figure 11.1, is given approximately by the simple power-law relation [Antonetti (1983)]:

$$\frac{Y}{\sigma} = 1.53 \left( \frac{P}{H_c} \right)^{-0.097} \tag{11.111}$$

The power-law relation shows that $Y/\sigma$ is a relatively weak function of the relative contact pressure. Using this relation, the joint conductance may be expressed as

$$h_j = \frac{k_g}{\sigma \cdot \left( \dfrac{Y}{\sigma} \right)} = \frac{k_g}{1.53 \, \sigma (P/H_c)^{-0/097}} \tag{11.112}$$

which clearly shows how the geometric, physical, and thermal parameters influence the joint conductance. The relation for the specific joint resistance is

$$A_a R_j = \frac{1}{h_j} = 1.53 \left( \frac{\sigma}{k_g} \right) \left( \frac{P}{H_c} \right)^{-0.097} \tag{11.113}$$

In general, if the metals work-harden, the relative contact pressure $P/H_c$ is obtained from the relationship:

$$\frac{P}{H_c} = \left[ \frac{P}{c_1 (1.62 \, \sigma/m)^{c_2}} \right]^{\frac{1}{1+0.071c_2}} \tag{11.114}$$

where the coefficients $c_1, c_2$ are obtained from Vickers micro-hardness tests. The Vickers micro-hardness coefficients are related to the Brinell hardness $H_B$ for a wide range of metals. The units of $\sigma$ in the above relation must be given in microns (μm). The units of $P$ and $c_1$ must be consistent.

The approximation of Hegazy (1985) for micro-hardness is recommended:

$$H_c = (12.2 - 3.54 H_B)(\sigma/m)^{-0.26} \tag{11.115}$$

where $H_c$, the effective contact micro-hardness, and $H_B$, the Brinell hardness, are in GPa, and the effective surface parameter $(\sigma/m)$ is in microns.

If the softer metal does not work-harden, then $H_c \approx H_B$. Since $H_B < H_c$, then if we set $H_c = H_B$ in the specific joint resistance relationship, this will give a lower bound for the joint resistance or an upper bound for the joint conductance.

The simple grease model for joint conductance or specific joint resistance was compared against the specific joint resistance data reported by Prasher (2001). The surface roughness parameters of

**Table 11.11** Surface roughness and grease thermal conductivity.

| Test | Roughness $\sigma_1 = \sigma_2$ μm | Conductivity $k_g$ W/(m K) |
|------|------|------|
| 1 | 0.12 | 3.13 |
| 2 | 1 | 3.13 |
| 3 | 3.5 | 3.13 |
| 4 | 1 | 0.4 |
| 5 | 3.5 | 0.4 |
| 6 | 3.5 | 0.25 |
| 7 | 3.5 | 0.22 |

Source: Adapted from Prasher (2001).

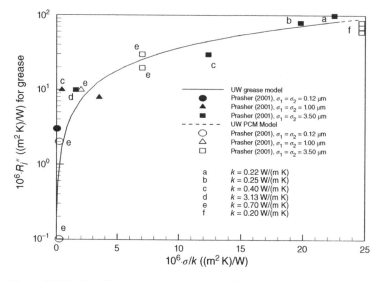

**Figure 11.22** Specific joint resistance versus $\sigma/k$ for grease. Source: Adapted from Prasher (2001).

the bounding copper surfaces and the grease thermal conductivities are given in Table 11.11. All tests were conducted at an apparent contact pressure of one atmosphere, and in a vacuum.

Prasher (2001) reported his data as specific joint resistance $R_j'' = A_a R_j$ versus the parameter $\sigma/k_g$ where $k_g$ is the thermal conductivity of the grease as shown in Figure 11.22 along with model predictions.

**Example 11.5** TIMs such as polymers are used as inserts between two nominally flat rough surfaces. The surfaces of the metals and the polymers have microscopic roughness; therefore, micro-contacts and micro-gaps are formed at the two interfaces. The apparent contact pressures are in the light-to-moderate range. If there is grease in one of the interfaces, radiation heat transfer does not occur through the grease, and the gap conductance is much greater than the contact conductance. We will assume that there is negligible thermal resistance at this interface. If the second interface is in a vacuum, heat transfer occurs through the micro-contacts and across the micro-gaps by radiation. If radiation heat transfer is negligible relative to the conduction through the micro-gaps, the overall joint conductance of the joint is given by the following relation:

$$h_j = \left[ \frac{1}{h_c} + \frac{t}{k_2} \right]^{-1}$$

where $h_c$ is the contact conductance, and $t$ is the polymer thickness under load, and $k_2$ is the thermal conductivity of the polymer. The thickness under load is given by

$$t = t_0 \left( 1 - \frac{P}{E_2} \right)$$

where $t_0$ is the polymer thickness before load, and $E_2$ is the modulus of elasticity of the polymer.

There are three contact conductance models: (i) the elastic model of Mikic (1974), (ii) the elastic model of Mikic (1974) modified by Fuller and Marotta (2000), and (iii) the plastic model of Cooper et al. (1969).

Given the thermophysical properties of Al 6061 and Delrin, and the micro-geometric characteristics of the metal and polymer, calculate the values of $h_c$, $h_b = k_2/t$, and $h_j$ for three apparent contact pressures: $P = 275.8, 827.4$, and $2758$ kPa.

The Al 6061 thermophysical and geometric parameter values are

$$k_1 = 183 \text{ W/(m K)} \quad \sigma_1 = 0.51 \text{ μm} \quad m_1 = 0.05 \quad E_1 = 72.1 \text{ GPa} \quad v_1 = 0.32$$

The Delrin thermophysical and geometric parameter values are

$$k_2 = 0.38 \text{ W/(m K)} \quad \sigma_2 = 2.19 \text{ μm} \quad m_2 = 0.26 \quad E_2 = 3.59 \text{ GPa} \quad v_2 = 0.38$$

The contact micro-hardness for Delrin is $H_c = 0.37$ GPa. The initial thickness of Delrin is $t_0 = 1.27$ mm. The experimental values of $h_j$ are $h_j = 78.5, 163.5$, and $251.6$ W/(m$^2$ K) at the following three contact pressures [Fuller (2000)]: $P = 275.8, 827.4$, and $2758$ kPa.

To compare the three approaches, we use the following equations:

$$C_c = 1.54 \left( \frac{P}{H_e} \right)^{0.94} \qquad \text{Mikic (M) Model}$$

$$C_c = 1.49 \left( \frac{P}{H_{ep}} \right)^{0.935} \qquad \text{Fuller and Marotta (FM) Model}$$

$$C_c = 1.25 \left( \frac{P}{H_p} \right)^{0.95} \qquad \text{Cooper–Mikic–Yovanovich (CMY) Model}$$

where $C_c = h_c \sigma / k_s m$. The equivalent roughness $\sigma$ and surface slope $m$ are

$$\sigma = \sqrt{\sigma_1^2 + \sigma_2^2} = 2.25 \text{ μm}$$

$$m = \sqrt{m_1^2 + m_2^2} = 0.265$$

While the respective hardness for each model is defined according to

$$H_e = mE'/\sqrt{2} = 0.747 \text{ GPa}$$

$$H_{ep} = mE_p/2.3 = 0.414 \text{ GPa}$$

$$H_p = 0.370 \text{ GPa}$$

The results for $h_c$ and $h_j$ for each model are tabulated below assuming $h_b = k_2/t \approx 299\,\text{W/(m}^2\,\text{K)}$ (which does not change appreciably with load here):

| P (kPa) | $h_c$ (W/(m$^2$ K)) | | | $h_j$ (W/(m$^2$ K)) | | | $h_{j,exp}$ (W/(m$^2$ K)) |
|---|---|---|---|---|---|---|---|
| | (M) | (FM) | (CMY) | (M) | (FM) | (CMY) | |
| 275.8 | 81.5 | 142.7 | 119.2 | 64.1 | 96.6 | 85.3 | 78.5 |
| 827.4 | 229.0 | 398.5 | 338.6 | 129.8 | 171.0 | 158.9 | 163.5 |
| 2758 | 710.2 | 1228.5 | 1062.7 | 210.6 | 240.7 | 233.6 | 251.6 |

Both the models of Fuller–Marotta and Cooper–Mikic–Yovanovich provide a good prediction of the experimental results for joint conductance reported by Fuller (2000).

### 11.10.5 Phase Change Materials (PCM)

PCM are being used to reduce thermal joint resistance in microelectronic systems. The PCM may consist of a substrate such as an aluminum foil which supports the PCM such as paraffin. In some applications, the paraffin may be filled with some solid particles to increase the "effective" thermal conductivity of the paraffin.

At some temperature $T_m$ above room temperature, the PCM melts, then "flows" through the micro-gaps, "expels" the air, and then "fills" the voids completely. After the temperature of the joint falls below $T_m$, the PCM solidifies. Depending on the level of surface roughness and out-of-flatness, the thickness of the PCM, a complex joint is formed. Thermal tests reveal that the specific joint resistance is very small relative to the bare joint resistance with air occupying the micro-gaps, Figure 11.22.

Because of the complex nature of a joint with PCM, there are no simple models available for the several types of joints that can be formed when a PCM is used.

## 11.11 Thermal Resistance at Bolted Joints

Bolted joints are frequently found in aerospace systems and less often in microelectronics systems. The bolted joints are complex because of their geometric configurations, the materials used, and the number of bolts and washers used. The pressure distributions near the location of the bolts are not uniform, and the region influenced by the bolts is difficult to predict. A number of papers are available to provide information on measured thermal resistances and to provide some models to predict the thermal resistance under various conditions.

Madhusudana (1996) and Johnson (1985) present material on the thermal and mechanical aspects of bolted joints.

For bolted joints used in satellite thermal design, the publications of Mantelli and Yovanovich (1996, 1998a,b), Song et al. (1993a) are recommended.

## References

Antonetti, V.W., *On the Use of Metallic Coatings to Enhance Thermal Contact Conductance*, Ph.D. Thesis, University of Waterloo, Waterloo, Ontario, Canada, 1983.

Antonetti, V.W., and Yovanovich, M.M., "Using Metallic Coatings to Enhance Thermal Contact Conductance of Electronic Packages", *Heat Transfer in Electronic Equipment*, ASME-HTD-28, ASME, New York, pp. 71–77, 1983.

Antonetti, V.W., and Yovanovich, M.M., "Enhancement of Thermal Contact Conductance by Metallic Coatings: Theory and Experiments", *Journal of Heat Transfer*, vol. 107, pp. 513–519, 1985.

Antonetti, V.W., and Yovanovich, M.M., "Using Metallic Coatings to Enhance Thermal Contact Conductance of Electronic Packages", *Journal of Heat Transfer Engineering*, vol. 9, no. 3, pp. 85–92, 1988.

Antonetti, V.W., Whittle, T.D., and Simons, R.E., "An Approximate Thermal Contact Conductance Correlation", *Experimental/Numerical Heat Transfer in Combustion and Phase Change*, ASME-HTD-170, ASME, New York, 1991.

Bahrami, M., *Modeling of Thermal Joint Resistance for Sphere-Flat Contacts in a Vacuum*, Ph.D. Thesis, University of Waterloo, Waterloo, Ontario, Canada, 2004.

Bahrami, M., Yovanovich, M.M., and Culham, J.R., "Thermal Joint Resistances of Conforming Rough Surfaces with Gas Filled Gaps", *Journal of Thermophysics and Heat Transfer*, vol. 18, no. 3, pp. 318–325, 2004a.

Bahrami, M., Yovanovich, M.M., and Culham, J.R., "Modeling Thermal Contact Resistance: A Scale Analysis Approach", *Journal of Heat Transfer*, vol. 126, pp. 896–905, 2004b.

Braun, D., and Frohn, A., "Heat Transfer in Simple Monatomic Gases and in Binary Mixtures of Monatomic Gases", *International Journal of Heat and Mass Transfer*, vol. 19, pp. 1329–1335, 1976.

Bush, A.W., and Gibson, R.D., "A Theoretical Investigation of Thermal Contact Conductance", *Journal of Applied Energy*, vol. 5, pp. 11–22, 1979.

Bush, A.W., Gibson, R.D., and Thomas, T.R., "The Elastic Contact of a Rough Surface", *Wear*, vol. 35, pp. 87–111, 1975.

Cetinkale, T.N., and Fishenden, M., "Thermal Conductance of Metal Surfaces in Contact", in *Proceedings of the General Discussion on Heat Transfer*, Institute of Mechanical Engineers, London, pp. 271–275, 1951.

Cooper, M.G., Mikic, B.B., and Yovanovich, M.M., "Thermal Contact Conductance", *International Journal of Heat and Mass Transfer*, vol. 12, pp. 279–300, 1969.

Fletcher, L.S., "Recent Developments in Contact Conductance Heat Transfer", *Journal of Heat Transfer*, vol. 110, pp. 1059–1070, 1988.

Fletcher, L.S., "A Review of Thermal Enhancement Techniques for Electronic Systems", *IEEE Transactions on Components Hybrids and Manufacturing Technology*, vol. 13, no. 4, pp. 1012–1021, 1990.

Fuller, J.J., *Thermal Contact Conductance of Metal/Polymer Joints: An Analytical and Experimental Investigation*, M.S. Thesis, Department of Mechanical Engineering, Clemson University, Clemson, SC, 2000.

Fuller, J.J., and Marotta, E.E., "Thermal Contact Conductance of Metal/Polymer Joints", *Journal of Thermophysics and Heat Transfer*, vol. 14, no. 2, pp. 283–286, 2000.

Garnier, J.E., and Begej, S., "Ex-reactor Determination of Gap and Contact Conductance Between Uranium Dioxide: Zircaloy-4 Interfaces", Report, *Nuclear Regulatory Commission*, Washington, DC, 1979.

Greenwood, J.A., "The Area of Contact Between Rough Surfaces and Flats", *Journal of Lubrication Technology*, vol. 81, pp. 81–91, 1967.

Greenwood, J.A., and Tripp, J.H., "The Elastic Contact of Rough Spheres", *Journal of Applied Mechanics* vol. 89, no. 1, pp. 153–159, 1967.

Greenwood, J.A., and Tripp, J.H., "The Contact of Two Nominally Flat Rough Surfaces", *Proceedings of the Institution of Mechanical Engineers*, vol. 185, pp. 625–633, 1970.

Greenwood, J.A., and Williamson, J.B.P., "Contact of Nominally Flat Surfaces", *Proceedings of the Royal Society of London*, A295, pp. 300–319, 1966.

Hartnett, J.P., "A Survey of Thermal Accommodation Coefficients", in *Rarefied Gas Dynamics*, ed. L. Talbot, Academic Press, New York, pp. 1–28, 1961.

Hegazy, A.A., *Thermal Joint Conductance of Conforming Rough Surfaces: Effect of Surface Micro-hardness Variation*, Ph.D. Thesis, University of Waterloo, Waterloo, Ontario, Canada, 1985.

Johnson, K.L., *Contact Mechanics*, Cambridge University Press, Cambridge, UK, 1985.

Lloyd, W.R., Wilkins, D.R., and Hill, P.R., "Heat Transfer in Multicomponent Mon-atomic Gases in the Low, Intermediate and High Pressure Regime", *Proceedings of the Nuclear Thermonics Conference*, 1973.

Loyalka, S.K., "A Model for Gap Conductance in Nuclear Fuel Rods", *Nuclear Technology*, vol. 57, pp. 220–227, 1982.

Madhusudana, C.V., *Thermal Contact Conductance*, Springer-Verlag, New York, 1996.

Madhusudana, C.V., and Fletcher, L.S., "Contact Heat Transfer : The Last Decade", *AIAA Journal*, Vol. 24, no. 3, pp. 510–523, 1986.

Mantelli, M.B.H., and Yovanovich, M.M., "Experimental Determination of the Overall Thermal Resistance of Satellite Bolted Joints", *Journal of Thermophysics and Heat Transfer*, vol. 10, no. 1, pp. 177–179, 1996.

Mantelli, M.B.H., and Yovanovich, M.M., "Parametric Heat Transfer Study of Bolted Joints", *Journal of Thermophysics and Heat Transfer*, vol. 12, no. 3, pp. 382–390, 1998a.

Mantelli, M.B.H., and Yovanovich, M.M., "Compact Analytical Model for Overall Thermal Resistance of Bolted Joints", *International Journal of Heat and Mass Transfer*, vol. 41, no. 10, pp. 1255–1266, 1998b.

Marotta, E.E., and Fletcher, L.S., "Thermal Contact Resistance of Selected Polymeric Materials", *Journal of Thermophysics and Heat Transfer*, vol. 10, no. 2, pp. 334–342, 1996.

McWaid, T.H., and Marschall, E., "Applications of the Modified Greenwood and Williamson Contact Model for Prediction of Thermal Contact Resistance", *Wear*, vol. 152, pp. 263–277, 1992a.

McWaid, T.H., and Marschall, E., "Thermal Contact Resistance Across Pressed Metal Contacts in a Vacuum Environment", *International Journal of Heat and Mass Transfer*, vol. 35, pp. 2911–2920, 1992b.

Mikic, B.B., "Thermal Contact Conductance: Theoretical Considerations", *International Journal of Heat Mass Transfer*, vol. 17, pp. 205–214, 1974.

Mikic, B.B., and Rohsenow, W.M., "Thermal Contact Resistance", Mechanical Engineering Report, DSR 38 74542-41, MIT, Cambridge, MA, 1966.

Milanez, F.H., Yovanovich, M.M., and Culham, J.R., "Effect of Surface Asperity Truncation on Thermal Contact Conductance," *IEEE Transactions on Components and Packaging Technologies*, vol. 26. no. 1, pp. 48–54, 2003.

Negus, K.J., and Yovanovich, M.M., "Correlation of Gap Conductance Integral for Conforming Rough Surfaces", *Journal of Thermophysics and Heat Transfer*, vol. 12, pp. 279–281, 1988.

Nho, K.M., *Experimental Investigation of Heat Flow Rate and Directional Effect on Contact Resistance of Anisotropic Gound/Lapped Interfaces*, Ph.D. Thesis, Department of Mechanical Engineering, University of Waterloo, Waterloo, Ontario, Canada, 1989.

Prasher, R.S., "Surface Chemistry and Characteristics Based Model for the Thermal Contact Resistance of Fluidic Interstitial Thermal Interface Materials", *Journal of Heat Transfer*, vol. 123, pp. 969–975, 2001.

Rapier, A.C., Jones, T.M., and McIntosh, J.E., "The Thermal Conductance of Uranium Dioxide/ Stainless Steel Interfaces", *International Journal of Heat and Mass Transfer*, vol. 6, pp. 397–416, 1963.

Savija, I., Culham, J.R., Yovanovich, M.M., and Marotta, E.E., "Review of Thermal Conductance Models for Joints Incorporating Enhancement Materials", AIAA-2002-0494, *40th AIAA Aerospace Sciences Meeting and Exhibit*, January 14–17, Reno, NV, USA, 2002a.

Savija, I., Yovanovich, M.M., Culham, J.R., and Marotta, E.E., "Thermal Joint Resistance Models for Conforming Rough Surfaces with Grease Filled Gaps", AIAA-2002-0495, *40th AIAA Aerospace Sciences Meeting and Exhibit*, January 14–17, Reno, NV, USA, 2002b.

Semyonov, Y.G., Borisov, S.E., and Suetin, P.E., "Investigation of Heat Transfer in Rarefied Gases over a Wide Range of Knudsen Numbers", *International Journal of Heat and Mass Transfer*, vol. 27, pp. 1789–1799, 1984.

Shlykov, Y.P., "Calculation of Thermal Contact Resistance of Machined Metal Sufaces", *Teploenergetika*, vol. 12, no. 10, pp. 79–83, 1965.

Song, S., *Analytical and Experimental Study of Heat Transfer through Gas Layers of Contact Interfaces*, Ph.D. Thesis, University of Waterloo, Waterloo, Ontario, Canada, 1988.

Song, S., and Yovanovich, M.M., "Correlation of Thermal Accommodation Coefficients for Engineering Surfaces", *ASME-HTD-69*, ASME, New York, pp. 107–116, 1987.

Song, S., and Yovanovich, M.M., "Relative Contact Pressure: Dependence on Surface Roughness and Vickers Microhardness", *Journal of Thermophysics and Heat Transfer*, vol. 2, no. 1, pp. 43–47, 1988.

Song, S., Park, C., Moran, K.P., and Lee, S., "Contact Area of Bolted Joint Interface: Analytical, Finite Element Modeling, and Experimental Study", *Computer Aided Design in Electronics Packaging*, vol. 3, pp. 73–81, 1992a.

Song, S., Yovanovich, M.M., and Nho, K., "Thermal Gap Conductance: Effects of Gas Pressure and Mechanical Load", *Journal of Thermophysics and Heat Transfer*, vol. 6, no. 1, pp. 62–68, 1992b.

Song, S., Moran, K.P., Augi, R., and Lee, S., "Experimental Study and Modeling of Thermal Contact Resistance across Bolted Joints", AIAA-93-0844, *31st Aerospace Sciences Meeting and Exhibit*, Reno, NV, USA, January 11–14, 1993a.

Song, S., Yovanovich, M.M., and Goodman, F.O., "Thermal Gap Conductance of Conforming Rough Surfaces in Contact", *Journal of Heat Transfer*, vol. 115, pp. 533–540, 1993b.

Sridhar, M.R., *Elastoplastic Contact Models for Sphere-Flat and Conforming Rough Surface Applications*, Ph.D. Dissertation, University of Waterloo, Waterloo, Ontario, Canada, 1994.

Sridhar, M.R., and Yovanovich, M.M., "Thermal Contact Conductance of Tool Steel and Comparison with Model", *International Journal of Heat and Mass Transfer*, vol. 39, no. 4, pp. 831–839, 1996a.

Sridhar, M.R., and Yovanovich, M.M., "Empirical Methods to Predict Vickers Micro-hardness", *Wear*, vol. 193, pp. 91–98, 1996b.

Sridhar, M.R., and Yovanovich, M.M., "Elastoplastic Contact Model for Isotropic Conforming Rough Surfaces and Comparison with Experiments", *Journal of Heat Transfer*, vol. 118, no. 1, pp. 3–9, 1996c.

Sridhar, M.R., and Yovanovich, M.M., "Elastoplastic Constriction Resistance Model for Sphere-Flat Contacts", *Journal of Heat Transfer*, vol. 118, no. 1, pp. 202–205, 1996d.

Tabor, D., *The Hardness of Metals*, Oxford University Press, London, 1951.

Teagan, W.P., and Springer, G.S., "Heat-Transfer and Density Distribution Measurements between Parallel Plates in the Transition Regime", *Physics of Fluids*, vol. 11, no. 3, pp. 497–506, 1968.

Thomas, L.B., *Rarefied Gas Dynamics*, Academic Press, New York, 1967.

Thomas, T.R, *Rough Surfaces*, Longman Group, London, 1982.

Thomas, L.B., and Loyalka, S.K., "Determination of the Thermal Accommodation Coefficients of Helium, Argon and Xenon on a Surface of Zircaloy-2 at About 2 C", *Nuclear Technology*, vol. 57, pp. 213–219, 1982.

Veziroglu, T.N., "Correlation of Thermal Contact Conductance Experimental Results", *AIAA Thermophysics Specialist Conference*, New Orleans, LA, USA, 1967.

Wachman, H.Y., "The Thermal Accommodation Coefficient: A Critical Survey", *ARS Journal*, vol. 32, pp. 2–12, 1962.

Wesley, D.A., and Yovanovich, M.M., "A New Gaseous Gap Conductance Relationship", *Nuclear Technology*, vol. 72, pp. 70–74, 1986.

Whitehouse, D.J., and Archard, J.F., "The Properties of Random Surfaces of Significance in Their Contact", *Proceedings of the Royal Society of London*, A316, pp. 97–121, 1970.

Wiedmann, M.L., and Trumpler, P.R., "Thermal Accommodation Coefficients", *Transactions of the ASME*, vol. 68, pp. 57–64, 1946.

Yovanovich, M.M., "Effect of Foils on Joint Resistance: Evidence of Optimum Foil Thickness", AIAA-72-283, *AIAA 7th Thermophysics Conference*, San Antonio, TX, USA, 1972.

Yovanovich, M.M., "Thermal Contact Correlations", in *Progress in Astronautics and Aeronautics: Spacecraft Radiative Transfer and Temperature Control*, vol. 83, ed. T.E. Horton, AIAA, New York, pp. 83–95, 1982.

Yovanovich, M.M., "Recent Developments in Thermal Contact, Gap and Joint Conductance: Theories and Experiments", *Proceedings of the 8th International Heat Transfer Conference*, San Francisco, CA, USA, Vol. 1, pp. 35–45, 1986.

Yovanovich, M.M., and Antonetti, V.W., "Application of Thermal Contact Resistance Theory to Electronic Packages", in *Advances in Thermal Modeling of Electronic Components and Systems*, Vol. 1, A. Bar-Cohen and A.D. Kraus, Hemisphere Publishing, New York, pp. 79–128, 1988.

Yovanovich, M.M., and Hegazy, A., "An Accurate Universal Contact Conductance Correlation for Conforming Rough Surfaces with Different Micro-Hardness Profiles", *18th Thermophysics Conference*, Montreal, Canada, AIAA Paper no. 83-1434, 1983.

Yovanovich, M.M., Hegazy, A.A., and DeVaal, J., "Surface Hardness Distribution Effects upon Contact, Gap and Joint Conductances", AIAA-82-0887, *AIAA/ASME 3rd Joint Thermophysics, Fluids, Plasma and Heat Transfer Conference*, St. Louis, MO, USA, June 7–11, 1982a.

Yovanovich, M.M., DeVaal, J., and Hegazy, A.A., "A Statistical Model to Predict Thermal Gap Conductance Between Conforming Rough Surfaces", AIAA-82-0888, *AIAA/ASME 3rd Joint Thermophysics, Fluids, Plasma and Heat Transfer Conference*, St. Louis, MO, USA, June 7–11, 1982b.

# 12

# Contact of Nonconforming Smooth Solids

In this chapter, we examine the contact of nonconforming smooth solids and present thermal resistance models for specific forms of mechanical contact, i.e. point contact and line contact. Classic elastic contact theory is used to predict the contact region size and then combined with thermal constriction models to predict the contact resistance. Models for gap resistance are also developed which allows the joint resistance to be calculated.

A comprehensive review of the Hertz elastic contact model applicable for smooth nonconforming surfaces is given. The contacting surfaces are associated with ellipsoids such as nonaligned long circular cylinders, sphere/flat contact, and a cylinder/flat contact. In the general case, the contact is a small elliptical area with semiaxes $a$ and $b$ with $b \leq a$. The semiaxes are related to the so-called "Hertz parameters" $m$ and $n$. The associated gap has thickness $\delta$ which is complex and nonsymmetric. The local gap thickness depends on the approach of the contacting surfaces under loading conditions and on local displacements of the bounding surfaces.

For the special case of a sphere-flat contact, the contact area is circular with radius $a$ and the associated gap has thickness $\delta$ which is axi-symmetric. The thermal contact resistance of an isothermal contact area is given in terms of the semiaxes $a, b$ and harmonic mean thermal conductivity of the contacting solids $k_1$ and $k_2$. For heat transfer across, the gas-filled gap an elasto-gap resistance model is presented. This model shows how the complex gap resistance depends on the gas thermal conductivity, rarefaction effects at the solid surface/gas interface as well as the local temperature drop across the gap. A complex model is also given for radiative heat transfer across the gap and a radiative overall resistance called the joint radiative resistance is presented.

The special case of a sphere-flat contact is presented because it occurs frequently in applications such as microelectronics cooling. The proposed models for a sphere-flat in a vacuum environment when only conduction across the contact area and radiation across the gap are presented in some detail. Then the model for gap resistance when a gas is present in the gap is presented and compared with experimental data. The special case of thermal contact resistance of elastic contact between a smooth sphere and smooth layered substrate in a vacuum environment is given.

The joint resistance of an elastic–plastic contact between a smooth hemisphere and smooth flat in a vacuum is considered. The elastic–plastic contact model is compared with conduction data obtained with several metals. Agreement between data and model is very good. The special case of the overall thermal resistance of instrument bearings in a vacuum is given. Models for determining the elasto-constriction (EC) resistance in ball bearings are also considered.

A line contact model applicable for elastic contact of a smooth circular cylinder and a smooth flat with fluid in the gap is then presented. This model is applicable for the nuclear industry.

*Thermal Spreading and Contact Resistance: Fundamentals and Applications*, First Edition.
Yuri S. Muzychka and M. Michael Yovanovich.
© 2023 John Wiley & Sons, Inc. Published 2023 by John Wiley & Sons, Inc.

Finally, we conclude with a comprehensive model for the contact of smooth nonconforming surfaces and rough conforming surfaces in a vacuum is presented. This complex contact model is based on scaling analyses.

## 12.1 Joint Resistances of Nonconforming Smooth Solids

The "elasto-constriction" and "elasto-gap" resistance models [Yovanovich (1986)] are based on the Boussinesq point load model [Timoshenko and Goodier (1970)] and the Hertz distributed load model [Hertz (1896), Timoshenko and Goodier (1970), Walowit and Anno (1975), Johnson (1985)]. Both models assume the bodies have "smooth" surfaces, they are perfectly elastic, and the applied load is static and normal to the plane of the contact area. In the general case, the contact area will be elliptical having semimajor and semiminor axes $a$, $b$ respectively. These dimensions are much smaller than the dimensions of the contacting bodies. The circular contact area produced when two spheres or a sphere and a flat are in contact are two special cases of the elliptical contact. Also, the rectangular contact area, produced when two ideal circular cylinders are in line contact or an ideal cylinder and a flat are in contact are special cases of the elliptical contact area.

Figure 12.1 shows the contact between two elastic bodies having physical properties (Young's modulus and Poisson's ratio): $E_1, v_1$ and $E_2, v_2$, respectively. One body is a smooth flat and the other body may be a sphere or a circular cylinder having radius $D/2$. The contact $2a$ is the diameter of a circular contact area for the sphere/flat contact, and the width of the contact strip for the cylinder/flat contact.

A gap is formed adjacent to the contact area, and its local thickness is characterized by $\delta$.

Heat transfer across the joint can take place by (i) conduction by means of the contact area, (ii) conduction through the substance in the gap, and (iii) by radiation across the gap if the substance is "transparent," or by radiation if the contact is formed in a vacuum.

The thermal joint resistance model to be presented below was given by Yovanovich (1971, 1986). It was developed for the elastic contact of paraboloids, i.e. the elastic contact formed by a ball and the inner and outer races of an instrument bearing.

## 12.2 Point Contact Model

When two curved surfaces are brought into contact with each other, the general shape of the contact area is an ellipse with semiaxes $a$ and $b$, and area $A = \pi ab$. The semiaxes are given by the

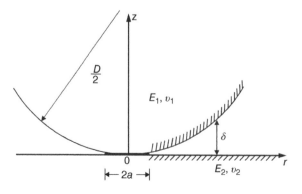

**Figure 12.1** Joint formed by elastic contact of sphere or cylinder with smooth flat. Source: Adapted from Kitscha (1982).

relationships [Timoshenko and Goodier (1970)]:

$$a = m \left[ \frac{3F\Delta}{2(A+B)} \right]^{1/3} \quad \text{and} \quad b = n \left[ \frac{3F\Delta}{2(A+B)} \right]^{1/3} \tag{12.1}$$

where $F$ is the total normal load acting upon the contact area, and $\Delta$ is a physical parameter defined by

$$\Delta = \frac{1}{2} \left[ \frac{(1-\nu_1^2)}{E_1} + \frac{(1-\nu_2^2)}{E_2} \right] \tag{12.2}$$

when dissimilar materials form the contact. The physical parameters are Young's modulus: $E_1$ and $E_2$ and Poisson's ratio: $\nu_1$ and $\nu_2$. The geometric parameters $A$ and $B$ are related to the radii of curvature of the two contacting solids [Timoshenko and Goodier (1970)]:

$$2(A+B) = \frac{1}{\rho_1} + \frac{1}{\rho_1'} + \frac{1}{\rho_2} + \frac{1}{\rho_2'}$$

$$= \frac{1}{\rho^\star} \tag{12.3}$$

where the local radii of curvature of the contacting solids are denoted as $\rho_1, \rho_1', \rho_2$, and $\rho_2'$. The second relationship between $A$ and $B$ is

$$2(B-A) = \left[ \left( \frac{1}{\rho_1} - \frac{1}{\rho_1'} \right)^2 + \left( \frac{1}{\rho_2} - \frac{1}{\rho_2'} \right)^2 \right.$$
$$\left. + 2 \left( \frac{1}{\rho_1} - \frac{1}{\rho_1'} \right) \left( \frac{1}{\rho_2} - \frac{1}{\rho_2'} \right) \cos 2\phi \right]^{1/2} \tag{12.4}$$

The parameter $\phi$ is the angle between the principal planes which pass through the contacting solids.

The dimensionless parameters $m$ and $n$ which appear in the equations for the semiaxes are called the Hertz elastic parameters. They are determined by means of the following Hertz relationships [Timoshenko and Goodier (1970)]:

$$m = \left[ \frac{2}{\pi} \frac{E(k')}{k^2} \right]^{1/3} \quad \text{and} \quad n = \left[ \frac{2}{\pi} k E(k') \right]^{1/3} \tag{12.5}$$

where $E(k')$ is the complete elliptic integral of the second kind of modulus $k'$ [Abramowitz and Stegun (1965), Byrd and Friedman (1971)], and

$$k' = \sqrt{1-k^2} \quad \text{with} \quad k = \frac{n}{m} = \frac{b}{a} \le 1 \tag{12.6}$$

The additional parameters $k$ and $k'$ are solutions of the transcendental equation [Timoshenko and Goodier (1970)]:

$$\frac{B}{A} = \frac{(1/k^2)E(k') - K(k')}{K(k') - E(k')} \tag{12.7}$$

where $K(k')$ and $E(k')$ are complete elliptic integrals of the first and second kind of modulus $k'$.

The Hertz solution requires the calculation of $k$, the ellipticity, $K(k')$ and $E(k')$. This requires a numerical solution of the transcendental equation which relates $k, K(k')$ and $E(k')$ to the local geometry of the contacting solids through the geometric parameters $A$ and $B$. This is usually accomplished by an iterative numerical procedure.

To this end, additional geometric parameters have been defined [Timoshenko and Goodier (1970)]:

$$\cos \tau = \frac{B - A}{B + A} \quad \text{and} \quad \omega = \frac{A}{B} \leq 1 \tag{12.8}$$

Computed values of $m$ and $n$, or $m/n$, and $n$ are presented with $\tau$ or $\omega$ as the independent parameter. Table 12.1 shows how $k$, $m$, and $n$ depend on the parameter $\omega$ over a range of values that should cover most practical contact problems.

The parameter $k'$ may be computed accurately and efficiently by means of the Newton–Raphson iteration method applied to the following relationships [Yovanovich (1986)]:

$$k'_{new} = k' + \frac{N(k')}{D(k')} \tag{12.9}$$

where

$$N(k') = k^2 \frac{E(k')}{K(k')} \left[ k^2 + \frac{A}{B} \right] - k^4 \left[ 1 + \frac{A}{B} \right] \tag{12.10}$$

and

$$D(k') = \frac{E(k')}{K(k')} \left[ k'k^2 - 2k' \left( \frac{A}{B} \right) \right] + \left( \frac{A}{B} \right) k^2 k' \tag{12.11}$$

**Table 12.1**  Hertz contact parameters and elastoconstriction parameter.

| $\omega$ | $k$ | $m$ | $n$ | $\psi^\star$ |
|---|---|---|---|---|
| 0.001 | 0.0147 | 14.316 | 0.2109 | 0.2492 |
| 0.002 | 0.0218 | 11.036 | 0.2403 | 0.3008 |
| 0.004 | 0.0323 | 8.483 | 0.2743 | 0.3616 |
| 0.006 | 0.0408 | 7.262 | 0.2966 | 0.4020 |
| 0.008 | 0.0483 | 6.499 | 0.3137 | 0.4329 |
| 0.010 | 0.0550 | 5.961 | 0.3277 | 0.4581 |
| 0.020 | 0.0828 | 4.544 | 0.3765 | 0.5438 |
| 0.040 | 0.1259 | 3.452 | 0.4345 | 0.6397 |
| 0.060 | 0.1615 | 2.935 | 0.4740 | 0.6994 |
| 0.080 | 0.1932 | 2.615 | 0.5051 | 0.7426 |
| 0.100 | 0.2223 | 2.391 | 0.5313 | 0.7761 |
| 0.200 | 0.3460 | 1.813 | 0.6273 | 0.8757 |
| 0.300 | 0.4504 | 1.547 | 0.6969 | 0.9261 |
| 0.400 | 0.5441 | 1.386 | 0.7544 | 0.9557 |
| 0.500 | 0.6306 | 1.276 | 0.8045 | 0.9741 |
| 0.600 | 0.7117 | 1.1939 | 0.8497 | 0.9857 |
| 0.700 | 0.7885 | 1.1301 | 0.8911 | 0.9930 |
| 0.800 | 0.8618 | 1.0787 | 0.9296 | 0.9972 |
| 0.900 | 0.9322 | 1.0361 | 0.9658 | 0.9994 |
| 1.000 | 1.0000 | 1.0000 | 1.0000 | 1.0000 |

If the initial guess for $k'$ is based upon the following correlation of the results given in Table 12.1, the convergence will occur within two to three iterations:

$$k' = \left\{ 1 - \left[ 0.9446 \left( \frac{A}{B} \right)^{0.6135} \right]^2 \right\}^{1/2} \tag{12.12}$$

Polynomial approximations of the complete elliptic integrals [Abramowitz and Stegun (1965)] may be used to evaluate them with an absolute error less than $10^{-7}$ over the full range of $k'$.

## 12.3 Local Gap Thickness

The local gap thickness is required for the "elasto-gap" resistance model developed by Yovanovich (1986). The general relationship for the gap thickness can be determined by means of the following surface displacements [Johnson (1985), Timoshenko and Goodier (1970)]:

$$\delta(x, y) = \delta_0 + w(x, y) - w_0 \tag{12.13}$$

where $\delta_0(x, y)$ is the local gap thickness under zero load conditions, $w(x, y)$ is the total local displacement of the surfaces of the bodies outside the loaded area, and $w_0$ is the approach of the contact bodies due to loading.

The total local displacement of the two bodies is given by

$$\frac{3F\Delta}{2\pi} \int_\mu^\infty \left( 1 - \frac{x^2}{a^2 + t} - \frac{y^2}{b^2 + t} \right) \frac{dt}{\{(a^2 + t)(b^2 + t)t\}^{1/2}} \tag{12.14}$$

where $\mu$ is the positive root of the equation

$$\frac{x^2}{(a^2 + \mu)} + \frac{y^2}{(b^2 + \mu)} = 1 \tag{12.15}$$

When $\mu > 0$, the point of interest lies outside the elliptical contact area

$$\frac{x^2}{a^2} + \frac{y^2}{b^2} = 1 \tag{12.16}$$

When $\mu = 0$, the point of interest lies inside the contact area, and when $x = y = 0$, $w(0, 0) = w_0$ the total approach of the contacting bodies is

$$w_0 = \frac{3F\Delta}{2\pi} \int_0^\infty \frac{dt}{\{(a^2 + t)(b^2 + t)t\}^{1/2}} \tag{12.17}$$
$$= \frac{3F\Delta}{\pi a} K(k')$$

The relationships for the semiaxes and the local gap thickness will be used in Sections 12.4–12.7 to develop the general relationships for the contact and gap resistances.

## 12.4 Contact Resistance of Isothermal Elliptical Contact Area

The general spreading/constriction resistance model, as proposed by Yovanovich (1971, 1986), is based upon the assumption that both bodies forming an elliptical contact area can be taken to be a conducting half-space. This approximation of actual bodies is reasonable because the dimensions of the contact area are very small relative to the characteristic dimensions of the contacting bodies.

If the free (noncontacting) surfaces of the contacting bodies are adiabatic, then the total "ellipsoidal" spreading/constriction resistance of an isothermal elliptical contact area with $a \geq b$ is [Yovanovich (1971, 1986)]:

$$R_c = \frac{\psi}{2k_s a} \tag{12.18}$$

where $a$ is the semimajor axis, $k_s$ is the harmonic mean thermal conductivity of the joint:

$$k_s = \frac{2k_1 k_2}{k_2 + k_2} \tag{12.19}$$

and $\psi$ is the spreading/constriction parameter of the isothermal elliptical contact area developed in the section for spreading resistance of an isothermal elliptical area on an isotropic half-space:

$$\psi = \frac{2}{\pi} K(k') \tag{12.20}$$

in which $K(k')$ is the complete elliptic integral of the first kind of modulus $k'$, and it is related to the semiaxes:

$$k' = \left[ 1 - \left( \frac{b}{a} \right)^2 \right]^{1/2}$$

The complete elliptic integral can be computed accurately by means of accurate polynomial approximations and by Computer Algebra Systems. This important special function can also be approximated by means of the following simple relationships:

$$\begin{aligned} K(k') &= \quad \ln(4a/b) \qquad & 0 \leq k < 0.1736 \\ K(k') &= \frac{2\pi}{\left[ 1 + \sqrt{b/a} \right]^2} \qquad & 0.1736 < k \leq 1 \end{aligned} \tag{12.21}$$

These approximations have a maximum error less than 0.8% which occurs at $k = 0.1736$. The "ellipsoidal" spreading/constriction parameter approaches the value of one when $a = b$, the circular contact area.

When the results of the Hertz elastic deformation analysis are substituted into the results of the ellipsoidal constriction analysis, one obtains the "elasto-constriction" resistance relationship developed by [Yovanovich (1971, 1986)]:

$$k_s \left[ 24F\Delta\rho^\star \right]^{1/3} R_c = \frac{2}{\pi} \left[ \frac{K(k')}{m} \right] \equiv \psi^\star \tag{12.22}$$

where the effective radius of the ellipsoidal contact is defined as $\rho^\star = [2(A + B)]^{-1}$. The left hand side is a dimensionless group consisting of the known total mechanical load $F$, the effective thermal conductivity $k_s$ of the joint, the physical parameter $\Delta$, and the isothermal, elliptical spreading/constriction resistance $R_c$. The right hand side is defined to be $\psi^\star$ which is called the thermal "elasto-constriction" parameter [Yovanovich (1971, 1986)]. Typical values of $\psi^\star$ for a range of values of $\omega$ are given in Table 12.1. The "elasto-constriction" parameter $\psi^\star \to 1$ when $k = b/a = 1$, the case of the circular contact area.

## 12.5 Elastogap Resistance Model

The thermal resistance of the gas-filled gap depends on three local quantities: (i) the local gap thickness, (ii) the thermal conductivity of the gas, and (iii) the temperature difference between the bounding solid surfaces.

The gap model is based on the subdivision of the gap into elemental heat flow channels (flux tubes) having isothermal upper and lower boundaries, and adiabatic sides [Yovanovich and Kitscha (1974)]. The heat flow lines in each channel (tube) are assumed to be straight and perpendicular to the plane of contact.

If the local "effective" gas conductivity, $k_g(x, y)$, in each elemental channel is assumed to be uniform across the local gap thickness $\delta(x, y)$, then the differential gap heat flow rate is

$$dQ_g = \frac{k_g(x, y)\Delta T_g(x, y)}{\delta(x, y)} dx\, dy \tag{12.23}$$

The total gap heat flow rate is given by the double integral $Q_g = \iint_{A_g} dQ_g$ where the integration is performed over the entire effective gap area $A_g$.

The thermal resistance of the gap, $R_g$, is defined in terms of the overall joint temperature drop, $\Delta T_j$ [Yovanovich and Kitscha (1974)]:

$$\frac{1}{R_g} = \frac{Q_g}{\Delta T_j} = \iint_{A_g} \frac{k_g(x, y)\Delta T_g(x, y)}{\delta(x, y)\Delta T_j} dA_g \tag{12.24}$$

The local gap thickness in the general case of two bodies in elastic contact forming an elliptical contact area is given above.

The local "effective" gas conductivity is based on a model for the "effective" thermal conductivity of a gaseous layer bounded by two infinite, isothermal, parallel plates. Therefore, for each heat flow channel (tube), the effective thermal conductivity is approximated by the relation [Yovanovich and Kitscha (1974)]:

$$k_g(x, y) = \frac{k_{g,\infty}}{\left[1 + \dfrac{\alpha\beta\Lambda}{\delta(x, y)}\right]} \tag{12.25}$$

where $k_{g,\infty}$ is the gas conductivity under continuum conditions at standard temperature and pressure (STP). The accommodation parameter, $\alpha$, is defined as

$$\alpha = \frac{(2 - \alpha_1)}{\alpha_1} + \frac{(2 - \alpha_2)}{\alpha_2} \tag{12.26}$$

where $\alpha_1$ and $\alpha_2$ are the accommodation coefficients at the solid–gas interfaces [Wiedmann and Trumpler (1946), Hartnett (1961), Wachman (1962), Thomas (1982), Kitscha and Yovanovich (1975), Madhusudana (1975, 1996), Semyonov et al. (1984), Wesley and Yovanovich (1986), Song and Yovanovich (1987a,b), Song (1988), Song et al. (1993a,b)]. The fluid property parameter, $\beta$, is defined by

$$\beta = \frac{2\gamma}{\left[(\gamma + 1)/Pr\right]} \tag{12.27}$$

where $\gamma$ is the ratio of the specific heats, and $Pr$ is the Prandtl number. The mean free path $\Lambda$ of the gas molecules is given in terms of $\Lambda_{g,\infty}$, the mean free path at STP, as follows:

$$\Lambda = \Lambda_{g,\infty}\left(\frac{T_g}{T_{g,\infty}}\right)\left(\frac{P_{g,\infty}}{P_g}\right) \tag{12.28}$$

Two models for determining the local temperature difference, $\Delta T_g(x, y)$, are proposed [Yovanovich and Kitscha (1974)]. In the first model, it is assumed that the bounding solid surfaces are isothermal at their respective contact temperatures; hence,

$$\Delta T_g(x, y) = \Delta T_j \tag{12.29}$$

This is called the thermally "decoupled" model [Yovanovich and Kitscha (1974)], since it assumes that the surface temperature at the solid–gas interface is independent of the temperature field within each solid.

In the second model [Yovanovich and Kitscha (1974)], it is assumed that the temperature distribution of the solid–gas interface is induced by the conduction through the solid–solid contact, under vacuum conditions. This temperature distribution is approximated by the temperature distribution immediately below the surface of an insulated half-space that receives heat from an isothermal elliptical contact. Solving for this temperature distribution, using ellipsoidal coordinates, it was found that

$$\frac{\Delta T_g(x,y)}{\Delta T_j} = 1 - \frac{F(k',\psi)}{K(k')} \tag{12.30}$$

where $F(k',\psi)$ is the incomplete elliptic integral of the first kind of modulus $k'$ and amplitude angle $\psi$ [Abramowitz and Stegun (1965), Byrd and Friedman (1971)]. The modulus, $k'$, is given above, and the amplitude angle is

$$\psi = \sin^{-1}\left[\frac{a^2}{(a^2+\mu)}\right]^{1/2} \tag{12.31}$$

where the parameter $\mu$ is defined above. It ranges between $\mu = 0$, the edge of the elliptical contact area, to $\mu = \infty$, the distant points within the half-space.

Since the solid–gas interface temperature is coupled to the interior temperature distribution, it is called the "coupled" half-space model temperature drop.

The general "elasto-gap" model has not been solved. Two special cases of the general model have been examined. They are the sphere-flat contact studied by [Yovanovich and Kitscha (1974), Kitscha and Yovanovich (1975)], and the cylinder-flat contact was studied by McGee et al. (1985). The two special cases are discussed below.

## 12.6 Joint Radiative Resistance

If the joint is in a vacuum, or the gap is filled with a "transparent" substance such as dry air, then there is heat transfer across the gap by radiation. It is difficult to develop a general relationship that would be applicable for all point contact problems because the radiation heat transfer occurs in a complex enclosure that consists of at least three nonisothermal convex surfaces. The two contacting surfaces are usually metallic, and the third surface forming the enclosure is frequently a reradiating surface such as insulation.

Yovanovich and Kitscha (1974) examined an enclosure that was formed by the contact of a metallic hemisphere and a metallic circular disk of diameter $D$. The third boundary of the enclosure was a nonmetallic circular cylinder of diameter $D$ and height $D/2$. The metallic surfaces were assumed to be isothermal at temperatures $T_1$ and $T_2$ with $T_1 > T_2$. These temperature correspond to the "extrapolated" temperatures from temperatures measured on both sides of the joint. The joint temperature was defined as $T_j = (T_1 + T_2)/2$. The dimensionless radiation resistance was found to have the relationship:

$$Dk_s R_r = \frac{k_s}{\pi \sigma D T_j^3}\left[\frac{1-\epsilon_2}{\epsilon_2} + \frac{1-\epsilon_1}{2\epsilon_1} + 1.103\right] \tag{12.32}$$

where $\sigma = 5.67 \times 10^{-8}$ W/(m$^2$ K$^4$) is the Stefan–Boltzmann constant, $\epsilon_1, \epsilon_2$ are the emissivities of the hemisphere and disk, respectively, and $k_s$ is the effective thermal conductivity of the joint.

## 12.7 Joint Resistance of Sphere-Flat Contact

The contact, gap, radiative, and joint resistances of the sphere-flat contact shown in Figure 12.1 are presented here. The contact radius $a$ is much smaller than the sphere diameter $D$. Assuming an isothermal contact area the general "elasto-constriction" resistance model [Yovanovich (1971, 1986), Yovanovich and Kitscha (1974)] becomes

$$R_c = \frac{1}{2k_s a} \tag{12.33}$$

where $k_s = 2k_1 k_2/(k_1 + k_2)$ is the harmonic mean thermal conductivity of the contact, and the contact radius is obtained from the Hertz elastic model [Timoshenko and Goodier (1970)]:

$$\frac{2a}{D} = \left[\frac{6F\Delta}{D^2}\right]^{1/3} \tag{12.34}$$

where $F$ is the mechanical load at the contact, and $\Delta$ is the joint physical parameter defined above.

The general coupled "elastogap" resistance model for point contacts reduces, for the sphere-flat contact, to [Yovanovich and Kitscha (1974), Yovanovich (1986)]

$$\frac{1}{R_g} = \left(\frac{D}{L^2}\right) k_{g,0} I_{g,p} \tag{12.35}$$

where $L = D/(2a)$ is the relative contact size. The gap integral for point contacts proposed by Yovanovich and Kitscha (1974) and Kitscha and Yovanovich (1975) is defined as

$$I_{g,p} = \int_1^L \frac{2x \tan^{-1}\sqrt{x^2 - 1}}{\frac{2\delta}{D} + \frac{2M}{D}} \, dx \tag{12.36}$$

The local gap thickness $\delta$ is obtained from the relationship:

$$\begin{aligned}\frac{2\delta}{D} &= 1 - \left(1 - \left(\frac{x}{L}\right)^2\right)^{1/2} \\ &+ \frac{1}{\pi L^2}\left[(2 - x^2)\sin^{-1}\left(\frac{1}{x}\right) + \sqrt{x^2 - 1}\right] - \frac{1}{L^2}\end{aligned} \tag{12.37}$$

where $x = r/a$ and $1 \leq x \leq L$. The gap gas rarefaction parameter is defined as

$$M = \alpha\beta\Lambda \tag{12.38}$$

where the gas parameters $\alpha$, $\beta$, and $\Lambda$ have been defined above.

### 12.7.1 Joint Resistance for Sphere-Flat in Vacuum

The joint resistance for a sphere-flat contact in a vacuum is [Yovanovich and Kitscha (1974), Kitscha and Yovanovich (1975)]

$$\frac{1}{R_j} = \frac{1}{R_c} + \frac{1}{R_r} \tag{12.39}$$

The proposed models were verified by experiments conducted by Kitscha (1982). The test conditions were sphere diameter is $D = 25.4$ mm; vacuum pressure is $P_g = 10^{-6}$ Torr; mean interface temperature range: $316 \leq T_m \leq 321$ K; harmonic mean thermal conductivity of sphere-flat contact is $k_s = 51.5$ W/(m K); emissivities of very smooth sphere and lapped flat (RMS roughness is $\sigma = 0.13$ μm) are $\epsilon_1 = 0.2$ and $\epsilon_2 = 0.8$, respectively; elastic properties of sphere and flat: $E_1 = E_2 = 206$ GPa and $v_1 = v_2 = 0.3$.

**Table 12.2** Dimensionless load, constriction, radiative and joint resistances.

| F (N) | L (D/2a) | $T_m$ (K) | $R_r^\star$ (model) | $R_j^\star$ (model) | $R_j^\star$ (test) |
|---|---|---|---|---|---|
| 16.0 | 115.1 | 321 | 1155 | 104.7 | 107.0 |
| 22.2 | 103.2 | 321 | 1155 | 94.7 | 99.4 |
| 55.6 | 76.0 | 321 | 1155 | 71.3 | 70.9 |
| 87.2 | 65.4 | 320 | 1164 | 61.9 | 61.9 |
| 195.7 | 50.0 | 319 | 1177 | 48.0 | 48.8 |
| 266.9 | 45.1 | 318 | 1188 | 43.4 | 42.6 |
| 467.0 | 37.4 | 316 | 1211 | 36.4 | 35.4 |

Source: Kitscha (1982)/William Kitscha.

The dimensionless joint resistance is given by the relationship:

$$\frac{1}{R_j^\star} = \frac{1}{R_c^\star} + \frac{1}{R_r^\star} \tag{12.40}$$

where

$$R_j^\star = Dk_s R_j, \quad R_c^\star = Dk_s R_c = L, \quad R_r^\star = 1415\left(\frac{300}{T_m}\right)^3 \tag{12.41}$$

The model and vacuum data are compared for a load range in Table 12.2. The agreement between the joint resistance model and the data is excellent over the full range of the tests.

### 12.7.2 Effect of Gas Pressure on Joint Resistance of a Sphere-Flat Contact

According to the model of Yovanovich and Kitscha (1974), the dimensionless joint resistance with a gas in the gap is given by

$$\frac{1}{R_j^\star} = \frac{1}{R_c^\star} + \frac{1}{R_r^\star} + \frac{1}{R_g^\star} \tag{12.42}$$

where

$$R_g^\star = Dk_s R_g = \frac{k_s L^2}{k_{g,\infty} I_{g,p}} \tag{12.43}$$

The joint model for the sphere-flat contact is compared against data obtained for the following test conditions: sphere diameter is $D = 25.4$ mm; load is 16 newton; dimensionless load is $L = 115.1$; mean interface temperature range: $309 \leq T_m \leq 321$ K; harmonic mean thermal conductivity of sphere-flat contact is $k_s = 51.5$ W/(m K); emissivities of smooth sphere and lapped flat are $\epsilon_1 = 0.2$ and $\epsilon_2 = 0.8$, respectively.

The load was fixed such that $L = 115.1$ for all tests, while the air pressure was varied from 400 mmHg down to a vacuum. The dimensionless resistances are given in Table 12.3. It can be seen that the dimensionless radiative resistance was relatively large with respect to the dimensionless gap and contact resistances. The dimensionless gap resistance values varied greatly with the gas pressure. The agreement between the joint resistance model and the data is very good for all test points.

**Table 12.3** Effect of gas pressure on gap and joint resistances for air.

| $T_m$ (K) | $P_g$ (mmHg) | $R_g^\star$ (model) | $R_r^\star$ (model) | $R_j^\star$ (model) | $R_j^\star$ (test) |
|---|---|---|---|---|---|
| 309 | 400.0 | 77.0 | 1295 | 44.5 | 46.8 |
| 310 | 100.0 | 87.6 | 1282 | 47.9 | 49.6 |
| 311 | 40.0 | 97.4 | 1270 | 50.7 | 52.3 |
| 316 | 4.4 | 138.3 | 1211 | 59.7 | 59.0 |
| 318 | 1.8 | 168.9 | 1188 | 64.7 | 65.7 |
| 321 | 0.6 | 231.3 | 1155 | 72.1 | 73.1 |
| 322 | 0.5 | 245.9 | 1144 | 73.4 | 74.3 |
| 325 | 0.2 | 352.8 | 1113 | 80.5 | 80.3 |
| 321 | Vacuum | $\infty$ | 1155 | 104.7 | 107.0 |

**Example 12.1** A joint consists of the elastic contact of a smooth sphere of diameter $D$ and a smooth flat of diameter $D = 2b$ under a static, external, axial force $F$. The elastic properties of the sphere and flat are $E_1$, $v_1$, and $E_2$, $v_2$, respectively, and their thermal conductivities are $k_1$ and $k_2$, respectively. The sphere-flat joint is under vacuum conditions. The thermophysical properties of the joint are listed below:

$$k_1 = 50.2\,\text{W/(m K)}, \quad E_1 = 206\,\text{GPa}, \quad v_1 = 0.3$$

$$k_2 = 52.8\,\text{W/(m K)}, \quad E_2 = 206\,\text{GPa}, \quad v_2 = 0.3$$

The diameter of the sphere is $D = 25.4\,\text{mm}$, and the flat is much larger than this dimension.

(a) Calculate the elasto-constriction resistance of the joint if the force is $F = 16.0\,\text{N}$.

To find the elasto-constriction resistance of the joint, we must first find $a$ which requires the use of the mechanical contact equations:

$$\Delta = \frac{1}{2}\left[\frac{(1 - v_1^2)}{E_1} + \frac{(1 - v_2^2)}{E_2}\right] = 8.835 \times 10^{-12}\,\text{GPa}^{-1}$$

$$\frac{2a}{D} = \left[\frac{6F\Delta}{D^2}\right]^{1/3} = 0.008695$$

which gives a contact spot radius of $a = 0.0001104\,\text{m}$. The harmonic mean effective thermal conductivity is

$$k_s = \frac{2k_1 k_2}{k_1 + k_2} = 51.47\,\text{W/(m K)}$$

Finally, we determine the constriction resistance for the joint to be

$$R_c = \frac{1}{2k_s a} = \frac{1}{2 \cdot 51.47 \cdot 0.0001104} = 87.98\,\text{K/W}$$

(b) The dimensionless radiation resistance of the sphere-flat-insulation enclosure is given by

$$R_r^\star = Dk_s R_r = C\left(\frac{300}{T_m}\right)^3$$

where $C = 1900$ and the mean absolute temperature of the joint is $T_m = 325$ K. Calculate values of $R_r^\star$ and $R_r$.

Using the above equation and parameters, we find

$$R_r^\star = 1900\left(\frac{300}{325}\right)^3 = 1494.4$$

which gives

$$R_r = \frac{1494.4}{0.0254 \cdot 51.47} = 1143.5 \text{ K/W}$$

(c) Calculate the dimensionless joint resistance $R_j^\star = Dk_sR_j$.
The joint resistance for a vacuum is determined from

$$\frac{1}{R_j} = \frac{1}{R_c} + \frac{1}{R_r}$$

This gives a value of $R_j = 81.69$ K/W and dimensionless joint resistance of $R_j^\star = 106.8$.

(d) If the force is increased to $F = 467.0$ N and the mean joint temperature falls to $T_m = 305$ K, calculate new values of $R_c$, $R_r$, and $R_j$.
Repeating the calculations for the new load of $F = 467$ N we find $a = 0.0003400$ m and

$$R_c = 28.57 \text{ K/W}$$

$$R_r = 1383.1 \text{ K/W}$$

$$R_j = 27.99 \text{ K/W}$$

**Example 12.2** In this problem, we will consider the elasto-gap resistance $R_g$ of the macro-gap of the joint described in Example 12.1 when the gas in the macro-gap is helium. This problem is related to the IBM Thermal Conduction Module (TCM) which is used to cool chips in the IBM servers. For this problem, use the helium properties listed below:

$$\alpha_1 = 0.55, \quad \alpha_2 = 0.60, \quad \gamma = 1.67, \quad Pr = 0.67, \quad \Lambda_0 = 186 \text{ nm}$$

The reference gas temperature and gas pressure are $T_0 = 288$ K and $P_{g,0} = 760$ Torr. The thermal conductivity of helium is given by

$$k_g = 0.145 + 3.24 \times 10^{-4} T_g$$

for the temperature range: $27 °C \le T_g \le 400 °C$.

(a) Calculate the value of the macro-gap integral $I_{g,p}$ for $F = 10$ N, $D = 25.4$ mm, $T_g = 330$ K, and the gas pressures: $P_g = 400, 40$, and $4$ Torr. Use the thermophysical properties given in Example 12.1.
For the physical properties in the previous example, we find the contact spot radius:

$$\frac{2a}{D} = \left[\frac{6F\Delta}{D^2}\right]^{1/3} = 0.007434$$

or

$$a = 0.00009441 \text{ m}$$

Next, we need to calculate the following parameters:

$$\alpha = \frac{(1-\alpha_1)}{\alpha_1} + \frac{(1-\alpha_2)}{\alpha_2} = 4.97$$

$$\beta = \frac{2\gamma}{(\gamma + 1)Pr} = 1.867$$

$$\Lambda = \Lambda_0 \left(\frac{T_g}{288}\right)\left(\frac{760}{P_g}\right)$$

We must now use Eqs. (12.35)–(12.38) to find the gap conductance integral which must be evaluated numerically. The results are provided in the table below:

| $P_g$ (Torr) | $I_{g,p}$ | $R_g^{\star}$ | $R_g$ (K/W) |
|---|---|---|---|
| 400 | 378563.8 | 15.09 | 11.54 |
| 40 | 265849.0 | 21.49 | 16.44 |
| 4 | 148102.6 | 38.57 | 29.5 |

(b) Calculate the values of the macro-gap resistance $R_g$ and the dimensionless macro-gap resistance defined as

$$R_g^{\star} = \left(\frac{k_s}{k_g}\right)\left(\frac{L^2}{I_{g,p}}\right)$$

We now use the above equation (Eq. (12.43)) to find the dimensionless gap resistance and the dimensional gap resistance. The dimensionless load from part (a) is $L = D/2a = 134.5$. The results are provided in the table above.

(c) Calculate the joint resistance for the gas pressures given in (a). The dimensionless radiation resistance is

$$R_g^{\star} = 1800\left(\frac{300}{T_m}\right)^3$$

with $T_m = T_g$.
The total joint resistance is found from

$$\frac{1}{R_j} = \frac{1}{R_c} + \frac{1}{R_g} + \frac{1}{R_c}$$

Using the above equation for the radiation resistance and the previous results for $R_c$ and $R_g$, we find the values for $R_j$ tabulated below.

| $P_g$ (Torr) | $R_j^{\star}$ | $R_j$ (K/W) |
|---|---|---|
| 400 | 13.43 | 10.27 |
| 40 | 18.28 | 13.98 |
| 4 | 29.32 | 22.43 |

## 12.8 Joint Resistance for Contact of a Sphere and Layered Substrate

Figure 12.2 shows three joints: (a) contact between a hemisphere and a substrate, (b) contact between a hemisphere and a layer of finite thickness bonded to a substrate, and (c) contact between a hemisphere and a very thick layer where $t/a \gg 1$.

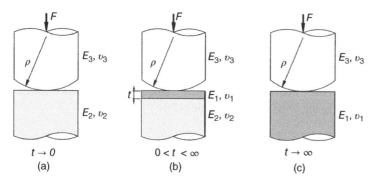

**Figure 12.2** Contact between hemisphere and layer on substrate. (a) Uncoated substrate, (b) coated substrate (thin layer), (c) coated substrate (thick layer). Source: Modified from Stevanović et al. (2002).

In the general case contact is between an elastic hemisphere of radius $\rho$ and elastic properties: $E_3, v_3$ and an elastic layer of thickness $t$ and elastic properties: $E_1, v_1$ which is bonded to an elastic substrate of elastic properties: $E_2, v_2$. The axial load is $F$. It is assumed that $E_1 < E_2$ for layers that are less rigid than the substrate.

The contact radius $a$ is much smaller than the dimensions of the hemisphere and the substrate. The solution for arbitrary layer thickness is complex because the contact radius depends on several parameters, i.e. $a = f(F, \rho, t, E_i, v_i)$, $i = 1, 2, 3$. The contact radius lies in the range: $a_S \le a \le a_L$, where $a_S$ corresponds to the very thin layer limit, $t/a \to 0$ (Figure 12.2a) and $a_L$ corresponds to the very thick layer limit, $t/a \to \infty$ (Figure 12.2c).

For the general case, a contact in a vacuum, and if there is negligible radiation heat transfer across the gap, the joint resistance is equal to the contact resistance which is equal to the sum of the spreading/constriction resistances in the hemisphere and layer-substrate, respectively.

The joint resistance is given by [Stevanović et al. (2002)]:

$$R_j = R_c = \frac{1}{4k_3 a} + \frac{\psi_{12}}{4k_2 a} \tag{12.44}$$

where $a$ is the contact radius. The first term on the right-hand side represents the constriction resistance in the hemisphere, and $\psi_{12}$ is the spreading resistance parameter in the layer-substrate. The thermal conductivities of the hemisphere and the substrate appear in the first and second terms, respectively. The layer-substrate spreading resistance parameter depends on two dimensionless parameters: $\tau = t/a$ and $\kappa = k_1/k_2$. This parameter was presented above under spreading resistance in a layer on a half-space. To calculate the joint resistance, the contact radius must be found.

A special case arises when the rigidity of the layer is much smaller than the rigidity of the hemisphere and the layer. This corresponds to "soft" metallic layers such as indium, lead, and tin, or nonmetallic layers such as rubber or elastomers. In this case, since $E_1 \ll E_2$ and $E_1 \ll E_3$, the hemisphere and substrate may be modeled as perfectly rigid while the layer deforms elastically.

The dimensionless numerical values for $a/a_L$ obtained from the elastic contact model of Chen and Engel (1972) according to Stevanović et al. (2002) are plotted in Figure 12.3 for a wide range of relative layer thickness $\tau = t/a$, and for a range of values of the layer Young's modulus, $E_1$.

The contact model which is represented by the correlation equation of the numerical values is [Stevanović et al. (2002)]:

$$\frac{a}{a_L} = 1 - c_3 \exp\left(c_1 \tau^{c_2}\right) \tag{12.45}$$

**Figure 12.3** Comparison of data and model for contact between rigid hemisphere and elastic layer on rigid substrate. Source: Adapted from [Stevanović et al. (2002)].

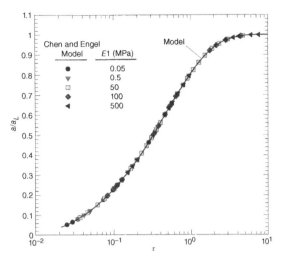

with correlation coefficients: $c_1 = -1.73$, $c_2 = 0.734$, $c_3 = 1.04$. The reference contact radius is $a_L$ which corresponds to the very thick layer limit given by

$$a_L = \left[\frac{3}{4}\frac{F\rho}{E_{13}}\right]^{1/3} \quad \text{for} \quad \frac{t}{a} \to \infty \tag{12.46}$$

The maximum difference between the correlation equation and the numerical values obtained from the model of Chen and Engel (1972) is approximately 1.9% for $\tau = 0.02$. The following relationship, based on the Newton–Raphson method, is recommended for calculation of the contact radius [Stevanović et al. (2002)]:

$$a_{n+1} = \frac{a_n - a_L(1 - 1.04\exp\left[-1.73(t/a_n)^{0.734}\right])}{1 + 1.321\left(a_L/a_S\right)\left(t/a_n\right)^{0.734}\exp\left[-1.73(t/a_n)^{0.734}\right]} \tag{12.47}$$

If the first guess is $a_0 = a_L$, then fewer than six iterations are required to give eight-digit accuracy.

In the general case where the hemisphere, layer, and substrate are elastic, then the contact radius lies in the range: $a_S \leq a \leq a_L$ for $E_2 < E_1$. The two limiting values of $a$ are according to Stevanović et al. (2002):

$$a = \begin{cases} a_S = \left(\dfrac{3}{4}\dfrac{F\rho}{E_{23}}\right)^{1/3} & \text{for} \quad \dfrac{t}{a} \to 0 \\[4mm] a_L = \left(\dfrac{3}{4}\dfrac{F\rho}{E_{13}}\right)^{1/3} & \text{for} \quad \dfrac{t}{a} \to \infty \end{cases} \tag{12.48}$$

where the effective Young's modulus for the two limits are defined as

$$E_{13} = \left[\frac{1-v_1}{E_1} + \frac{1-v_3}{E_3}\right]^{-1} \qquad E_{23} = \left[\frac{1-v_2}{E_2} + \frac{1-v_3}{E_3}\right]^{-1} \tag{12.49}$$

The dimensionless contact radius and dimensionless layer thickness were defined as [Stevanović et al. (2002)]:

$$a^{\star} = \frac{a - a_S}{a_L - a_S} \quad \text{where} \quad 0 < a^{\star} < 1 \tag{12.50}$$

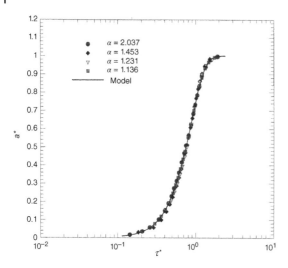

**Figure 12.4** Comparison of data and model for elastic contact between hemisphere and layer on substrate. Source: [Stevanović et al. (2002)]/with permission of IEEE.

and

$$\tau^{\star} = \left[\frac{t}{a}\sqrt{\alpha}\right]^{1/3} \quad \text{where} \quad \alpha = \frac{a_L}{a_S} = \left[\frac{E_{23}}{E_{13}}\right]^{1/3} \tag{12.51}$$

The dimensionless numerical values obtained from the full model of Chen and Engel (1972) for values of $\alpha$ in the range: $1.136 \leq \alpha \leq 2.037$ are shown in Figure 12.4.

The correlation equation is [Stevanović et al. (2002)]:

$$\frac{a - a_S}{a_L - a_S} = 1 - \exp\left[-\pi^{1/4}\left(\frac{t\sqrt{\alpha}}{a}\right)^{\pi/4}\right] \tag{12.52}$$

Since the unknown contact radius $a$ appears on both sides, the numerical solution of the correlation equation requires an iterative method (Newton–Raphson method) to find its root. For all metal combinations, the following solution is recommended [Stevanović et al. (2002)]:

$$a = a_S + (a_L - a_S)\left\{1 - \exp\left[-\pi^{1/4}\left(\frac{t\sqrt{\alpha}}{a_0}\right)^{\pi/4}\right]\right\} \tag{12.53}$$

where

$$a_0 = a_S + (a_L - a_S)\left\{1 - \exp\left[-\pi^{1/4}\left(\frac{2t\sqrt{\alpha}}{a_S + a_L}\right)^{\pi/4}\right]\right\} \tag{12.54}$$

## 12.9 Joint Resistance for Elastic–Plastic Contact of Hemisphere and Flat in Vacuum

A model is available for calculating the joint resistance of an elastic–plastic contact of a portion of a hemisphere whose radius of curvature is $\rho$ attached to a cylinder whose radius is $b_1$ and a cylindrical flat whose radius is $b_2$.

The elastic properties of the hemisphere are $E_1$, $v_1$, and the elastic properties of the flat are $E_2$, $v_2$. The thermal conductivities are $k_1$ and $k_2$, respectively.

If the contact strain is very small, the contact is elastic, and the Hertz model can be used to predict the elastic contact radius denoted as $a_e$. On the other hand, if the contact strain is very large, then plastic deformation may occur in the flat which is assumed to be fully *work hardened* and the plastic contact radius is denoted $a_p$. Between the fully elastic and fully plastic contact regions, there is a transition called the elastic–plastic contact region, which is very difficult to model. In the region, the contact radius is denoted as $a_{ep}$, the elastic–plastic contact radius. The relationship between $a_e$, $a_p$ and $a_{ep}$ is $a_e \leq a_{ep} \leq a_p$.

The elastic–plastic radius is related to the elastic and plastic contact radii by means of the composite model based the method of Churchill and Usagi (1972) for combining asymptotes [Sridhar and Yovanovich (1994)]:

$$a_{ep} = \left(a_e^n + a_p^n\right)^{1/n} \tag{12.55}$$

where $n$ is the combination parameter which is found empirically to have the value $n = 5$. The elastic and plastic contact radii may be obtained from the relationships [Sridhar and Yovanovich (1994)]:

$$a_e = \left(\frac{3}{4}\frac{F\rho}{E'}\right)^{1/3} \quad \text{and} \quad a_p = \left(\frac{F}{\pi H_B}\right)^{1/2} \tag{12.56}$$

with the effective modulus:

$$\frac{1}{E'} = \frac{1 - v_1^2}{E_1} + \frac{1 - v_2^2}{E_2} \tag{12.57}$$

The plastic parameter is the Brinell hardness $H_B$ of the flat. The elastic–plastic deformation model assumes that the hemispherical solid is harder than the flat. The static axial load is $F$.

The joint resistance for a smooth hemispherical solid in elastic–plastic contact with smooth flat is given by [Sridhar and Yovanovich (1994)]:

$$R_j = \frac{\psi_1}{4k_1 a_{ep}} + \frac{\psi_2}{4k_2 a_{ep}} \tag{12.58}$$

The spreading/constriction resistance parameters for the hemisphere and flat are

$$\psi_1 = \left(1 - \frac{a_{ep}}{b_1}\right)^{1.5} \quad \text{and} \quad \psi_2 = \left(1 - \frac{a_{ep}}{b_2}\right)^{1.5} \tag{12.59}$$

### 12.9.1  Alternative Constriction Parameter for Hemisphere

The following spreading/constriction parameter can be derived from the hemisphere solution:

$$\psi_1 = 1.0014 - 0.0438\epsilon - 4.0264\epsilon^2 + 4.968\epsilon^3 \tag{12.60}$$

where $\epsilon = a/b_1$.

If the contact is in a vacuum, and the radiation heat transfer across the gap is negligible, the $R_j = R_c$. Also, if $b_1 = b_2 = b$, then

$$\psi_1 = \psi_2 = \left(1 - \frac{a}{b}\right)^{1.5} \tag{12.61}$$

The joint and dimensionless joint resistances for this case becomes

$$R_j = \frac{\psi}{2k_s a} \quad \text{and} \quad R_j^\star = 2bk_s R_j = \frac{(1 - a/b)^{1.5}}{a/b} \tag{12.62}$$

where $k_s = 2k_1 k_2/(k_1 + k_2)$.

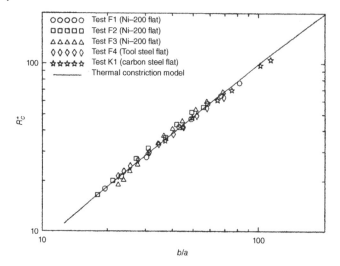

**Figure 12.5** Comparison of data and model for elastic–plastic contact between hemisphere and flat. Source: [Sridhar and Yovanovich (1994)]/with permission of American Society of Mechanical Engineers.

Sridhar and Yovanovich (1994) compared the dimensionless joint resistance against data obtained for contacts between a carbon steel ball and several flats of Ni 200, carbon steel, and tool steel. The nondimensional data and the dimensionless joint resistance model are compared in Figure 12.5 for a range of values of the reciprocal contact strain $b/a$.

The agreement between the model and the data over the entire range $20 < b/a < 120$ is very good. The points near $b/a \approx 100$ are in the elastic contact region, and the points near $b/a = 20$ are close to the plastic contact region. In between, the points are in the transition region (TR) called the elastic–plastic contact region.

If the material of the flat "work-hardens" as the deformation takes place, the model for predicting the contact radius is much more complex as described by Sridhar and Yovanovich (1994) and Johnson (1985). This case will not be given here.

**Example 12.3** In this problem, the elasto–plastic contact model will be used to find the joint resistance $R_j$ of a smooth sphere-smooth flat joint. The sphere is carbon steel, and the flat is Ni 200 which is "softer" than the sphere. The thermophysical properties of the sphere and flat are given below. The Brinell hardness of the flat is denoted as $H_B = 1.01$ GPa. Assume that the joint is in a vacuum and neglect radiation heat transfer across the macro-gap.

$$k_1 = 78.5 \, \text{W/(m K)}, \quad E_1 = 207 \, \text{GPa}, \quad v_1 = 0.3$$

$$k_2 = 45.8 \, \text{W/(m K)}, \quad E_2 = 204 \, \text{GPa}, \quad v_2 = 0.3$$

The diameter of the sphere is $D = 25.4$ mm, and the diameter of the flat is $2b = 25.4$ mm.

(a) Calculate the elastic, plastic, and elasto–plastic contact radii denoted as $a_e$, $a_p$, and $a_{ep}$ by means of the relations given in this section, for three loads: $F = 10, 100, 1000$ N.
   We begin by determining the effective elastic modulus $E'$:

$$\frac{1}{E'} = \left[ \frac{(1 - v_1^2)}{E_1} + \frac{(1 - v_2^2)}{E_2} \right]$$

or $E' = 112.9$ GPa. Next, the elastic and plastic contact parameters are found from (with $\rho = D/2$):

$$a_e = \left(\frac{3}{4}\frac{F\rho}{E'}\right)^{1/3} \quad \text{and} \quad a_p = \left(\frac{F}{\pi H_B}\right)^{1/2}$$

and

$$a_{ep} = \left(a_e^5 + a_p^5\right)^{1/5}$$

The results are provided in the table below:

| F (N) | $a_e$ (μm) | $a_p$ (μm) | $a_{ep}$ (μm) |
|-------|-----------|-----------|--------------|
| 10    | 94        | 156       | 95.4         |
| 100   | 204       | 178       | 221.4        |
| 1000  | 439       | 561       | 590.6        |

(b) Calculate the values of the joint resistance based on the elastic contact model for the three loads.

The constriction resistance (since we have a vacuum and negligible radiation) is determined from:

$$R_c = \frac{\psi}{2k_s a}$$

where

$$k_s = \frac{2k_1 k_2}{k_1 + k_2} = 57.85 \text{ W/(m K)}$$

and

$$\psi = (1 - a/b)^{3/2}$$

Using the calculated contact spot size for elastic contact $a = a_e$, and the radius of the flat $b = D/2$, we find the values for each load tabulated in the table below.

(c) Calculate the values of the joint resistance based on the elasto–plastic contact model for the three loads.

Repeating the calculations from part (b) with $a = a_{ep}$, we obtain the results provided in the table below.

(d) Calculate the ratio of the resistance values obtained in (b) and (c).

The ratio of the two joint resistances are also provided in the table below.

| F (N) | $R_{j,e}$ (K/W) | $R_{j,ep}$ (K/W) | $R_{j,e}/R_{j,ep}$ |
|-------|-----------------|------------------|---------------------|
| 10    | 90.9            | 89.5             | 1.016               |
| 100   | 41.3            | 38.0             | 1.087               |
| 1000  | 18.6            | 13.6             | 1.368               |

## 12.10 Ball Bearing Resistance

Models have been presented [Yovanovich (1967, 1971)] for calculating the overall thermal resistance of slowly rotating instrument bearings which consist of many very smooth balls contained by very smooth inner and outer races. The thermal resistance models for bearings are based on elastic contact of the balls with the inner and outer races, and spreading and constriction resistances in the balls and in the inner and outer races. For each ball, there are two elliptical contact areas, one at the inner race and one at the outer race. The local thickness of the adjoining gap is very complex to model. There are four spreading/constriction zones associated with each ball. The full "elasto-constriction" resistance model must be used to obtain the overall thermal resistance of the bearing.

Since these are complex systems, the contact resistance models are also complex; therefore, they will not be presented here. The above references should be consulted for the development of the contact resistance models and other pertinent references.

## 12.11 Line Contact Models

If a long smooth circular cylinder with radius of curvature $\rho_1 = D_1$, length $L_1$, and elastic properties: $E_1, v_1$ makes contact with another long smooth circular cylinder with radius of curvature $\rho_2 = D_2/2$, length $L_1$, and elastic properties: $E_2, v_2$, then in general, if the axes of the cylinders are not aligned (i.e. they are crossed), an elliptical contact area is formed with semi-axes: $a, b$ where it is assumed that $a < b$. If the cylinder axes are aligned, then the contact area becomes a strip of width $2a$, and the larger axes is equal to the length of the cylinder. The general Hertz model presented above may be used to find the semiaxes and the local gap thickness if the axes are not aligned. For aligned axes, the general equations reduce to simple relationships which are given below.

### 12.11.1 Contact Strip and Local Gap Thickness

If the two cylinder axes are aligned, then the contact area is a strip of width $2a$ and length $L_1$ where [Timoshenko and Goodier (1970), Walowit and Anno (1975)]:

$$a = 2 \left[ \frac{2F \rho \Delta}{\pi L_1} \right]^{1/2} \tag{12.63}$$

where the effective curvature is

$$\frac{1}{\rho} = \frac{1}{\rho_1} + \frac{1}{\rho_2} \tag{12.64}$$

and the contact parameter is

$$\Delta = \frac{1}{2} \left[ \frac{1 - v_1^2}{E_1} + \frac{1 - v_2^2}{E_2} \right] \tag{12.65}$$

The contact pressure is maximum along the axis of the contact strip, and it is given by the relationship:

$$P_0 = \frac{2}{\pi} \frac{F}{a L_1} = \left[ \frac{F}{2\pi L_1 \rho \Delta} \right]^{1/2} \tag{12.66}$$

and the pressure distribution has the form [Timoshenko and Goodier (1970), Walowit and Anno (1975)]:

$$P(x) = P_0 \sqrt{1 - \left(\frac{x}{a}\right)^2} \quad \text{for} \quad 0 \le x \le a \tag{12.67}$$

The mean contact area pressure is

$$P_m = \frac{F}{2aL_1} = \frac{4P_0}{\pi} \tag{12.68}$$

The normal approach of the two aligned cylinders is [Timoshenko and Goodier (1970), Walowit and Anno (1975)]

$$\alpha = \frac{2F'}{\pi} \left\{ \frac{1 - v_1^2}{E_1} \left[ \ln\left(\frac{4\rho_1}{a}\right) - \frac{1}{2} \right] + \frac{1 - v_2^2}{E_2} \left[ \ln\left(\frac{4\rho_2}{a}\right) - \frac{1}{2} \right] \right\} \tag{12.69}$$

where $F' = F/L_1$ is the load per unit cylinder length. The general local gap thickness relationship is [Timoshenko and Goodier (1970), Walowit and Anno (1975)]:

$$\frac{2\delta}{\rho} = (1 - 1/L^2)^{1/2} - (1 - \xi^2/L^2)^{1/2}$$
$$+ \left[ \xi(\xi^2 - 1)^{1/2} - \cosh^{-1}(\xi) - \xi^2 + 1 \right] 2L \tag{12.70}$$

where

$$L = \frac{\rho}{2a}, \quad \xi = \frac{x}{L}, \quad \text{and} \quad 1 \le \xi \le L \tag{12.71}$$

If a single circular cylinder of diameter $D$ or ($\rho_1 = D_1/2 = D/2$) is in elastic contact with a flat ($\rho_2 = \infty$), then put $\rho = D/2$ in the above relationships.

### 12.11.2  Contact Resistance at Line Contact

The thermal contact resistance for the very narrow contact strip of width $2a$ formed by the elastic contact of a long smooth circular cylinder of diameter $D$ and a smooth flat whose width is $2b$, and its length $L_1$ is identical to the cylinder length, is given by the approximate relationship [McGee et al. (1985)]:

$$R_c = \frac{1}{\pi L_1 k_1} \left[ \ln\left(\frac{4}{\epsilon_1}\right) - \frac{\pi}{2} \right] + \frac{1}{\pi L_1 k_2} \ln\left(\frac{2}{\pi \epsilon_2}\right) \tag{12.72}$$

where the thermal conductivities of the cylinder and flat are $k_1$ and $k_2$, respectively. The contact parameters are $\epsilon_1 = 2a/D$ for the cylinder and $\epsilon_2 = a/b$ for the flat. For elastic contacts, $2a/D \ll 1$ and $2a/b \ll 1$ for most engineering applications. The approximate relationship for $R_c$ becomes more accurate for very narrow strips.

The width of the flat relative to the cylinder diameter may be (i) $2b > D$, (ii) $2b = D$, or (iii) $2b < D$. McGee et al. (1985) proposed the use of the dimensionless form of the contact resistance:

$$R_c^\star = L_1 k_s R_c = \frac{1}{2\pi} \frac{k_s}{k_1} \ln\left(\frac{\pi}{F^\star}\right) - \frac{k_s}{2k_1} + \frac{1}{2\pi} \frac{k_s}{k_2} \ln\left(\frac{1}{4\pi F^\star}\right) \tag{12.73}$$

where

$$F^\star = \frac{F\Delta}{L_1 D} \quad \text{and} \quad k_s = \frac{2k_1 k_2}{k_1 + k_2} \tag{12.74}$$

### 12.11.3 Gap Resistance at Line Contact

The general "elasto-gap" resistance model for line contacts proposed by Yovanovich (1986) reduces for the circular cylinder/flat contact to

$$\frac{1}{R_g} = \frac{4aL_1}{D} k_{g,\infty} I_{g,l} \tag{12.75}$$

where $k_{g,\infty}$ is the gas thermal conductivity, and the line contact "elasto-gap" integral is defined as [Yovanovich (1986)]:

$$I_{g,l} = \frac{2}{\pi} \int_1^L \frac{\cosh^{-1}(\xi)d\xi}{2\delta/D + M/D} \tag{12.76}$$

where

$$L = \frac{D}{2a}, \quad \xi = \frac{x}{L}, \quad \text{and} \quad 1 \le \xi \le L \tag{12.77}$$

This is the "coupled elasto-gap" model. Numerical integration is required to calculate values of $I_{g,l}$. The gas rarefaction parameter which appears in the gap integral is

$$M = \alpha\beta\Lambda \tag{12.78}$$

where the accommodation parameter and other gas parameters are defined as

$$\alpha = \frac{(2-\alpha_1)}{\alpha_1} + \frac{(2-\alpha_2)}{\alpha_2} \qquad \beta = \frac{2\gamma}{(\gamma+1)\,\text{Pr}} \qquad \gamma = \frac{C_p}{C_v} \tag{12.79}$$

and the molecular mean free path is

$$\Lambda = \Lambda_{g,\infty} \left(\frac{T_g}{T_{g,\infty}}\right) \cdot \left(\frac{P_{g,\infty}}{P_g}\right) \tag{12.80}$$

where $\Lambda_{g,\infty}$ is the value of the molecular mean free path at the reference temperature $T_{g,\infty}$ and gas pressure $P_{g,\infty}$.

### 12.11.4 Joint Resistance at Line Contact

The joint resistance at a line contact, neglecting radiation heat transfer across the gap, is

$$\frac{1}{R_j} = \frac{1}{R_c} + \frac{1}{R_g} \tag{12.81}$$

McGee et al. (1985) examined the accuracy of the above contact, gap, and joint resistance relationships for helium and argon for gas pressures between $10^{-6}$ Torr and one atmosphere. The effect of contact load was investigated for mechanical loads between 80 Newtons and 8000 Newtons on specimens fabricated from Keewatin tool steel, Type 304 stainless steel, and Zircaloy-4. The experimental data were compared with the model predictions, and good agreement was obtained over a limited range of experimental parameters. Discrepancies were observed at the very light mechanical loads due to slight amounts of form error (crowning) along the contacting surfaces.

## 12.12   Joint Resistance of Nonconforming Rough Surfaces

There is ample empirical evidence that surfaces may not be conforming and rough as shown in Figure 10.5c,f. The surfaces may be both nonconforming and rough as shown in Figure 10.5b,e, where a smooth hemispherical surface is in contact with a flat, rough surface.

If surfaces are nonconforming and rough, then the joint that is formed is more complex from the standpoint of defining the micro- and macro-geometry before load is applied, and the definition of the micro- and macro-contacts that are formed after load is applied. The deformation of the contacting asperities may be elastic, plastic, or elastic–plastic. The deformation of the bulk may also be elastic, plastic, or elastic–plastic. The mode of deformation of the micro- and macro-geometry are closely connected under conditions that are not understood today.

The thermal joint resistance of such a contact is complex because heat can cross the joint by conduction through the micro-contacts and the associated micro-gaps and by conduction across the macro-gap. If the temperature level of the joint is sufficiently high, there may be significant radiation across the micro-gaps and macro-gap.

Clearly, this type of joint represents complex thermal and mechanical problems that are coupled. Much vacuum data have been reported [Clausing and Chao (1965), Burde (1977), Kitscha (1982)] that shows that the presence of roughness can alter the joint resistance of a nonconforming surface under light mechanical loads, and have negligible effects at higher loads. Also, the presence of out-of-flatness can have significant effect on the joint resistance of a rough surface under vacuum conditions.

It is generally accepted that the joint resistance under vacuum conditions may be modeled as the superposition of microscopic and macroscopic resistance [Clausing and Chao (1965), Greenwood and Tripp (1967), Holm (1967), Yovanovich (1969), Burde and Yovanovich (1978), Lambert (1995), Lambert and Fletcher (1997)]. The joint resistance can be modeled as

$$R_j = R_{mic} + R_{mac} \tag{12.82}$$

The microscopic resistance is given by the relationship:

$$R_{mic} = \frac{\psi_{mic}}{2k_s N a_S} \tag{12.83}$$

where $\psi_{mic}$ is the average spreading/constriction resistance parameter, $N$ is the number of micro-contacts that are distributed in some complex manner over the "contour" area of radius $a_L$, and $a_S$ represents some average micro-contact spot radius, and the harmonic mean thermal conductivity of the joint is $k_s = 2k_1 k_2/(k_1 + k_2)$.

The macroscopic resistance is given by the relationship:

$$R_{mac} = \frac{\psi_{mac}}{2k_s a_L} \tag{12.84}$$

where $\psi_{mac}$ is the spreading/constriction resistance parameter for the contour area of radius $a_L$.

The mechanical model should be capable of predicting the contact parameters: $a_S, a_L$, and $N$. These parameters are also required for the determination of the thermal spreading/constriction parameters: $\psi_{mic}, \psi_{mac}$.

At this time, there is no simple mechanical model available for prediction of the geometric parameters required in the microscopic and macroscopic resistance relationships. There are publications (e.g. Greenwood and Tripp (1967), Holm (1967), Burde and Yovanovich (1978), Johnson (1985), Lambert and Fletcher (1997), Marotta and Fletcher (2001)) that deal with various aspects of this very complex problem.

## 12.13 System for Nonconforming Rough Surface Contact

The thermal and mechanical models are developed for nonconforming rough (CR) surfaces as shown in Figure 12.6. The system consists of two solid cylindrical rods of radius $b_L$ making contact Bahrami et al (2005a, 2006). The nonconforming surface is smooth $\sigma = 0$ and polished with radius of curvature $\rho > 0$. The polished surface is a hemisphere. The deformation of the hemispherical surface is assumed to be elastic and therefore, the effective Young's modulus is

$$\frac{1}{E'} = \frac{1 - v_1^2}{E_1} + \frac{1 - v_2^2}{E_2} \tag{12.85}$$

The Young's modulus of the rough flat surface is $E_2$, and it undergoes elastic deformation under an axial load $F$. The parameters $v_1$ and $v_2$ are called Poisson's ratios.

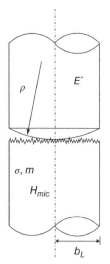

**Figure 12.6**
Schematic of the non-conforming smooth – rough surface contact.

When the surfaces are smooth, the contact with be elastic and the contact area will be circular with radius $a_H < b_L$. The Hertz theory of elastic contact can be used to predict the radius $a_H$ will depend on the parameters: $(F, \rho, E_1, E_2, v_1, v_2)$, and the contact radius has the scale $a_H \sim F^{1/3}$. When surface roughness is present, then the mechanical contact becomes significantly more complex because the contacting asperities undergo plastic deformation while the hemispherical surface undergoes elastic deformation simultaneously.

The Hertz contact theory predicts that the contact pressure has a parabolic profile with the maximum value at the center of the contact area, and it is zero at the edge of the contact area.

The presence of surface roughness modifies the contact pressure distribution. The maximum value decreases, and the pressure is now distributed over a larger area such that the effective macro-contact radius is $a_L > a_H < b_L$.

Heat transfer across the joint formed by the contact of a rough flat and a smooth nonconforming surface is complex because it consists of the micro-spreading resistance which is associated with the micro-contacts and the macro-spreading resistance which is associated with the macro-contact geometry. The micro- and macro-resistances are thermally connected in series and therefore the overall joint resistance is the sum of the micro- and macro-resistances, i.e. $R_j = R_s + R_L$.

Simple scaling analysis will be employed in Sections 12.12.1 and 12.13.2 to find the pertinent mechanical and thermal relations for the micro-geometry for conforming rough surface plastic contact. Subsequently, relations for the nonconforming rough surface contact will be obtained.

### 12.13.1 Vickers Micro-hardness Model

Whenever a polished flat surface is glass bead-blasted under certain conditions, a rough flat surface is produced that is characterized by its asperity roughness denoted as $\sigma$ and its asperity mean slope denoted as $m$. The parameter $\sigma/m$ is the effective surface roughness and it appears in micro-geometric models that give several parameters such as the mean contact spot $a_s$, the contact spot density $n$, the relative real area of contact $A_r/A_a$, and the relative mean plane separation denoted as $\lambda = Y/\sigma$, where $Y$ is the mean plane separation. These parameters are given in Chapter 11 under the Sections 11.2–11.4 elastic, plastic, and elasto–plastic contact.

For plastic contact, the micro-hardness is developed from the Vickers micro-hardness measurements which are given in terms of the Vickers diagonal $d_V$. Hegazy (1985) correlated the Vickers micro-hardness for several metals as

$$H_V = c_1 \left( \frac{d_V}{d_0} \right)^{c_2} \tag{12.86}$$

where $d_0$ represents some convenient reference value for the average diagonal, and $c_1$ and $c_2$ are the correlation coefficients. It is conventional to set $d_0 = 1\,\mu m$. $d_V$ can also be related to the mean radius of micro-contacts $a_s$ such that $d_V = \sqrt{2\pi}a_s$, the micro-hardness has the following scale:

$$H_{mic} \sim H^\star = c_1 \left( \frac{\sigma}{\sigma_0 m} \right)^{c_2} \tag{12.87}$$

where $\sigma_0 = 1\,\mu m$.

Sridhar and Yovanovich (1996) developed correlation coefficients for the Vickers micro-hardness coefficients:

$$\frac{c_1}{3178} = 4.0 - 5.77\kappa + 4.0\kappa^2 - 0.61\kappa^3 \tag{12.88}$$

and

$$c_2 = -0.370 + 0.442 \left( \frac{H_B}{c_1} \right) \tag{12.89}$$

where $H_B$ is the Brinell hardness [Johnson (1985), Tabor (1951)] and $\kappa = H_B/3178$. The correlations are valid for the Brinell hardness range 1300–7600 MPa. The foregoing correlations were developed for a range of metal types (e.g. Ni200, SS304, Zr alloys, Ti alloys, and tool steel). Sridhar and Yovanovich (1996) also proposed a correlation equation that relates the Brinell hardness number to the Rockwell C hardness number:

$$BHN = 43.7 + 10.92\,HRC - \frac{HRC^2}{5.18} + \frac{HRC^3}{340.26} \tag{12.90}$$

for the range: $20 \le HRC \le 65$. The foregoing correlations may be used to calculate the Vickers micro-hardness coefficients and, therefore, the micro hardness of many metals.

### 12.13.2 Scale Analysis Results

Bahrami (2004) and Bahrami et al. (2004a) under took a simple scale analysis of conforming rough surfaces which are characterized by the effective surface roughness $\sigma$ and the effective surface asperity slope $m$. The results of the scaling analyses are presented next.

The mean contact spot which is modeled as an equivalent circle of radius $a_s$ has the following scale:

$$a_s \sim \frac{\sigma}{m} \tag{12.91}$$

Assuming plastic deformation of the contacting asperities the effective micro-hardness is obtained from the Vickers micro-hardness measurements and scaled as

$$H_{mic} \sim c_1 \left( \frac{\sigma}{m\sigma_0} \right)^{c_2} \tag{12.92}$$

where $\sigma_0 = 1\,\mu m$ and $c_1$ and $c_2$ are correlation coefficients of the Vickers micro-hardness test. A force balance for the contacting asperities $F = A_r H_{mic}$, where $A_r$ is the total real contact area yields the following scale:

$$F \sim \pi n_s \left( \frac{\sigma}{\sigma_0 m} \right)^2 H_{mic} \tag{12.93}$$

where $n_s$ is the total number of contact spots over the total apparent area $A_a$. The total real contact area has the following scale:

$$A_r \sim \pi n_s \left( \frac{\sigma}{\sigma_0 m} \right)^2 \tag{12.94}$$

If the distance between adjacent contact spots is very large, then the thermal spreading resistance of the entire set of micro-contacts is modeled as micro-contacts on a half-space which is thermally connected in parallel and the total spreading resistance has the scale $R_s \sim 1/2k_s n_s a_s$, where $k_s = 2k_1 k_2/(k_1 + k_2)$ is the effective thermal conductivity of the contact. Thus, the total spreading resistance of the micro-contacts has the following scale:

$$R_s \sim \frac{1}{2 k_s n_s \left( \frac{\sigma}{\sigma_0 m} \right)} \tag{12.95}$$

If the distance between adjacent contact spots is comparable to the radius of a typical contact spot, then the spreading resistance of a typical contact spot must be obtained by means of a flux tube solution and, therefore, the spreading resistance of each micro-contact spot must be multiplied by the spreading resistance parameter $\psi(1 - \epsilon_s)^{1.5}$ were $\epsilon_s = \sqrt{A_r/A_a} = \sqrt{F/\pi b_L^2 H_{mic}} = \sqrt{P^\star}$.

The total spreading resistance of closely spaced micro spots has the scale:

$$R_s(\epsilon) \sim \frac{\psi(\epsilon_s)}{2 k_s n_s \left( \frac{\sigma}{\sigma_0 m} \right)} \tag{12.96}$$

The total dimensionless spreading resistance is defined as

$$R_s^\star = 2 k_s L R_s \tag{12.97}$$

with the characteristic length $L = b_L^2/(\sigma/\sigma_0 m)$ here $b_L$ is the radius of the apparent contact area $A_a$.

The dimensionless apparent contact pressure is defined as $P^\star = P/H_{mic} = F/(\pi b_L^2 H_{mic})$.

The dimensionless total spreading resistance has two scales: one for closely spaced micro spots corresponding to high contact pressures and one for very light contact pressures. The two relations are (i) for the flux tube model

$$R_s^\star \sim \frac{\left( 1 - \sqrt{P^\star} \right)^{1.5}}{P^\star} \qquad P^\star < 10^{-4} \tag{12.98}$$

and (ii) for the half-space model

$$R_s^\star \sim \frac{1}{P^\star} \qquad P^\star > 10^{-4} \tag{12.99}$$

If we use the solution for isothermal circular contact spots on a half-space, the effective micro-spreading resistance becomes

$$R_s^\star = \frac{C}{P^\star} \tag{12.100}$$

where $C$ is an empirical parameter to be found from extensive experimental data.

In dimensional form, we have for the total spreading resistance

$$R_s = \frac{C\pi \left( \sigma/\sigma_0 m \right) H_{mic}}{2 k_s F} \tag{12.101}$$

and for the contact conductance,

$$h_s = \frac{1}{C} \frac{2}{\pi} \left( \frac{\sigma}{\sigma_0 m} \right) \frac{P}{H_{mic}} \tag{12.102}$$

Five hundred experimental data of Hegazy (1985) for Ni200, SS304, Zr-2.5 wt%Nb, and Zircaloy 4 were plotted as $R_s^\star$ versus $P^\star$ in the range $10^{-4}$ to $10^{-2}$, and it was found that $C = 0.36$ gives good agreement between the model and data.

### 12.13.3 Contact of Smooth Hemisphere and Rough Flat

A smooth ($\sigma = 0, m = 0$) hemisphere whose curvature is $\rho$ placed in mechanical contact with a rough flat with curvature $\rho \to \infty$ and surface roughness $\sigma > 0$ and $m > 0$ contact. A brief review of the Hertz contact of the hemisphere and flat is necessary before obtaining the approximate model for the smooth hemisphere-smooth flat contact.

The Hertz model for perfectly smooth surfaces yields the pressure distribution $P_H(r/a_H)$ and Hertz contact radius $a_H$. The pressure distribution is

$$P_H(r/a_H) = P_{0,H} \sqrt{1 - (r/a_H)^2} \tag{12.103}$$

where

$$P_{0,H} = \frac{3F}{2\pi a_H^2}, \qquad a_H = \left( \frac{3F\rho}{3E'} \right)^{1/3} \tag{12.104}$$

where $F$ is the total axial load, $P_{0,H}$ is the maximum contact pressure at the center of the contact area and the effective Young's modulus is

$$\frac{1}{E'} = \frac{1 - \nu_1^2}{E_1} + \frac{1 - \nu_2^2}{E_2} \tag{12.105}$$

with Young's moduli ($E_1, E_2$) and Poisson's ratios ($\nu_1, \nu_2$).

The presence of surface roughness in the flat surface distorts the pressure distribution such that the maximum contact pressure at the axis decreases and the pressure distribution occurs over a larger radius of contact $a_L > a_H$.

Bahrami (2004) developed a contact model based on the following geometric parameter:

$$\alpha = \frac{\sigma \rho}{a_H^2} \tag{12.106}$$

and another geometric parameter based on the hemisphere curvature and the Hertz contact radius

$$\tau = \frac{\rho}{a_H} \tag{12.107}$$

The complex analysis yielded the following correlation for the modified Hertz radius:

$$a' = \frac{a_L}{a_H} = \frac{1.80}{\tau^{0.028}} \sqrt{\alpha + 0.31\tau^{0.056}} \tag{12.108}$$

For hemispheres with relatively large radius of curvature, the following approximate relation is recommended:

$$a' = 1.5\sqrt{\alpha + 0.45} \tag{12.109}$$

### 12.13.4 General Micro–Macro Spreading Resistance Model

The macro-spreading resistance model is based on the flux tube spreading resistance model which gives

$$R_L = \frac{\left(1 - \frac{a_L}{b_L}\right)^{1.5}}{2k_s a_L} \tag{12.110}$$

The effective micro-spreading resistance for nonconforming rough surface contacts is given by

$$R_s = \frac{CH_{mic}\left(\frac{\sigma}{\sigma_0 m}\right)}{4k_s}\left[\int_0^{a_L} P(r) r \, dr\right]^{-1} \tag{12.111}$$

A force balance over the entire contact area gives

$$F = 2\pi \int_0^{a_L} P(r) r \, dr \tag{12.112}$$

which is used to give the following relation for the effective micro-thermal spreading resistance for nonconforming surfaces

$$R_s = \frac{C\pi(\sigma/\sigma_0 m)H_{mic}}{2k_s F} \tag{12.113}$$

which is identical to the relation for the micro-spreading resistance of conforming rough surfaces. These results show that the micro-thermal spreading resistance is independent of surface curvature $\rho$ and independent of the pressure distribution profile.

Finally, the overall thermal joint resistance of nonconforming rough surfaces has the form

$$R_j = \frac{0.36\pi\left(\sigma/\sigma_0 m\right)H_{mic}}{2k_s F} + \frac{\left(1 - \frac{a_L}{b_L}\right)^{1.5}}{2k_s a_L} \tag{12.114}$$

with the empirical constant $C = 0.36$ which is based on 500 vacuum data points of Hegazy (1985).

Combining all constants, we write the total spreading resistance

$$R_j = \frac{0.565\left(\sigma/\sigma_0 m\right)H_{mic}}{k_s F} + \frac{\left(1 - \frac{a_L}{b_L}\right)^{1.5}}{2k_s a_L} \tag{12.115}$$

It can be seen that in the conforming rough surface limit, where $a_L \to 0$, the macro resistance $R_L \to 0$. Also, in the elasto-constriction limit, where $\sigma \to 0, a_L \to a_H$, and the micro-spreading resistance $R_s \to 0$.

Diving both sides by the micro resistance, we obtain

$$\frac{1.77k_s F}{H_{mic}(\sigma/\sigma_0 m)} R_j = 1 + \Theta \tag{12.116}$$

where $\Theta$ is the ratio of the macro- to micro-thermal resistances

$$\Theta = \frac{F(1 - a_L/b_L)^{1.5}}{1.13 H_{mic}(\sigma/\sigma_0 m)a_L} \tag{12.117}$$

The nondimensional parameter $\Theta$ includes the applied load, the macro- and micro-geometrical parameters $(\sigma, m, \rho)$ as well as the elastic and plastic mechanical properties of the contacting solids $E'$ and $H_{mic}$. Based on this nondimensional parameter, a criterion can be defined for the elasto-constriction and conforming rough limits:

$$\begin{cases} \Theta \ll 1 & \text{conforming rough limit} \\ \Theta \gg 1 & \text{elasto-constriction limit} \end{cases} \tag{12.118}$$

The parameter $\Theta$ is independent of the thermal conductivities of the contacting solids.

The relation for $R_j$ can be nondimensionalized with respect to the length $L$ to give

$$R_j^{\star} = 2k_s L R_j = \frac{0.36}{P^{\star}} + \frac{L(1 - a_L/b_L)^{1.5}}{a_L} \tag{12.119}$$

where $L = b_L^2/(\sigma/\sigma_0 m)$ and $P^{\star} = F/(\pi b_L^2 H_{mic})$.

### 12.13.5 Comparisons of Nonconforming Rough Surface Model with Vacuum Data

The test parameters for comparisons of the nonconforming rough surface model and vacuum data are given in Table 12.4 for three tests which are denoted as $T1$, $T2$, and $T3$. Tests $T1$ and $T3$ have almost identical micro ($\sigma, m$) and macro ($\rho$) geometric characteristics. The hemisphere and flat are fabricate from stainless steel. The Vickers micro-hardness coefficients $c_1$ and $c_2$ are similar. Test $T2$ has the smallest surface roughness and smallest radius of curvature. The temperature level of the joints are such that radiative heat transfer across the joint are negligible.

The parameter $\Theta$ has three ranges: for $T1$ the range is $1.34 \leq \Theta \leq 6.21$, for $T2$ the range is $1.13 \leq \Theta \leq 8.96$ and for $T3$ the range is $0.14 \leq \Theta \leq 6.04$. The range of parameters for all tests is given in Table 12.5.

**Table 12.4** Parameter values for T1–T3 tests for non-conforming rough model.

| Test | $\sigma$ (µm) | $m$ (–) | $\rho$ (m) | $c_1$ (GPa) | $c_2$ (–) | $k_s$ (W/(mK)) |
|------|------|------|------|------|------|------|
| T1 | 2.04 | 0.087 | 0.95 | 6.23 | −0.23 | 18.83 |
| T2 | 2.78 | 0.199 | 0.45 | 6.55 | −0.12 | 18.82 |
| T3 | 2.04 | 0.087 | 0.95 | 6.23 | −0.23 | 18.46 |

**Table 12.5** Ranges of geometric, mechanical and thermal parameters of experiments.

| Parameter | Range |
|------|------|
| $\sigma$ | 0.12–13.94 (µm) |
| $m$ | 0.04–0.34 (–) |
| $\rho$ | 0.0127–120 (m) |
| $b_L$ | 7.15–14.28 (mm) |
| $E'$ | 25.64–114.0 (GPa) |
| $F$ | 7.72–16764 (N) |
| $k_s$ | 16.6–227.2 (W/(m K)) |
| $\bar{T}_c$ | 60–195 (°C) |

**Table 12.6** Comparison between model and test results *T*1.

| F | α | τ | $a_L/b_L$ | $R_s$ | $R_L$ | $R_j$ | Θ | $R_{exp}$ |
|---|---|---|---|---|---|---|---|---|
| (N) | (−) | (−) | (−) | (K/W) | (−) | (K/W) | (−) | (K/W) |
| 210.9 | 1.71 | 527.1 | 0.152 | 9.34 | 11.12 | 20.46 | 1.19 | 21.50 |
| 303.6 | 1.34 | 466.8 | 0.157 | 6.44 | 10.61 | 17.05 | 1.65 | 17.50 |
| 445.7 | 1.04 | 410.8 | 0.162 | 4.41 | 10.17 | 14.58 | 2.31 | 13.80 |
| 883.6 | 0.66 | 326.9 | 0.176 | 2.20 | 9.04 | 11.25 | 4.11 | 10.90 |
| 1116.9 | 0.56 | 302.4 | 0.182 | 1.72 | 8.52 | 10.24 | 4.96 | 10.10 |
| 2577.4 | 0.32 | 228.9 | 0.210 | 0.73 | 6.86 | 7.59 | 9.42 | 7.10 |

**Table 12.7** Comparison between model and test results *T*2.

| F | α | τ | $a_L/b_L$ | $R_s$ | $R_L$ | $R_j$ | Θ | $R_{exp}$ |
|---|---|---|---|---|---|---|---|---|
| (N) | (−) | (−) | (−) | (K/W) | (−) | (K/W) | (−) | (K/W) |
| 373.2 | 1.10 | 715.1 | 0.198 | 5.84 | 7.83 | 13.67 | 1.34 | 14.08 |
| 495.3 | 0.91 | 650.7 | 0.204 | 4.41 | 7.53 | 11.94 | 1.71 | 12.19 |
| 869.4 | 0.62 | 539.4 | 0.219 | 2.52 | 6.81 | 9.33 | 2.70 | 9.42 |
| 1381.5 | 0.46 | 462.3 | 0.236 | 1.57 | 6.07 | 7.64 | 3.87 | 7.62 |
| 1740.9 | 0.39 | 427.9 | 0.245 | 1.22 | 5.60 | 6.82 | 4.60 | 6.71 |
| 2656.1 | 0.36 | 371.7 | 0.266 | 0.77 | 4.80 | 5.57 | 6.21 | 5.53 |

The test results are compared with the model in Figures 12.7–12.9 and in Tables 12.6–12.8. The thermal contact resistance *TCR* given as K/W are plotted versus the total mechanical load *F* (N) for loads from 31.3 N up to 2656.1 N.

The dimensionless joint resistance is $2k_sLR_j$, where $R_j$ is the experimental value or it is the model prediction shown as a solid line. The horizontal axis is the total resistance $0.36/P^\star + L(1 - a_L/b_L)^{1.5}/a_L$. The model parameters are the dimensionless contact pressure $P^\star = F/(\pi b_L^2 H^\star)$ and $L = b_L^2(\sigma/m)$. The dimensionless micro-hardness expressed in terms of the Vickers micro-hardness coefficients ($c_1, c_2$) is written as $H^\star = c_1(\sigma/\sigma_0 m)^{c_2}$.

### 12.13.6 General Model Obtained from Scaling Analysis and Data

When the joint is formed by conforming rough surfaces the hemispherical surface is also flat because $\rho \gg 1$ and the joint is said to be conforming and rough. The scaling analysis yields a relation between the micro-spreading resistance and the applied load. The macro-spreading resistance is negligible $R_L \to 0$ and $R_j = R_s$ when radiative heat transfer is negligible. The dimensionless spreading resistance is given by

$$R_s^\star = 2k_sLR_s = \frac{C}{P^\star} \tag{12.120}$$

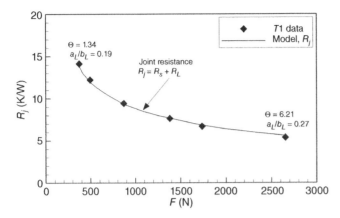

**Figure 12.7** Comparison between model and T1 test results. Source: Bahrami (2004)/Majid Bahrami.

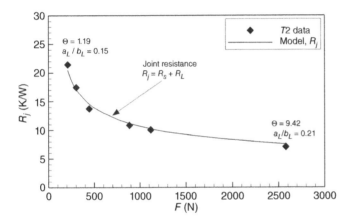

**Figure 12.8** Comparison between model and T2 test results. Source: Bahrami (2004)/Majid Bahrami.

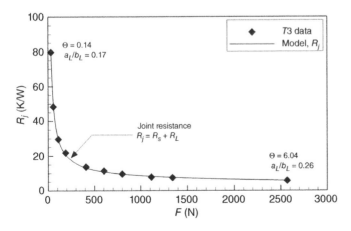

**Figure 12.9** Comparison between model and T3 test results. Source: Bahrami (2004)/Majid Bahrami.

**Table 12.8** Comparison between model and test results $T3$.

| $F$ | $\alpha$ | $\tau$ | $a_L/b_L$ | $R_s$ | $R_L$ | $R_j$ | $\Theta$ | $R_{exp}$ |
|---|---|---|---|---|---|---|---|---|
| (N) | (–) | (–) | (–) | (K/W) | (–) | (K/W) | (–) | (K/W) |
| 31.3 | 5.73 | 1634.4 | 0.169 | 71.85 | 9.89 | 81.74 | 0.14 | 79.90 |
| 55.6 | 3.90 | 1349.1 | 0.173 | 40.16 | 9.55 | 49.71 | 0.24 | 48.50 |
| 110.1 | 2.47 | 1074.2 | 0.179 | 20.15 | 9.06 | 29.21 | 0.45 | 29.60 |
| 189.2 | 1.72 | 896.7 | 0.186 | 11.91 | 8.77 | 20.68 | 0.74 | 21.80 |
| 409.3 | 1.03 | 693.4 | 0.200 | 5.36 | 7.76 | 13.12 | 1.45 | 13.80 |
| 600.5 | 0.80 | 610.2 | 0.209 | 3.64 | 7.26 | 10.90 | 2.00 | 11.50 |
| 795.8 | 0.66 | 555.6 | 0.217 | 2.74 | 6.88 | 9.63 | 2.51 | 9.70 |
| 1110.2 | 0.53 | 497.2 | 0.227 | 1.97 | 6.42 | 8.38 | 3.26 | 7.90 |
| 1338.6 | 0.47 | 467.1 | 0.234 | 1.61 | 6.07 | 7.69 | 3.77 | 7.50 |
| 2561.5 | 0.30 | 376.3 | 0.264 | 0.84 | 5.05 | 5.89 | 6.04 | 5.80 |

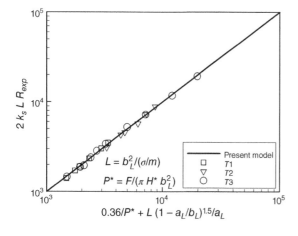

**Figure 12.10** Comparison of $T1$–$T3$ test data and model. Source: Bahrami (2004)/Majid Bahrami.

with the dimensionless contact pressure $P^\star = P/H^\star$ and the micro-hardness is based on the Vickers micro-hardness coefficients $c_1$ and $c_2$. The dimensionless spreading resistance is expressed as

$$R_s^\star = \frac{C}{P^\star} \tag{12.121}$$

which is plotted as a straight line in Figure 12.10. Data from many studies are also shown in Figure 12.10 over almost two decades of the independent parameter $P^\star$. The data except for those in the range $10^{-4} < P^\star < 10^{-2}$ are within the $\pm 15\%$ bounds when the fitting parameter has the value $C = 0.36$. There are many data which were obtained at very light contact pressures which lie even further below the model predictions. These are said to be due to truncation effects. The best agreement between data and the model occur at higher contact pressure where $P^\star \approx 10^{-3}$.

When the general model for nonconforming rough surfaces are compared with all data as shown in Figure 12.11, the dimensionless total spreading resistance is based on the micro- and

**Figure 12.11** Comparison of the scale analysis model and data for the conforming rough limit. Source: Bahrami et al. (2004a)/with permission of American Society of Mechanical Engineers.

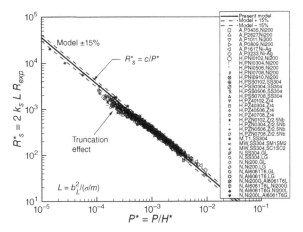

macro-spreading resistances as $R_j = R_s + R_L$ which can be written in full as

$$R_j^\star = \frac{0.36}{P^\star} + \frac{L\left(1 - a_L/b_L\right)^{1.5}}{a_L}$$

(12.122)

When the dimensionless total spreading resistance from the experimental data $2k_sLR_{exp}$ and plotted against $R_j^\star$ one obtains the plot shown in Figure 12.12. The data fall on a straight line with a 45° slope, and the data are within ±15% of the straight line.

Finally, Bahrami (2004) chose to present all data as $R_{exp}/R_s$ versus the parameter $\Theta = R_L/R_s$ and when plotted over five decades of $\Theta$ three regions are clearly seen in Figure 12.13. The conforming rough (CR) limit, the transition region (TR), and the elasto-constriction (EC) are clearly seen. In the conforming rough region, the micro-spreading resistance is the dominant mode of heat transfer, in the transition region, the plastic macro-resistance dominates, and in the elasto-constriction region, the macro-spreading resistance dominates. The data are within ±15% of the model over the entire range of $\Theta$.

**Figure 12.12** Comparison of general model with all data. Source: Bahrami et al. (2004a)/with permission of American Society of Mechanical Engineers.

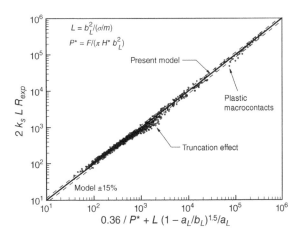

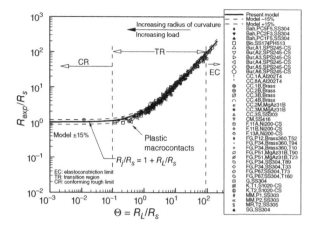

**Figure 12.13** Comparison between general model and non-conforming rough data. Source: Bahrami et al. (2004a)/with permission of American Society of Mechanical Engineers.

## 12.14 Joint Resistance of Nonconforming Rough Surface and Smooth Flat Contact

The geometry of two nonconforming rough contacts is shown in Figure 12.14 with radius $b_L$ that is pressed against each other with an external load $F$. The central portion of the contact plane consists of contacting asperities and associated gaps which are filled with a gas. The contact region is characterized by the radius $a_L$. In the macro-gap region ($a_L \leq r \leq b_L$), heat transfer across the joint occurs by gas conduction only because radiation heat transfer is assumed to be zero.

In the micro-gap region ($0 \leq r \leq a_L$), heat transfer occurs by conduction via the solid micro-contacts and the micro-gaps as shown in the insert in the left of Figure 12.14. This region is very narrow because the spreading resistance zone on both sides of the micro-contacts is related to the radius of the micro-contact spots ($t \approx 4a$), where $a$ is the micro-spot radius.

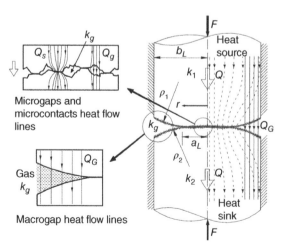

**Figure 12.14** Contact of non-conforming rough surfaces with presence of interstitial gas. Source: Adapted from Bahrami et al. (2004b).

The overall joint resistance depends on four resistances which are thermally connected in parallel and series. It is given by following thermal network:

$$R_j = \left[ 1 / \left[ (1/R_s + 1/R_g)^{-1} + R_L \right] + 1/R_G \right]^{-1} \tag{12.123}$$

The micro-solid and micro-gap resistances are denoted as $R_s$ and $R_g$, and the macro-solid and macro-gap resistances are denoted as $R_L$ and $R_G$. For contacts in a vacuum, $R_g \to \infty$ and $R_G \to \infty$.

Models for the micro- and macro-solid and gap resistances were proposed by Bahrami et al. (2004d), and the salient points are summarized in Sections 12.14.1 and 12.14.2.

With the assumption of plastic deformation of the contacting asperities, simple correlations were proposed [Bahrami et al. (2004b,c)] for predicting the micro-contact resistance $R_s$ and the macro-contact resistance $R_L$ in a vacuum with negligible radiation heat transfer across the gaps.

The relationships were given as follows:

$$R_S = \frac{0.565 H'(\sigma'/m)}{k_s F} \tag{12.124}$$

and

$$R_L = \frac{(1 - a_L/b_L)}{2 k_s a_L} \tag{12.125}$$

where $k_s = 2 k_1 k_2 / (k_1 + k_2)$ and $H' = c_1 (\sigma'/m)^{c_2}$ with $\sigma' = \sigma/\sigma_0$ where $\sigma_0 = 1\,\mu m$ and $c_1, c_2, k_s$ and, $F$ are the Vickers micro-hardness coefficients, the harmonic mean thermal conductivity, and the applied load, respectively.

### 12.14.1 Micro-gap Thermal Resistance

The overall micro-gap heat transfer rate is given by

$$Q_g = \iint \frac{k_g (T_{i,1} - T_{i,2})}{Y(r) - M} dA_g \tag{12.126}$$

where the local micro-gap temperature drop is $(T_{i,1} - T_{i,2})$. The definition of the total thermal gap resistance $R_g = (T_{i,1} - T_{i,2})/Q_g$ leads to the following relation for the micro-gap thermal resistance:

$$R_g = \frac{1}{2\pi k_g} \left[ \int_0^{a_L} \frac{r\,dr}{Y(r) + M} \right]^{-1} \tag{12.127}$$

because $dA_g = dA_a = 2\pi r\,dr$. To determine $R_g$, the local plane separation $Y(r)$ is required.

It can be shown that the local normalized mean plane separation is

$$\lambda = \frac{Y}{\sqrt{2}\sigma} = \text{erfc}^{-1}\left( \frac{2P}{H'} \right) \tag{12.128}$$

where $\text{erfc}^{-1}(\cdot), H' = c_1 (1.62\sigma'/m)^{c_2}, P = F/A_a$ and $A_a$ are the inverse complementary error function, the effective micro-hardness, nominal contact pressure, and the apparent macro-contact area, respectively. Bahrami et al. (2004b) presented the following approximate relationship for the inverse complementary error function:

$$\text{erfc}^{-1}(x) = \frac{1}{0.218 + 0.735 x^{0.173}} \tag{12.129}$$

for the range: $10^{-9} \leq x \leq 0.02$. Song and Yovanovich (1987a,b) proposed the following approximation:

$$\text{erfc}^{-1}(x) = 0.9638 [-\ln(2.795x)]^{1/2} \tag{12.130}$$

for the range: $10^{-6} \leq x \leq 2 \times 10^{-2}$. The maximum difference between the correlations is about 3% at $x = 0.02$. Otherwise, the differences are much smaller.

It can be shown that

$$\frac{2P(\xi)}{H'} = \text{erfc}\,(\lambda(\xi)) \tag{12.131}$$

where $\lambda(\xi) = Y(\xi)/\sqrt{2}\sigma$ and $\xi = r/a_L$ are the nondimensional mean plane separation and the nondimensional radial position.

Bahrami (2004) reported the local mean plane separation has the form

$$\lambda(\xi) = a_1 + a_2\xi^2 \tag{12.132}$$

with coefficients

$$a_1 = \text{erfc}^{-1}\left(\frac{2P_0}{H'}\right) \quad \text{and} \quad a_2 = \text{erfc}^{-1}\left(\frac{0.03P_0}{H'}\right) - a_1 \tag{12.133}$$

with

$$P_0 = \frac{P_{0,H}}{1 + 1.37\alpha\tau^{-0.075}} \tag{12.134}$$

and $\alpha = \sigma\rho/a_H^2$ and $\tau = \rho/a_H$.

The micro-gap thermal resistance was found to have the form

$$R_g = \frac{\sqrt{2}\sigma a_1}{\pi k_g a_L^2 \ln[1 + a_2/(a_1 + M/\sqrt{2}\sigma)]} \tag{12.135}$$

In the conforming rough limit $\rho \to \infty$ and $a_L \to b_L$. The contact pressure distribution becomes uniform over the macro-contact area and $P = P_0 = F/\pi b_L^2$. Consequently, the mean plane separation becomes uniform over the apparent contact area. Thus, $a_2 = 0$ and $a_1 = \text{erfc}^{-1}(2P/H')$. The micro-gap resistance becomes

$$R_g = \frac{Y + M}{k_g A_a} \tag{12.136}$$

which is the conforming rough micro-gap resistance relation developed earlier. For conforming rough contacts, $\text{erfc}^{-1}(2P/H') = \lambda = Y/\sqrt{2}\sigma$.

### 12.14.2 Macro-gap Thermal Resistance

The total heat transfer rate across the macro-gap area $A_G$ is given by

$$Q_G = \iint \frac{k_g(T_1 - T_2)}{D(r) + M} dA_G \tag{12.137}$$

where $(T_1 - T_2)$ represents the effective temperature drop across the macro-gap, $k_g$ is the gas thermal conductivity, $M$ is the gas parameter that accounts for temperature jump at the gas–solid boundary, and the local gap thickness is denoted $D(r)$.

The definition $R_G = (T_1 - T_2)/Q_G$ gives the following relation for the macro-gap thermal resistance:

$$R_G = \frac{1}{2\pi k_g}\left[\frac{rdr}{D(r) + M}\right]^{-1} \tag{12.138}$$

**Figure 12.15** Micro-gap geometry. Source: Adapted from Bahrami et al. (2004b).

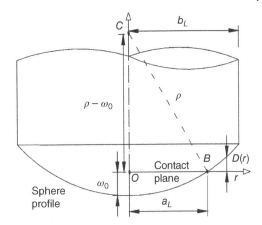

Figure 12.15 shows the geometry for the macro-gap contact. The radius of curvature is $\rho$, the radius of the nonconforming cylinder is $b_L$, the approach of the loaded cylinder is $\omega_0$ under the load $F$. From the right-triangle $OBC$, we find

$$\omega_0 = \frac{a_L^2}{2\rho} \tag{12.139}$$

The local macro-gap thickness is found to be

$$D(r) = \rho - \omega_0 - \sqrt{\rho^2 - r^2} \tag{12.140}$$

where $a_L \leq r \leq b_L$.

After substitution and integration, one obtains the relation for the macro-gap resistance:

$$2\pi k_g R_G = \frac{1}{\left[ S \ln \left[ \dfrac{S-B}{S-A} \right] + B - A \right]} \tag{12.141}$$

where $A = \sqrt{\rho^2 - a_L^2}$, $B = \sqrt{\rho^2 - a_L^2}$ and $S = \rho - \omega_0 + M$. In the conforming rough surface limit, $\rho \to \infty$, $a_L \to b_L$, and $R_G \to \infty$.

Having established the four thermal resistances ($R_s$, $R_g$, $R_L$, $R_G$), Bahrami (2004) conducted parametric studies to find how the joint resistance varied over different ranges of the load $F$, surface micro-roughness $\sigma$, gas pressure $P_g$, and the curvature of the smooth nonconforming surface.

The parametric studies were based on the following typical system: ($\rho = 20\,\text{mm}$, $\sigma = 4.24\,\mu\text{m}$, $F = 100\,\text{N}$, $P_g = 200\,\text{Torr}$). The solids were stainless steel and the gas was nitrogen. Values of the parameters used in the study are listed in Table 12.9.

In the first study, the fixed system parameters values were $\rho = 20\,\text{mm}$, $P_g = 200\,\text{Torr}$, $\sigma = 4.24\,\mu\text{m}$, while the mechanical load was varied over the range $10\,\text{N} \leq F \leq 1000\,\text{N}$. The overall joint resistance $R_j$ fell from about 30 to 10 K/W.

In the second study, the system parameters values were $\rho = 20\,\text{mm}$, $P_g = 200\,\text{Torr}$, $F = 100\,\text{N}$, while surface asperity roughness $\sigma$ was varied over the range $0.01\,\mu\text{m} \leq \sigma \leq 10\,\mu\text{m}$. The overall joint resistance $R_j$ fell from about 20 to 18 K/W.

In the third study, the system parameter values were $\rho = 20\,\text{mm}$, $F = 100\,\text{N}$, $\sigma = 4.24\,\mu\text{m}$, while the gas pressure $P_g$ was varied over the range $10^{-6}$ to $10^3$ Torr. The overall joint resistance $R_j$ fell from about 80 to 20 K/W.

During the fourth study, the system parameter values were $\sigma = 4.24\,\mu\text{m}$, $F = 100\,\text{N}$, $P_g = 200\,\text{Torr}$, while the surface curvature $\rho$ was varied over the range $\rho = 0.01$ to 10 m. The overall joint resistance $R_j$ fell dramatically from about 15 to 0.5 K/W.

**Table 12.9** Values of parameters in parametric studies.

| Parameter | Value |
|---|---|
| $\alpha(\text{SS—N}_2)$ | 0.78 |
| $b_L$ | 12 mm |
| $\rho$ | 20 mm |
| $\sigma$ | 4.24 µm |
| $m$ | 0.19 |
| $P_g$ | 200 Torr |
| $F$ | 100 N |
| $k_g$ | 0.026 W/(m K) |
| $k_s$ | 20 W/(m K) |
| $T_g$ | 300 K |
| $\Lambda_0$ | 62.5 nm |
| $c_1, c_2$ | 4 GPa, 0 |

Bahrami (2004) compared the predictions of the nonconforming rough surface model with the data of Kitscha (1982) which correspond to heat transfer across the contact formed by a smooth hemisphere and a smooth flat. Both surfaces were highly polished and the surface roughness for both surfaces was $\sigma \approx 0$. Tests were performed with carbon steel and steel-1020 with air and argon in the gaps. Kitscha reported 110 data. The gas pressure was varied, and the radiation heat transfer was not negligible especially at light mechanical loads. The data were shown to fall between the $\pm 15\%$ bounds of the model predictions. Bahrami (2004) determined that the RMS relative difference between the model and data were approximately 7.3%.

The proposed nonconforming rough surface model gave acceptable agreement with the data of Kitscha (1982).

# References

Abramowitz, M., and Stegun, I.A., *Handbook of Mathematical Functions*, Dover, New York, 1965.

Bahrami, M., *Modeling of Thermal Joint Resistance for Rough Sphere-Flat Contact in a Vacuum*, Ph.D. Thesis, University of Waterloo, Dept. of Mech. Eng., Waterloo, Ontario, Canada, 2004.

Bahrami, M., Yovanovich, M.M., and Culham, J.R., "Modeling Thermal Contact Resistance: A Scale Analysis Approach", *Journal of Heat Transfer*, Vol. 126, pp. 896–905, 2004a.

Bahrami, M., Culham, J.R., Yovanovich, M.M., and Schneider, G.E., "Thermal Contact Resistance of Nonconforming Rough Surfaces, Part 1: Contact Mechanics Model", *Journal of Thermophysics and Heat Transfer*, Vol. 18, no. 2, pp. 209–217, 2004b.

Bahrami, M., Culham, J.R., Yovanovich, M.M., and Schneider, G.E., "Thermal Contact Resistance of Nonconforming Rough Surfaces, Part 2: Thermal Model", *Journal of Thermophysics and Heat Transfer*, Vol. 18, no. 2, pp. 218–227, 2004c.

Bahrami, M., Yovanovich, M.M., and Culham, J.R., "Thermal Joint Resistances of Nonconforming Rough Surfaces with Gas-Filled Gaps", *Journal of Thermophysics and Heat Transfer*, Vol. 18, no. 3, pp. 326–332, 2004d.

Bahrami, M., Yovanovich, M.M., and Culham, J.R., "A Compact Model for Spherical Rough Contacts", *Journal of Tribology*, Vol. 127, pp. 884–889, 2005a.

Bahrami, M., Yovanovich, M.M., and Culham, J.R., "Thermal Contact Resistance: Effect of Elastic Deformation", *21st IEEE SEMI-THERM Symposium*, 9 pages, 2005b.

Bahrami, M., Culham, J.R., Yovanovich, M.M., and Schneider, G.E., "Review of Thermal Joint Resistance Models for Non-conforming Rough Surfaces", *Applied Mechanics Reviews*, Vol. 59/1, pp. 1–12, 2006.

Burde, S.S., *Thermal Contact Resistance Between Smooth Spheres and Rough Flats*, Ph.D. Dissertation, Department of Mechanical Engineering, University of Waterloo, Waterloo, Ontario, Canada, 1977.

Burde, S.S., and Yovanovich, M.M., "Thermal Resistance at Smooth Sphere / Rough Flat Contacts: Theoretical Analysis", AIAA 78-871, *2nd AIAA/ASME Thermophysics and Heat Transfer Conference*, Palo Allto, CA, 1978.

Byrd, P.F., and Friedman, M.D., *Handbook of Elliptic Integrals for Engineers and Scientists*, 2nd Edition, Springer-Verlag, New York, 1971.

Chen, W.T., and Engel, P.A.,"Impact and Contact Stress Analysis in Multilayer Media", *International Journal of Solids and Structures*, Vol. 8, pp. 1257–1281, 1972.

Churchill, S.W., and Usagi, R., "A General Expression for the Correlation of Rates of Transfer and Other Phenomena", *AIChE Journal*, Vol. 18, pp. 1121–1132, 1972.

Clausing, A.M., and Chao, B.T., "Thermal Contact Resistance in a Vacuum Environment", *Journal of Heat Transfer*, Vol. 87, pp. 243–251, 1965.

Greenwood, J .A., and Tripp, J.H., "The Elastic Contact of Rough Spheres", *Journal of Applied Mechanics*, Vol. 89, no. 1, pp. 153–159, 1967.

Hartnett, J.P., "A Survey of Thermal Accommodation Coefficients", in *Rarefied Gas Dynamics*, ed. L. Talbot, Academic Press, New York, pp. 1–28, 1961.

Hegazy, A., *Thermal Joint Resistance of Conforming Rough Surfaces: Effect of Surface Micro-hardness Variation*, Ph.D. Thesis, University of Waterloo, Waterloo, Ontario, Canada, 1985.

Hertz, H.R., *Miscellaneous Papers*, English Translation, Macmillan, London, 1896.

Holm, R., *Electric Contacts: Theory and Applications*, Springer-Verlag, New York, 1967.

Johnson, K.L., *Contact Mechanics*, Cambridge University Press, Cambridge, 1985.

Kitscha, W.W., *Thermal Resistance of Sphere-Flat Contacts*, M.A.Sc. Thesis, Department of Mechanical Engineering, University of Waterloo, Waterloo, Ontario, Canada, 1982.

Kitscha, W.W., and Yovanovich, M.M., "Experimental Investigation on the Overall Thermal Resistance of Sphere-Flat Contacts", in *Progress in Astronautics and Aeronautics: Heat Transfer with Thermal Control Applications*, Vol. 39, MIT Press, Cambridge, MA, pp. 93–110, 1975.

Lambert, M.A., *Thermal Contact Conductance of Spherical Rough Surfaces*, Ph.D. Thesis, Texas AçM University, College Station, TX, 1995.

Lambert, M.A., and Fletcher, L.S., "Thermal Contact Conductance of Spherical Rough Surfaces", *Journal of Heat Transfer*, Vol. 119, pp. 684–690, 1997.

Madhusudana, C.V.,"The Effect of Interface Fluid on Thermal Contact Conductance", *International Journal of Heat and Mass Transfer*, Vol. 18, pp. 989–991, 1975.

Madhusudana, C.V., *Thermal Contact Conductance*, Springer-Verlag, New York, 1996.

McGee, G.R., Schankula, M.H., and Yovanovich, M.M., "Thermal Resistance of Cylinder-Flat Contacts: Theoretical Analysis and Experimental Verification of a Line- Contact Model", *Nuclear Engineering Design*, Vol. 86, pp. 369–381, 1985.

Marotta, E., and Fletcher, L.S., "Thermal Contact Resistance Modelling of Non-Flat, Roughened Surfaces with Non-Metallic Coatings", *Journal of Heat Transfer*, Vol. 123, pp. 11–23, 2001.

Semyonov, Yu.G., Borisov, S.E., and Suetin, P.E., "Investigation of Heat Transfer in Rarefied Gases over a Wide Range of Knudsen Numbers", *International Journal of Heat and Mass Transfer*, Vol. 27, pp. 1789–1799, 1984.

Song, S., *Analytical and Experimental Study of Heat Transfer through Gas Layers of Contact Interfaces*, Ph.D. Thesis, University of Waterloo, Waterloo, Ontario, Canada, 1988.

Song, S., and Yovanovich, M.M., "Correlation of Thermal Accommodation Coefficients for Engineering Surfaces", *ASME-HTD-69*, ASME, New York, pp. 107–116, 1987a.

Song, S., and Yovanovich, M.M., "Explicit Relative Contact Pressure: Dependence Upon Surface Roughness Parameters and Vickers Micro-hardness Coefficients", *AIAA Paper No. 87-0152*, 25th Aerospace Sciences Meeting, Reno, NV, USA, 1987b.

Song, S., Moran, K.P., Augi, R., and Lee, S., "Experimental Study and Modeling of Thermal Contact Resistance across Bolted Joints", *AIAA-93-0844, 31st Aerospace Sciences Meeting and Exhibit*, Reno, NV, USA, January 11–14, 1993a.

Song, S., Yovanovich, M.M., and Goodman, F.O., "Thermal Gap Conductance of Conforming Rough Surfaces in Contact", *Journal of Heat Transfer*, Vol. 115, pp. 533–540, 1993b.

Sridhar, M.R., and Yovanovich, M.M., "Review of Elastic and Plastic Contact Conductance Models: Comparison with Experiment", *Journal of Thermophysics and Heat Transfer*, Vol. 8, no. 4, pp. 633–640, 1994.

Sridhar, M.R., and Yovanovich, M.M., "Elasto-plastic Constriction Resistance Model for Sphere/Flat Contacts", *Journal of Heat Transfer*, Vol. 118, pp. 202–205, 1996.

Stevanović, M., Yovanovich, M.M., and Culham, J.R., "Modeling Thermal Contact Resistance Between Elastic Hemisphere and Elastic Layer on Bonded Elastic Substrate", *Proceedings of the 8th Intersociety Conference on Thermal and Thermomechanical Phenomena in Electronic Systems*, San Diego, CA, USA, May 29-June 1, 2002.

Tabor, D., *Hardness of Metals*, Oxford University Press, London, 1951.

Thomas, T.R., *Rough Surfaces*, Longman Group, London, 1982.

Timoshenko, S.P., and Goodier, J.N., *Theory of Elasticity*, 3rd Edition, McGraw-Hill, New York, 1970.

Wachman, H.Y., "The Thermal Accommodation Coefficient: A Critical Survey", *ARS Journal*, Vol. 32, pp. 2–12, 1962.

Walowit, J.A., and Anno, J.N., *Modern Developments in Lubrication Mechanics*, Applied Science Publishers, Barking, Essex, England, 1975.

Wesley, D.A., and Yovanovich, M.M., "A New Gaseous Gap Conductance Relationship", *Nuclear Technology*, Vol. 72, pp. 70–74, 1986.

Wiedmann, M.L., and Trumpler, P.R., "Thermal Accommodation Coefficients", *Transactions of the ASME*, Vol. 68, pp. 57–64, 1946.

Yovanovich, M.M., "Thermal Contact Resistance across Elastically Deformed Spheres", *Journal of Spacecraft and Rockets*, Vol. 4, pp. 119–122, 1967.

Yovanovich, M.M., "Overall Constriction Resistance Between Contacting Rough Wavy Surfaces", *International Journal of Heat and Mass Transfer*, Vol. 12, pp. 1517–1520, 1969.

Yovanovich, M.M., "Thermal Constriction Resistance between Contacting Metallic Paraboloids: Application to Instrument Bearings", in *Progress in Astronautics and Aeronautics: Heat Transfer and Spacecraft Control*, Vol. 24, ed. J.W. Lucas, AIAA, New York, pp. 337–358, 1971.

Yovanovich, M.M., "Recent Developments in Thermal Contact, Gap and Joint Conductance: Theories and Experiments", *Proceedings of the 8th International Heat Transfer Conference*, San Francisco, CA, USA, Vol. 1, pp. 35–45, 1986.

Yovanovich, M.M., and Kitscha, W.W., "Modeling the Effect of Air and Oil upon the Thermal Resistance of a Sphere-Flat Contact", in *Progress in Astronautics and Aeronautics: Thermophysics and Spacecraft Control*, Vol. 35, ed. R.G. Hering, AIAA, New York, pp. 293–319, 1974.

# Appendix A

# Special Functions

Special functions arise in the solution of ordinary and partial differential equations. These functions are defined in terms of integrals and/or infinite series. Of particular interest to engineers are the Gamma and Beta functions, the Gaussian error function, the Bessel functions, Legendre functions, and elliptic integrals. Other functions which appear in mathematical physics which we will not consider include those related to the Bessel functions: spherical Bessel, Hankel, Struve, Airy, and Kelvin functions. Those related to the elliptic integrals: the Jacobi elliptic functions, other integral functions such as the Fresnel integrals, the family of orthogonal polynomials: Hermite, Laguerre, Jacobi, and Chebyshev, and the hypergeometric and confluent hypergeometric functions. The latter being quite useful for defining many special functions in terms of a generalized series. There are many others not included in the above list. The vast majority can be found discussed in the Handbook of Mathematical Functions, by Abramowitz and Stegun (1961) and other references Andrews (1998), Bowman (1958), Davis (1962), Farell and Bertram (1963), Gray and Matthews (1966), Hildebrand (1976), Hochstadt (1986), Lebedev (1972), MacMillan (1958), and Sneddon (1955) included at the end of the Appendix.

## A.1 Gamma and Beta Function

We begin our examination of special functions by considering the Gamma and Beta functions given their close relationship to one another.

### A.1.1 Gamma Function

The Gamma function is one of the simplest special functions and also quite important since it appears in the definitions of several other special functions and can be used to define many definite integrals.

The Gamma function is defined by the following limit:

$$\Gamma(x) = \lim_{n \to \infty} \frac{n! n^x}{x(x+1)(x+2)\cdots(x+n)} \tag{A.1}$$

From the above definition, it is clear that $\Gamma(x)$ cannot be defined at $x = 0, -1, -2, \ldots$, since the limit becomes infinite for any of these values, see Figure A.1. It also has the special result $\Gamma(1) = 1$,

*Thermal Spreading and Contact Resistance: Fundamentals and Applications*, First Edition.
Yuri S. Muzychka and M. Michael Yovanovich.
© 2023 John Wiley & Sons, Inc. Published 2023 by John Wiley & Sons, Inc.

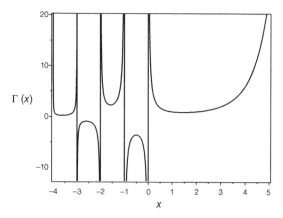

**Figure A.1** The Gamma function.

which can be deduced from the definition above. We may also define the Gamma function as an infinite product series:

$$\Gamma(z) = \frac{1}{z} \prod_{n=1}^{\infty} \frac{\left(1 + \frac{1}{n}\right)^z}{\left(1 + \frac{z}{n}\right)} \tag{A.2}$$

where $z = x + iy$ is a complex variable. This is Euler's product for $\Gamma(z)$. The function converges everywhere except at $z = 0, -1, -2, -3, \dots$. The Gamma function is one of the few mathematical functions which are not a direct solution to a differential equation. However, it plays a significant role in the solution to many equations and special functions.

The Gamma function also has the following recurrence relationship which is useful for determining values of $\Gamma(x)$ for values greater than one, that is

$$\Gamma(x + 1) = x\Gamma(x) \qquad x > 0 \tag{A.3}$$

The Gamma function can also be related to the factorial function $n!$, since

$$\begin{aligned}
\Gamma(1) &= 1 = 0! \\
\Gamma(2) &= 1 \cdot \Gamma(1) = 1 = 1! \\
\Gamma(3) &= 2 \cdot \Gamma(2) = 2 \cdot 1! = 2! \\
\Gamma(4) &= 3 \cdot \Gamma(3) = 3 \cdot 2! = 3! \\
&\vdots \\
\Gamma(n + 1) &= n! \qquad n = 0, 1, 2, \dots
\end{aligned} \tag{A.4}$$

which can be easily deduced, making it a generalization of the factorial function $n!$.

The Gamma function can also be defined as an infinite integral:

$$\Gamma(x) = \int_0^{\infty} e^{-t} t^{x-1} \, dt \qquad x > 0 \tag{A.5}$$

although it is limited to only positive values of $x$, the integral form is the most common way in which $\Gamma(x)$ is defined.

The integral form also allows us to define the derivative of the Gamma function, since we may differentiate under the integral sign and obtain:

$$\Gamma'(x) = \int_0^{\infty} e^{-t} t^{x-1} \ln t \, dt \qquad x > 0 \tag{A.6}$$

and

$$\Gamma''(x) = \int_0^\infty e^{-t} t^{x-1} (\ln t)^2 \, dt \qquad x > 0 \tag{A.7}$$

By and large the Gamma function is most useful for solving definite integrals. For example, if the following change of variables are employed, we obtain the following alternate forms: for example, if $t = u^2$, then we obtain:

$$\Gamma(x) = 2 \int_0^\infty e^{-u^2} u^{2x-1} \, du \tag{A.8}$$

or if $t = \ln(1/u)$, then

$$\Gamma(x) = \int_0^1 \left( \ln \frac{1}{u} \right)^{x-1} \, du \tag{A.9}$$

A more complicated but very useful form can be obtained by using Eq. (A.8) in the following form:

$$\Gamma(x)\Gamma(y) = 2 \int_0^\infty e^{-u^2} u^{2x-1} \, du \cdot 2 \int_0^\infty e^{-v^2} v^{2y-1} \, dv \tag{A.10}$$

or

$$\Gamma(x)\Gamma(y) = 4 \int_0^\infty \int_0^\infty e^{-(u^2+v^2)} u^{2x-1} v^{2y-1} \, du \, dv \tag{A.11}$$

Further if we introduce $u = r\cos\theta$ and $v = r\sin\theta$, we can show (left as an exercise) that

$$\Gamma(x)\Gamma(y) = 2\Gamma(x+y) \int_0^{\pi/2} \cos^{2x-1}\theta \sin^{2y-1}\theta \, d\theta \tag{A.12}$$

or

$$\frac{\Gamma(x)\Gamma(y)}{2\Gamma(x+y)} = \int_0^{\pi/2} \cos^{2x-1}\theta \sin^{2y-1}\theta \, d\theta \qquad x > 0, y > 0 \tag{A.13}$$

Equation (A.13) proves quite useful in evaluating a host of trigonometric integrals.

Finally, we conclude with the *incomplete* Gamma function and *complementary incomplete* Gamma functions. These are defined as follows:

$$\gamma(a,x) = \int_0^x e^{-t} t^{a-1} \, dt \qquad a > 0 \tag{A.14}$$

and

$$\Gamma(a,x) = \int_x^\infty e^{-t} t^{a-1} \, dt \qquad a > 0 \tag{A.15}$$

such that

$$\gamma(a,x) + \Gamma(a,x) = \Gamma(a) \tag{A.16}$$

These functions find uses in probability theory.

### A.1.2 Beta Function

The Beta function is defined by the following integral relationship:

$$B(x,y) = \int_0^1 t^{x-1}(1-t)^{y-1}\, dt \qquad x > 0, \quad y > 0 \tag{A.17}$$

It is related to the Gamma function, but due to its importance and frequency of appearance, it is given its own designation. Alternate forms of the Beta function integral are obtained with following change of variables. First, if we introduce $t = u/(1+u)$ in Eq. (A.17), we obtain:

$$B(x,y) = \int_0^\infty \frac{u^{x-1}}{(1+u)^{x+y}}\, du \qquad x > 0, \quad y > 0 \tag{A.18}$$

Further if we introduce the change of variable $t = \cos^2\theta$ in Eq. (A.13), we obtain:

$$B(x,y) = 2 \int_0^{\pi/2} \cos^{2x-1}\theta \sin^{2y-1}\theta\, d\theta \qquad x > 0, \quad y > 0 \tag{A.19}$$

which when compared with Eq. (A.12) yields:

$$B(x,y) = \frac{\Gamma(x)\Gamma(y)}{\Gamma(x+y)} \qquad x > 0, \quad y > 0 \tag{A.20}$$

from which we also observe the following symmetry property:

$$B(x,y) = B(y,x) \tag{A.21}$$

The Beta function (and hence, the Gamma function) is quite useful in representing definite integrals. One class of definite integrals which arise in engineering analysis, are those involving, areas, moments, and volumes. These integrals can be generalized through the definition of the Dirichlet integrals. If $V$ denotes the closed region in the first octant bounded by the planes $x = 0$, $y = 0$, and $z = 0$, and the surface:

$$\left(\frac{x}{a}\right)^\alpha + \left(\frac{y}{b}\right)^\beta + \left(\frac{z}{c}\right)^\gamma = 1 \tag{A.22}$$

and if all constants are positive, we define the Dirichlet integrals by

$$I = \int\int\int_V x^{p-1}y^{q-1}z^{r-1}\, dx\, dy\, dz = \frac{a^p b^q c^r}{\alpha\beta\gamma} \frac{\Gamma(p/\alpha)\Gamma(q/\beta)\Gamma(r/\gamma)}{\Gamma(1 + p/\alpha + q/\beta + r/\gamma)} \tag{A.23}$$

These integrals as we shall see are useful in evaluating the multiple integrals associated with the evaluation of areas, volumes, and moments of many simple and complex geometries. In the case of area integrals, bounded by the axes $x = 0$ and $y = 0$, and the curve:

$$\left(\frac{x}{a}\right)^\alpha + \left(\frac{y}{b}\right)^\beta = 1 \tag{A.24}$$

one would simplify Eq. (A.23) as

$$I = \int\int_A x^{p-1}y^{q-1}\, dx\, dy = \frac{a^p b^q}{\alpha\beta} \frac{\Gamma(p/\alpha)\Gamma(q/\beta)}{\Gamma(1 + p/\alpha + q/\beta)} \tag{A.25}$$

## A.2 Error Function

The Gaussian error function appears in many physical and mathematical applications such as heat conduction, mass diffusion, probability theory, and the theory of errors. It is defined by the integral:

$$\operatorname{erf} x = \frac{2}{\sqrt{\pi}} \int_0^x e^{-t^2}\, dt \tag{A.26}$$

**Figure A.2** The Gaussian error function.

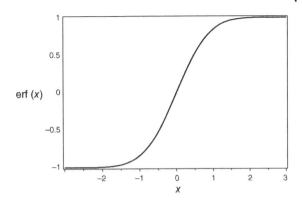

By means of a series expansion of the exponential function, we may obtain the following infinite series formulation:

$$\text{erf } x = \frac{2}{\sqrt{\pi}} \sum_{n=0}^{\infty} \frac{(-1)^n x^{2n+1}}{n!(2n+1)} \qquad |x| < \infty \tag{A.27}$$

The error function is plotted in Figure A.2. From Eq. (A.26), we may deduce that the error function is an odd function having the property:

$$\text{erf}(-x) = -\text{erf}(x) \tag{A.28}$$

Additional properties include $\text{erf}(0) = 0$ and $\text{erf}(\infty) = 1$. The complementary error function denoted as erfc $x$ occurs frequently in diffusion problems. It is defined as

$$\text{erfc } x = \frac{2}{\sqrt{\pi}} \int_x^{\infty} e^{-t^2} \, dt \tag{A.29}$$

such that

$$\text{erfc } x = 1 - \text{erf } x \tag{A.30}$$

The error function and complementary error function can also be expressed in terms of the incomplete Gamma functions. It will be left as an exercise for you to verify that

$$\text{erf } x = \frac{2}{\sqrt{\pi}} \int_0^x e^{-t^2} \, dt = \frac{1}{\sqrt{\pi}} \gamma(1/2, x^2) \tag{A.31}$$

and

$$\text{erfc } x = \frac{2}{\sqrt{\pi}} \int_x^{\infty} e^{-t^2} \, dt = \frac{1}{\sqrt{\pi}} \Gamma(1/2, x^2) \tag{A.32}$$

As mentioned earlier, the error function or complementary error function appears in the solution of many diffusion problems involving semi-infinite domains. One such problem which we shall examine involves the following ordinary differential equation:

$$\frac{d^2\phi}{d\eta^2} + 2\eta \frac{d\phi}{d\eta} = 0 \tag{A.33}$$

subject to $\phi = 1$ when $\eta = 0$ and $\phi \to 0$ when $\eta \to \infty$. The solution can be found by means of the method of *reduction of order*. That is we will define $w = d\phi/d\eta$ to obtain

$$\frac{dw}{d\eta} + 2\eta w = 0 \tag{A.34}$$

Separating variables allows us to write

$$\frac{dw}{w} = -2\eta d\eta \tag{A.35}$$

or after integrating

$$\ln w = -\eta^2 + \ln C_1 \tag{A.36}$$

or

$$\ln \frac{w}{C_1} = -\eta^2 \tag{A.37}$$

which gives

$$w = \frac{d\phi}{d\eta} = C_1 e^{-\eta^2} \tag{A.38}$$

Integrating once more, yields

$$\phi = C_1 \int_0^\eta e^{-\eta^2} \, d\eta + C_2 \tag{A.39}$$

Finally, application of the two boundary conditions yields

$$1 = C_1 \int_0^0 e^{-\eta^2} \, d\eta + C_2 = C_2 \tag{A.40}$$

and

$$0 = C_1 \int_0^\infty e^{-\eta^2} \, d\eta + C_2 \tag{A.41}$$

or

$$C_1 = \frac{-1}{\int_0^\infty e^{-\eta^2} \, d\eta} = -\frac{1}{\sqrt{\pi}/2} \tag{A.42}$$

The above equation may be evaluated by means of Eqs. (A.8) and (A.13). Again, it will be left as an exercise for you to show that

$$\int_0^\infty e^{-\eta^2} \, d\eta = \frac{\sqrt{\pi}}{2} \tag{A.43}$$

Returning to our solution, we obtain

$$\phi = 1 - \frac{2}{\sqrt{\pi}} \int_0^\eta e^{-\eta^2} \, d\eta = 1 - \text{erf } \eta = \text{erfc } \eta \tag{A.44}$$

## A.3   Bessel Functions

The family of functions denoted as Bessel functions have many applications in applied physics and engineering. We are specifically interested the functions denoted as $J_n(x)$, $Y_n(x)$, $I_n(x)$, and $K_n(x)$. These appear frequently in engineering problems related to heat conduction, electric fields, buckling, and dynamics. Bessel functions are also commonly referred to as cylinder functions due to their association with solutions for cylindrical domains.

### A.3.1 Bessel Functions of the First and Second Kind

We begin our examination of the Bessel functions by examining Bessel's equation:

$$x^2 \frac{d^2y}{dx^2} + x\frac{dy}{dx} + (x^2 - n^2)y = 0 \qquad n \geq 0 \tag{A.45}$$

The general solution to this equation can be found using the series form through the application of the method of Frobenius. Provided that $n$ is not an integer, the general solution is of the form:

$$y(x) = C_1 J_n(x) + C_2 J_{-n}(x) \qquad n \neq 0, 1, 2, \ldots \tag{A.46}$$

where

$$J_n(x) = \sum_{k=0}^{\infty} \frac{(-1)^k (x/2)^{2k+n}}{k!\Gamma(k+n+1)} \tag{A.47}$$

and

$$J_{-n}(x) = \sum_{k=0}^{\infty} \frac{(-1)^k (x/2)^{2k-n}}{k!\Gamma(k-n+1)} \tag{A.48}$$

$J_n(x)$ is referred to as a Bessel function of the first kind. It can be shown that $J_n(x)$ and $J_{-n}(x)$ are not linearly independent solutions when $n$ is an integer. If $n$ is an integer, we find that

$$J_{-n}(x) = (-1)^n J_n(x) \qquad n = 0, 1, 2, \ldots \tag{A.49}$$

For cases when $n$ is equal to an integer, only $J_n(x)$ can be assumed to be the solution to Eq. (A.45). A second linearly independent solution can be found of the form:

$$Y_n(x) = \begin{cases} \dfrac{\cos(n\pi)J_n(x) - J_{-n}(x)}{\sin(n\pi)} & n \neq 0, 1, 2, \ldots \\[2ex] \lim_{p \to n} \dfrac{\cos(p\pi)J_p(x) - J_{-p}(x)}{\sin(p\pi)} & n = 0, 1, 2, \ldots \end{cases} \tag{A.50}$$

This is referred to as a Bessel function of the second kind. Since both $J_n(x)$ and $Y_n(x)$ are linearly independent solutions, they are valid for *all* values of $n$, and hence, the general solution to Eq. (A.45) is taken as

$$y(x) = C_1 J_n(x) + C_2 Y_n(x) \tag{A.51}$$

The functions $J_n(x)$ and $Y_n(x)$ are shown in Figures A.3 and A.4 for the values of $n = 0, 1$, and 2. In problems where $x = 0$ is part of the solution domain, the $Y_n(x)$ must be excluded since they are undefined at $x = 0$.

The Bessel functions of the first and second kind have several special characteristics, particularly when the order is *half integral*. In these cases, the functions reduce to elementary forms of sine and cosine functions. For example if $n = 1/2$, we have

$$J_{1/2}(x) = \sqrt{\frac{2}{\pi x}} \sin(x) \tag{A.52}$$

$$J_{-1/2}(x) = \sqrt{\frac{2}{\pi x}} \cos(x) \tag{A.53}$$

$$Y_{1/2}(x) = -\sqrt{\frac{2}{\pi x}} \cos(x) \tag{A.54}$$

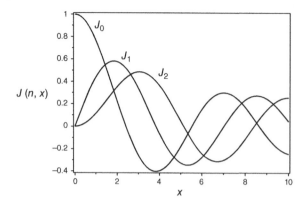

**Figure A.3**   The Bessel function $J_n(x)$.

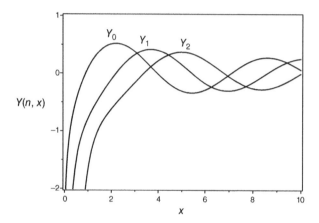

**Figure A.4**   The Bessel function $Y_n(x)$.

$$Y_{-1/2}(x) = \sqrt{\frac{2}{\pi x}}\,\sin(x) \tag{A.55}$$

Some useful recurrence relationships are

$$J_{n+1}(x) = \frac{2n}{x}J_n(x) - J_{n-1}(x) \tag{A.56}$$

$$\frac{d}{dx}J_n(x) = \frac{1}{2}[J_{n-1}(x) - J_{n+1}(x)] \tag{A.57}$$

$$x\frac{d}{dx}J_n(x) = xJ_{n-1}(x) - nJ_{n+1}(x) = nJ_n(x) - xJ_{n+1}(x) \tag{A.58}$$

$$\frac{d}{dx}x^n J_n(x) = x^n J_{n-1}(x) \tag{A.59}$$

$$\frac{d}{dx}x^{-n}J_n(x) = -x^{-n}J_{n+1}(x) \tag{A.60}$$

The above equations are also valid if $J_n(x)$ is replaced by $Y_n(x)$.

**Table A.1** Roots of the Bessel functions.

| $i$ | $J_0(x) = 0$ | $J_1(x) = 0$ | $Y_0(x) = 0$ | $Y_1(x) = 0$ |
|---|---|---|---|---|
| 1 | 2.4048 | 3.8317 | 0.8936 | 2.1971 |
| 2 | 5.5201 | 7.0156 | 3.9577 | 5.4297 |
| 3 | 8.6537 | 10.1735 | 7.0861 | 8.5960 |
| 4 | 11.7915 | 13.3237 | 10.2223 | 11.7492 |
| 5 | 14.9309 | 16.4706 | 13.3611 | 14.8974 |

### A.3.2 Zeroes of the Bessel Functions

The Bessel functions of the first and second kind are harmonic and as such contain repeated zeroes or roots. The roots of these functions are tabulated in the various mathematical handbooks and are also easily computed using root-finding solvers in the various mathematical packages. Maple and Mathematica also have special functions to represent the zeroes of the Bessel functions. As an example, the first five roots of the Bessel functions of the first and second kind are given in Table A.1. Additional roots may be approximated from

$$x_0^{i+1} = x_0^i + \pi \tag{A.61}$$

### A.3.3 Modified Bessel Functions of the First and Second Kind

Another differential equation related to Bessel's equation is the modified Bessel equation which takes the form:

$$x^2 \frac{d^2y}{dx^2} + x\frac{dy}{dx} - (x^2 + n^2)y = 0 \qquad n \geq 0 \tag{A.62}$$

The solution to the above equation is in the form of

$$y(x) = C_1 I_n(x) + C_2 I_{-n}(x) \qquad n \neq 0, 1, 2, \dots \tag{A.63}$$

where

$$I_n(x) = i^{-n} J_n(ix) = \sum_{k=0}^{\infty} \frac{(x/2)^{2k+n}}{k!\Gamma(k+n+1)} \tag{A.64}$$

and

$$I_{-n}(x) = i^n J_{-n}(ix) = \sum_{k=0}^{\infty} \frac{(x/2)^{2k-n}}{k!\Gamma(k-n+1)} \tag{A.65}$$

where $I_n(x)$ is the modified Bessel function of the first kind.

However, it can be shown that $I_n(x) = I_{-n}(x)$ when $n = 0, 1, 2, \dots$, otherwise, $I_n(x)$ and $I_{-n}(x)$ are linearly independent. As with the Bessel's equation, a second linearly independent solution can be developed which satisfies Bessel's modified equation. The function is defined analogously in a manner similar to the Bessel function of the second kind:

$$K_n(x) = \begin{cases} \dfrac{\pi}{2\sin(n\pi)}[I_{-n}(x) - I_n(x)] & n \neq 0, 1, 2, \dots \\ \lim\limits_{p \to n} \dfrac{\pi}{2\sin(p\pi)}[I_{-p}(x) - I_p(x)] & n = 0, 1, 2, \dots \end{cases} \tag{A.66}$$

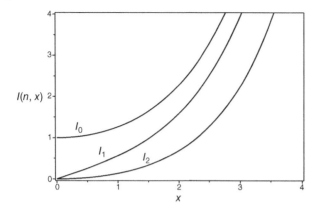

**Figure A.5** The Bessel function $I_n(x)$.

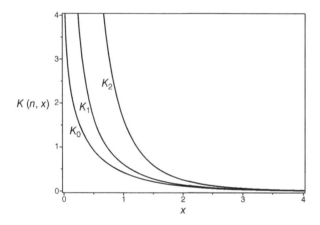

**Figure A.6** The Bessel function $K_n(x)$.

This is referred to as a modified Bessel function of the second kind. Since both $I_n(x)$ and $K_n(x)$ are linearly independent solutions, they are valid for *all* values of $n$, and hence, the general solution to Eq. (A.62) is taken as

$$y(x) = C_1 I_n(x) + C_2 K_n(x) \tag{A.67}$$

The functions $I_n(x)$ and $K_n(x)$ are shown in Figures A.5 and A.6 for the values of $n = 0, 1$, and 2. In problems where $x = 0$ is part of the solution domain, the $K_n(x)$ must be excluded since they are undefined at $x = 0$.

The modified Bessel functions of the first and second kind have several special characteristics, particularly when the order is *half-integral*. In these cases, the functions reduce to elementary forms of hyperbolic sine and cosine functions. For example, if $n = 1/2$, we have

$$I_{1/2}(x) = \sqrt{\frac{2}{\pi x}} \sinh(x) \tag{A.68}$$

$$I_{-1/2}(x) = \sqrt{\frac{2}{\pi x}} \cosh(x) \tag{A.69}$$

$$K_{1/2}(x) = K_{-1/2}(x) = \sqrt{\frac{\pi}{2x}} \exp(-x) \tag{A.70}$$

Some useful recurrence relationships for the modified Bessel functions are

$$I_{n+1}(x) = I_{n-1}(x) - \frac{2n}{x} I_n(x) \tag{A.71}$$

$$\frac{d}{dx} I_n(x) = \frac{1}{2} [I_{n-1}(x) + I_{n+1}(x)] \tag{A.72}$$

$$x \frac{d}{dx} I_n(x) = x I_{n-1}(x) - n I_n(x) = x I_{n+1}(x) + n I_n(x) \tag{A.73}$$

$$\frac{d}{dx} x^n I_n(x) = x^n I_{n-1}(x) \tag{A.74}$$

$$\frac{d}{dx} x^{-n} I_n(x) = x^{-n} I_{n+1}(x) \tag{A.75}$$

and

$$K_{n+1}(x) = K_{n-1}(x) + \frac{2n}{x} K_n(x) \tag{A.76}$$

$$\frac{d}{dx} K_n(x) = -\frac{1}{2} [K_{n-1}(x) + K_{n+1}(x)] \tag{A.77}$$

$$x \frac{d}{dx} K_n(x) = -x K_{n-1}(x) - n K_n(x) = n K_n(x) - x K_{n+1}(x) \tag{A.78}$$

$$\frac{d}{dx} x^n K_n(x) = -x^n K_{n-1}(x) \tag{A.79}$$

$$\frac{d}{dx} x^{-n} K_n(x) = -x^{-n} K_{n+1}(x) \tag{A.80}$$

These expressions only cover a number of useful relationships involving the Bessel functions and modified Bessel functions. The reader is referred to the various references and handbooks for additional expressions involving integrals of Bessel functions, namely the book by Gradshteyn and Ryzhik (2007). Of course, the mathematical packages Maple and Mathematica can also assist you quite well.

## A.4 Elliptic Integrals

Elliptic integrals are another class of functions defined by integrals which have no simple algebraic form. There are three principal forms, and each may be defined as either an incomplete or a complete elliptic integral.

We begin by examining the *elliptic integral of the first kind* which takes the form:

$$F(x, k) = \int_0^x \frac{dx}{\sqrt{(1 - x^2)(1 - k^2 x^2)}} \qquad k^2 < 1 \tag{A.81}$$

An alternate form which is more frequently used is obtained by introducing the change of variable $x = \sin \phi$:

$$F(\phi, k) = \int_0^\phi \frac{d\phi}{\sqrt{(1 - k^2 \sin^2 \phi)}} \qquad k^2 < 1 \tag{A.82}$$

The parameter $k$ is called the modulus of the elliptic integral and the parameter $k' = \sqrt{1 - k^2}$, the complementary modulus with respect to $k$. When the variable $x = 1$ or $\phi = \pi/2$, the elliptic integral is referred to as the *complete elliptic integral of the first kind* and is denoted as

$$K = K(k) = \int_0^{\pi/2} \frac{d\phi}{\sqrt{(1 - k^2 \sin^2 \phi)}} \tag{A.83}$$

When the complementary modulus $k'$ is the argument of the complete elliptic integral, it is frequently denoted as

$$K' = K(k') = \int_0^{\pi/2} \frac{d\phi}{\sqrt{(1 - k'^2 \sin^2 \phi)}} \tag{A.84}$$

The *elliptic integral of the second kind* is defined as

$$E(x, k) = \int_0^x \sqrt{\frac{1 - k^2 x^2}{1 - x^2}} \, dx \qquad k^2 < 1 \tag{A.85}$$

or more frequently as

$$E(\phi, k) = \int_0^\phi \sqrt{1 - k^2 \sin^2 \phi} \, d\phi \qquad k^2 < 1 \tag{A.86}$$

Once again, if the variable $x = 1$ or $\phi = \pi/2$, the elliptic integral is referred to as the *complete elliptic integral of the second kind* and is denoted as

$$E = E(k) = \int_0^{\pi/2} \sqrt{1 - k^2 \sin^2 \phi} \, d\phi \tag{A.87}$$

When the complementary modulus $k'$ is the argument of the complete elliptic integral, it is frequently denoted as

$$E' = E(k') = \int_0^{\pi/2} \sqrt{1 - k'^2 \sin^2 \phi} \, d\phi \tag{A.88}$$

Finally, we define the *elliptic integral of the third kind* which has the form:

$$\Pi(x, n, k) = \int_0^x \frac{dx}{(1 + nx^2)\sqrt{(1 - x^2)(1 - k^2 x^2)}} \qquad k^2 < 1 \tag{A.89}$$

or more frequently as

$$\Pi(\phi, n, k) = \int_0^\phi \frac{d\phi}{(1 + n\sin^2 \phi)\sqrt{1 - k^2 \sin^2 \phi}} \qquad k^2 < 1 \tag{A.90}$$

Once again, if the variable $x = 1$ or $\phi = \pi/2$, the elliptic integral is referred to as the *complete elliptic integral of the third kind* and is denoted as

$$\Pi = \Pi(n, k) = \Pi(1, n, k) = \Pi(\pi/2, n, k) \tag{A.91}$$

Care must be taken to ensure which form is used, e.g. the algebraic form $(x)$ or the trigonometric form $(\phi)$ when using computer algebra systems such as Maple or Mathematica. In Maple, the default form is the algebraic formulation.

Some useful identities related to complete elliptic integrals of the first and second kind, are

$$\frac{dE}{dk} = \frac{1}{k}(E - K) \tag{A.92}$$

and

$$\frac{dK}{dk} = \frac{1}{kk'^2}(E - k'^2 K) \tag{A.93}$$

Related to the elliptic integral of the first kind are the Jacobi elliptic functions. These are inverse forms. We will limit ourselves to the elliptic integrals for this appendix.

## A.5  Legendre Functions

The Legendre functions arise in the solution of many problems in spherical coordinates. They are sometimes referred to as spherical harmonics. Legendre's equation is represented by the following differential equation:

$$(1 - x^2)\frac{d^2y}{dx^2} - 2x\frac{dy}{dx} + n(n+1)y = 0 \qquad n \geq 0, \quad |x| < 1 \tag{A.94}$$

or

$$\frac{d}{dx}\left[(1 - x^2)\frac{dy}{dx}\right] + n(n+1)y = 0 \qquad n \geq 0, \quad |x| < 1 \tag{A.95}$$

The solution of Eq. (A.94) is found in terms of

$$y(x) = C_1 P_n(x) + C_2 Q_n(x) \tag{A.96}$$

where $P_n(x)$ and $Q_n(x)$ are Legendre functions of the first and second kind of order $n$. When $n = 0, 1, 2, 3, \ldots$, the solutions are obtained in terms of the Legendre polynomials. The Legendre polynomial $P_n(x)$ is defined as

$$P_n(x) = \frac{1}{2^n n!}\frac{d^n}{dx^n}(x^2 - 1)^n \tag{A.97}$$

or

$$\begin{aligned}
P_0(x) &= 1 \\
P_1(x) &= x \\
P_2(x) &= \tfrac{1}{2}(3x^2 - 1) \\
P_3(x) &= \tfrac{1}{2}(5x^3 - 3x) \\
&\vdots
\end{aligned} \tag{A.98}$$

The following recurrence relationship is obtained:

$$P_{n+1}(x) = \frac{2n+1}{n+1}x P_n(x) - \frac{n}{n+1}P_{n-1}(x) \qquad n \geq 1 \tag{A.99}$$

The Legendre function of the second kind for the case when $n = 0, 1, 2, 3, \ldots$ is defined by

$$Q_n(x) = \frac{1}{2}P_n(x)\ln\frac{1+x}{1-x} - \sum_{m=1}^{n}\frac{1}{m}P_{m-1}(x)P_{n-m}(x) \tag{A.100}$$

or

$$\begin{aligned}
Q_0(x) &= \tfrac{1}{2}\ln\tfrac{1+x}{1-x} \\
Q_1(x) &= P_1(x)Q_0(x) - 1 \\
Q_2(x) &= P_2(x)Q_0(x) - \tfrac{3}{2}x \\
Q_3(x) &= P_3(x)Q_0(x) - \tfrac{5}{2}x^2 + \tfrac{3}{2} \\
&\vdots
\end{aligned} \tag{A.101}$$

It is clear that singularities at $x = \pm 1$ exist in the first term of the $Q_n(x)$ and hence must be avoided if these points exist in the solution domain of an application. The first five Legendre polynomials are plotted in Figures A.7 and A.8 for $P_n(x)$ and $Q_n(x)$. For the general case of $n$ not being an integer, the Legendre functions may be defined in terms of the hypergeometric function. We will discuss these at the conclusion of the Section A.6.

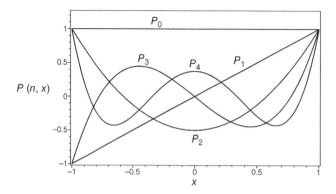

**Figure A.7** The Legendre function $P_n(x)$.

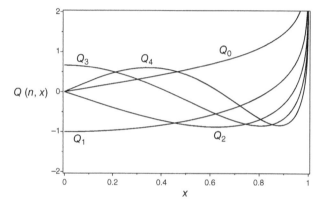

**Figure A.8** The Legendre function $Q_n(x)$.

## A.6 Hypergeometric Function

We conclude with a short discussion of the hypergeometric function. This product series can be used to define many of the special functions we have considered thus far. The hypergeometric function is defined by the product series:

$$F(a, b; c; x) = \frac{\Gamma(a)\Gamma(b)}{\Gamma(c)} \sum_{n=0}^{\infty} \frac{\Gamma(a+n)\Gamma(b+n)}{\Gamma(c+n)} \frac{x^n}{n!} \qquad |x| < 1, \quad c \neq 0, -1, -2, \ldots \qquad \text{(A.102)}$$

or

$$F(a, b; c; x) = \sum_{n=0}^{\infty} \frac{(a)_n (b)_n}{(c)_n} \frac{x^n}{n!} \qquad |x| < 1, \quad c \neq 0, -1, -2, \ldots \qquad \text{(A.103)}$$

where the operation defined by $(\delta)_n$ is used to represent in a shorthand fashion the following:

$$(\delta)_n = \frac{\Gamma(\delta + n)}{\Gamma(\delta)} \qquad \text{(A.104)}$$

It is referred to as the Pochhammer symbol.

The hypergeometric function has the following property:

$$F(a, b; c; x) = F(b, a; c; x) \qquad \text{(A.105)}$$

Also, note that some books utilize a notation that denotes $F(a, b; c; x) = {}_2F_1(a, b; c; x) = F([a, b]; [c]; x)$, where the 2 and 1 denote the number of numerator and denominator variables. Clearly, this is indicative that the hypergeometric function itself is a special case of a more generalized hypergeometric series.

The hypergeometric function is one solution to the following differential equation:

$$x(1 - x)\frac{d^2y}{dx^2} + [c - (a + b + 1)x]\frac{dy}{dx} - aby = 0 \tag{A.106}$$

Another linearly independent solution can be found yielding the following general solution:

$$y(x) = C_1 F(a, b; c; x) + C_2 x^{1-c} F(a - c + 1, b - c + 1; 2 - c; x) \tag{A.107}$$

A general property of the hypergeometric function is

$$\frac{d^k}{dx^k} F(a, b; c; x) = \frac{(a)_k (b)_k}{(c)_k} F(a + k, b + k; c + k; x) \tag{A.108}$$

Finally, the hypergeometric function also has a useful integral representation:

$$F(a, b; c; x) = \frac{\Gamma(c)}{\Gamma(b)\Gamma(c - b)} \int_0^1 t^{b-1} (1 - t)^{c-b-1} (1 - xt)^{-a} \, dt \tag{A.109}$$

### A.6.1 Relationship to Other Functions

The hypergeometric is most useful at representing other mathematical functions. The following is a short list of functions that may be evaluated using the hypergeometric function:

$$B(p, q) = \frac{1}{p} F(p, 1 - q; 1 + p; 1) \tag{A.110}$$

$$J_v(x) = \frac{(x/2)^v}{\Gamma(v + 1)} F\left([\ ]; v + 1; -\frac{x^2}{4}\right) \tag{A.111}$$

$$K(k) = \frac{\pi}{2} F\left(\frac{1}{2}, \frac{1}{2}; 1; k^2\right) \tag{A.112}$$

$$E(k) = \frac{\pi}{2} F\left(-\frac{1}{2}, \frac{1}{2}; 1; k^2\right) \tag{A.113}$$

$$P_n(x) = \frac{\Gamma(2n + 1)}{2^n [\Gamma(n + 1)]^2} x^n F\left(-\frac{n}{2}, \frac{1 - n}{2}; \frac{1}{2} - n; \frac{1}{x^2}\right) \tag{A.114}$$

$$Q_n(x) = \frac{\Gamma\left(\frac{1}{2}\right)\Gamma(n + 1)}{2^{n+1}\Gamma\left(n + \frac{3}{2}\right)} x^{-n-1} F\left(1 + \frac{n}{2}, \frac{1 + n}{2}; n + \frac{3}{2}; \frac{1}{x^2}\right) \tag{A.115}$$

This is just a small sampling of such relationships. Additional functions may be defined on the basis of other hypergeometric functions such as the *confluent* hypergeometric functions of the first and second kind.

## References

Abramowitz, M., and Stegun, I., *Handbook of Mathematical Functions*, Dover Publications, 1961.

Andrews, L.C., *Special Functions of Mathematics for Engineers*, SPIE Press, 1998.

Bowman, F., *Introduction to Bessel Functions*, Dover Publications, 1958.

Davis, H.T., *Introduction to Non-Linear Differential and Integral Equations*, Dover Publications, 1962.

Farrell, O.J., and Bertram, R., *Solved Problems in Analysis as Applied to Gamma, Beta, Legendre, and Bessel Functions*, Dover Publications, 1963.

Gradshteyn, I.S., and Ryzhik, I.M., *Tables of Integrals, Series, and Products*, Academic Press, 2007.

Gray, A., and Mathews, G.B., *A Treatise on Bessel Functions and Their Applications to Physics*, Dover Publications, 1966.

Hildebrand, F.B., *Advanced Calculus for Applications*, Prentice-Hall, 1976.

Hochstadt, H., *The Functions of Mathematical Physics*, Dover Publications, 1986.

Lebedev, N.N., *Special Functions and Their Applications*, Dover Publications, 1972.

MacMillan, W.D., *The Theory of the Potential*, Dover Publications, 1958.

Sneddon, I.N., *Special Functions of Mathematical Physics and Chemistry*, Oliver and Boyd, 1955.

# Appendix B

# Hardness

## B.1 Micro- and Macro-hardness Indenters

Several micro and macrohardness indenters will be reviewed in this appendix based on Yovanovich (2006). The macro-hardness indentation testers are the Brinell (Meyer) and Rockwell, and the microhardness indenters are the Berkovich, Knoop, and Vickers. There are six types of hardness indentation tests used to determine the macro-hardness and micro-hardness of materials such a metals, ceramics, and plastics. The indenters are much harder than the specimen, and they are smooth balls, cones with hemispherical tips, and pyramidal having three or four faces. A brief description of the various types of indenters are given below. More details regarding the nano, micro, and macro-indenters can be found in the texts [Mott (1956), Tabor (1951), Johnson (1985), McColm (1990), Fisher-Cripps (2000)].

### B.1.1 Brinell and Meyer Macrohardness

The Brinell and Meyer macro-hardnesses are determine by the same indentation test. A hard smooth ball of diameter $D$ (mm) is pressed into a smooth flat surface under a known load $F$ (N) for a duration of 30–60 seconds, depending on the whether the metal is hard or soft as shown in Figure B.1. After removal of the ball, the diameter of the indentation is measured by means of an optical microscope. The diameter of the indentation $d$ (mm) is the average value of two measurements, i.e. $d = (d_1 + d_2)/2$, where $d_1$ and $d_2$ are the measured diameters which are perpendicular to each other.

#### Brinell Hardness Number
The Brinell hardness number (BHN) is expressed as the load divided by the *actual* area of the indentation. Therefore,

$$BHN = \frac{2F}{\pi D \left( D - \sqrt{D^2 - d^2} \right)} = \frac{F}{\pi D t} \tag{B.1}$$

where the penetration depth, defined as the distance from the original surface to the maximum indentation depth, is

$$t = \frac{D - \sqrt{D^2 - d^2}}{2} \tag{B.2}$$

as shown in Figure B.1.

The relative indentation size is recommended to lie in the range: $d/D = 0.25$–$0.6$. If the load is given in kgf and the indentation diameter is given as mm, then the Brinell hardness has units of

*Thermal Spreading and Contact Resistance: Fundamentals and Applications*, First Edition.
Yuri S. Muzychka and M. Michael Yovanovich.
© 2023 John Wiley & Sons, Inc. Published 2023 by John Wiley & Sons, Inc.

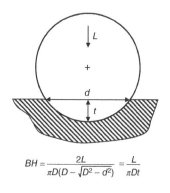

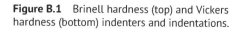

**Figure B.1**  Brinell hardness (top) and Vickers hardness (bottom) indenters and indentations.

$$BH = \frac{2L}{\pi D(D - \sqrt{D^2 - d^2})} = \frac{L}{\pi D t}$$

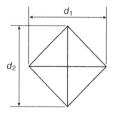

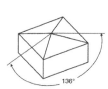

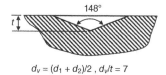

$d_v = (d_1 + d_2)/2 \,, \; d_v/t = 7$

$\text{kgf/mm}^2$. These are the units used in handbooks and older texts. Today, its more common to give the Brinell hardness in units of MPa or GPa. In this text, the Brinell hardness will be denoted as $H_B$.

**Meyer Hardness Number**

The Meyer hardness number (MHN) is based on the same indentation test; however, the Meyer hardness number is expressed as the indentation load divided by the *projected* area of the indentation. Therefore,

$$MHN = \frac{4F}{\pi d^2} = P_m \tag{B.3}$$

where $P_m$ is the mean contact pressure. The Meyer hardness is said to be a true representation of the hardness of the material. Here, the Meyer hardness will be denoted as $H_M$.

The Brinell and Meyer hardnesses are related as

$$\frac{H_B}{H_M} = \frac{(d/D)^2}{2\left[1 - \sqrt{1 - (d/D)^2}\right]} \tag{B.4}$$

Table B.1 shows the relationship between the numerical values of the Brinell and Meyer hardnesses for typical values of $d/D$ which is sometimes called the contact strain.

The difference between the Brinell hardness and the Meyer hardness is about 10% at the largest recommended value of $d/D$.

The Brinell hardness number is favored by certain engineers because there is an empirical relationship between it and the ultimate tensile strength of the material. The Meyer hardness which

**Table B.1** Ratio of Brinell to Meyer hardness versus contact strain.

| $d/D$ | $H_B/H_M$ |
| --- | --- |
| 0.20 | 0.9899 |
| 0.30 | 0.9770 |
| 0.40 | 0.9583 |
| 0.50 | 0.9330 |
| 0.60 | 0.9000 |

is based on the projected area is preferable since it gives the mean pressure beneath the indenter which opposes the applied force.

## B.1.2  Rockwell Macro-hardness

The Rockwell hardness test is a static indentation test similar to the Brinell indentation test. It differs in that it measures the permanent increase in the indentation depth from the depth reached under an initial load of 98.1 N, due to the application of an additional load. Measurement is made after recovery which takes place following the removal of the additional load. The Rockwell hardness number is a direct reading (in units of 0.002 mm) from the dial gauge which is attached to the Rockwell machine while the initial minor load is still imposed.

According to the material being tested, the indenter may be a 120° diamond cone with a blended spherical apex of 0.2 mm radius or a steel ball indenter. The steel ball indenter is normally 1.588 mm in diameter; however, larger diameters such as 3.175, 6.350, or 12.7 mm may be used for soft materials.

The Rockwell hardness testers are constructed to apply a fixed minor load of 98.1 N which is used to establish the measurement datum. This is followed by an additional load, within two to eight seconds, which may be 0.49, 0.88, or 1.37 kN. The combination of three loads and five indenters gives 15 conditions of test; each has its own hardness scale.

There is no Rockwell hardness number value designated by a number alone because it is necessary to specify which indenter and load have been used in an indentation test. Therefore, a prefix letter is employed to designate the scale and test conditions. Of the several scales available, the B and C scales are the most widely used. For the B scale, a 0.88 kN additional load with a 1.588 mm diameter steel ball indenter is used. For the C scale, a 1.37 kN additional load a conical indenter is used.

The Rockwell scales are divided into 100 divisions, each equivalent to 0.002 mm of recovered indentation. Since the scales are reversed, the number is higher, the harder the material, as shown by the following relations which define the Rockwell B and C hardness numbers.

$$\left. \begin{array}{l} \text{Rockwell B } = RB = 130 - \dfrac{\text{depth of penetration (mm)}}{0.002} \\[2mm] \text{Rockwell C } = RC = 130 - \dfrac{\text{depth of penetration (mm)}}{0.002} \end{array} \right\} \tag{B.5}$$

The Brinell (Meyer) hardness tests and the Rockwell tests give essentially identical macro-hardness values which are constant with respect to the indentation load. The hardness values correspond to the resistance of the bulk material to the penetration of the indenter. Brinell hardness values are frequently reported in the materials handbooks.

### B.1.3 Knoop Micro-hardness Indenter and Test

The Knoop ($H_K$) hardness indenter and test procedure were developed at the National Bureau of Standards (now NIST) in 1939. The indenter used is a rhombic-based pyramidal diamond that produces an elongated diamond-shaped indent. The angles from the opposite faces of a Knoop indenter are $172°30'$ and $130°$.

The Knoop indenter is a diamond ground to a pyramidal form that produces a diamond-shaped indentation having approximate ratio between long and short diagonals of 7 to 1. The depth of indentation is about 1/30 of the diagonal length.

The Knoop indenter is particularly useful for the study of highly brittle materials due to the small depth of penetration for a given indenter load. Also, due to the unequal lengths of the diagonals, it is very useful for investigating anisotropy of the surface of the specimen.

Knoop tests are mainly done at test forces from 10 to 1000 g, so a high-powered microscope is necessary to measure the indent size. Because of this, Knoop tests have mainly been known as micro-hardness tests. The newer standards more accurately use the term "micro-indentation" tests. The magnifications required to measure Knoop indents dictate a highly polished test surface. To achieve this surface, the samples are normally mounted and metallurgically polished; therefore, Knoop is almost always a destructive test.

The indenter is pressed into the polished surface of a sample by an accurately controlled test force which is maintained for a specific dwell time, normally 10–15 seconds. After the dwell time is complete, the indenter is removed leaving an elongated diamond-shaped indent in the sample. The size of the indent is determined optically by measuring the longest diagonal $d$ of the diamond shaped indent. The length of the smaller diagonal is $d/7$. The Knoop hardness number (KHN) is defined as the ratio of the test force divided by the *projected* area of the indent. The Knoop hardness number is calculated from

$$KHN = \frac{2F}{d^2 \left[ \cot \frac{172.5°}{2} \tan \frac{130°}{2} \right]} = 14.240 \frac{F}{d^2} \tag{B.6}$$

where the units of $d$ are mm and the load is kgf. The Knoop micro-hardness will be denoted as $H_K$. The conventional units are kgf/mm$^2$. Typical values of $H_K$ are in the range from 100 to 1000 kgf/mm$^2$. Table B.2 lists nominal values for four materials.

**Table B.2** Values of Knoop hardness number for selected materials.

| Material | $H_K$ |
|---|---|
| Gold foil | 69 |
| Quartz | 820 |
| Silicon carbide | 2480 |
| Diamond | 8000 |

### B.1.4 Vickers Micro-hardness Indenter and Test

In the Vickers micro-hardness test, a diamond indenter, in the form of a square-based pyramid with an angle of 136° between the opposite faces at the vertex, is pressed into the polished surface of the test specimen using a prescribed force $F$ as shown in Figure B.1. After the force has been removed, the diagonal lengths of the indentation $d_1$ and $d_2$ are measured with an optical microscope. The time for the initial application of the force is 2–8 seconds, and the test force is maintained for 10–15 seconds. The applied loads vary from 1 to 120 kgf, and the standard loads are 5, 10, 20, 30, 50, 100, and 120 kgf.

The Vickers hardness number (VHN) is defined as the test force divided by the *actual* area of the residual indent. It is given by

$$VHN = \frac{2F}{d^2} \sin \frac{136°}{2} = 1.854 \frac{F}{d^2} \tag{B.7}$$

where the mean diagonal is $d = (d_1 + d_2)/2$ and its units are mm. The units of load is kgf. The Vickers hardness number is smaller than the mean contact pressure by about 7%.

The Vickers micro-hardness will be denoted as $H_V$. The units of Vickers micro-hardness are frequently reported as kgf/mm². There is now a trend toward reporting Vickers micro-hardness in SI units (MPa or GPa). To convert Vickers micro-hardness values from kgf/mm² to MPa multiply by 9.807.

The Vickers micro-hardness depends on the load applied to the indenter. As the load increases, the diagonal and corresponding penetration depth increase. The Vickers micro-hardness can be related to the diagonal $d_V$ or the penetration depth $t$ which are related as $d_V = 7t$. The Vickers contact area and the penetration depth are related as $A_V = 24.5\ t^2$.

The Vickers micro-hardness test is reliable for measuring the micro-hardness of metals, polymers, and ceramics.

### B.1.5 Berkovich Micro and Nano Hardness Indenter and Nano Hardness Tests

The Berkovich diamond indenter is a triangular pyramid with a true point since only three sides have to meet. The angle between the vertical axis and each of the faces is 65°. The Berkovich diamonds are cut with an angle of 142° between any two of the planes along the line so that the surface areas of indents are the same as the Vickers indent for the same depth of penetration. This means that isotropic micro-hardness values are the same for a given material when it is indented by the Berkovich and the Vickers indenters.

The Berkovich indenter is used to study the micro and sub-micron indentations of various materials. Equipment for this purpose consists of an instrumented loading device that records the indenter load $P$ in mN and indenter displacements $h$ in micrometers or nanometers. Estimates of the elastic modulus $E$ and micro-hardness $H$ of the specimen are obtained from the load versus penetration measurements.

Rather than the direct measurements of the size of residual impressions, which require optical microscopes, contact areas are calculated from the depth measurements together with a knowledge of the indenter shape. This is in contrast to the procedures employed for the macro-indentation tests, where the lateral dimensions (diagonals for the Vickers and Knoop tests, and diameter for the Brinell and Rockwell tests), rather than the depth of penetration of the residual impression, are used to calculate the micro-hardness and the macro-hardness.

## B.2 Micro- and Macro-hardness Tests and Correlations

The Vickers, Brinell, and Rockwell B micro- and macro-hardness test results and correlation equations for SS 304, Ni 200, Zr-4, and Zr-Nb will be presented in this section. The data are given in several papers [Yovanovich et al. (1982a,b, 1983), Yovanovich and Hegazy (1983)] and the dissertation of Hegazy (1985). Figures B.3 and B.4 show the Vickers, Brinell, and Rockwell B results for SS 304 and Ni 200, respectively. In order to show the micro- and macro-hardness values on the same plot, its necessary to plot the hardness versus the penetration depth $t$ which varies from about $t = 1$ μm to approximately $t = 800$ μm. The largest penetration depths of about 800 μm correspond to the Brinell tests which were based on a 10 mm diameter steel ball forced into the surface of the specimens under a 29.43 kN load. In the Rockwell B tests a 1.59 mm diameter steel ball was forced into the surface of the specimens under a 981 N load. The penetration depths were about 100 μm.

The Brinell and Rockwell B macro-hardness values denoted as $H_b$ for SS 304 were identical, $H_b = 150$ kg/mm$^2$ (Figure B.2). The Brinell and Rockwell B macro-hardness values for Ni 200 were identical, $H_b = 170$ kg/mm$^2$ (Figure B.3).

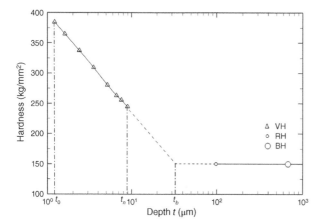

**Figure B.2** Vickers, Brinell, and Rockwell Hardness versus indentation depths for SS 304. Source: Adapted from after Yovanovich and Hegazy (1983).

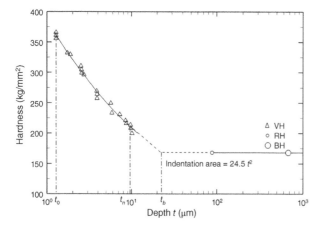

**Figure B.3** Vickers, Brinell, and Rockwell Hardness versus indentation depths for Ni 200. Source: Adapted from after Yovanovich and Hegazy (1983).

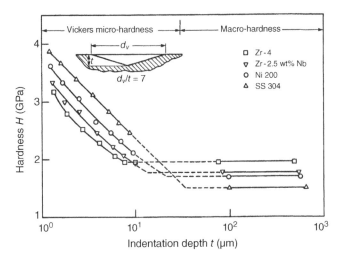

**Figure B.4** Vickers, Brinell, and Rockwell Hardness versus indentation depths for four metals and alloys. Source: Hegazy (1985).

The Vickers micro-hardness values $H$ for both metals are dependent on the penetration depth $t$ as shown in Figures B.3 and B.4. The points shown in Figure B.3 represent the average value of five test values at each load.

The maximum values $H = H_{max}$ were measured at the minimum depths $t = t_0$, and the smallest values $H = H_{min}$ were measured at the maximum test depths $t = t_n$. The Vickers micro-hardness values were correlated and the extrapolated curve intersected the bulk values $H_b$ at the penetration depth $t_b$. The Vickers micro-hardness values were correlated with the following general set of equations:

$$\left.\begin{array}{ll} H = H_{max} = \text{constant} & t \le t_0 \\ H = H(t) = c_1 t^{c_2} + c_3 & t_0 \le t \le t_n \\ H = H_b = \text{constant} & t \ge t_b \end{array}\right\} \tag{B.8}$$

The correlation coefficients are $c_1, c_2$ and $c_3$. The values for SS 304 are

$$c_1 = 3049.6 \qquad c_2 = -0.024 \qquad c_3 = -2649.8 \tag{B.9}$$

The maximum and minimum hardness values and the corresponding penetration depths are

$$\left.\begin{array}{ll} H_{max} = 385 \text{ kg/mm}^2 & t_0 = 1.2 \text{ μm} \\ H_b = 150 \text{ kg/mm}^2 & t_b = 34 \text{ μm} \end{array}\right\} \tag{B.10}$$

The values for Ni 200 are

$$c_1 = 377.27 \qquad c_2 = -0.274 \qquad c_3 = 7.79 \tag{B.11}$$

The maximum and minimum hardness values and the corresponding penetration depths are

$$\left.\begin{array}{ll} H_{max} = 362.3 \text{ kg/mm}^2 & t_0 = 1.24 \text{ μm} \\ H_b = 170.4 \text{ kg/mm}^2 & t_b = 21.56 \text{ μm} \end{array}\right\} \tag{B.12}$$

In order to implement the correlation equations, it is necessary to convert the penetration depths to microns from μm. For example, if the Vickers penetration depth is $5 \times 10^{-6}$ m for both metals, then $t = 5 \times 10^{-6}/10^{-6} = 5$ μm, and the micro-hardness values are $H = 284.3$ kg/mm$^2$ for SS 304 and $H = 250.5$ kg/mm$^2$ for Ni 200.

A model is required to calculate the contact micro-hardness given the correlation equation and the effective surface roughness $\sigma/m$ and the apparent contact pressure $P$. Several models were presented [Yovanovich et al. (1982a,b, 1983), Yovanovich and Hegazy (1983)]. A simple direct method for calculation of the effective contact micro-hardness will be presented next.

### B.2.1  Direct Approximate Method

The Direct Approximate Method (DAM) is based on the observation that the geometric parameters of the Cooper–Mikic–Yovanovich (CMY) model [Cooper et al. (1969)] change slowly as the apparent contact pressure is varied over a wide range. This model is based on the equivalence of the Vickers indentation area $A_V = d_V^2/2 = 49t^2/2$ and the mean contact spot area $A_c = \pi a^2$, where $a$ is the mean contact spot radius. The area equivalence leads to the following relationship between the indentation depth and the contact spot radius:

$$t = \left(\frac{\pi a^2}{24.5}\right)^{1/2} = 0.358\, a \tag{B.13}$$

The contact micro-hardness can be related to the mean contact spot radius as

$$H_c = c_1(0.358\, a)^{c_2} + c_3 \tag{B.14}$$

It was shown by Yovanovich (1981) that the following simple explicit relationship:

$$a = 0.99\left(\frac{\sigma}{m}\right)\left[-\ln\left(3.132\,\frac{P}{H}\right)\right]^{-0.547} \tag{B.15}$$

can be used to calculate the mean contact spot radius when $\sigma/m$, $P$, and $H$ are known.

For a particular metal such as Ni 200, we have after substitution the implicit relationship:

$$H = 501.3\left(\frac{m}{\sigma}\right)^{0.274}\left[-\ln\left(3.132\,\frac{P}{H}\right)\right]^{0.150} \tag{B.16}$$

where the units of $\sigma/m$ must be microns, and the units of $P$ and $H$ must be consistent. Since the unknown $H$ appears on both sides, a numerical method to find its root can be used, or an iterative approach can be employed to calculate the value of $H$ beginning with an initial guess. It is found that starting with the lowest value $H = H_b$, only two to three iterations are required to calculate an accurate value of $H$.

The following explicit relationship based on $H = H_b$ substituted on the right-hand side gives approximate values for the effective contact micro-hardness which is denoted as $H_c$:

$$H_c = 501.3\left(\frac{m}{\sigma}\right)^{0.274}\left[-\ln\left(3.132\,\frac{P}{H_b}\right)\right]^{0.150} \tag{B.17}$$

This relationship shows clearly how the contact micro-hardness is related to the effective surface roughness of the joint, the apparent contact pressure, and the bulk hardness.

For a given metal, the contact micro-hardness decreases with increasing surface roughness and increasing contact pressure.

Two examples will be given to illustrate the use of the correlation equation. For a Ni 200 joint, where $\sigma/m = 1.21/0.139 = 8.71$ microns, $P_{min} = 622$ kg/mm$^2$, $P_{max} = 3510$ kg/mm$^2$, and $H_b = 170.4$ kg/mm$^2$. The mean contact pressure is $P_m = (P_{min} + P_{max})/2 = 2066$ kg/mm$^2$.

Substitution of these values in the correlation equation gives the contact micro-hardness value: $H_c = 366.0$ kg/mm$^2$ which is 2.15 times greater than the bulk hardness.

For the second example, the contact parameters are $\sigma/m = 8.48/0.344 = 24.65$ microns, $P_{min} = 571$ kg/mm$^2$, $P_{max} = 3433$ kg/mm$^2$, and $H_b = 170.4$ kg/mm$^2$. The mean contact pressure is $P_m = (P_{min} + P_{max})/2 = 2002$ kg/mm$^2$.

Substitution of these values in the correlation equation gives the contact micro-hardness value: $H_c = 277.3$ kg/mm$^2$ which is 1.63 times greater than the bulk hardness.

### B.2.2   Vickers Micro-hardness Correlation Equations

The material in this section is obtained from the work of Hegazy (1985). The development of Vickers micro-hardness correlation equations are based on Vickers micro-hardness and Brinell hardness measurements. Seven loads from 0.147 to 4.9 N were used in the Vickers micro-hardness measurements. For each load, five indentations were used to calculate the value of $H_V$. Each value of $H_V$ is based on the average of the measured diagonals for a particular load, i.e.

$$H_V = \frac{1.854\,F}{d_V^2}\ \text{GPa} \tag{B.18}$$

where the applied load is $F$ in Newtons, and

$$d_V = \frac{(d_1 + d_2)}{2} \tag{B.19}$$

and $d_1$ and $d_2$ are the measured diagonals in μm.

For any load, the maximum percent difference in the calculated values of $H_V$ was less than 5 %. The Vickers micro-hardness values $H_V$ in GPa for four metals: Ni 200, SS 304, Zr-4, and Zr-Nb are plotted against the average values of the Vickers indentation diagonal $d_V$ in μm as shown in Figures B.4 and B.5.

The Vickers micro-hardness values are correlated by the power–law relation:

$$H_V = c_1 \left( \frac{d_V}{d_0} \right)^{c_2} \quad \text{with} \quad d_0 = 1\ \text{μm} \tag{B.20}$$

The parameter $d_0$ was introduced to make the ratio $d_V/d_0$ dimensionless. The Vickers micro-hardness correlation coefficients are $c_1$ and $c_2$. The units of $H_V$ and $c_1$ are GPa, and

**Figure B.5**   Vickers, Brinell, and Rockwell Hardness versus indentation diagonals for four metals and alloys. Source: Hegazy (1985).

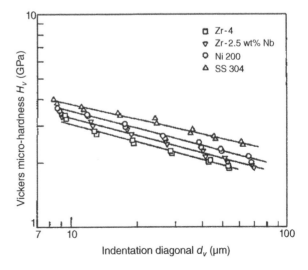

the size index $c_2$ is dimensionless. The correlation equation is based on the average Vickers diagonal $d_V$. The correlation coefficient $c_1$ is the Vickers micro-hardness when $d_V = d_0$.

The bulk or material hardness which is denoted as $H_m$ was determined by Brinell and Rockwell B indentation tests which gave the same value although the indentations depths are significantly different as seen in Figure B.4.

The values of $H_m$, $c_1$, and $c_2$ are given in Table B.3 for the four metals.

The maximum and RMS percent differences between the measured values and the predicted values are given. Except for the maximum percent difference of 10.2 % for Zr-Nb, the maximum percent differences are below 5 %, and the RMS percent differences are below 3 %.

The average value of $c_2$ is −0.260. Since the variation in the values of $c_2$ is relatively small, this value was selected for the four metals to develop an alternative Vickers micro-hardness correlation equation of the form:

$$H_V = \xi \left( \frac{d_V}{d_0} \right)^{\eta} \quad \text{with} \quad d_0 = 1 \ \mu m \tag{B.21}$$

where $\eta = -0.260$, fixed for all metals. The Vickers micro-hardness values were used to find values for the correlation coefficients $\xi$. These values are given in Table B.4. The alternative correlation equation with the fixed value of $\eta = -0.260$ and the corresponding values of $\xi$ have maximum and RMS percent differences which are comparable with the original correlation equation. All values of $H_V / \xi$ when plotted against $d_V$ fall on a single curve as seen in Figure B.6. This confirms that there is a close relationship between $H_V$ and $\xi$ for the four metals.

The values of $\xi$ when plotted against values of $H_m$ fall on a straight line as shown in Figure B.7. The following correlation equation for $\xi$ versus $H_m$ is obtained

$$\xi = 12.04 - 3.49 \ H_m \quad \text{for} \quad 1.472 \le H_m \le 1.913 \tag{B.22}$$

The units of $\xi$ and $H_m$ are GPa. The comparisons between the actual and predicted values of $\xi$ are given in Table B.5.

The RMS percent difference is about 2.3 %.

**Table B.3** Vickers correlation coefficients for four metals.

| Metal | $H_m$ (GPa) | $c_1$ (GPa) | $c_2$ | Max % diff. | RMS % diff. |
|---|---|---|---|---|---|
| Ni 200 | 1.668 | 6.304 | −0.264 | 4.8 | 1.8 |
| SS 304 | 1.472 | 6.271 | −0.229 | 4.2 | 1.4 |
| Zr-4 | 1.913 | 5.677 | −0.278 | 3.4 | 1.7 |
| Zr-Nb | 1.727 | 5.884 | −0.267 | 10.2 | 2.7 |

**Table B.4** Modified Vickers correlation coefficients for four metals.

| Metal | $H_m$ (GPa) | $\xi$ (GPa) | $\eta$ | Max % diff. | RMS % diff. |
|---|---|---|---|---|---|
| Ni 200 | 1.668 | 6.217 | −0.260 | 5.2 | 1.8 |
| SS 304 | 1.472 | 6.906 | −0.260 | 5.9 | 2.4 |
| Zr-4 | 1.913 | 5.367 | −0.260 | 3.9 | 1.8 |
| Zr-Nb | 1.727 | 5.750 | −0.260 | 9.7 | 2.7 |

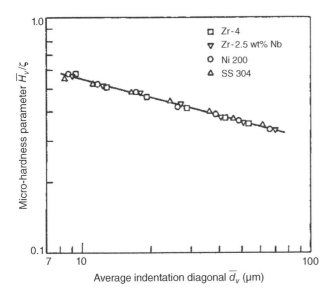

**Figure B.6** Normalized Vickers micro-hardness versus Vickers diagonals for four metals and alloys. Source: Hegazy (1985).

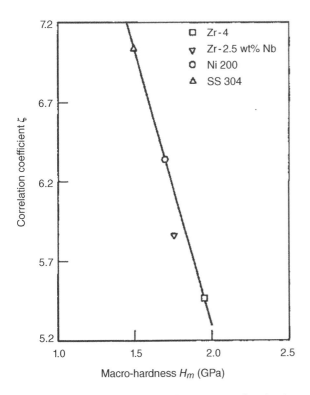

**Figure B.7** Plot of Vickers coefficients versus micro-hardness. Source: Hegazy (1985).

**Table B.5** Comparisons of modified correlations with Vickers hardness values.

| Metal | $H_m$ (GPa) | $\xi$ (GPa) | Corr. eq. | % diff. |
|-------|------------|------------|-----------|---------|
| Ni 200 | 1.668 | 6.217 | 6.219 | 0.03 |
| SS 304 | 1.472 | 6.906 | 6.903 | −0.04 |
| Zr-4 | 1.913 | 5.367 | 5.364 | −0.06 |
| Zr-Nb | 1.727 | 5.750 | 6.013 | 4.60 |

A new general Vickers micro-hardness correlation equation for the four metals and any other metal whose bulk hardness lies in the range: $1.472 \leq H_m \leq 1.913$ is

$$H_V = \left(12.04 - 3.49\, H_m\right) \left(\frac{d_V}{d_0}\right)^{-0.260} GPa \tag{B.23}$$

where the units of $H_m$ must be GPa.

## B.3 Correlation Equations for Vickers Coefficients

When Vickers indentation tests were done on harder metals and alloys, it was observed that the simple general correlation equation developed for the four metals whose bulk hardness values lie in the range: $(1.33 \leq H_m \leq 1.91)$ GPa predicts values of $H_V$ which have large errors.

When a titanium alloy with $H_B = 3.07$ GPa, and untreated and heat-treated tool steel with bulk hardness values in the range: $(1.98 \leq H_B \leq 7.57)$ GPa, the simple general correlation equation gives poor agreement with the measured Vickers micro-hardness measurements. Sridhar (1994) conducted extensive Vickers micro-hardness tests, and Brinell and Rockwell C macro-hardness tests, and he found new relationships between the Vickers correlation coefficients $c_1$ and $c_2$ and the bulk hardness $H_B$. The development of the new correlation equations are given in Sridhar (1994) and Sridhar and Yovanovich (1994); only the final results will be presented.

For Brinell hardness values in the range 1.30–7.60 GPa, a least-squares cubic fit was used to obtain correlation equations for $c_1$ and $c_2$. The correlation equation for $c_1$ is as derived by Sridhar (1994) and Sridhar and Yovanovich (1994):

$$\frac{c_1}{H_{BGM}} = 4.0 - 5.77\, Z + 4.0\, Z^2 - 0.61\, Z^3 \tag{B.24}$$

with

$$Z = \frac{H_B}{H_{BGM}}$$

where $H_{BGM} = 3.178$ GPa which is the geometric mean of the minimum and maximum values of $H_B$ for the test materials.

The correlation equation for $c_2$ is as derived by Sridhar (1994) and Sridhar and Yovanovich (1994):

$$c_2 = -0.57 + \frac{Z}{1.22} - \frac{Z^2}{2.42} + \frac{Z^3}{16.58} \tag{B.25}$$

The maximum percent difference and RMS % difference between the $c_1$ values and the correlation equation are −11.0% and 5.3%, respectively. The maximum percent difference and RMS %

difference between the $c_2$ values and correlation equation are $-41.5\%$ and $20.8\%$, respectively. The relatively large percent differences between the $c_2$ values and the correlation values are less important than the good agreement between the $c_1$ values and the correlation equation. The values of $c_2$ for the very hard heat-treated tool steel lie in the range: $[-0.040 \; to - 0.129]$.

An alternative correlation equation for $c_2$ was given by Sridhar and Yovanovich (1996a,b):

$$c_2 = -0.370 + 0.442 \, \frac{H_B}{c_1} \tag{B.26}$$

The percent difference and RMS percent difference between all data and the correlation equation are $25.8\%$ and $10.9\%$, respectively.

The Brinell hardness tests were performed with a standard steel ball for materials with hardness less than $4.40 \, GPa$ and with a carbide ball with hardness greater than $4.40 \, GPa$ as recommended by ASTM E10. A load of $3000 \, kgf$ was used to make Brinell hardness tests so that included the entire range $(1.30–7.60) \, GPa$

Rockwell C hardness tests were also performed on the heat-treated tool steel specimens to compliment the Brinell hardness tests. The Brinell hardness values and the Rockwell C hardness values in the range $(15 \leq HRC \leq 65)$ were in good agreement. The following correlation equation between the Brinell hardness number and the Rockwell C hardness number was obtained using a third-order polynomial:

$$BHN = 43.7 + 10.92 \, HRC - \frac{(HRC)^2}{5.18} + \frac{(HRC)^3}{340.26} \tag{B.27}$$

which is valid for $(20 \leq HRC \leq 65)$. This correlation equation is a useful relationship between Brinell hardness number ($BHN$) and Rockwell C hardness number ($HRC$) for tool steel (01). It is not possible to perform accurate Rockwell C hardness tests below $HRC = 20$.

## B.4 Temperature Effects on Vickers and Brinell Hardness

Thermophysical properties of all materials depend on temperature level. Since the temperature level of most joints is above room temperature, it is important to conduct Vickers micro-hardness and Brinell hardness test at elevated temperatures.

### B.4.1 Temperature Effects on Yield Strength and Vickers Micro Hardness of SS 304L

The effects of temperature level on the values of yield strength $S_y$ and Vickers micro-hardness $H_V$ are given [Nho (1990)] in Table B.6.

The values of yield strength and Vickers micro-hardness are strongly dependent on temperature for temperatures from $(200 \; to \; 800) \, °C$. The temperature dependence of $S_y$ and $H_V$ with temperature are similar.

### B.4.2 Temperature Effect on Brinell Hardness

Temperature effects on the Brinell hardness of SS 304, Ni 200, and Al 6061 T6 were found by Nho (1990). The load was $1500 \, kg$, and the ball diameter was $10 \, mm$ for the Brinell test on SS 304 and Ni 200. Five measurements of the indentation diameter were made at the reported temperatures. The average values for SS 304 are given in Table B.7.

**Table B.6** Temperature effects on yield strength and Vickers hardness.

| $T(°C)$ | $S_y$ (MPa) | $S_y(T)/S_y(25)$ | $H_V$ (MPa) | $H_V(T)/H_V(25)$ |
|---------|-------------|------------------|-------------|-------------------|
| 25 | 274 | 1.00 | 1570 | 1.00 |
| 200 | 223 | 0.81 | 1180 | 0.75 |
| 400 | 198 | 0.72 | 1090 | 0.69 |
| 600 | 157 | 0.57 | 863 | 0.55 |
| 800 | 78.9 | 0.29 | 392 | 0.25 |

**Table B.7** Temperature effect on Brinell hardness of SS 304.

| $T(°C)$ | $d$(mm) | $H_B$ (MPa) | $H_B(T)/H_B(23.8)$ |
|---------|---------|-------------|---------------------|
| 23.8 | 3.68 | 1335 | 1.00 |
| 55.6 | 3.79 | 1256 | 0.94 |
| 89.9 | 3.89 | 1189 | 0.89 |
| 123.0 | 4.00 | 1122 | 0.84 |
| 152.4 | 4.07 | 1082 | 0.81 |
| 186.1 | 4.15 | 1039 | 0.78 |

The correlation equation is

$$\frac{H_B(T)}{H_B(23.8)} = \exp\left[C_T\left(T - 23.8\right)\right] \tag{B.28}$$

where the correlation coefficient is $C_T = -1.51 \times 10^{-3} \ C^{-1}$.

The average values for Ni 200 are given in Table B.8.

The correlation equation for Ni 200 is

$$\frac{H_B(T)}{H_B(23.1)} = \exp\left[C_T\left(T - 23.1\right)\right] \tag{B.29}$$

where the correlation coefficient is $C_T = -1.86 \times 10^{-3} \ C^{-1}$.

**Table B.8** Temperature effect on Brinell hardness of Ni 200.

| $T(°C)$ | $d$(mm) | $H_B$ (MPa) | $H_B(T)/H_B(23.1)$ |
|---------|---------|-------------|---------------------|
| 23.1 | 3.51 | 1472 | 1.00 |
| 58.5 | 3.62 | 1381 | 0.94 |
| 95.1 | 3.70 | 1320 | 0.90 |
| 121.8 | 3.80 | 1249 | 0.85 |
| 159.0 | 3.91 | 1177 | 0.80 |
| 209.3 | 4.17 | 1028 | 0.70 |

**Table B.9** Temperature effect on Brinell hardness of Al 6061 T-6.

| $T(°C)$ | $d$(mm) | $H_B$ (MPa) | $H_B(T)/H_B(22.3)$ |
|---------|---------|-------------|---------------------|
| 22.3 | 2.59 | 915 | 1.00 |
| 48.9 | 2.61 | 901 | 0.98 |
| 71.1 | 2.64 | 880 | 0.96 |
| 102.3 | 2.68 | 854 | 0.93 |
| 124.3 | 2.72 | 828 | 0.91 |
| 140.8 | 2.77 | 798 | 0.87 |
| 164.3 | 2.81 | 775 | 0.85 |
| 195.6 | 2.86 | 748 | 0.82 |

For the temperature tests for Al 6061 T6, the load was 500 kg, and the ball diameter was 10 mm. Five measurements of the indentation diameter were made at the reported temperatures. The average values are Al 6061 T6 are given in Table B.9. The correlation equation is

$$\frac{H_B(T)}{H_B(22.3)} = \exp\left[C_T\left(T - 22.3\right)\right] \tag{B.30}$$

where the correlation coefficient is $C_T = -1.20 \times 10^{-3}\ C^{-1}$.

### B.4.3 Temperature Effect on Vickers Micro-hardness and Correlation Coefficients

The hot Vickers micro-hardness tests and correlations of Nho (1990). He conducted tests with Ni 200, SS 304, and Al 6061-T6. He obtained indentation data from three tests and reported the average values. Six loads were used: $(F = 25, 50, 100, 200, 300, 500)$ kg. Tests were conducted at room temperature and four higher temperature levels.

The Vickers micro-hardness results were correlated with the power–law relation:

$$H_V = c_1\left(\frac{d_V}{d_0}\right)^{c_2} \quad \text{with} \quad d_0 = 1\ \mu m \tag{B.31}$$

For the three metals tested, the correlation coefficient $c_1(T)$ was dependent on the test temperature while the size index $c_2$ was independent of the temperature. For Ni 200, the values of $c_2$ lie in the narrow range of $-0.209$ to $-0.237$. The average value $c_2 = -0.226$ was selected, and a new set of values for $c_1(T)$ were calculated.

For SS 304, the values of $c_2$ lie in the narrow range of $-0.265$ to $-0.289$. The average value $c_2 = -0.279$ was selected, and a new set of values for $c_1(T)$ were calculated.

For Al 6061 T-6 the values of $c_2$ lie in the narrow range of $-0.00643$ to $-0.0117$. The average value $c_2 = -0.0079$ was selected, and a new set of values for $c_1(T)$ were calculated.

The values of $c_1(T)$ for each metal were correlated, and the following general relationship was given

$$\frac{c_1(T)}{c_1(T_{rm})} = \exp\left[C_T\left(T - T_{rm}\right)\right] \quad 25 \leq T \leq T_{max} \tag{B.32}$$

The correlation parameters $c_1(T_{rm})$, $C_T$, and $T_{max}$ are given in Table B.10.

**Table B.10** Temperature coefficients for Brinell hardness of three metals.

| Metal | $c_1(T_{rm})$ (MPa) | $10^3 C_T (C^{-1})$ | $T_{max}(°C)$ |
|---|---|---|---|
| Ni 200 | 6636 | −1.372 | 186 |
| SS 304 | 7339 | −1.675 | 190 |
| Al 6061 T-6 | 1123 | −1.190 | 183 |

The room temperatures for the Vickers microhardness tests for Ni 200, SS 304, and Al 6061 T-6 were $T_{rm}$ = 22.9, 25.2, 23.7°C, respectively.

For Ni 200 with $c_2$ = −0.226, the re-correlated values are listed in Table B.11. For SS 304 with $c_2$ = −0.279, the re-correlated values are listed in Table B.12. For Al 6061 T-6 with $c_2$ = −0.0079, the re-correlated values are listed in Table B.13.

**Table B.11** Temperature effect on coefficients for Ni 200.

| T(°C) | $c_1$ (MPa) | $c_1(T)/c_1(T_{rm})$ | Max.% diff. |
|---|---|---|---|
| 22.9 | 6636 | 1.00 | +2.6 |
| 48.4–45.9 | 6282 | 0.95 | −2.7 |
| 95.6–82.3 | 6115 | 0.92 | −2.3 |
| 156.5–138.0 | 5613 | 0.85 | −2.5 |
| 185.9–179.7 | 5258 | 0.79 | −2.9 |

**Table B.12** Temperature effect on coefficients for SS 304.

| T(°C) | $c_1$ (MPa) | $c_1(T)/c_1(T_{rm})$ | Max.% diff. |
|---|---|---|---|
| 25.2 | 7339 | 1.00 | −1.5 |
| 61.6–54.3 | 6716 | 0.92 | −2.7 |
| 114.3–92.8 | 6321 | 0.86 | −2.9 |
| 150.8–146.1 | 5898 | 0.80 | +2.3 |
| 197.2–182.4 | 5495 | 0.75 | +2.7 |

**Table B.13** Temperature effect on coefficients for Al 6061 T-6.

| T(°C) | $c_1$ (MPa) | $c_1(T)/c_1(T_{rm})$ | Max.% diff. |
|---|---|---|---|
| 23.7 | 1123 | 1.00 | +1.8 |
| 52.3–48.1 | 1104 | 0.98 | −1.8 |
| 92.4–86.4 | 1072 | 0.95 | −1.4 |
| 157.0–156.4 | 1013 | 0.90 | +1.8 |
| 183.0–177.9 | 909 | 0.81 | −1.8 |

## B.5 Nanoindentation Tests

A comprehensive review of nano-indentation tests and nano-hardness is beyond the scope of this paper. Therefore, a relatively short review of the important features of nano-indentation tests will be presented. Details are available in the text of Fisher-Cripps (2000) and in the papers by Oliver and Pharr (1992), Pharr et al. (1992), Hendrix (1995), and Hay and Pharr (2000).

The indenter of choice is the Berkovich (three-sided) indenter, although the Vickers indenter has been used.

As the indenter is slowly forced into the surface of the specimen, both elastic and plastic deformation processes occur, and a contact area is produced that conforms to the shape of the indenter. The displacement $h$ and the load $P$ are continuously monitored until the maximum load $P_{max}$ is reached and the corresponding penetration depth is $h_{max}$. As the indenter is slowly withdrawn, only the elastic portion of the displacement is recovered. The unloading load and displacement are continuously monitored until the load is zero and the final or residual penetration depth $h_f$ measured. The loading and unloading curves are shown schematically in Figure B.8. The load range is typically from zero to $P_{max} = 200$ mN and the penetration depth is from zero to $h_{max} = 1000$ nm. Figure B.9 shows unloading nano-indentation curves for six different materials [Pharr et al. (1992)]. The horizontal axis is the relative displacement $(h - h_f)$ which shows clearly the different unloading trends. The aluminum and tungsten unloading curves are very similar.

The slope of the upper portion of the unloading curve is denoted as $S = dP/dh$ as shown in Figure B.10. The parameter $S$ is called the elastic contact stiffness [Fisher-Cripps (2000), Oliver and Pharr (1992), Pharr et al. (1992), Hendrix (1995), Hay and Pharr (2000)]. The elastic modulus $E$ and the nano- or micro-hardness $H$ are derived from these quantities by a set of relationships

**Figure B.8** Typical loading and unloading curves for nano-indentation tests. Source: Adapted from Oliver and Pharr (1992).

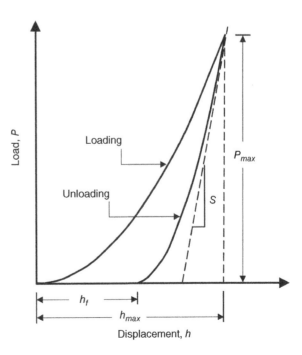

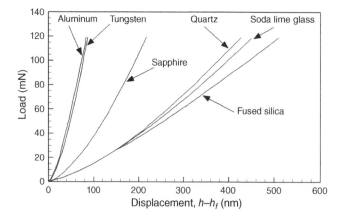

**Figure B.9** Final unloading curves versus reduced displacements for six materials. Source: Adapted from Oliver and Pharr (1992).

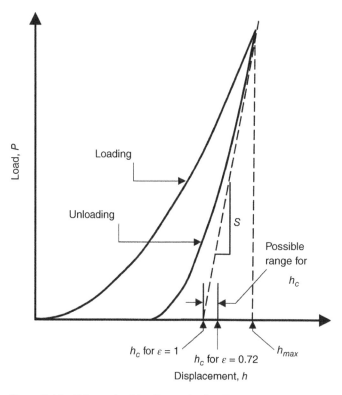

**Figure B.10** Schematic of loading and unloading curves versus displacements showing quantities used in analysis. Source: Adapted from Oliver and Pharr (1992).

based on elasticity theory. Figure B.10 also shows the contact penetration depth $h_c$ for an ideal circular punch where $\epsilon = 1$, and the contact depth for the Berkovich indenter with $\epsilon = 0.72$.

The fundamental relationships from which $H$ and $E$ are determined are

$$H = \frac{P}{A} \tag{B.33}$$

where $P$ is the load and $A$ is the *projected* contact area at that load, and

$$E_r = \frac{\sqrt{\pi}}{2\beta} \frac{S}{\sqrt{A}} \tag{B.34}$$

where $E_r$ is the reduced elastic modulus of the contact, and $\beta$ is a constant that depends on the geometry of the indenter. The reduced modulus is defined as

$$\frac{1}{E_r} = \frac{1 - v^2}{E} + \frac{1 - v_i^2}{E_i} \tag{B.35}$$

where $E$ and $v$ are the elastic modulus and Poisson's ratio of the test specimen, and $E_i$ and $v_i$ are the elastic modulus and Poisson's ratio of the indenter. The reduced modulus is used to account for the fact that elastic displacements occur in the specimen and the indenter. For a diamond indenter, the values $E_i = 1140\,\text{GPa}$ and $v_i = 0.07$ are frequently used. The reduced modulus requires the Poisson's ratio of the specimen which is unknown. However, the value $v = 0.25$ produces a 5% uncertainty in the calculated value of $E$ for most materials.

The equation for the reduced elastic modulus is based on the classical problem of the axisymmetric contact of a smooth, rigid, circular punch with an isotropic elastic half-space whose elastic properties $E$ and $v$ are constants. For this type of indenter, the geometric parameter value is $\beta = 1$. However, it has been shown that the equation can be applied to indenters whose geometry is not axisymmetric, provided appropriate values of $\beta$ are used. For indenters with square cross sections such as the Vickers pyramid, $\beta = 1.01$, and for the triangular cross section such as the Berkovich pyramid, $\beta = 1.034$.

In order to implement the method to calculate $H$ and $E$ accurately from the indentation load-displacement data, one must have an accurate measurement of the elastic contact stiffness $S$ and the *projected* contact area $A$ under the load $P$.

The widely used method of establishing the contact area was proposed by Oliver and Pharr (1992) which expands on the ideas by others [Fisher-Cripps (2000)] and in a few papers [Oliver and Pharr (1992), Pharr et al. (1992), Hendrix (1995), Hay and Pharr (2000)]. The first step of the analysis procedure consists in fitting the unloading part of the load-displacement data to the power–law relation derived from the elastic contact theory:

$$P = B(h - h_f)^m \tag{B.36}$$

where $B$ and $m$ are empirically determined fitting parameters, and $h_f$ is the final displacement after complete unloading. It is also determined from the curve fit.

The second step in the analysis consists of finding the contact stiffness $S$ by differentiating the unloading curve fit, and evaluating the result at the maximum depth of penetration, $h = h_{max}$. This gives

$$S = \left(\frac{dP}{dh}\right)_{h=h_{max}} = Bm(h_{max} - h_f)^{m-1} \tag{B.37}$$

The third step in the procedure is to determine the contact depth $h_c$ which for elastic contact is smaller than the total depth of penetration. The contact depth is estimated according to

$$h_c = h - \epsilon \frac{P}{S} \tag{B.38}$$

where $\epsilon$ is another constant that depends on the indenter geometry. This geometric parameter has the value $\epsilon = 0.75$ for spherical indenters, and $\epsilon = 0.72$ for conical indenters. Experiments with Berkovich indenters (three-sided pyramidal indenters) have shown that $\epsilon = 0.75$ and works well even when there is elastic–plastic deformation during the unloading. It should be noted that the correction for $h_c$ should be used with some caution because it is not valid in the case of material pile-up around an indent. Therefore, inspection of the residual impression using a scanning electron microscope (SEM) or an atomic force microscope (AFM) is useful.

If we assume that the Berkovich indenter is ideal, then the relation

$$\left.\begin{aligned} A &= 3\sqrt{3} \tan^2 65.3^\circ\, h_c^2 \\ &= 24.5\, h_c^2 \end{aligned}\right\} \tag{B.39}$$

is valid, and fitting the upper 25–50% of the unloading curve is sufficient.

If the indenter tip is blunted or it has other defects, then the following procedure is recommended.

In the last step in the analysis, the *projected* contact area is calculated by evaluating an empirically determined indenter area function $A = f(d)$ at the contact depth $h_c$ such that

$$A = f(h_c) \tag{B.40}$$

The area function $A = f(h_c)$ is also called the shape function or tip function because it relates the cross-sectional area of the indenter $A$ to the distance $h_c$ from its tip.

A general polynomial is used [Pharr et al. (1992)]:

$$A = 24.5 h_c^2 + \sum_{i=1}^{n} C_i\, h_c^{1/2^i} \tag{B.41}$$

The leading term of the polynomial fit corresponds to the ideal Berkovich indenter, and the remaining terms account for deviations from the ideal geometry due to indenter tip rounding. The number of terms is chosen to give a good fit over the entire range of depths as assessed by comparing a log–log plot of the fit with the data. Because the data are often obtained over more than one order of magnitude in depth, a weighted fitting procedure should be used to assure that data from all depths have equal importance. The fitting parameters $C_i$ can be obtained by performing nano-indentation tests on materials with known elastic modulus.

The procedure described above is essentially the one used by many researchers to obtain values of the elastic modulus $E$ and the nano-hardness $H$ of many different materials. It is also used to

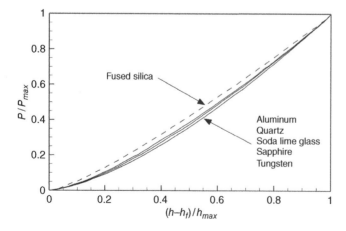

**Figure B.11** Normalized final unloading curves versus displacements for six materials showing similar trends. Source: Adapted from Oliver and Pharr (1992).

**Table B.14** Parameter values for power–law fits of unloading curve.

| Material | B ($mN/nm^m$) | m |
|---|---|---|
| Aluminum | 0.2650 | 1.38 |
| Fused silica | 0.0500 | 1.25 |
| Quartz | 0.0215 | 1.43 |
| Sapphire | 0.0435 | 1.47 |
| Soda-lime glass | 0.0279 | 1.37 |
| Tungsten | 0.141 | 1.51 |

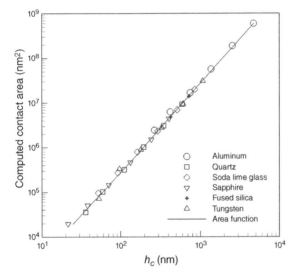

**Figure B.12** Calculated contact areas versus contact depths for six materials. Source: Adapted from Oliver and Pharr (1992).

determine these important physical properties of thin films and thin films bonded to substrates. The procedure continues to be modified for different applications.

The normalized unloading curves for the six materials are shown in Figure B.11.

The procedure given above was used to obtain $H$ and $E$ for six materials. The power–law fit correlation coefficients [Pharr et al. (1992)] are listed in Table B.14.

The good agreement between the calculated values of the contact area and the correlation equation is shown in Figure B.12.

Results for three materials are listed below. The values were obtained digitizing the values given in Tables B.15–B.17. The values for the first load were observed to lie below the values for the second load. Therefore, the digitized values were normalized with the second load values to show the trends of the data as the load increases.

The values of hardness and elastic modulus for all materials show a definite size effect.

**Table B.15** Values of hardness and elastic modulus for aluminum.

| F (mN) | H (GPa) | Ratio | F (mN) | E (GPa) | Ratio |
|--------|---------|-------|--------|---------|-------|
| 0.504 | 0.271 | 0.886 | 0.504 | 77.7 | 0.970 |
| 1.50 | 0.306 | 1.000 | 1.53 | 80.1 | 1.000 |
| 4.53 | 0.295 | 0.964 | 4.62 | 74.7 | 0.933 |
| 13.5 | 0.258 | 0.843 | 14.0 | 72.1 | 0.900 |
| 41.0 | 0.231 | 0.755 | 40.7 | 71.9 | 0.898 |
| 119 | 0.202 | 0.661 | 118 | 70.9 | 0.885 |

**Table B.16** Values of hardness and elastic modulus for tungsten.

| Load (mN) | Hardness (GPa) | Load (mN) | Modulus (GPa) |
|-----------|----------------|-----------|---------------|
| 0.512 | 5.86 | 0.499 | 372.0 |
| 1.49 | 5.67 | 1.48 | 495.0 |
| 4.53 | 5.21 | 4.49 | 427.0 |
| 13.2 | 4.55 | 13.6 | 403.0 |
| 40.9 | 3.88 | 41.1 | 401.0 |
| 119 | 3.75 | 120 | 400.0 |

**Table B.17** Values of hardness and elastic modulus for quartz.

| Load (mN) | Hardness (GPa) | Load (mN) | Modulus (GPa) |
|-----------|----------------|-----------|---------------|
| 0.506 | 12.9 | 0.515 | 119.0 |
| 1.50 | 13.8 | 1.50 | 122.0 |
| 4.46 | 13.3 | 4.63 | 119.0 |
| 13.5 | 13.2 | 13.8 | 123.0 |
| 40.2 | 12.5 | 40.8 | 121.0 |
| 117 | 12.4 | 121 | 125.0 |

# References

Cooper, M.G., Mikic, B.B., and Yovanovich, M.M., "Thermal Contact Conductance," *International Journal of Heat and Mass Transfer*, Vol. 12, pp. 279–300, 1969.

Fisher-Cripps, A.C., *Introduction to Contact Mechanics*, Springer, New York, 2000.

Hay, J.L., and Pharr, G.M., "Instrumented Indentation Testing", ASM Handbook Volume 08: Mechanical Testing and Evaluation, ASM International, Materials Park, Ohio, USA, pp. 231–242, 2000.

Hegazy, A.A., *Thermal Joint Conductance of Conforming Rough Surfaces: Effect of Surface Microhardness Variation*, Ph.D. Thesis, University of Waterloo, Canada, 1985.

Hendrix, B.C., "The Use of Shape Correction Factors for Elastic Indentation Measurements", *Journal of Materials Research*, Vol. 10, no. 2, pp. 255–257, 1995.

Johnson, K.L., *Contact Mechanics*, Cambridge University Press, Cambridge, 1985.

McColm, I.J., *Ceramic Hardness*, Plenum Press, New York, 1990.

Mott, M.A., *Micro-Indentation Hardness Testing*, Butterworths Scientific Publications, London, 1956.

Nho, K.M., *Experimental Investigation of Heat Flow Rate and Directional Effect on Contact Conductance of Anisotropic Ground/Lapped Interfaces*, Ph.D. Thesis, University of Waterloo, Canada, 1990.

Oliver, W.C., and Pharr, G.M., "An Improved Technique for Determining Hardness and Elastic Modulus Using Load and Displacement Sensing Indentation Experiments", *Journal of Materials Research*, Vol. 7, no. 6, pp. 1564–1583, 1992.

Pharr, G.M., Oliver, W.C., and Brotzen, F.R., "On the Generality of the Relationship Among Contact Stiffness, Contact Area, and Elastic Modulus During Indentation", *Journal of Materials Research*, Vol. 7, no. 3, pp. 613–617, 1992.

Sridhar, M.R., *Elastoplastic Contact Models for Sphere-Flat and Conforming Rough Surface Applications*, Ph.D. Thesis, University of Waterloo, Canada, 1994.

Sridhar, M.R., and Yovanovich, M.M., "Review of Elastic and Plastic Contact Conductance Models: Comparison with Experiment", *Journal of Thermophysics and Heat Transfer*, Vol. 8, no. 4, pp. 633–640, 1994.

Sridhar, M.R., and Yovanovich, M.M., "Empirical Methods to Predict Vickers Microhardness", *Wear*, Vol. 193, pp. 91–98, 1996a.

Sridhar, M.R., and Yovanovich, M.M., "Elastoplastic Contact Model for Isotropic Conforming Rough Surfaces and Comparison with Experiments", *Journal of Heat Transfer*, Vol. 118, no. 1, pp. 3–9, 1996b.

Tabor, D., *Hardness of Metals*, Oxford University Press, London, 1951.

Yovanovich, M.M., "New Contact and Gap Conductance Correlations for Conforming Rough Surfaces", AIAA Paper No. 81-1164, *AIAA 16th Thermophysics Conference*, Palo Alto, CA, USA, June 23–25. Published in AIAA Progress in Astronautics and Aeronautics, *Spacecraft Radiative Transfer and Temperature Control*, Vol. 83, pp. 83–95, 1982. Editor Dr. T.E. Horton, 1981.

Yovanovich, M.M., "Micro and Macro Hardness Measurements, Correlations, and Contact Models", AIAA 2006-979, *44th AIAA Aerospace Sciences Meeting and Exhibit*, Reno, NV, USA, January 9–12, 2006.

Yovanovich, M.M., and Hegazy, A., "An Accurate Universal Contact Conductance Correlation for Conforming Rough Surfaces with Different Micro-Hardness Profiles", AIAA 83-1434, *AIAA 18th Thermophysics Conference*, Montreal, Quebec, June 1–3, 1983.

Yovanovich, M.M., DeVaal, J., and Hegazy, A., "A Statistical Model to Predict Thermal Gap Conductance Between Conforming Rough Surfaces", AIAA 82-0888, *AIAA/ASME 3rd Joint Thermophysics, Fluids, Plasma and Heat Transfer Conference*, St. Louis, MO, USA, June 7–ll, 1982a.

Yovanovich, M.M., Hegazy, A.A., and DeVaal, J., "Surface Hardness Distribution Effects Upon Contact, Gap and Joint Conductances", AIAA 82-0887, *AIAA/ASME 3rd Joint Thermophysics, Fluids, Plasma and Heat Transfer Conference*, St. Louis, MO, USA, June 7–11, 1982b.

Yovanovich, M.M., Hegazy, A., and Antonetti, V.W., "Experimental Verification of Contact Conductance Models Based Upon Distributed Surface Micro-Hardness", AIAA-83-0532, *AIAA 21st Aerospace Sciences Meeting*, Reno, NV, USA, January 1–13, 1983.

# Appendix C

## Thermal Properties

In this appendix, we report thermal conductivity for solids, pastes and greases, interface materials, and gases.

**Table C.1**  Thermal properties of metals at 300 K.

| Material | $k$ (W/(m K)) | $\rho$ (kg/m$^3$) | $c_p$ (J/(kg K)) |
|---|---|---|---|
| Aluminum | 237 | 2702 | 903 |
| Cadmium | 96.8 | 8650 | 231 |
| Copper | 401 | 8933 | 385 |
| Germanium | 59.9 | 5360 | 322 |
| Gold | 317 | 19,300 | 129 |
| *Iron* | 80.2 | 7870 | 447 |
| Armco | 72.7 | 7870 | 447 |
| Lead | 35.3 | 11,340 | 129 |
| Molybdenum | 138 | 10,240 | 251 |
| Nickel | 90.7 | 8900 | 444 |
| Platinum | 71.6 | 21,450 | 133 |
| Silver | 429 | 10,500 | 235 |
| *Stainless steel* | | | |
| AISI 302 | 15.1 | 8055 | 480 |
| AISI 304 | 14.9 | 7900 | 477 |
| AISI 316 | 13.3 | 8238 | 468 |
| AISI 347 | 14.2 | 7978 | 480 |
| Tin | 66.6 | 7310 | 227 |
| Titanium | 21.9 | 4500 | 522 |
| Tungsten | 174 | 19,300 | 132 |
| Zinc | 116 | 7140 | 389 |
| Zirconium | 22.7 | 6570 | 278 |

*Thermal Spreading and Contact Resistance: Fundamentals and Applications*, First Edition.
Yuri S. Muzychka and M. Michael Yovanovich.
© 2023 John Wiley & Sons, Inc. Published 2023 by John Wiley & Sons, Inc.

**Table C.2** Thermal properties of nonmetals at 300 K.

| Material | $k$ (W/(m K)) | $\rho$ (kg/m³) | $c_p$ (J/(kg K)) |
|---|---|---|---|
| *Aluminum oxide* | | | |
| Sapphire | 46.0 | 3970 | 765 |
| Polycrystalline | 36.0 | 3970 | 765 |
| *Carbon* | | | |
| Amorphous | 1.60 | 1950 | — |
| Diamond type IIa insulator | 2300 | 3500 | 509 |
| Graphite pyrolytic | | | |
| ‖ to layers | 1950 | 2210 | 709 |
| ⊥ to layers | 5.70 | 2210 | 709 |
| Silicon | 148 | 2320 | 712 |
| Silicon carbide | 490 | 3160 | 675 |
| *Silicon dioxide* | | | |
| Crystalline | | | |
| ‖ to layers | 10.4 | 2650 | 745 |
| ⊥ to layers | 6.2 | 2650 | 745 |
| *Silicon dioxide* | | | |
| Polycrystalline | 1.38 | 2220 | 745 |
| Silicon nitride | 16.0 | 2400 | 691 |
| *Titanium dioxide* | | | |
| Polycrystalline | 8.4 | 4157 | 710 |

Source: Data obtained from Bergman et al. (2011).

## C.1 Thermal Properties of Solids

Thermal conductivity of some pure metals, alloys, and nonmetals are reported in Tables C.1 and C.2 to assist in computations of thermal spreading resistance.

## C.2 Thermal Conductivity of Gases

A common approximation for the conductivity of gases is the power–law relation:

$$\frac{k}{k_0} = \left(\frac{T}{T_0}\right)^n \tag{C.1}$$

where $T_0, k_0$ are reference values. Table C.3 lists the values of $n$ for several gases and the accuracy obtained for a given temperature range.

Another approximation for the thermal conductivity is based on the Sutherland formula:

$$\frac{k}{k_0} = \left(\frac{T}{T_0}\right)^{3/2} \frac{T_0 + S}{T + S} \tag{C.2}$$

**Table C.3** Power–law and Sutherland formula conductivity parameters for gases.

| Gas | $T_0$ (K) | $k_0$ (W/(m K)) | $n$ (%) | Range (K) | $S$ | Range (K) |
|---|---|---|---|---|---|---|
| Air | 273 | 0.0241 | 0.81 ± 3 | 210–2000 | 194 | 160–2000 |
| Argon | 273 | 0.0163 | 0.73 ± 4 | 210–1800 | 170 | 150–1800 |
| $CO_2$ | 273 | 0.0146 | 1.30 ± 2 | 180–1700 | 180 | 180–1700 |
| CO | 273 | 0.0232 | 0.82 ± 2 | 210–800 | 180 | 200–800 |
| $N_2$ | 273 | 0.0242 | 0.74 ± 3 | 210–1200 | 150 | 200–1200 |
| $O_2$ | 273 | 0.0244 | 0.84 ± 2 | 220–1200 | 240 | 200–1200 |
| $H_2$ | 273 | 0.168 | 0.72 ± 2 | 200–1000 | 120 | 200–1000 |
| Steam | 300 | 0.0181 | 1.35 ± 2 | 300–900 | 2200 | 300–700 |

Source: Data obtained from White (1988).

**Table C.4** Specific thermal resistance of greases and pastes.

| Material | $R''_{TIM}$ ((cm$^2$ K)/W) |
|---|---|
| Greases | 0.2–1 |
| Phase change materials | 0.3–0.7 |
| Gels | 0.4–0.8 |
| Solder | ≤0.05 |

**Table C.5** Specific thermal resistance of interface materials.

| Material | $R''_{TIM}$ ((cm$^2$ K)/W) |
|---|---|
| Elastomeric pads | 1–3 |
| Thermal tapes | 1–4 |
| Thermally conductive adhesives | 0.15–1 |

Source: Data are obtained from Blazej (2003). Fletcher (1972) presented a detailed analysis of many common interfacial materials. A graphical summary of dimensional measured contact conductance versus contact pressure is provided in Figure C.1.

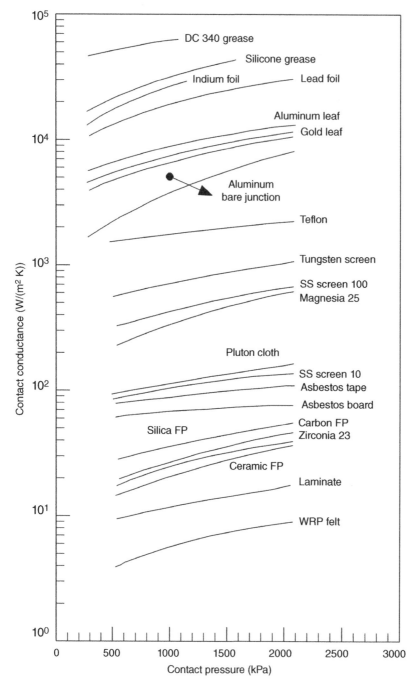

**Figure C.1** Contact conductance of interface materials. Source: Adapted from Fletcher (1972).

where $S$ is an effective temperature, called the Sutherland constant, which is a characteristic of the gas. Values of $S$ for several gases are also listed in the Table C.3. The accuracy of the Sutherland formula is slightly better than the power–law relation.

## C.3   Resistance of Thermal Interface Materials (TIMs)

Thermal interface materials (TIMs) are used to reduce contact resistance between contacting surfaces. A wide array of materials such as pastes, phase change materials, greases, gels, tapes, elastomers, and adhesives are often used between surfaces. Their thermal conductivity typically falls in the range of $1 < k < 10\,\mathrm{W/(m\,K)}$ and most often, the resistance of the actual interface layer is reported rather than thermal conductivity. We only report typical ranges for these materials in Tables C.4 and C.5 as the numbers of products are too numerous to tabulate.

## References

Bergman, T.L., Lavine, A.S, Incropera, F.P., and DeWitt, D.P., *Fundamentals of Heat and Mass Transfer*, 7th Edition, Wiley, 2011.

Blazej, D., "Thermal Interface Materials", *Electronics Cooling*, Vol. 9, no. 4, pp. 14–20, 2003.

Fletcher, L.S., "A Review of Thermal Control Materials for Metallic Junctions", AIAA Paper No. 72-284, *AIAA 7th Thermophysics Conference*, San Antonio, TX, USA, 1972.

White, F.M., *Viscous Fluid Flow*, McGraw-Hill, 1988.

# Index

Note: Page number followed by "*f*" denotes figures and page number followed by "*t*" denotes tables respectively.

*Thermal Spreading and Contact Resistance: Fundamentals and Applications*, First Edition.
Yuri S. Muzychka and M. Michael Yovanovich.
© 2023 John Wiley & Sons, Inc. Published 2023 by John Wiley & Sons, Inc.

Printed and bound by CPI Group (UK) Ltd, Croydon, CR0 4YY

16/04/2025

14658353-0004